Elements of Unit Operations and Chemical Reactors

Fabio Montagnaro • Marco Balsamo
Francesca Di Lauro

Elements of Unit Operations and Chemical Reactors

Springer

Fabio Montagnaro
Department of Chemical Sciences
University of Naples Federico II
Naples, Italy

Marco Balsamo
Department of Chemical Sciences
University of Naples Federico II
Naples, Italy

Francesca Di Lauro
Department of Chemical, Materials,
and Industrial Production Engineering
University of Naples Federico II
Naples, Italy

This work was supported by EU Recovery Plan.

ISBN 978-3-032-18045-2 ISBN 978-3-032-18046-9 (eBook)
https://doi.org/10.1007/978-3-032-18046-9

This Springer imprint is published by the registered company Springer Nature Switzerland AG
The registered company address is: Gewerbestrasse 11, 6330 Cham, Switzerland

To all readers and students
“Scrivi il tuo nome
Su qualcosa che vale”
(“Write your name
On something worthwhile”)
(Lucio Battisti, 1982)

Preface

The first word of this book is *elements*. It (in Ancient Greek, *Stoikheîa*) recalls the foundations of a composite subject. And the subject of unit operations and chemical reactors is certainly composite. As the readers know, the literature is dense with textbooks (some of them historically significant) on these topics, which represent some of the cornerstones that students (in industrial chemical engineering or similar subjects) face in their education. This literature, we believe, is currently largely sufficient for anyone wishing to delve deeper into the topics in question.

The idea that guided the composition of this manuscript was therefore focused on the fundamentals of the matter, that is, to prepare a book that would appeal to those who are approaching this subject *for the first time* (and, naturally, we intend to address with the same attention those who undertake the delicate task of teaching these topics to those who are engaging with them for the first time). This idea carries with it challenging aspects, linked to the fine task of presenting the topics in a way simple enough to understand, yet rigorous enough to provide with the tools to analyse the matter in detail later.

Considering that the approach just outlined is less prevalent in the literature, this book aims to cover the *rudiments* of unit operations and chemical reactors (including related transport phenomena) in a single "object". Its stated goal is not to be exhaustive, but to present the elements of the subject, with the assumptions and simplifications that will be recalled from time to time, but without losing rigor in its approach and methodology. The theoretical concepts are interspersed with worked exercises, limited in number but deemed sufficient to cover the aspects developed in theory with practical case studies.

This textbook has been prepared in the year 2025. We are aware that we stand on the verge of a revolution that has already begun, namely that involving the use of forms of artificial intelligence in all fields of knowledge (and their applications) and in everyday life. We realise the importance of the arrival of artificial intelligence in our time (in fact, for future readers this will be such obvious that what we have just said will be superfluous) and of the irreversibility of this "new present", which will bring with it undeniable advantages, presenting us with delicate challenges in ethical, social, and professional matters (challenges that, we trust, future generations will be able to address as part of the natural laws of evolution). Even with this awareness, we wanted to proceed in a manner that likely would not have been possible just five years from now: to prepare this book in a completely "artisanal" manner, a true *manuscript* ("hand-written") in which no artificial intelligence was used. This choice is not based on a lack of confidence in artificial media (which, we repeat, will become increasingly central in the future), but rather on the desire to offer readers a product crafted in a traditional-fashioned way, as if it were a placeholder in a transitional era. While we hope readers will appreciate, our approach inevitably leads to errors and inaccuracies: we would be happy to hear from you about this, and we thank you in advance for your feedback and suggestions/corrections.

* * *

This manuscript is dedicated by the authors to every reader and would not have been possible without the transfer of expertise we have absorbed over the years from our *Alma Mater*, the University of Naples Federico II (Italy), to which we owe our education and where the textbook has been prepared. While the list of mentors to whom the authors, belonging to different generations, are indebted would be long, we wish to mention here the names of Professor Piero Salatino and Professor Luciano Santoro, who have contributed (in more or less recent times) to the formation and illuminated transmission of the concepts regarding chemical reactors and unit operations. Finally, this manuscript is personally dedicated by Fabio to his wife Alessandra for her tender and irreplaceable support; by Marco to his son Diego, whose bright spirit is the engine driving him forward; and by Francesca to her husband Mattia and her son-to-be Andrea, who have driven her to be the best version of herself.

Naples, Italy — Fabio Montagnaro
Naples, Italy — Marco Balsamo
Naples, Italy — Francesca Di Lauro

Competing Interests The authors have no competing interests to declare that are relevant to the content of this manuscript.

Contents

1 Distillation . . . 1

1.1 Generalities . . . 3

1.2 Liquid-Vapour Equilibria for Binary Systems . . . 3

1.2.1 Relative Volatility . . . 3

1.2.2 Temperature-Composition Phase Diagrams . . . 6

1.3 Flash Distillation . . . 11

1.4 Continuous Distillation with Reflux: Operating Principles and General Scheme . . . 16

1.5 Operating Lines . . . 18

1.5.1 Assumption of Constant Molar Flow Rates . . . 18

1.5.2 Rectifying Section . . . 19

1.5.3 Stripping Section . . . 21

1.5.4 Feed Line . . . 22

1.5.5 Relation Among the Three Operating Lines . . . 25

1.6 The McCabe-Thiele's Method . . . 26

1.7 Effect of the Reflux Ratio . . . 28

1.7.1 Total Reflux . . . 28

1.7.2 Minimum Reflux Ratio . . . 30

1.7.3 Optimal Reflux Ratio . . . 31

1.8 Thermal Aspects for Condenser and Reboiler . . . 32

1.9 The Murphree's Efficiency . . . 33

1.10 Background Information . . . 38

References . . . 38

2 Liquid-Liquid Extraction . . . 39

2.1 Generalities . . . 40

2.2 Principles of Liquid-Liquid Extraction . . . 41

2.2.1 Triangular Diagrams . . . 41

2.2.2 Liquid-Liquid Equilibria . . . 43

2.3 Single-Stage Process . . . 46

2.3.1 Mass Balances . . . 46

2.3.2 Operating Limits of Extracting Solvent Flow Rate . . . 49

2.4 Multistage Crossflow Process . . . 53

2.5 Multistage Countercurrent Flow Process . . . 57

2.5.1 Mass Balances and Operating Lines . . . 57
2.5.2 Effect of the Extracting Solvent Flow Rate . . . 60
2.6 Background Information . . . 65
References . . . 66

3 Momentum Transport . . . 67
3.1 General Characteristics . . . 68
3.2 Flow in Conduits . . . 69
3.3 Laminar and Turbulent Flow . . . 70
3.4 The Viscosity Law for Newtonian Fluids . . . 71
3.4.1 Shear . . . 71
3.4.2 General Law of Molecular Transport . . . 71
3.4.3 Newtonian Fluids . . . 72
3.4.4 Viscosity . . . 73
3.5 The Reynolds Number . . . 75
3.6 Mean Velocity . . . 76
3.7 Laminar Flow . . . 77
3.7.1 Balance of Forces . . . 77
3.7.2 Radial Profile for the Shear Stress . . . 78
3.7.3 Radial Profile for the Axial Velocity . . . 79
3.8 Turbulent Flow . . . 81
3.8.1 Conceptual Presentation . . . 81
3.8.2 Quantitative Analysis . . . 83
3.9 Friction Factor . . . 89
3.10 Background Information . . . 93
References . . . 94

4 Energy Transport . . . 95
4.1 Mechanisms of Energy Transport . . . 97
4.2 Thermal Conduction . . . 97
4.2.1 General Characteristics . . . 97
4.2.2 The Fourier's Law on Conduction . . . 98
4.2.3 Thermal Conductivity . . . 98
4.2.4 Thermal Diffusivity and Prandtl Number . . . 100
4.3 Conduction Through Walls . . . 102
4.3.1 Flat Wall . . . 102
4.3.2 Cylindrical Wall . . . 104
4.4 Thermal Convection . . . 106
4.4.1 General Characteristics . . . 106
4.4.2 Nusselt Number . . . 107
4.5 Convection and Conduction Through Walls . . . 108
4.5.1 Flat Wall . . . 108
4.5.2 Cylindrical Wall . . . 112
4.6 Background Information . . . 116
References . . . 116

5 Heat Exchangers . . . 117
5.1 Double Pipe Exchangers . . . 119
5.2 Counterflow Operation . . . 121
5.3 Energy Balance . . . 122
5.4 Operating Lines . . . 124
5.5 Overall Heat Transfer Coefficients . . . 128
5.6 Local Heat Transfer Coefficients . . . 130
5.6.1 General Considerations . . . 130
5.6.2 Equivalent Diameter for Heat Exchangers Annulus . . . 131
5.6.3 Non-Dimensional Numbers . . . 131
5.6.4 Evaluation of the Local Coefficients . . . 132
5.7 Design Equations . . . 133
5.8 Background Information . . . 143
References . . . 144

6 Evaporators . . . 145
6.1 Generalities . . . 146
6.2 Boiling Point Elevation . . . 147
6.3 Single-Effect Evaporator . . . 148
6.4 Multiple-Effect Evaporator . . . 151
6.5 Background Information . . . 152
References . . . 152

7 Mass Transport . . . 155
7.1 Generalities . . . 156
7.2 Mass Diffusion . . . 157
7.2.1 Definition . . . 157
7.2.2 The Fick's Law on Diffusion . . . 157
7.2.3 Diffusion Coefficient . . . 158
7.2.4 Schmidt Number . . . 159
7.3 Mass Transfer by Convection . . . 161
7.3.1 Mass Transfer Coefficients . . . 161
7.3.2 Sherwood Number . . . 162
7.4 Convective Mass Transfer Between Phases . . . 163
7.4.1 General Presentation . . . 163
7.4.2 Main Equations . . . 165
7.4.3 Local and Global Mass Transfer Coefficients . . . 166
7.4.4 Relation Between Local and Global Mass Transfer Coefficients . . . 166
7.4.5 Resistances and Overall Transfer Coefficients . . . 168
7.5 Background Information . . . 171
References . . . 172

8 Absorption Columns . . . 175
8.1 Gas-Liquid Absorption . . . 177
8.2 Absorption Packed Towers . . . 178

8.3 Mass Balance . . . 180
8.4 Operating Lines . . . 180
8.5 Design of the Absorber Cross-Section . . . 183
8.6 Absorption Rate, Mass Transfer Coefficients and Interfacies Conditions . . . 186
8.7 Design Equations . . . 189
8.8 Height of Transfer Unit . . . 192
8.9 Number of Transfer Units . . . 194
8.9.1 Local Gas Side . . . 194
8.9.2 Local Liquid Side . . . 196
8.9.3 Global Gas Side . . . 197
8.9.4 Global Liquid Side . . . 197
8.10 Overall View . . . 198
8.11 Background Information . . . 203
References . . . 204

9 Humidification Processes . . . 205
9.1 Generalities . . . 207
9.2 Hygrometric Quantities and Psychrometric Charts . . . 208
9.3 Design of Cooling Towers . . . 210
9.3.1 Section of Natural Draught Tower . . . 210
9.3.2 Operating and Equilibrium Conditions . . . 211
9.3.3 Minimum Gas Flux . . . 213
9.3.4 Packing Height: Design Equation . . . 214
9.3.5 Interface Conditions and Gas Temperature Profile . . . 216
9.4 Background Information . . . 219
References . . . 220

10 Ideal Chemical Reactors . . . 221
10.1 General Definitions and Concepts . . . 222
10.1.1 Reaction Rate . . . 223
10.1.2 Degree of Conversion and Effect of Volume Changes . . . 226
10.1.3 Damköhler Number . . . 231
10.2 Batch (Discontinuous) Reactor with Perfect Mixing (STR = Stirred Tank Reactor) . . . 233
10.2.1 Productivity for STR . . . 240
10.3 Continuous Reactors . . . 245
10.3.1 PFR = Plug Flow Reactor (Perfect Axial Segregation) . . . 245
10.3.2 CSTR = Continuous Stirred Tank Reactor (Perfect Mixing) . . . 252
10.4 Productivity for Continuous Reactors . . . 258
10.4.1 The Differential Reactor . . . 260

10.5 Background Information 263
References 263

11 Comparison Between Chemical Reactors 265
11.1 PFR Vs. CSTR 266
11.2 Autocatalytic Reactions 273
11.2.1 Kinetics 273
11.2.2 Reactor Choice 276
11.3 Background Information 285
References 286

12 Combination of Chemical Reactors, and Recycle Reactor 287
12.1 General Premise 288
12.2 Reactors in Parallel 289
12.3 Reactors in Series 296
12.4 The Role of Mixing 304
12.5 PFR with Recycle 311
12.5.1 Illustration 311
12.5.2 Design Equation 314
12.5.3 Optimisation of the Recycle Ratio 316
12.6 Background Information 321
References 322

13 Extensions to Ideal Reactors: Non-Isothermal Reactions 323
13.1 Introductory Considerations on Non-Isothermal Rcacting Systems 325
13.1.1 Design Equations 325
13.1.2 Kinetics 325
13.1.3 Thermodynamics 327
13.2 Thermokinetic Maps 329
13.2.1 Irreversible Reactions 330
13.2.2 Reversible Reactions 331
13.3 Optimal Temperature Profiles 335
13.3.1 Irreversible Reactions; Endothermic Reversible Reactions 335
13.3.2 Exothermic Reversible Reactions 337
13.4 Mass and Energy Balances for Adiabatic Reactors 340
13.5 Optimisation for Adiabatic Reactors 347
13.5.1 Optimal Temperature Profiles (Operating Lines) 347
13.5.2 Autothermal Reactors 355
13.5.3 Reactor Choice 360
13.6 Background Information 369
References 370

14 Extensions to Ideal Reactors: Reaction Networks 371
14.1 General Picture 372
14.2 Approach 376

14.3 Considerations on the Role of Temperature 394
14.4 Background Information 396
References 396

15 Non-Ideal Reactors: Fluid Dynamic Aspects 397
15.1 Premise 398
15.2 Residence Time Distribution 399
15.2.1 General Presentation and Properties 399
15.2.2 Residence Time Distribution for PFR 402
15.3 Tracing of Non-Ideal Reactors to Find the RTD Function 404
15.3.1 Tracing 404
15.3.2 Pulse Tracing 406
15.3.3 Step Tracing 408
15.4 Interpretation of Non-Ideal Reactors Via Compartment Models 414
15.5 Background Information 424
References 424

16 Non-Ideal Reactors: Conceptualisation and Design 425
16.1 Maximum Segregation/Mixing Reactor—General Approach . . . 426
16.2 Maximum Segregation Reactor 428
16.3 Maximum Mixing Reactor 429
16.3.1 Premise 429
16.3.2 Presentation of the System 430
16.3.3 Design Equation 432
16.4 Actual Conversion Degree 434
16.5 Axial Dispersion Reactor 437
16.5.1 Presentation of the System 437
16.5.2 Design Equation 438
16.6 Laminar Flow Reactor 443
16.7 Background Information 446
References 448

Distillation

1

Abstract

This chapter inaugurates the discussion of unit operations, by presenting the fundamental principles of distillation in binary mixtures. We start with the analysis of liquid-vapour equilibria through relative volatility, phase diagrams and equilibrium curve. These concepts provide the foundation for practical applications such as flash distillation and continuous distillation with reflux. The McCabe-Thiele's graphical method, based on assumptions of constant molar flux and ideal equilibrium stage, is presented for the design of distillation columns. The role of operating and feed lines, reflux ratio, and energy balances in condenser and reboiler are discussed. Finally, the chapter addresses deviations from the ideal stage model by introducing the concept of Murphree's efficiency.

Nomenclature

B	Molar flow rate of bottom residue [moles/time]
$\mathbb{C}_L$	Average specific heat of liquid [energy/(moles·temperature)]
$\mathbb{C}_V$	Average specific heat of vapour [energy/(moles·temperature)]
D	Molar flow rate of distillate [moles/time]
F	Molar flow rate of feed [moles/time]
f_L	Liquid fraction (molar basis) of the feed [−]
H_F	Molar enthalpy of feed [energy/moles]
H_L	Molar enthalpy of saturated liquid [energy/moles]
H_V	Molar enthalpy of saturated vapour [energy/moles]
L	Molar flow rate of liquid (generic; in flash distillation; in the rectifying section) [moles/time]
$\overline{L}$	Molar flow rate of liquid in the stripping section [moles/time]
N	Total number of stages [−]
N_{min}	Minimum number of stages [−]

F. Montagnaro et al., *Elements of Unit Operations and Chemical Reactors*,
https://doi.org/10.1007/978-3-032-18046-9_1

P^0	Vapour pressure of the pure component [mass/(length·time^2)]
p_i	Partial pressure of the i component [mass/(length·time^2)]
q	Parameter related to the thermal condition of the feed [−]
q_c	Thermal power at the condenser [energy/time]
q_r	Thermal power at the reboiler [energy/time]
R	Reflux ratio [−]
R_{min}	Minimum reflux ratio [−]
R_{opt}	Optimal reflux ratio [−]
R_∞	Infinite reflux ratio [−]
$Rec\%$	Recovery of the more volatile species in distillate [−]
T	Temperature [temperature]
T_F	Feed temperature [temperature]
T_b	Bubble point temperature [temperature]
T_d	Dew point temperature [temperature]
T_{flash}	Temperature of flash distillation [temperature]
T_r	Reduced temperature [−]
V	Molar flow rate of vapour (generic; in flash distillation; in the rectifying section) [moles/time]
$\overline{V}$	Molar flow rate of vapour in the stripping section [moles/time]
v_i	Volatility of the i component [−]
x	Molar fraction of the more volatile species in liquid phase [−]
x^+	Molar fraction of the more volatile species in liquid phase corresponding to the pinch point [−]
x_B	Molar fraction of the more volatile species in bottom residue [−]
x_D	Molar fraction of the more volatile species in distillate [−]
y	Molar fraction of the more volatile species in vapour phase [−]
y^+	Molar fraction of the more volatile species in vapour phase corresponding to the pinch point [−]
y^*	Equilibrium molar fraction of the more volatile species in vapour phase (for tray efficiency <1) [−]
z_F	Molar fraction of the more volatile species in feed [−]
α	Relative volatility [−]
η_M	Murphree's efficiency [−]
$\lvert\lambda\rvert$	Latent heat of liquid-vapour phase transition [energy/moles]
$\lvert\lambda_F\rvert$	Latent heat of vaporisation of feed [energy/moles]
π	Total pressure [mass/(length·time^2)]
Σ	Molar flow rate of the mixing point [moles/time]

1.1 Generalities

Distillation is a process that separates a mixture of two or more components into its constituents or fractions of different composition from that of the initial mixture. It relies on mass and energy transfer between liquid and vapour phases (arguments detailed in later chapters), and separation occurs due to the different volatility of the mixture components. Main industrial applications of distillation include, e.g., petroleum refining, production of alcoholic beverages, air separation and waste management/recycling. In the following discussion, *the analysis will be limited to binary mixtures only*, while extensions to multicomponent distillation are covered in specialised textbooks (Blackadder and Nedderman 1971; McCabe et al. 2001; Seader and Henley 2006; Treybal 1981). Distillation will be considered as *stationary*, i.e. its parameters (e.g., flow rates, compositions) are constant in time. Moreover, the operation will be considered as being non-reactive (i.e., no chemical reaction occurs during the separation).

1.2 Liquid-Vapour Equilibria for Binary Systems

As a general consideration for the reader, it is emphasised that the concepts of liquid-vapour equilibrium discussed here are directly useful for developing the unit operations as presented in this elements textbook. Reference is made to specific literature (e.g., Atkins et al. 2022; Smith et al. 2020) for further information on the topic beyond the scope of this discussion.

1.2.1 Relative Volatility

Let us consider a binary system constituted by A and B components, with the former characterised by lower boiling temperature and defined as *low-boiling* or equivalently *more volatile* (conversely, B is the *high-boiling* or *less volatile* component). The *volatility* υ for each species can be defined as:

$$\begin{cases} \upsilon_A = \dfrac{y}{x} \\ \upsilon_B = \dfrac{(1-y)}{(1-x)} \end{cases} \tag{1.1}$$

where y and x represent the *molar fraction* of the low-boiling in the vapour and liquid phases, respectively, at thermodynamic equilibrium. An important parameter affecting the components separation by distillation is the *relative volatility* α of A with respect to B:

$$\boxed{\alpha = \frac{v_A}{v_B} = \frac{y(1-x)}{x(1-y)}} \tag{1.2}$$

For *ideal solutions* (i.e. characterised by both enthalpy of mixing and volume change on mixing equal to zero, as in the case of, e.g., a benzene/toluene binary system), α depends only on temperature T at fixed total pressure π of the system, and it is, by definition, $\alpha > 1$.

Raoult's law applies to ideal solutions:

$$\begin{cases} p_A = P_A^0 x \\ p_B = P_B^0 (1-x) \end{cases} \tag{1.3}$$

where p_i represents the partial pressure of component i in the vapour phase, and P_i^0 is the vapour pressure of the pure component i at the same temperature of the binary system. The dependence of P^0 on temperature is generally expressed by the *Antoine's equation*:

$$\log_{10}\left(P^0\right) = \mathcal{A} - \frac{\mathcal{B}}{T + \mathcal{C}} \tag{1.4}$$

where $\mathcal{A}$, $\mathcal{B}$, and $\mathcal{C}$ are constants tabulated for each substance (Green and Perry 2008; NIST Chemistry WebBook 2025).

In ideal cases, *Dalton's law* applies:

$$\begin{cases} p_A = \pi y \\ p_B = \pi (1-y) \end{cases} \tag{1.5}$$

Combining the relations in Eqs. 1.3 and 1.5 for each component, we get:

$$\begin{cases} P_A^0 x = \pi y \\ P_B^0 (1-x) = \pi (1-y) \end{cases} \tag{1.6}$$

Dividing the two relations in Eq. 1.6 side by side, and taking Eq. 1.2 into account, yields:

$$\frac{P_A^0 x}{P_B^0 (1-x)} = \frac{y}{(1-y)} \Rightarrow \frac{y(1-x)}{x(1-y)} = \frac{P_A^0}{P_B^0} = \alpha \tag{1.7}$$

showing that, for ideal solutions, *the relative volatility equals the ratio of the vapour pressures of the pure components*. The greater the difference between the vapour pressures (or, equivalently, the farther α is from unity), the more efficient the separation by distillation. Under these conditions, in fact, the liquid and vapour phases differ markedly in composition, with the vapour phase significantly enriched in the more volatile component.

Equation 1.7 can be also rewritten as:

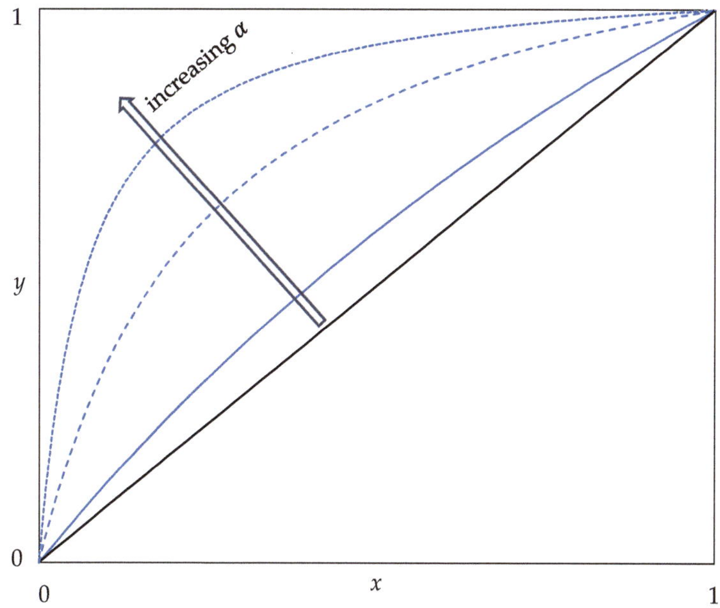

Fig. 1.1 Equilibrium curve for a binary system, parametric in relative volatility

$$\boxed{y = \frac{\alpha x}{1 + (\alpha - 1)x}} \tag{1.8}$$

providing an analytical relationship between the compositions of the liquid and vapour phases at equilibrium. This *equilibrium curve* or *y*-*x* diagram (Fig. 1.1) is fundamental for the methodologies described later in the chapter, when dealing with the design of distillation columns.

Example: Construction of the Equilibrium Curve for an Ideal Binary Mixture

Assume a binary mixture composed of acetonitrile and acetic acid, where acetonitrile is the more volatile component. Let the system be at a pressure of 1 atm. The mixture can be considered ideal, with $\alpha = 3.6$ from literature data. Therefore, it is possible to use Eq. 1.8 and construct the equilibrium curve as illustrated in Fig. 1.2.

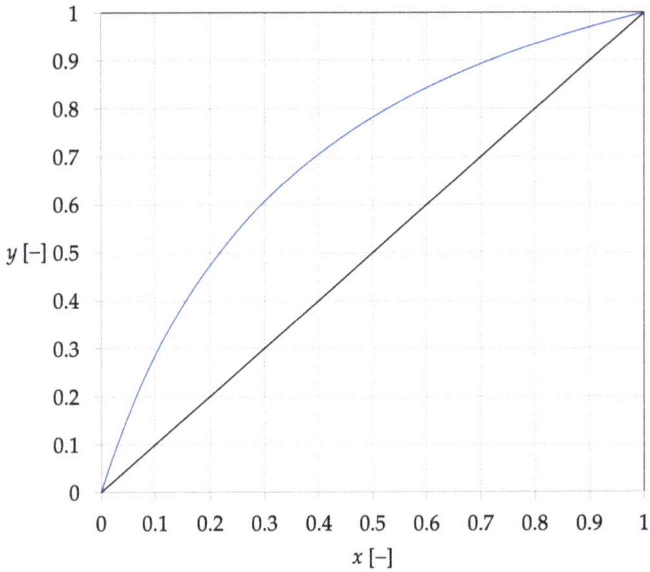

Fig. 1.2 Equilibrium curve for the binary system, at 1 atm, acetonitrile-acetic acid

1.2.2 Temperature-Composition Phase Diagrams

The determination of liquid-vapour equilibrium conditions for a binary mixture is typically carried out using *phase diagrams*, in which the temperature is plotted as a function of the composition of the more volatile component (at a fixed total pressure). In the case of ideal solutions, the diagram takes on the characteristic "lens" shape (Fig. 1.3). The lower and upper curves represent, respectively, the locus of points of bubble formation (initial boiling point) and dew formation (initial condensation temperature) of the binary mixture. These curves delimit the region of coexistence of *saturated liquid* and *saturated vapour* at equilibrium. Unlike pure compounds, at constant π, the *bubble point temperature* (denoted as T_b) and the *dew point temperature* (indicated as T_d) of a binary mixture are functions of the system composition. Actual systems whose temperature-composition pair lies below or above the lens represent a *subcooled liquid* or a *superheated vapour*, respectively.

For a deeper understanding of the phase diagram, consider, for example, isobaric (i.e. at fixed π) heating of a subcooled liquid stream represented by point L_O (where, in general, L and V also represent the total molar flow rate, [moles/time], of liquid and vapour, respectively) with coordinates (x_O;T_O) in Fig. 1.3:

- at temperature T_I, the liquid solution begins to boil (hence $T_I = T_b$) with the generation of the first vapour bubble V_I (composition y_I) on the dew curve in equilibrium with the liquid L_I on the bubble curve, the latter having a composition equivalent to that of L_O ($x_I = x_O$);

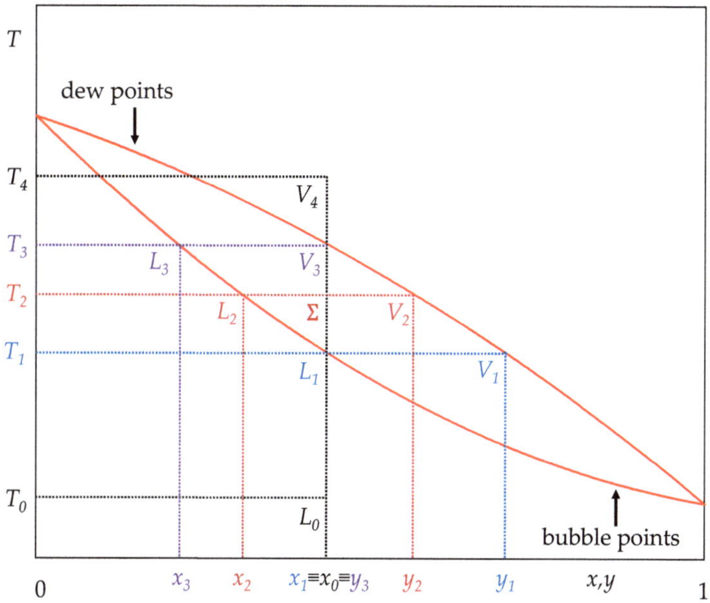

Fig. 1.3 Temperature-composition phase diagram for a binary ideal system at constant total pressure

- at temperature T_2, the two phases L_2 (composition x_2) and V_2 (composition y_2) are in equilibrium, with the liquid phase being depleted of the most volatile component compared to L_1 ($x_2 < x_1$);
- at temperature T_3 (dew point, $T_3 = T_d$), there is an infinitesimal flow of liquid L_3 (composition x_3) and a saturated vapour V_3 with composition $y_3 = x_0$;
- for $T_4 > T_3$, a superheated vapour V_4 is obtained with the same composition as the initial subcooled liquid.

It is highlighted the choice of referring to molar flow rates, dimensionally expressed in [moles/time], instead of molar quantities in [moles], for uniformity with the rest of the chapter, where continuous distillation operations will be dealt with. The diagram shows that, for T belonging to the interval $[T_b, T_d]$, it is possible to separate the initial solution into two phases, vapour and liquid, which are richer and poorer in the most volatile component, respectively, compared to the starting system.
The determination of the molar flow rates of saturated liquid and vapour in equilibrium at a given temperature can be obtained by applying materials balances in terms of [moles/time]. For instance, at temperature T_2 (Fig. 1.3), the overall material balance and the one for the most volatile species are expressed by, respectively:

$$L_0 = L_2 + V_2 \tag{1.9}$$

$$L_0 x_0 = L_2 x_2 + V_2 y_2 \tag{1.10}$$

Substituting Eq. 1.9 into Eq. 1.10 and performing simple algebraic manipulations, we obtain:

$$\frac{L_2}{V_2} = \frac{y_2 - x_0}{x_0 - x_2} \tag{1.11}$$

Let the point Σ in Fig. 1.3, described by coordinates (x_0;T_2), denote the *overall composition point* on the T-composition diagram (or *mixing point* if referring to its flow rate, with $\Sigma = L_0$ in terms of flow rate). It is, in terms of segment length, $y_2 - x_0 = \overline{\Sigma V_2}$ and $x_0 - x_2 = \overline{L_2 \Sigma}$; consequently, Eq. 1.11 can be rewritten as:

$$\boxed{\frac{L_2}{V_2} = \frac{\overline{\Sigma V_2}}{\overline{L_2 \Sigma}}} \tag{1.12}$$

also known as the *lever rule*. Therefore, the ratio between the molar flow rates of the liquid and vapour phases at equilibrium is equal to the reciprocal of the ratio of the distances of the liquid and vapour representative points from the overall composition point.

The analytical determination of the bubble point temperature of an ideal solution with composition x_0 can be obtained by applying Raoult's and Dalton's laws (Eqs. 1.3 and 1.5, respectively) and considering that the sum of the mole fractions of the two components in the vapour phase V_1 must, of course, be unity. So, it is (as, at the bubble point, a vapour of composition y_1 is in equilibrium with a liquid of composition x_0, cf. Fig. 1.3):

$$\boxed{y_1 + (1 - y_1) = \frac{p_A}{\pi} + \frac{p_B}{\pi} = \frac{P_A^0(T_b)}{\pi} x_0 + \frac{P_B^0(T_b)}{\pi}(1 - x_0) = 1} \tag{1.13}$$

Similarly, the dew point temperature of a superheated vapour of composition $y_0 = x_0$ can be determined by Raoult's and Dalton's laws and using the mole fraction congruence relation for the two components in the liquid phase L_3. Considering that, at the dew point, a vapour of composition y_0 is in equilibrium with a liquid of composition x_3, cf. Fig. 1.3, it is:

$$\boxed{x_3 + (1 - x_3) = \frac{p_A}{P_A^0(T_d)} + \frac{p_B}{P_B^0(T_d)} = \pi \left[\frac{y_0}{P_A^0(T_d)} + \frac{1 - y_0}{P_B^0(T_d)} \right] = 1} \tag{1.14}$$

Equations 1.13 and 1.14 are solved iteratively with respect to temperature (e.g., using numerical computation software), evaluating the vapour pressures via Antoine's equation (Eq. 1.4), so to obtain the values for T_b and T_d, respectively.

Table 1.1 Liquid-vapour equilibrium calculations for the ethyl acetate/toluene system

Pure species	$\mathcal{A}$	$\mathcal{B}$	C	T_b [K]
Ethyl acetate	4.228	1245.702	−55.189	349.8
Toluene	4.082	1346.382	−53.508	383.3
Binary systems				
	$P_A^0(T_b)$ [bar]	$P_B^0(T_b)$ [bar]	y_1 [−]	T_b [K]
Liquid solution ($x_0 = 0.3$)	1.81	0.65	0.54	369.0
	$P_A^0(T_d)$ [bar]	$P_B^0(T_d)$ [bar]	x_3 [−]	T_d [K]
Superheated vapour ($y_0 = 0.7$)	1.54	0.55	0.45	363.5

Example: Bubble and Dew Points for the System Ethyl Acetate/Toluene

Table 1.1 lists the values of Antoine's constants (NIST Chemistry WebBook 2025) for a reference binary system ethyl acetate/toluene at $\pi = 1$ bar. Firstly, application of Eq. 1.4 (with P^0 and T expressed in [bar] and [K], respectively) allows the evaluation of the normal T_b for pure substances, by letting $P^0 = \pi = 1$ bar. Since the bubble point temperature (the logarithm cancels out):

$$T_b = T = \frac{\mathcal{B}}{\mathcal{A}} - C \tag{1.15}$$

for ethyl acetate is lower (Table 1.1), this is the more volatile ("A") compound.

Let us now consider the determination of the bubble point of a liquid solution with $x_0 = 0.3$: we hypothesise a value for T_b, then we determine $P_A^0(T_b)$ and $P_B^0(T_b)$ by Antoine's Eq. 1.4:

$$P^0 = 10^{\mathcal{A} - \frac{\mathcal{B}}{T + C}} \tag{1.16}$$

and finally we verify that Eq. 1.13 is satisfied. The solution is reported in Table 1.1: it is $T_b = 369.0$ K and, by letting $y_1 = P_A^0(T_b)x_0/\pi$ as per Eq. 1.6, we obtain that the first vapour bubble has composition $y_1 = 54\%$ in equilibrium with the liquid at $x_1 = x_0 = 30\%$ (cf. Fig. 1.3). The vapour is richer in the more volatile compound.

Finally, let us calculate the dew point of a vapour solution with $y_0 = 0.7$: we hypothesise a value for T_d, then we determine $P_A^0(T_d)$ and $P_B^0(T_d)$ by Antoine's Eq. 1.4 (as rewritten in Eq. 1.16) and finally we verify that Eq. 1.14 is satisfied. The solution is reported in Table 1.1: it is $T_d = 363.5$ K and, by letting $x_3 = \pi y_0/P_A^0(T_d)$ as per Eq. 1.6, we obtain that the first liquid drop has composition $x_3 = 45\%$ in equilibrium with the vapour at $y_3 = y_0 = 70\%$ (cf. Fig. 1.3). The liquid is poorer in the more volatile compound.

The equilibrium curve can also be constructed from the temperature-composition diagram (apart from using the already presented Eq. 1.8 as in Fig. 1.1, if the relative volatility is known). Given, for example, the two phases L_1 (composition x_1) and V_1

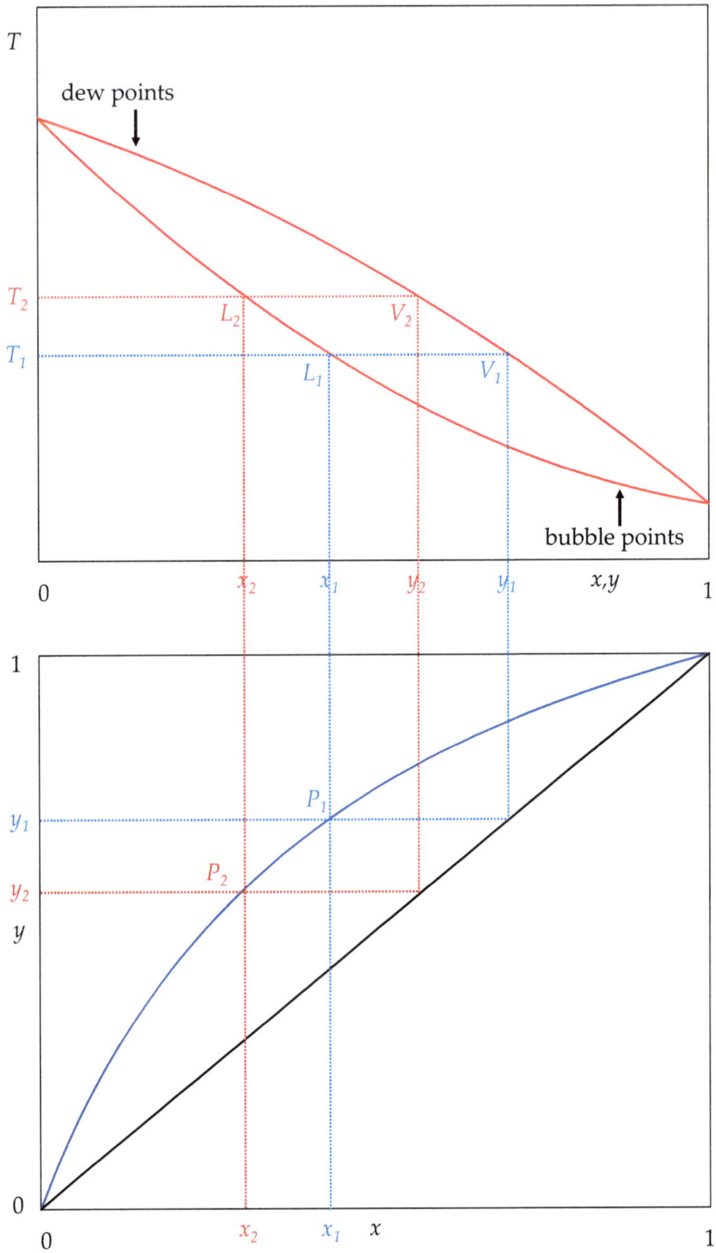

Fig. 1.4 Relationship between temperature-composition phase diagram and equilibrium curve for a binary ideal system at constant total pressure

(composition y_1) in equilibrium at temperature T_1, the point P_1 on the y-x diagram with coordinates (x_1;y_1) is obtained as follows (Fig. 1.4):

- draw the vertical line through x_I from the temperature-composition diagram to the abscissa of the y vs. x plane;
- draw the vertical line through y_I from the temperature-composition diagram to the diagonal ($y = x$) of the y vs. x plane, and from the resulting intersection point draw the horizontal line to the ordinate.

The procedure is applied for all the L_i-V_i equilibrium pairs to plot the overall equilibrium curve, as done in Fig. 1.4.

Finally, *non-ideal systems* show deviations from Raoult's law due to a nonzero enthalpy of mixing and a nonzero volume of mixing (related to the differences in interactions between the compounds in the mixture compared with those of the pure components). In particular, some systems (such as water/ethanol) form an *azeotrope*, i.e., a mixture that, at equilibrium, is characterised by a vapour phase and a liquid phase having the same composition (Fig. 1.5). At the azeotropic composition, the dew point and the boiling point coincide; therefore, the system behaves like a pure substance (actually, the word "azeotrope" derives from ancient Greek words that literally means "no change on boiling"). Minimum and maximum boiling point azeotropes have a boiling temperature lower and higher, respectively, than those of the pure components. From the above considerations, it follows that $\alpha = 1$ at the azeotropic composition, and the inequality between the volatilities of the two components is reversed when passing through the azeotrope (in other words, α can even become less than 1). Hence, for azeotropic systems, the equilibrium curve crosses the diagonal in the y vs. x plane (cf. Fig. 1.5).

1.3 Flash Distillation

Flash distillation is a continuous single-stage operation in which a binary liquid mixture undergoes partial vaporisation, resulting into a vapour and liquid phase in thermodynamic equilibrium. A typical flash distillation unit operates as follows (Fig. 1.6): the liquid mixture is pumped through a heat exchanger and then passes through a throttling valve, where a pressure drop occurs. This causes a partial and very rapid vaporisation of the fluid. The liquid-vapour mixture then enters a separator (flash drum), where the intimate contact between phases allows them to reach equilibrium (determined by the operating pressure and temperature). The liquid phase is separated by gravity at the bottom of the flash chamber, while the vapour phase, collected at the top, is subsequently sent to a condenser.

Flash distillation is particularly suitable when the relative volatility of the binary system is high (i.e., when the components have significantly different boiling points).

The solution of the flash distillation problem requires coupling the material balance equations with the knowledge of the phase equilibrium conditions of the system (the latter described through the equilibrium curve or the temperature-composition phase diagram). Let F, V, and L be the molar flow rates of the feed, the vapour leaving and liquid leaving the flash distillation unit, respectively. The

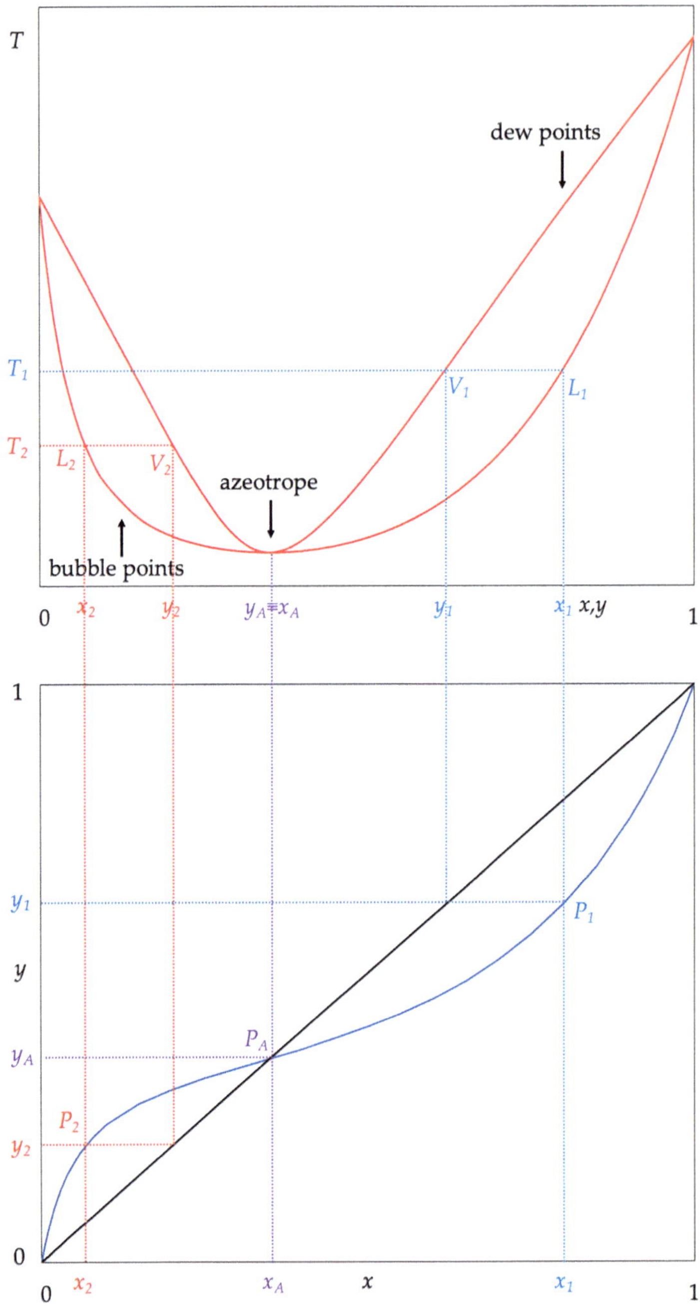

Fig. 1.5 Relationship between temperature-composition phase diagram and equilibrium curve for a binary non-ideal system at constant total pressure

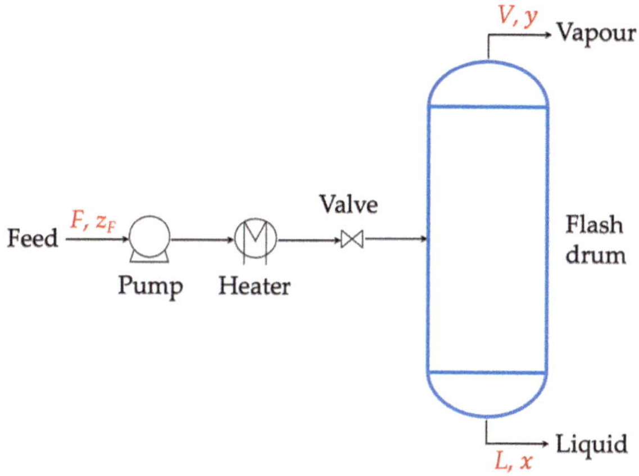

Fig. 1.6 Scheme of a flash distillation unit

corresponding mole fractions of the low-boiling component are z_F, y, and x. The overall material balance and the material balance for the more volatile species, expressed in [moles/time], are reported in Eqs. 1.17 and 1.18, respectively (all the balances in this chapter are written in terms of "*what enters is equal to what exits*", as the systems are stationary and no chemical reaction, or other phenomena leading to formation/disappearance of chemical species, are considered):

$$\boxed{F = L + V} \tag{1.17}$$

$$\boxed{Fz_F = Lx + Vy} \tag{1.18}$$

In other words, Eq. 1.17 states that the overall [moles/time] entering the unit are equal to the overall [moles/time] exiting (as both liquid and vapour) the unit. At the same time, Eq. 1.18 illustrates that the [moles/time] of the low-boiling species entering the unit are equal to the [moles/time] of the same species exiting (as both liquid and vapour) the unit. This approach will be followed also for other couples of balances, similar to Eqs. 1.17 and 1.18, in this chapter; while, later on in this textbook, the universal balance law will be presented for more general cases.

Given the phase diagram and the feed conditions (flow rate and composition), total pressure and temperature of the flash unit T_{flash}, the values of y and x can be read on the diagram (Fig. 1.7). Therefore, L and V can be computed by solving Eqs. 1.17 and 1.18. Alternatively, the lever rule can be applied directly on the diagram:

$$\frac{L}{V} = \frac{y - z_F}{z_F - x} = \frac{\overline{FV}}{\overline{LF}} \tag{1.19}$$

Once the L/V ratio is known, the flow rates can be determined from either one of the two material balance equations.

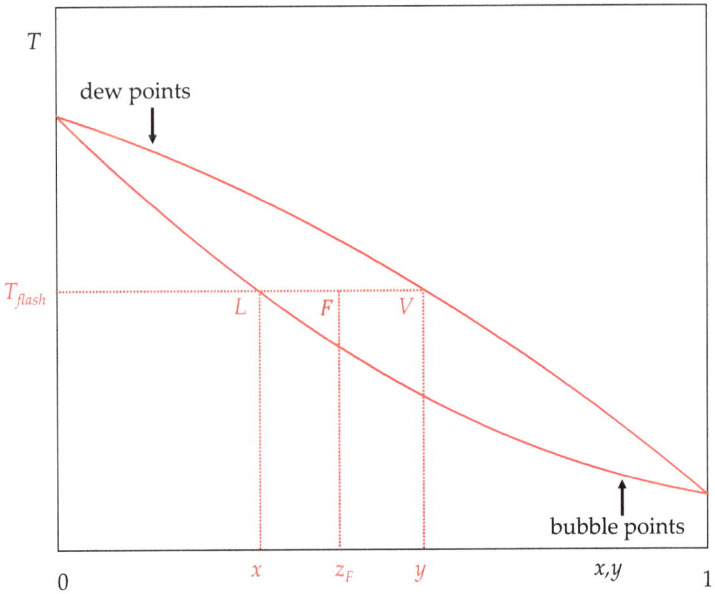

Fig. 1.7 Application of temperature-composition phase diagram to flash distillation

Example: Flash Distillation of the System Benzene/Toluene

Table 1.2 reports the equilibrium data (dew and bubble points) for the ideal system benzene/toluene (with benzene being the more volatile compound) at 1 bar total pressure (as obtained via Eqs. 1.13 and 1.14 using the Antoine's constants reported in NIST Chemistry WebBook 2025). The corresponding *T*-composition diagram is illustrated in Fig. 1.8.

Consider a liquid feed with $z_F = 0.5$, to be flashed at a temperature so that 23% of *F* is vaporised. Hence, it is $L/V = 0.77F/0.23F = 3.35$. Recalling the lever rule in Eq. 1.19, we get $L/V = \overline{FV}/\overline{LF} = 3.35$. The flash temperature is iteratively determined on the *T*-composition plot by identifying, on the vertical line with abscissa z_F, the summation point *F* for which $\overline{FV}/\overline{LF} = 3.35$. In this specific case, we obtain $T_{flash} = 93.5$ °C, and we read on the diagram the equilibrium compositions for the streams exiting the flash unit (cf. Fig. 1.6), $y = 0.67$ and $x = 0.45$.

Table 1.2 Liquid-vapour equilibrium data for benzene/toluene system ($\pi = 1$ bar)

T [°C]	110.6	107.7	106.0	102.0	98.3	96.6	92.0	89.2	83.0	80.2
x [−]	0	0.06	0.10	0.20	0.30	0.35	0.50	0.60	0.85	1
y [−]	0	0.13	0.21	0.37	0.51	0.57	0.71	0.79	0.94	1

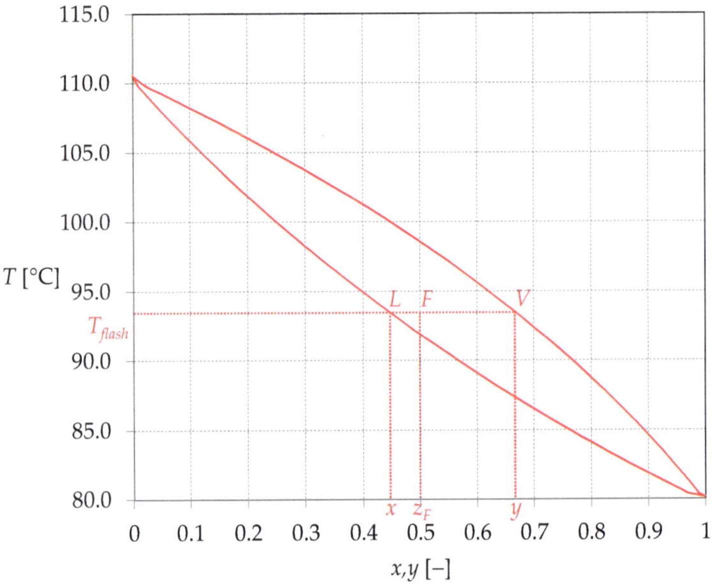

Fig. 1.8 Visualization on T-composition diagram of flash distillation of benzene/toluene system (molar fractions are referred to benzene)

1.4 Continuous Distillation with Reflux: Operating Principles and General Scheme

Continuous distillation with reflux (also termed *fractional distillation*) is applied for the separation of compounds with comparable volatility. In the following discussion, the operation is described with reference to *multistage equipment*, that is, systems composed of individual units (e.g., sieve trays; the detailed analysis of such units is left to specialised literature such as, e.g., Green and Perry 2008) that allow mass and energy transfer between two streams in *countercurrent flow*, through their intimate mixing and subsequent separation (Fig. 1.9). In this configuration, the liquid and vapour flow downward and upward, respectively, and the stages (or also termed trays in the following) are numbered in increasing order in the direction of the liquid phase flow. The subscripts of the streams (and their respective compositions) *refer to the stage from which they exit*. It is assumed that *the streams leaving each stage are at equilibrium conditions*. Therefore, such stages are defined as *equilibrium*, *theoretical* or *ideal stages*. In practice, the approach to equilibrium conditions depends on the effectiveness of mixing between the two phases and on the contact time between them. Deviations from equilibrium conditions will be considered later, through the evaluation of the stage efficiency.

Figure 1.9 depicts the general scheme of a fractionating distillation unit. The tray where the feed enters is called the *feed tray*, and the sections above and below it are

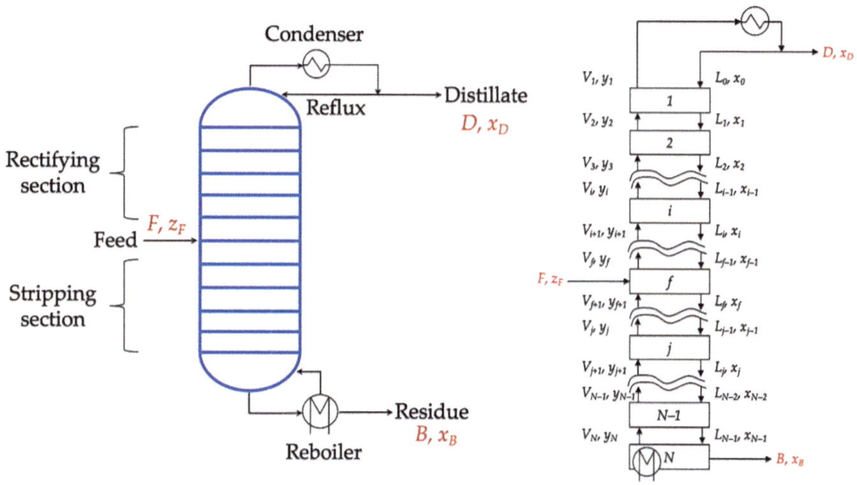

Fig. 1.9 Fractionating distillation unit. On the left: general operating scheme and main sections/streams. On the right: details of the multistage contact system with indication of streams and compositions

defined as the *rectifying section* and the *stripping section*, respectively. The vapour rising through the column becomes progressively richer in the more volatile component, while the liquid descending through the column becomes progressively depleted of the low-boiling species. The liquid flowing to the bottom of the column is sent to a reboiler, i.e. a heat exchanger (for example, a coil heated with steam) which generates both a vapour that is reintroduced into the column and a liquid that is collected as the *bottom product* (or *residue*). The reboiler is generally considered an equilibrium stage. The vapour at the top of the column is sent to a condenser, i.e. a heat exchanger typically cooled with water as the refrigerant, which usually condenses the overhead vapour completely. The condensed liquid is split into two streams: a *distillate* (or *overhead product*) collected as product, and a reflux stream returned to the top tray of the column.

With reference to the general scheme of Fig. 1.9, let F, D, and B be the molar flow rates of the feed, distillate, and bottom product, respectively. The corresponding mole fractions of the more volatile component are z_F, x_D, and x_B. The global mass balance and the mass balance on the low-boiling species, expressed in [moles/time], are given by Eqs. 1.20 and 1.21, respectively:

$$\boxed{F = D + B} \tag{1.20}$$

$$\boxed{F z_F = D x_D + B x_B} \tag{1.21}$$

Given the feed conditions and, for example, the specified purities of the bottom product and distillate, Eqs. 1.20 and 1.21 allow the two unknowns D and B to be determined. Alternatively, one may specify the purity and recovery of the more volatile component in the distillate ($Rec\%$) (in the case it is the target product of the separation process), defined as:

$$Rec\% = 100\frac{Dx_D}{Fz_F} \tag{1.22}$$

1.5 Operating Lines

1.5.1 Assumption of Constant Molar Flow Rates

The numerical analysis of a distillation column with trays requires, in addition to the knowledge of the equilibrium conditions between phases, appropriate relations between the compositions of the streams that cross the stages. These relations are defined as *operating conditions* and depend on the operating parameters of the equipment.

The McCabe-Thiele's method (McCabe and Thiele 1925) described later is based on the *constant molar flow rates* assumption, according to which, in each section of the equipment (rectifying and stripping), the molar flow rate of liquid, and the molar flow rate of vapour, both remain constant from stage to stage. More rigorous approaches accounting for energy balances in the stripping and rectifying sections, such as the Ponchon-Savarit's method, are described in devoted textbooks (Blackadder and Nedderman 1971; Seader and Henley 2006; Treybal 1981). The constant molar flow rates assumption is especially valid when the following conditions are satisfied:

- molar latent heats of vaporisation of the mixture components are comparable;
- sensible heats of the components, associated with their processing in the distillation unit, are negligible compared to their latent heats of vaporisation;
- heat losses to the surroundings are negligible.

The above conditions imply that, in each stage, every mole of high-boiling component that condenses from the vapour phase provides the energy to vaporise one mole of the more volatile component in the liquid phase. This, therefore, results in no change in the molar flow rates of liquid and vapour between the inlet and the outlet of the stage itself. Although, within each section (rectifying and stripping) of the column, the molar flow rates of liquid and vapour remain constant, they can differ between the rectifying and stripping sections as a consequence of the feed introduction. Therefore, in the tray-by-tray scheme of the distillation column, it is possible to omit subscripts for the molar flow rates of liquid and vapour (as, instead, reported in the general scheme of Fig. 1.9). In particular, the symbols L and V will be adopted for the rectifying section, and the symbols $\overline{L}$ and $\overline{V}$ for the stripping section (Fig. 1.10).

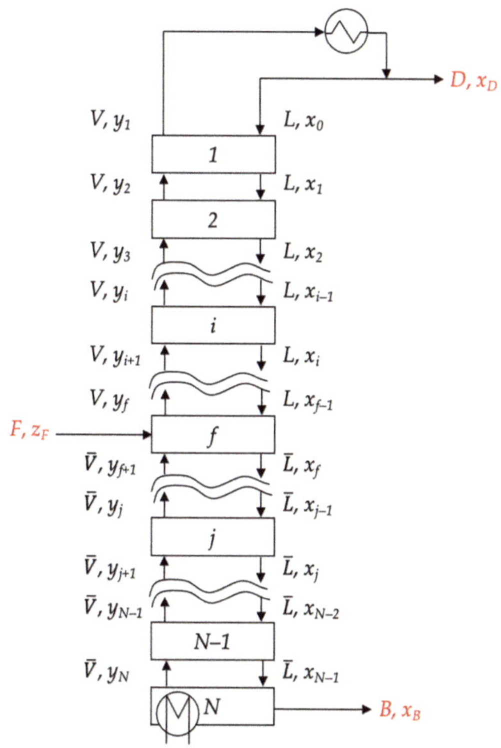

Fig. 1.10 Scheme of a fractional distillation column according to the constant molar flow rates assumption

1.5.2 Rectifying Section

To define the operating conditions in the rectifying section, it is necessary to write the material balances by considering the part of the column between its top and the generic i-th stage of the section under examination (see the dashed region in Fig. 1.11). The global and low-boiling mass balances are given in Eqs. 1.23 and 1.24, respectively (the balances are expressed in [moles/time] as usual, and with reference to the chosen control region):

$$V = L + D \tag{1.23}$$

$$V y_{i+1} = L x_i + D x_D \tag{1.24}$$

In particular, Eq. 1.23 translates into maths that the vapour flow rate leaving the first stage of the column, and entering the condenser, is subdivided into the reflux and the distillate. From Eq. 1.24 it follows that:

$$y_{i+1} = \frac{L}{V} x_i + \frac{D}{V} x_D \tag{1.25}$$

Substituting Eq. 1.23 into Eq. 1.24 yields an alternative expression for the operating line of the rectifying section:

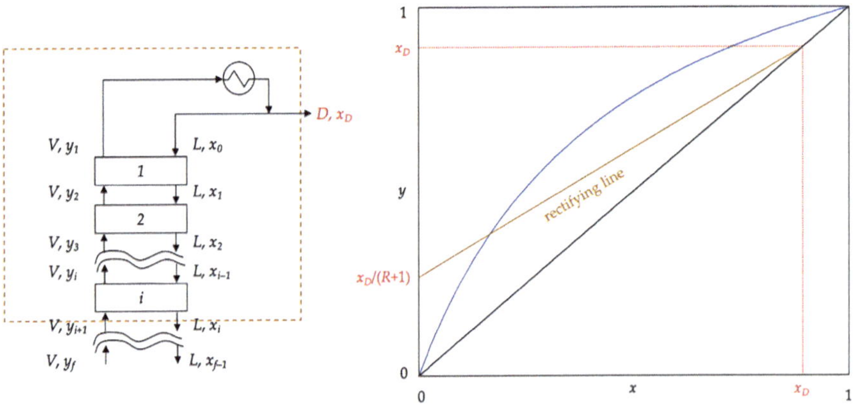

Fig. 1.11 Rectifying section. On the left: multistage scheme (the dashed line identifies the control surface for writing the material balances). On the right: operating line represented on the y vs. x plane

$$y_{i+1} = \frac{L}{L+D}x_i + \frac{D}{L+D}x_D \tag{1.26}$$

Equations 1.25, or 1.26, thus relate the vapour- and liquid-phase compositions crossing between two subsequent stages of the rectifying section (for example, for $i = 1$ we will have a relation between x_1 and y_2, and so on, cf. Fig. 1.11).

An important operating parameter used in the analysis of a fractionating column is the *reflux ratio R*, defined as the ratio between the liquid molar flow rate returned to the top of the column and the distillate molar flow rate withdrawn as product:

$$\boxed{R = \frac{L}{D}} \tag{1.27}$$

While a more extensive discussion on the reflux ratio will be developed later on, substituting $L = RD$ into Eq. 1.26 gives the following expression for the *operating line of the rectifying section* as a function of the reflux ratio:

$$\boxed{y_{i+1} = \frac{R}{R+1}x_i + \frac{x_D}{R+1}} \tag{1.28}$$

Equation 1.28 can be easily represented in the y vs. x plane when R and x_D are known. One point of the operating line lies on the diagonal: in fact, setting $x_i = x_D$ gives $y_{i+1} = x_D$; another point is its intercept with the y-axis ($x_i = 0$), equal to $x_D/(R + 1)$ (cf. Fig. 1.11, where a typical rectifying line is illustrated).

In most cases, fractionating distillation columns adopt a *total condenser*, which completely condenses the vapour at the top of the column in order to obtain both the reflux and the distillate in liquid phase. In this configuration, the streams V, L and D have the same composition ($y_1 = x_D = x_0$). In the case of a *partial condenser*, the distillate is obtained as vapour in equilibrium with the reflux liquid that is condensed. Under these conditions, the compositions of streams V, L and D are different, and in particular x_D and x_0 are equilibrium compositions (in the present treatment, we consider that the reflux is not subcooled). Therefore, the partial condenser represents in this case an additional ideal stage of the distillation column.

1.5.3 Stripping Section

To define the operating conditions in the stripping section, material balances are written considering as control surface the portion of the column enclosing the reboiler and the generic j-th stage of the column (see the dashed region in Fig. 1.12). Equations 1.29 and 1.30 report the total and more volatile component mass balances, respectively (the balances are expressed in [moles/time] as usual, and with reference to the chosen control region):

$$\overline{L} = \overline{V} + B \tag{1.29}$$

$$\overline{L}x_{j-1} = \overline{V}y_j + Bx_B \tag{1.30}$$

In particular, Eq. 1.29 translates into maths that the liquid flow rate leaving the last (except the reboiler) stage of the column, and entering the reboiler itself, is subdivided into a vapour refluxed into the column and the bottom product. From Eq. 1.30, the following expression for the *operating line of the stripping section* is obtained:

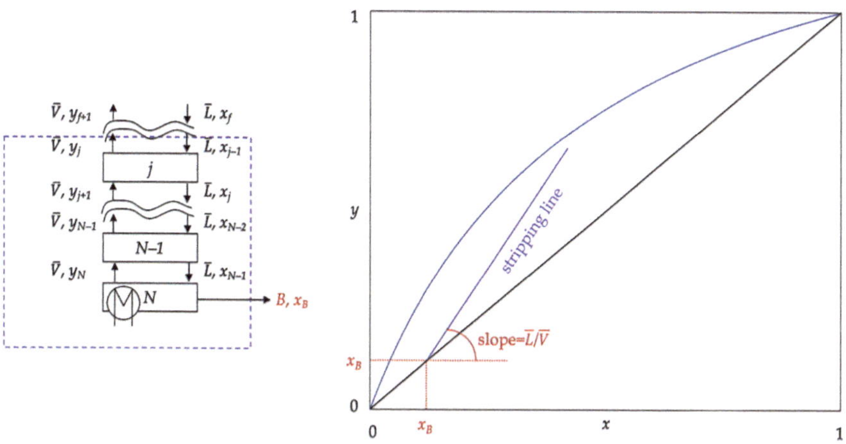

Fig. 1.12 Stripping section. On the left: multistage scheme (the dashed line identifies the control surface for writing the material balances). On the right: operating line represented on the y vs. x plane

$$y_j = \frac{\overline{L}}{\overline{V}} x_{j-1} - \frac{B}{\overline{V}} x_B \tag{1.31}$$

Equation 1.31 thus relate the vapour- and liquid-phase compositions crossing between two subsequent stages of the stripping section (cf. Fig. 1.12). Equation 1.31 can be easily represented in the y vs. x plane when the slope ($\overline{L}/\overline{V}$) and x_B are known. In particular, one point of the operating line lies on the diagonal: in fact, setting $x_{j-1} = x_B$ gives $y_j = x_B$ since, Eq. 1.30 and Fig. 1.12, $\overline{L} - B = \overline{V}$ (cf. Fig. 1.12, where a typical stripping line is illustrated).

1.5.4 Feed Line

At the feed stage of the distillation column, the flow rates of liquid and/or vapour may change, depending on the thermal condition of the feed (subcooled liquid, saturated liquid, liquid-vapour mixture, saturated vapour, superheated vapour). This effect is quantitatively analysed by introducing the q parameter (McCabe and Thiele 1925; McCabe et al. 2001; Treybal 1981) defined as the change in molar liquid flow rate at the feed stage per unit molar feed flow rate:

$$q = \frac{\overline{L} - L}{F} \tag{1.32}$$

Considering the global material balance around the feed plate (Fig. 1.13):

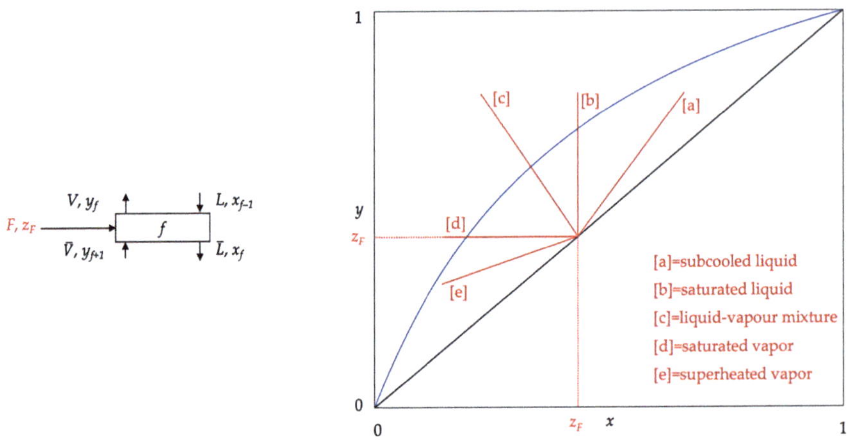

Fig. 1.13 Feed section. On the left: stage scheme. On the right: operating line as a function of the feed thermal state represented on the y vs. x plane

Table 1.3 Relevant quantities as a function of the feed thermal state

Thermal state of feed	relations for flow rates and enthalpies	q values	slope of the feed line $q/(q-1)$
Subcooled liquid	$\overline{L} > L + F$ $\overline{V} > V$ $H_F < H_L$	$q > 1$ (Eq. 1.36)	>1
Saturated liquid	$\overline{L} = L + F$ $\overline{V} = V$ $H_F = H_L$	$q = 1$	∞
Liquid-vapour mixture	$\overline{L} > L$ $V > \overline{V}$ $H_L < H_F < H_V$	$0 < q < 1$ (Eq. 1.37)	$f_L/(f_L-1) < 0$
Saturated vapour	$\overline{L} = L$ $V = \overline{V}+F$ $H_F = H_V$	$q = 0$	0
Superheated vapour	$\overline{L} < L$ $V > \overline{V} + F$ $H_F > H_V$	$q < 0$ (Eq. 1.38)	$\in]0, 1[$

$$F + L + \overline{V} = V + \overline{L} \Rightarrow \frac{\overline{L} - L}{F} = \frac{\overline{V} - V}{F} + 1 \tag{1.33}$$

we get:

$$\boxed{q = \frac{\overline{L} - L}{F} = \frac{\overline{V} - V}{F} + 1} \tag{1.34}$$

The q parameter is equivalently defined as (cf. Treybal 1981):

$$\boxed{q = \frac{H_V - H_F}{H_V - H_L} = \frac{H_V - H_F}{|\lambda_F|}} \tag{1.35}$$

where H_V, H_F and H_L are the molar enthalpies of the saturated vapour, of the feed and of the saturated liquid, respectively; λ_F is the molar latent heat of vaporisation of the feed stream. Hence, *q represents the ratio between the heat required to bring one mole of feed to the saturated vapour state and the molar latent heat of vaporisation of the feed.* Its values as a function of the feed thermal state are listed in Table 1.3.

We observe that for the feed stream in the state of saturated liquid, or saturated vapour, Eq. 1.35 implies that $q = 1$ (since $H_F = H_L$, that is $H_V - H_F = |\lambda_F|$), or $q = 0$ (since $H_F = H_V$), respectively. More in detail:

- when $q = 1$ (saturated liquid), elaboration of Eq. 1.34 gives $\overline{L} = L + F$ and $\overline{V} = V$, since the whole feed adds to L in determining $\overline{L}$ (Fig. 1.13);

- conversely, when $q = 0$ (saturated vapour), elaboration of Eq. 1.34 gives $\overline{L} = L$ and $V = \overline{V} + F$, since the whole feed adds to $\overline{V}$ in determining V (Fig. 1.13).

For the other three cases, i.e. subcooled liquid, liquid-vapour mixture and superheated vapour feed, q can be computed through Eqs. 1.36, 1.37 and 1.38, respectively:

$$q = \frac{\mathbb{C}_L(T_b - T_F) + |\lambda_F|}{|\lambda_F|} \tag{1.36}$$

$$q = \frac{f_L|\lambda_F|}{|\lambda_F|} = f_L \tag{1.37}$$

$$q = \frac{-\mathbb{C}_V(T_F - T_d)}{|\lambda_F|} \tag{1.38}$$

where T_b, T_d and T_F are the bubble point temperature of feed, dew point temperature of feed, and actual feed temperature, respectively. $\mathbb{C}_L$ and $\mathbb{C}_V$ represent the average specific heats (molar basis) of feed liquid and vapour, respectively. Finally, f_L is the liquid fraction (molar basis) of the feed. We observe that:

- for the case of subcooled liquid feed ($T_F < T_b$), Eq. 1.36, the difference $H_V–H_F$ (Eq. 1.35) is computed first by bringing the subcooled liquid to the state of saturated liquid, and then by considering its evaporation to the saturated vapour state. Since $H_F < H_L$, inspection of Eq. 1.35 (confirmed by Eq. 1.36) highlights that $q > 1$. Back to Eq. 1.34, it is $\overline{L} > L + F$ and $V < \overline{V}$, since the whole feed adds to L in determining $\overline{L}$ and, in addition, the feed is so "cold" that a fraction of $\overline{V}$ condenses at the feed stage (Fig. 1.13);
- for the case of liquid-vapour mixture (T_F comprised between T_b and T_d), Eq. 1.37, the difference $H_V - H_F$ (Eq. 1.35) is computed by considering that only a fraction f_L needs to be brought to the saturated vapour state. In this case, it is $0 < q < 1$. Back to Eq. 1.34, it is $\overline{L} > L$ and $V > \overline{V}$, since the feed partly (f_L) adds to L in determining $\overline{L}$ and partly ($1–f_L$) adds to $\overline{V}$ in determining V (Fig. 1.13);
- for the case of superheated vapour feed ($T_F > T_d$), Eq. 1.38, the difference $H_V–H_F$ (Eq. 1.35) is computed first by bringing back the superheated vapour to the state of saturated vapour. Since $H_F > H_V$, inspection of Eq. 1.35 (confirmed by Eq. 1.38) highlights that $q < 0$. Back to Eq. 1.34, it is $\overline{L} < L$ and $V > \overline{V} + F$, since the whole feed adds to $\overline{V}$ in determining V and, in addition, the feed is so "hot" that a fraction of L evaporates at the feed stage (Fig. 1.13).

The formulation of the feed line as a function of the q parameter derives from the material balance (in [moles/time]) on the more volatile component around the feed stage:

$$Fz_F + Lx + \overline{V}y = Vy + \overline{L}x \Rightarrow \left(V - \overline{V}\right)y = \left(L - \overline{L}\right)x + Fz_F \tag{1.39}$$

$$y = \frac{L - \overline{L}}{V - \overline{V}}x + \frac{F}{V - \overline{V}}z_F = \frac{-qF}{-(q-1)F}x + \frac{F}{-(q-1)F}z_F \tag{1.40}$$

$$\boxed{y = \frac{q}{q-1}x - \frac{z_F}{q-1}} \tag{1.41}$$

by using, for simplicity, the generic symbols y and x for the molar fractions in vapour and liquid phase (respectively), and with the help of Eq. 1.34. Equation 1.41 can be easily represented in the y vs. x plane when the slope ($q/(q-1)$) and z_F are known. In particular, one point of the feed line lies on the diagonal: in fact, setting $x = z_F$ gives $y = z_F$ (cf. Fig. 1.13, where typical feed lines are illustrated). More in detail:

- for saturated liquid ($q = 1$) we have a vertical line, while for saturated vapour ($q = 0$) we have a horizontal line;
- for liquid-vapour mixture ($0 < q < 1$), we have a feed line with negative slope;
- for superheated vapour ($q < 0$) the line has slope positive but not higher than 1 (lower slope than the diagonal line), while for subcooled liquid ($q > 1$) the line has slope > 1 (steeper than the diagonal).

1.5.5 Relation Among the Three Operating Lines

We confirm the use, for simplicity, of the generic symbols y and x for the molar fractions in vapour and liquid (respectively) phase. The operating line of the rectifying section has been obtained by the mass balance in Eq. 1.24:

$$Vy = Lx + Dx_D \tag{1.42}$$

The operating line of the stripping section has been obtained by the mass balance in Eq. 1.30:

$$\overline{V}y = \overline{L}x - Bx_B \tag{1.43}$$

Subtracting the two, and considering the overall balance in Eq. 1.21, it is:

$$\left(V - \overline{V}\right)y = \left(L - \overline{L}\right)x + Dx_D + Bx_B = \left(L - \overline{L}\right)x + Fz_F \tag{1.44}$$

This is the same as Eq. 1.39, that is the mass balance around the feed stage, that has led us to the operating feed line. Geometrically, this means that the three operating lines intersect at one point, as schematically illustrated in Fig. 1.14 (where, for simplicity, a vertical feed line typical of saturated liquid has been drawn, cf. Fig. 1.13).

A way to handle this result is: once got the rectifying line (knowing x_D and the y-axis intercept) and the feed line (on the basis of z_F and the feed thermal state), the

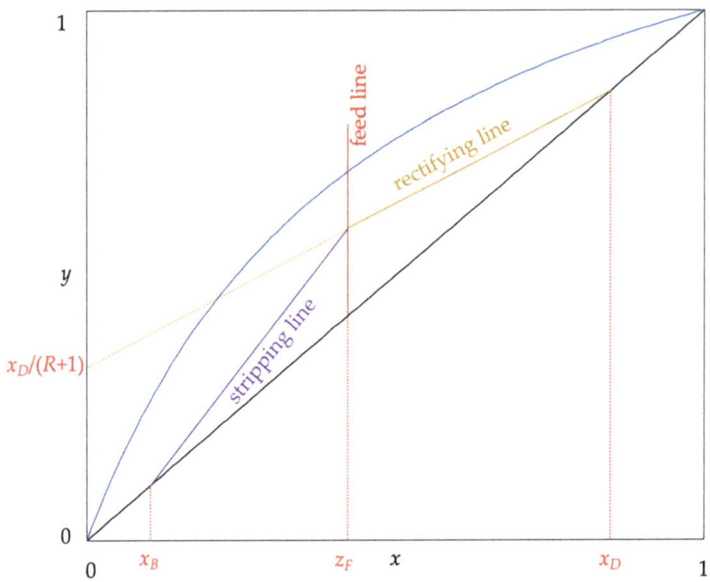

Fig. 1.14 The three operating lines intersect at one point (example of saturated liquid feed)

stripping line is obtained as the line starting from the diagonal at x_B, and passing through the intersection of the rectifying and feed lines.

1.6 The McCabe-Thiele's Method

The numerical evaluation of continuous distillation columns with reflux can be grouped into two main categories:

- *design problems*, in which it is generally necessary to determine the number of stages (N) required to achieve a desired separation of the feed mixture;
- *verification problems*, in which the number of stages of a distillation column is known, and the objective is to assess whether (and under which operating conditions) the desired separation can be achieved.

With the liquid-vapour equilibrium conditions for a binary system at a fixed pressure known, and having derived analytical expressions for the operating lines, the McCabe-Thiele's graphical method (under the constant molar fluxes assumption) can be applied to solve design and verification problems for a multistage distillation column.

A theoretical example follows to illustrate the McCabe-Thiele's method applied to a design problem, as illustrated in Fig. 1.15. Consider a feed stream with composition z_F and known thermal state (e.g., a saturated liquid characterised by a vertical

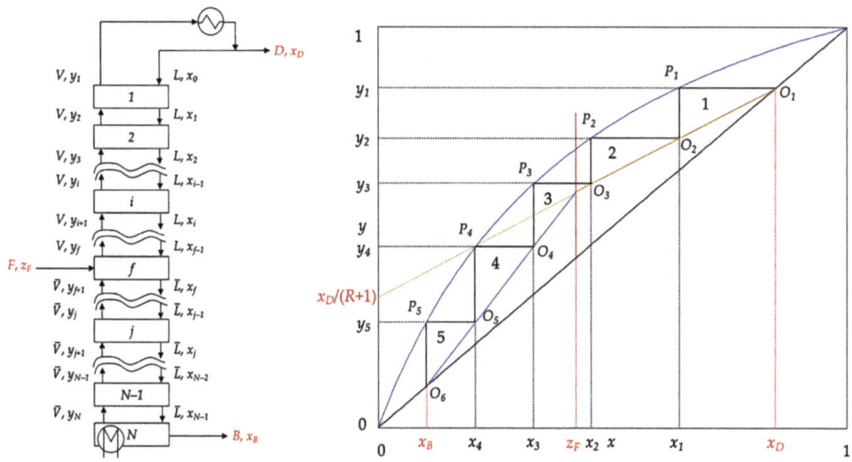

Fig. 1.15 McCabe-Thiele's graphical method. The scheme of the fractional distillation column is re-drawn here for clarity

operating line starting from the diagonal at z_F) to be separated into a distillate and bottom product with assigned compositions x_D and x_B. Let the reflux ratio R be fixed, and assume a total condenser at the column top. The goal is to determine the number of theoretical stages N required for the operation. With R and x_D fixed, the rectifying line is drawn and, consequently, the stripping line can be identified as explained above.

The number of stages is determined by a "stepwise" construction, alternating between equilibrium curve and operating lines. We can start either from the top or the bottom of the column; here, the top is chosen. For a total condenser, the vapour at the top, the reflux liquid, and the distillate all have the same composition ($y_1 = x_D = x_0$). The point $O_1(x_D;x_D)$ represents the operating condition at the column top. The liquid (composition x_1) and vapour (composition y_1) leaving the first stage are in equilibrium, so x_1 is obtained from the intersection of the equilibrium curve with the horizontal line $y_1 = x_D$. The point $P_1(x_1;y_1)$ represents the outlet streams of the first stage. Operating conditions for streams crossing between the first and second stage (i.e., the liquid at x_1 and the vapour at y_2) are determined by drawing a vertical from P_1 to the rectifying line (that relates streams crossing between stages in the rectifying section, see Eq. 1.28), giving y_2 and represented by point $O_2(x_1;y_2)$. The triangle $O_1P_1O_2$ represents the first stage. The horizontal from O_2 to the equilibrium curve gives x_2, with point $P_2(x_2;y_2)$ representing the outlet streams from the second stage. The vertical from P_2 to the rectifying line gives point $O_3(x_2;y_3)$, corresponding to the streams crossing between the second and third stage.

As the intersection of the three operating lines is approached, it is necessary to define the stage at which the feed is introduced. *The feed stage is located where internal stream compositions best match the feed composition, minimising N at fixed conditions.* Thus, the transition from the rectifying to the stripping line occurs at an

x value immediately greater than the abscissa of the intersection point of the operating lines (McCabe et al. 2001; Treybal 1981). The feed stage triangle, therefore, has one vertex on the rectifying line and one on the stripping line (of course, the third vertex must lie on the equilibrium curve), and straddles the intersection of the operating lines. Any alternative placement of the feed stage would increase the number of stages. In this example, the feed enters at the third stage.

The construction proceeds along the stripping section by using the relative operating line (that relates streams crossing between stages in the stripping section, see Eq. 1.31) until x approaches x_B (or slightly exceeds it, since compositions change discretely in this method). In our example, 4 stages plus the reboiler (counted as an equilibrium stage) are required ($N = 5$).

1.7 Effect of the Reflux Ratio

1.7.1 Total Reflux

The previous analysis showed that, once the feed conditions, the compositions of the distillate and the bottoms product, and the reflux ratio are fixed, it is possible to determine the number of ideal stages required to carry out the operation. Let us now consider the effect of increasing the reflux ratio at fixed specifications of the inlet and outlet streams of the distillation unit. Focusing on the rectifying operating line (Eq. 1.28), we observe that, as $R = L/D$ increases, its slope ($R/(R + 1)$) increases while its intercept with the y-axis ($x_D/(R + 1)$) decreases. *An increase in the reflux ratio leads to a larger distance between the equilibrium and operating conditions, i.e. an increase in the driving force of the separation process.* Consequently, a greater composition change per stage is obtained, which results in a *lower number of stages N* required to achieve a given separation (Fig. 1.16).

In the limiting case of infinite reflux ratio R_∞ ($R \rightarrow \infty$, *total reflux* condition), the *minimum number of stages* (N_{min}) required for the separation is obtained. Under this condition, the slope of the rectifying operating line tends to unity, and therefore the rectifying line coincides with the diagonal. This situation can occur for $D \rightarrow 0$, meaning that no distillate is withdrawn, and it means that $F = B = 0$ implying that also the stripping line lies on the diagonal (see Eq. 1.31 for $B = 0$). In other words, at R_∞ the operating lines of the two sections of the column do collapse on the diagonal. This operating mode corresponds, e.g., to pre-charging the column with a certain amount of material, with all the vapour reaching the condenser being returned to the column as liquid (total reflux, no distillate), and all the liquid reaching the reboiler being vaporised and rising back up the column (no bottom product). Under total reflux condition, N_{min} is determined by the stepwise construction between x_D and x_B using the diagonal as operating line (Fig. 1.16). The determination of N_{min} is important, for example, to preliminarily assess whether an existing distillation column can be used to carry out a new separation process (the minimum number of stages must be lower than the number of trays in the existing column). Of course, N_{min} is a theoretical, non-operating, value.

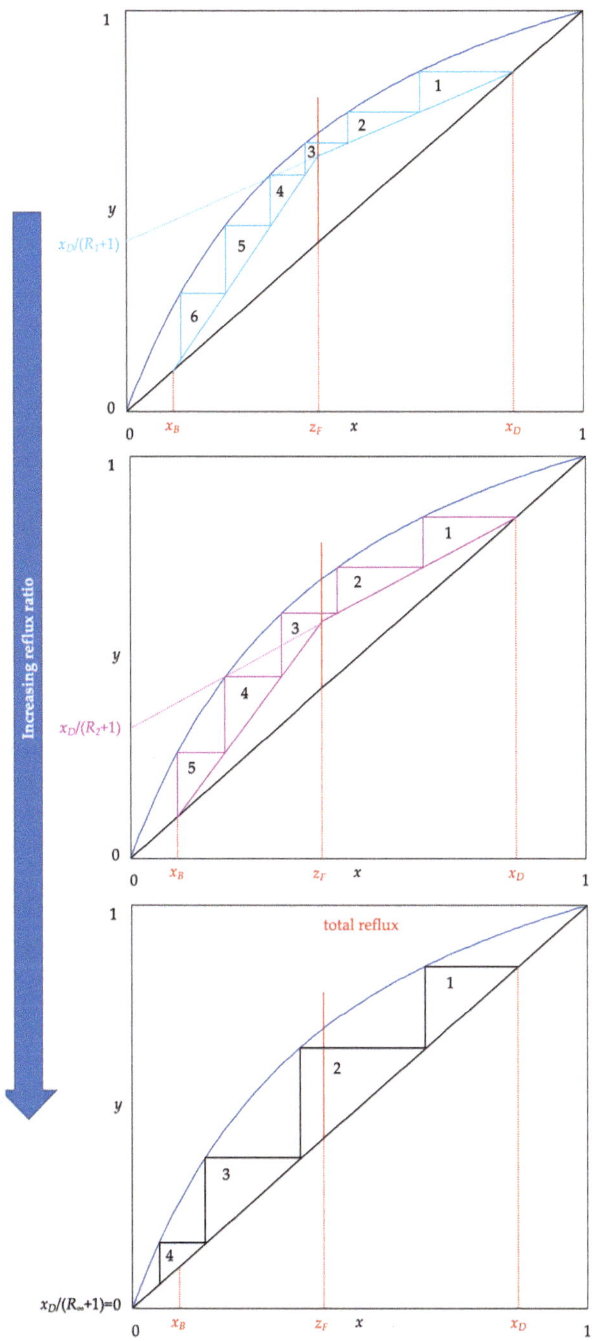

Fig. 1.16 Effect of the reflux ratio on the total number of stages at fixed feed properties, purity of distillate and bottom product

For ideal binary systems with constant relative volatility α, in some cases the Fenske's equation has been proposed to compute an estimation for N_{min} (Fenske 1932; McCabe et al. 2001):

$$N_{min} = \frac{\ln\left[\frac{x_D(1-x_B)}{x_B(1-x_D)}\right]}{\ln(\alpha)} - 1 \tag{1.45}$$

1.7.2 Minimum Reflux Ratio

Following the previous discussion, a decrease in the reflux ratio makes N, the number of stages, increase, due to a reduction in the driving force of the process. This happens up to a limiting condition (minimum R) in which a theoretically infinite number of stages would be required to overcome that intersection point, as the driving force vanishes (*pinch condition for which operating and equilibrium conditions coincide*; this concept will be also discussed other times in this textbook, with reference to other unit operations).

For ideal mixtures, where the equilibrium curve is concave downward, the condition of an infinite number of stages, corresponding to the *minimum reflux ratio* R_{min}, occurs when the intersection of the three operating lines lies on the equilibrium curve (Fig. 1.17). The value of R_{min} can be determined by substituting the coordinates (x^+;y^+) of the intersection point of the operating lines with the equilibrium curve into the rectifying line Eq. (1.28) or, equivalently, from the intercept with the y-axis of the rectifying line, which assumes the highest possible value, i.e. $x_D/(R_{min} + 1)$. For $R < R_{min}$, the intercept would be even higher, the operating lines would cross the equilibrium curve and this is, clearly, a non-operating condition.

Indeed, many real systems exhibit an inflection point in the equilibrium curve, so the minimum reflux ratio condition cannot be determined using the previous procedure. In such systems, the most limiting condition of zero driving force,

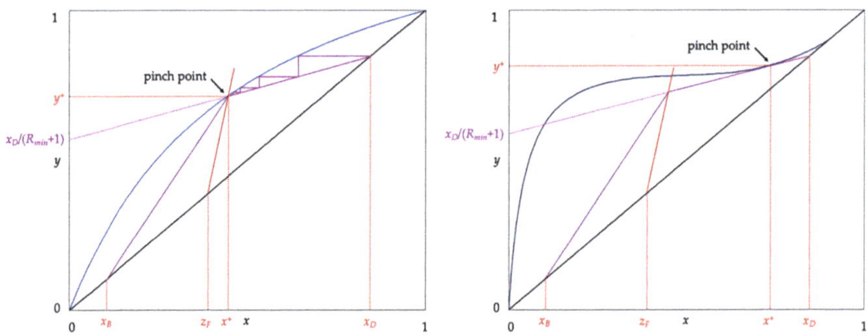

Fig. 1.17 Minimum reflux ratio. On the left: ideal system. On the right: non-ideal system

corresponding to R_{min}, can be determined by the enriching section line tangent to the equilibrium curve (Fig. 1.17). Again, even lower values for the reflux ratio would lead to operating lines above the equilibrium curve.

1.7.3 Optimal Reflux Ratio

From what we have seen so far, we have learned that the reflux ratio can vary between a minimum value (R_{min}, for which $N \rightarrow \infty$, a non-operating condition) and an infinite value (R_{∞}, for which N_{min} is obtained but it is again a non-operating condition). This requires us to search for an optimal reflux ratio, R_{opt}, whose evaluation derives from an economic optimisation of the process, that is, from minimising the total costs.

The total annual costs are the sum of the capital (or fixed) and operating costs. The capital costs are an increasing function of equipment size, while the operating costs are an increasing function of the energy duty to the reboiler, the cooling water flow to the condenser, and the pumping power for liquid circulation (e.g., reflux). These costs are schematically illustrated in Fig. 1.18 as a function of the reflux ratio. As for capital (fixed) costs, at the minimum reflux ratio they are infinite because the number of stages is infinite (infinite column height). An increase in R reduces the

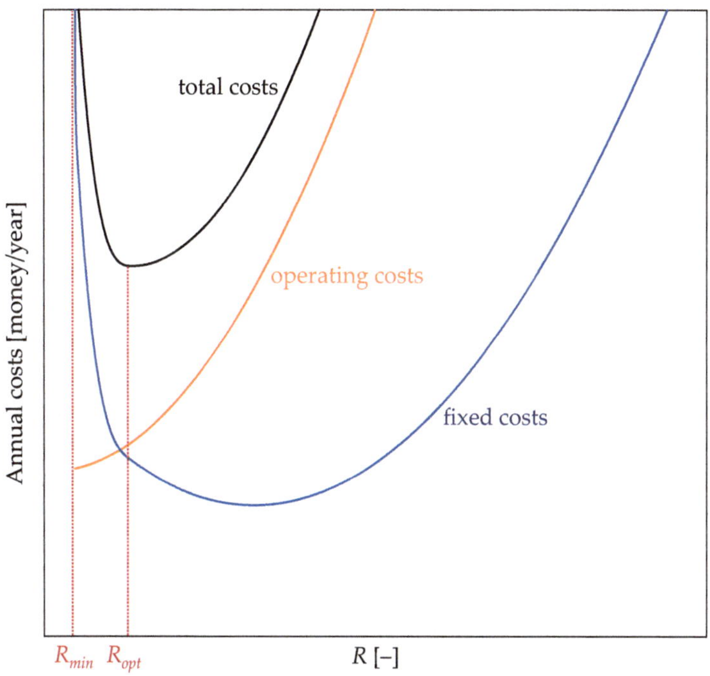

Fig. 1.18 Fixed, operating, and total costs as a function of the reflux ratio (schematic representation)

tower height but increases its cross-sectional area to accommodate higher liquid and vapour flow rates circulating in the column, together with an increase in the reboiler and condenser size. This implies that the capital cost curve has a minimum. The operating costs start from a minimum value at R_{min}, since heating, cooling, and circulation duties are lowest, and they increase monotonically with the increase of the reflux ratio. Therefore, the total costs curve exhibits a minimum, typically located at a reflux ratio value between 10% and 50% higher than the minimum reflux ratio ($R_{opt} = 1.1 - 1.5R_{min}$). The evaluation of R_{min} is therefore fundamental to calculate R_{opt}.

1.8 Thermal Aspects for Condenser and Reboiler

On industrial scale, heat losses in thermally insulated distillation columns can be considered negligible, and therefore the equipment is considered adiabatic (McCabe et al. 2001). The energy balance in the case of a total condenser (the stream V is entirely condensed), where the reflux liquid is at its boiling point, can be written as:

$$\boxed{q_c = |\lambda_{y_1,T_1}|V = |\lambda_{y_1,T_1}|(R+1)D} \tag{1.46}$$

where q_c is the thermal power (or heat duty) at the condenser, $|\lambda_{y_1,T_1}|$ is the molar latent heat of liquid-vapour phase transition of the vapour mixture leaving the top of the column and sent to the condenser, with composition y_1 and at temperature T_1 (see the general scheme in Fig. 1.10). The value of $|\lambda_{y_1,T_1}|$ can be calculated, given the composition and temperature of V, and given the molar latent heats of liquid-vapour phase transition of the pure components at the same temperature, $|\lambda_{A,T_1}|$ and $|\lambda_{B,T_1}|$ (recalling that A is the low-boiling one), as:

$$|\lambda_{y_1,T_1}| = y_1|\lambda_{A,T_1}| + (1-y_1)|\lambda_{B,T_1}| \tag{1.47}$$

where the $\lambda(T)$ functionality is (Green and Perry 2008):

$$|\lambda| = C_1(1-T_r)^{C_2+C_3T_r+C_4T_r^2} \tag{1.48}$$

with T_r is the reduced temperature, equal to the ratio between the actual temperature and the critical temperature (both expressed in [K]). Values for the critical temperature and for the constants $C_1 \ldots C_4$ as function of the substances are tabulated in literature (Green and Perry 2008).

For a partial condenser, instead, the energy balance can be written as:

$$\boxed{q_c = |\lambda_{x_0,T_0}|L} \tag{1.49}$$

where $|\lambda_{x_0,T_0}|$ is the molar latent heat of liquid-vapour phase transition of the liquid mixture leaving the condenser with composition x_0 and at temperature T_0 (in fact, in this case only a fraction of V entering the condenser, i.e. L, is actually condensed, while D remains in vapour phase).

Table 1.4 Constants for the evaluation of the molar latent heat of liquid-vapour phase transition

Species	C_1	C_2	C_3	C_4
Acetonitrile	$4.3511 \cdot 10^7$	0.34765	0	0
Acetic acid	$4.0179 \cdot 10^7$	2.6037	−5.0031	2.7069

Example: Evaluation of the Heat Duty at a Total Condenser

Consider a vapour stream having molar flow rate $V = 707.57$ kmol/h, exiting the first stage of a distillation column, and composed by 95% acetonitrile (low-boiling A compound, $y_1 = 0.95$) and 5% acetic acid (B compound) at the equilibrium temperature of $T_1 = 357$ K. This stream must be treated in a total condenser.

The critical temperature (Green and Perry 2008) is 545.5 K for A, and 591.95 K for B. Thus, the reduced temperature T_r is 0.654 for A and 0.603 for B. Table 1.4 lists the values of the constants for the application of Eq. 1.48, that yields the latent heat in [J/kmol].

We obtain $|\lambda_{A,T_1}| = 30.07 \cdot 10^3$ kJ/kmol and $|\lambda_{B,T_1}| = 23.71 \cdot 10^3$ kJ/kmol. Via Eq. 1.47, we get $|\lambda_{y_1,T_1}| = 29.75 \cdot 10^3$ kJ/kmol and, finally by Eq. 1.46, the total heat duty at the condenser, $q_c = 2.10 \cdot 10^7$ kJ/h.

As the reboiler is concerned (cf. Fig. 1.10: only a fraction of $\overline{L}$, entering the reboiler, is vaporised, and this corresponds to the $\overline{V}$ stream), the energy balance reads:

$$\boxed{q_r = |\lambda_{y_N,T_N}|\overline{V}} \tag{1.50}$$

where q_r is the thermal power at the reboiler, and $|\lambda_{y_N,T_N}|$ is the molar latent heat of liquid-vapour phase transition of the vapour mixture leaving the reboiler with composition y_N and temperature T_N (see Fig. 1.10). Again, $|\lambda_{y_N,T_N}|$ can be calculated using a similar approach as presented in Eqs. 1.47 and 1.48.

Of course, a cold (e.g., water) and a hot (e.g., steam) service fluid will be required at the condenser and reboiler, respectively. These aspects will be discussed in the chapter dedicated to heat exchangers.

1.9 The Murphree's Efficiency

In the previous discussion, it was assumed that each stage of a distillation column is *ideal*, implying that the liquid and vapour leaving each stage are in equilibrium. In practice, however, this condition is not achieved due to the finite liquid-vapour contact time, and because the composition of the vapour and liquid varies across the tray cross-section (McCabe et al. 2001). If the deviations from ideality are not negligible, a useful parameter for quantifying this behaviour is the *Murphree's*

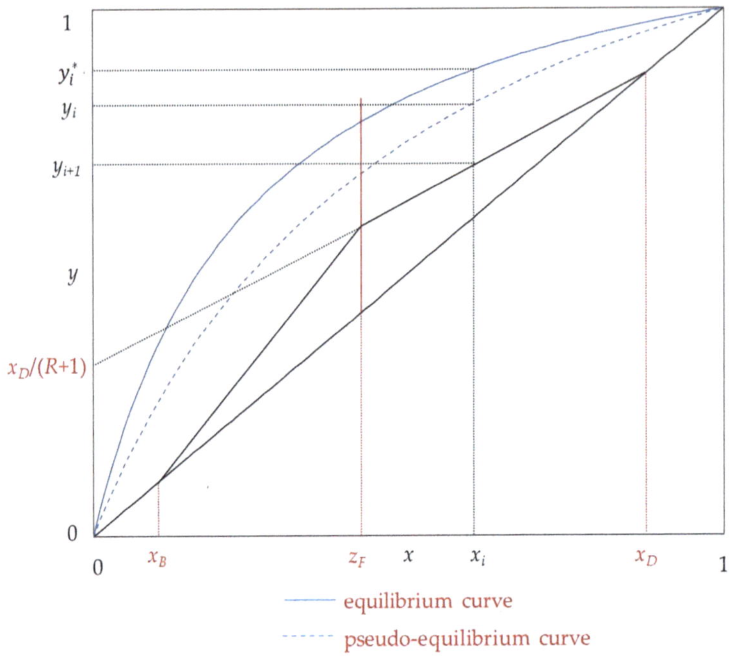

Fig. 1.19 Pseudo-equilibrium curve derived from the Murphree's efficiency of the distillation tray

efficiency η_M (alternative efficiencies for distillation columns are defined in McCabe et al. 2001), which for a generic *i*-th stage is (Murphree 1925):

$$\boxed{\eta_M = \frac{y_i - y_{i+1}}{y_i^* - y_{i+1}}} \tag{1.51}$$

where y_i is the actual mole fraction (of the more volatile component) in the vapour leaving tray *i*-th, y_{i+1} is the mole fraction in the vapour entering tray *i*-th, and y_i^* is the vapour composition that would be in equilibrium with the liquid leaving the same tray. It is evident that, for a real (non-ideal) tray, $y_i < y_i^*$ and therefore $\eta_M < 1$; in other words, the enrichment per tray of the vapour (in the more volatile component) is lower than in the ideal case. Starting from Eq. 1.51, it is possible to construct a *pseudo-equilibrium curve*. Take the example in Fig. 1.19 (please also refer to the scheme in Fig. 1.10): for the tray *i*, x_i is the composition of the liquid leaving the stage, y_{i+1} is the composition of the vapour entering the stage (known from the operating line), and y_i^* is the composition of the ideal vapour in equilibrium with x_i (known from equilibrium data). The denominator in Eq. 1.51, $y_i^* - y_{i+1}$, is known. Once the efficiency is given (often, from literature or semi-empirical correlations), the numerator in Eq. 1.51, $y_i - y_{i+1}$, is known, from which the actual value of y_i ($< y_i^*$) is calculated as:

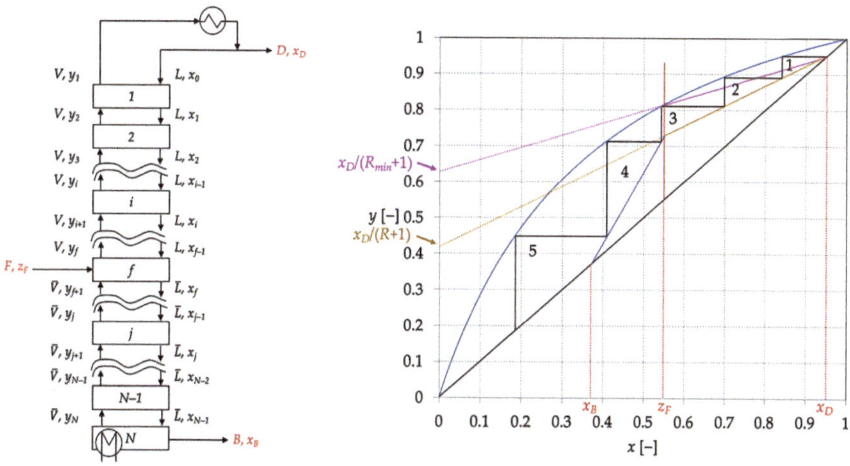

Fig. 1.20 McCabe-Thiele's method applied to the design of a fractionating column for acetonitrile/ acetic acid separation (molar fractions are referred to acetonitrile). The scheme of the fractional distillation column is re-drawn here for clarity

$$y_i = y_{i+1} + \eta_M \left(y_i^* - y_{i+1} \right) \tag{1.52}$$

Iterating the procedure (for a number of x-values as high as possible), the pseudo-equilibrium curve of Fig. 1.19 can be obtained. At this point, the McCabe-Thiele's method can be applied alternating between pseudo-equilibrium curve and operating lines. Of course, the reduced distance between operating lines and pseudo-equilibrium curve, with respect to the equilibrium one, diminishes the driving force, and ultimately increases the number of stages with respect to the ideal case.

Example: Design of a Fractionating Column

Consider the distillation of an ideal binary mixture composed of acetonitrile and acetic acid, where acetonitrile is the more volatile component. The system, at a pressure of 1 atm, presents an equilibrium curve already obtained in a previous example. Referring to the diagram in Fig. 1.20, let the feed be F = 1000 kmol/h (calculation basis) with $z_F = 0.55$. We wish to separate F into a distillate with $x_D = 0.95$ (therefore, richer in acetonitrile than F) and a bottoms product with $x_B = 0.37$ (therefore, richer in acetic acid than F). These data allow us to apply Eqs. 1.20 and 1.21, so to obtain $D = 310.34$ kmol/h and $B = 689.66$ kmol/h. This translates, Eq. 1.22, to a recovery $Rec\% = 53.6\%$.

Let the feed F in the physical state of a saturated liquid. Therefore, Table 1.3, $q = 1$ and, Fig. 1.13, the feed line will be a vertical meeting the diagonal in z_F (Fig. 1.20).

(continued)

In this case the equilibrium curve is concave downward, so the pinch condition (R_{min}) is determined by drawing the limiting rectifying line from (x_D;x_D) through the intersection point of the feed line with the equilibrium curve: we get the intercept on the y-axis $x_D/(R_{min} + 1) = 0.63$, and consequently $R_{min} = 0.51$ (Fig. 1.20). Assume an operating reflux ratio $R = 2.5R_{min} = 1.28$. The associated rectifying line has intercept on the y-axis $x_D/(R + 1) = 0.42$. We can also draw (Fig. 1.20) the stripping line passing through (x_B;x_B) and the intersection point of the other two (rectifying and feed) lines, graphically determined.

We can then apply the stepwise construction, alternating between equilibrium curve and operating lines, so to obtain a total number of stages $N = 5$ (including the reboiler) and optimal feed location at the third stage from the top of the column. In fact, working with 5 stages allows us to go even beyond the required specifications in terms of x_B; having $N = 5$ is still necessary, otherwise we would not obtain the required x_B value (Fig. 1.20).

Data also allow to determine the internal flow rates. In the rectifying section, it is $L = RD = 397.23$ kmol/h and $V = D + L = 707.57$ kmol/h. This also lead us to the evaluation of the thermal power at the condenser, which is $q_c = 2.10{\cdot}10^7$ kJ/h as already obtained in a previous example. To obtain the flow rates in the stripping section, we must use the fundamental Eq. 1.34. It is, in our case, $\overline{L} = L + qF = L + F = 1397.23$ kmol/h and $\overline{V} = V + (q-1)F = V = 707.57$ kmol/h. We can also confirm that, Fig. 1.20, $\overline{L} = \overline{V} + B$.

Example: Verification of a Non-ideal Fractionating Column

Consider the fractionation of the binary system acetonitrile/acetic acid already reported in the above example, starting now from a feed under saturated vapour conditions ($q = 0$). Assume the following specifications: $F = 1000$ kmol/h, $z_F = 0.55$, $x_D = 0.95$ and $x_B = 0.1$. The solution of the material balances in Eqs. 1.20 and 1.21 yields $D = 529.41$ kmol/h and $B = 470.59$ kmol/h.

The operation must be conducted in an existing tray column with $N = 8$ and Murphree's stage efficiency $\eta_M = 0.8$.

Firstly, Fig. 1.21, once known the equilibrium curve (same as above), we plot the horizontal feed line meeting the diagonal in z_F ($q = 0$; cf. Table 1.3 and Fig. 1.13). This problem can be solved by a trial-and-error procedure, according to the following algorithm:

- choose an operating value for R;
- draw the operating lines;

(continued)

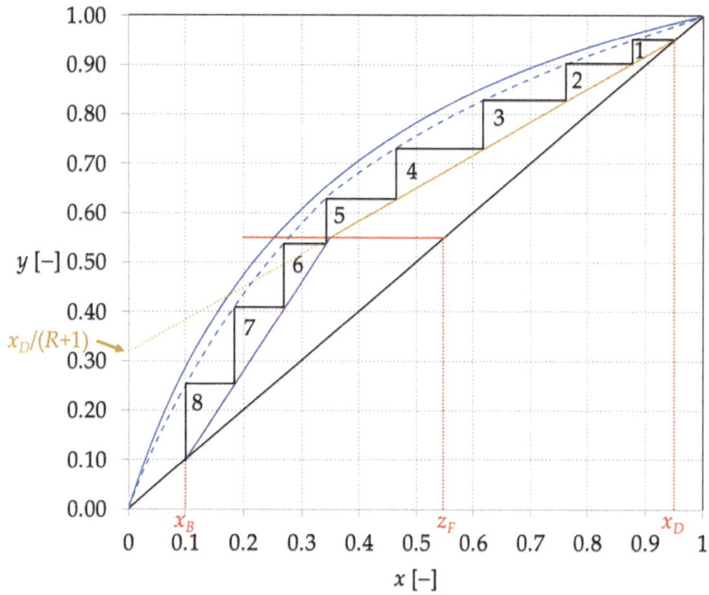

Fig. 1.21 McCabe-Thiele's method applied to the verification of a non-ideal fractionating column for acetonitrile/acetic acid separation (molar fractions are referred to acetonitrile)

- draw the pseudo-equilibrium curve adopting the assigned η_M;
- apply the McCabe-Thiele's method and verify $N = 8$.

We now report the solution of the trial-and-error procedure. For $R = 2$, we draw the rectifying line passing through (x_D;x_D) and the y-axis intercept $x_D/(R + 1) = 0.32$. The stripping line passes through (x_B;x_B) and the intersection point of the other two (rectifying and feed) lines. Now, we are able to build the pseudo-equilibrium curve (dashed line in Fig. 1.21, following the approach illustrated in Fig. 1.19 and Eq. 1.52). It is seen that the two (rectifying and stripping) operating lines fall below the pseudo-equilibrium curve, so the choice on R is conceptually valid. Now, applying the stepwise construction alternating between operating and pseudo-equilibrium conditions, we verify that the chosen $R = 2$ allows the desired separation with a total of 8 stages and with feed located at $N = 5$ from top of the column.

Concluding, data also allow to determine the internal flow rates. In the rectifying section, it is $L = RD = 1058.82$ kmol/h and $V = D + L = 1588.23$ kmol/h. To obtain the flow rates in the stripping section, use Eq. 1.34: $\overline{L} = L + qF = L = 1058.82$ kmol/h and $\overline{V} = V + (q-1)F = V - F = 588.23$ kmol/h. We can also confirm that $\overline{L} = \overline{V} + B$.

1.10 Background Information

- Liquid-vapour equilibria are described through relative volatility, Raoult's and Dalton's laws, the equilibrium curve (Eq. 1.8) and related diagrams (Fig. 1.1).
- Flash distillation is analysed using the temperature-composition diagram and material balances, with the lever rule applied for determining vapour and liquid compositions and flow rates (Eqs. 1.17–1.19 and Fig. 1.7).
- Continuous distillation with reflux (Fig. 1.10) is treated under the constant molar flow rates and ideal stage hypotheses, by defining rectifying (Eq. 1.28 and Fig. 1.11) and stripping (Eq. 1.31 and Fig. 1.12) operating lines.
- The feed conditions are represented by the parameter in Eq. 1.34 and Table 1.3, with the feed line illustrated in Fig. 1.13.
- The feed, rectifying, and stripping lines do intersect at one point (Fig. 1.14).
- The McCabe-Thiele's graphical method provides a design and verification tool to determine the number of stages and feed tray location, alternating between equilibrium and operating conditions (Fig. 1.15).
- The reflux ratio varies between a minimum value corresponding to a pinch condition (for which the number of stages would be infinite) and an infinite value (for which the number of stages would be minimum). Both conditions are non-operative and certify the existence of an optimal reflux ratio value, obtained by appropriately increasing the minimum value, on the basis on economic evaluations (Fig. 1.18).
- Deviations from the ideal stage model are accounted for through the Murphree's efficiency (Eq. 1.51 and Fig. 1.19).

References

Atkins P, de Paula J, Keeler J (2022) Atkins' physical chemistry. Oxford University Press, Oxford

Blackadder DA, Nedderman RM (1971) A handbook of unit operations. Academic, New York

Fenske MR (1932) Fractionation of straight-run Pennsylvania gasoline. Ind Eng Chem 24:482–485. https://doi.org/10.1021/ie50269a003

Green DW, Perry RH (2008) Perry's chemical engineers' handbook. McGraw-Hill, New York

McCabe WL, Smith JC, Harriott P (2001) Unit operations of chemical engineering. McGraw-Hill, Singapore

McCabe WL, Thiele EW (1925) Graphical design of fractionating columns. Ind Eng Chem 17:605–611. https://doi.org/10.1021/ie50186a023

Murphree EV (1925) Rectifying column calculations. Ind Eng Chem 17:747–750. https://doi.org/10.1021/ie50187a044

National Institute of Standards and Technology (NIST) Chemistry WebBook (2025) https://webbook.nist.gov/

Seader JD, Henley EJ (2006) Separation process principles. John Wiley & Sons, Hoboken

Smith JM, Van Ness HC, Abbott MM, Swihart MT (2020) Introduction to chemical engineering thermodynamics. McGraw-Hill, New York

Treybal RE (1981) Mass-transfer operations. McGraw-Hill, Singapore

Liquid-Liquid Extraction

2

Abstract

This chapter presents the fundamentals of the other stage unit operation chosen to open this book, i.e. liquid-liquid extraction, focusing on the transfer of a single solute from a carrier solvent into an extracting one under the assumption of ideal (equilibrium) stages. The discussion begins with graphical tools for representing ternary mixtures, with particular emphasis on triangular diagrams to solve mass balances. The principles of liquid-liquid equilibria are then presented, highlighting binodal curves, tie lines, distribution coefficients, and selectivity, all of which guide the choice of a suitable solvent. Attention then turns to the single-stage extraction process, where overall and component balances are treated graphically via the Hunter-Nash's method, and the operating limits of solvent flow rate are identified. The treatment is extended to multistage arrangements, including crossflow and countercurrent schemes. For each configuration, design and verification strategies are outlined, showing how multiple stages increase recovery efficiency, while overcoming the inherent constraints of single-contact operations.

Nomenclature

B	Mass flow rate of extracting solvent [mass/time]
$B^{\diamond}$	Mass flow rate of extracting solvent for which separation can be attained with a single stage [mass/time]
B_E	Mass flow rate of extracting solvent in extract [mass/time]
B_{max}	Maximum mass flow rate of extracting solvent [mass/time]
B_{min}	Minimum mass flow rate of extracting solvent [mass/time]
B_R	Mass flow rate of extracting solvent in raffinate [mass/time]
D_A	Distribution coefficient of the carrier solvent [–]
D_C	Distribution coefficient of the solute [–]

F. Montagnaro et al., *Elements of Unit Operations and Chemical Reactors*,
https://doi.org/10.1007/978-3-032-18046-9_2

E	Mass flow rate of extract [mass/time]
E'	Mass flow rate of extract on extracting solvent-free basis [mass/time]
E'^{*}	Mass flow rate of extract on extracting solvent-free basis with maximum purity of C in the extract [mass/time]
F	Mass flow rate of feed [mass/time]
N	Total number of stages [–]
R	Mass flow rate of raffinate [mass/time]
R'	Mass flow rate of raffinate on extracting solvent-free basis [mass/time]
$Rec\%$	Recovery of solute in extract [–]
x_A	Mass fraction of the carrier solvent [–]
x_B	Mass fraction of the extracting solvent [–]
x_C	Mass fraction of the solute [–]
$x_i(F)$	Mass fraction of the *i*-component in feed [–]
$x_i(R)$	Mass fraction of the *i*-component in raffinate [–]
$x_i(R')$	Mass fraction of the *i*-component in raffinate on extracting solvent-free basis [–]
$y_i(E)$	Mass fraction of the *i*-component in extract [–]
$y_i(E')$	Mass fraction of the *i*-component in extract on extracting solvent-free basis [–]
α	Selectivity [–]
Δ	Difference mass flow rate (net mass flow) between the extract and raffinate at the interstage [mass/time]
Δ_{min}	Difference mass flow rate (net mass flow) between the extract and raffinate at the interstage corresponding at B_{min} [mass/time]
Σ	Mass flow rate of the mixing point [mass/time]
$\Sigma^{\diamond}$	Mass flow rate of the mixing point corresponding at $B^{\diamond}$ [mass/time]
Σ^{*}	Mass flow rate of the mixing point corresponding at the maximum purity of C in the extract [mass/time]
Σ_{max}	Mass flow rate of the mixing point corresponding at B_{max} [mass/time]
Σ_{min}	Mass flow rate of the mixing point corresponding at B_{min} [mass/time]

2.1 Generalities

Liquid-liquid extraction (sometimes also termed LLE) is a separation operation in which a solute (multicomponent extraction is not considered in the following) is removed from a liquid phase, whose major constituent is usually referred to as the *carrier solvent*, by selective transfer into an *extracting solvent* (for further details not covered here, please refer to specialised literature such as Blackadder and Nedderman 1971; Green and Perry 2008; McCabe et al. 2001; Seader and Henley 2006; Treybal 1981). This process is based on the different solubility of the solute in the two (carrier and extracting) liquid phases. Examples of industrial applications of LLE include the separation of hydrocarbons in the petroleum industry, the purification of uranium in the nuclear industry, the recovery of antibiotics from fermentation broths obtained from organic substrates, and the extraction of flavours and fragrances in the food industry. Extraction will be considered as *stationary* and non-reactive. Thus, as in distillation, the material balances (here expressed in mass flow rates, i.e. [mass/time]) will be written in the simple form "IN = OUT".

Liquid-liquid extraction is generally applied as a separation technique instead of distillation in, e.g., the following cases:

- the compounds to be separated have a relative volatility close to unity or the solute is at very low concentration, so that fractional distillation would require a high number of stages and large energy input;
- the compounds to be separated are thermolabile;
- the formation of azeotropes limits the degree of separation achievable by distillation.

2.2 Principles of Liquid-Liquid Extraction

2.2.1 Triangular Diagrams

The representation of the compositions of ternary systems of interest for liquid-liquid extraction is usually carried out by means of *triangular diagrams*. The Hunter-Nash's graphical method, later described, makes use of triangular phase diagrams to solve design or verification problems of single-stage and multistage liquid-liquid extraction equipment (Hunter and Nash 1934). For alternative graphical methods applied to LLE, the scholar is referred to specific texts such as, for example, McCabe et al. 2001; Seader and Henley 2006.

The most used diagrams are *equilateral triangles* shown in Fig. 2.1. Let us consider a three-component system A, B, and C representing the carrier solvent (A), the extracting solvent (B), and the solute (C). The vertices of the triangle correspond to the pure components, while each side of the triangle represents the binary mixture of the two components located at the corresponding ends. Each interior point of the triangle represents a mixture containing all three components. In addition, on each axis the mass fraction (or mass percentage) of one component is reported, once a proper reading direction (clockwise or counterclockwise) is defined. So, x_A, x_B, and x_C represent the mass fractions of components A, B, and C, respectively. Let the point P be located inside the triangle, thus representing a ternary mixture. The composition relative to one component can be determined by drawing from point P a line parallel to the side opposite to the vertex representing that component, until it intersects the axis reporting its composition (see arrows in Fig. 2.1). For example, if x_A is desired, draw from P the line parallel to the side $\overline{BC}$ and read where this line intersect the x_A-axis. In our example, it is $x_A = 40\%$ (and, extending the concept, we obtain $x_B = 30\%$ and $x_c = 30\%$; of course, the *congruence condition* $x_A + x_B + x_c = 1$ must hold).

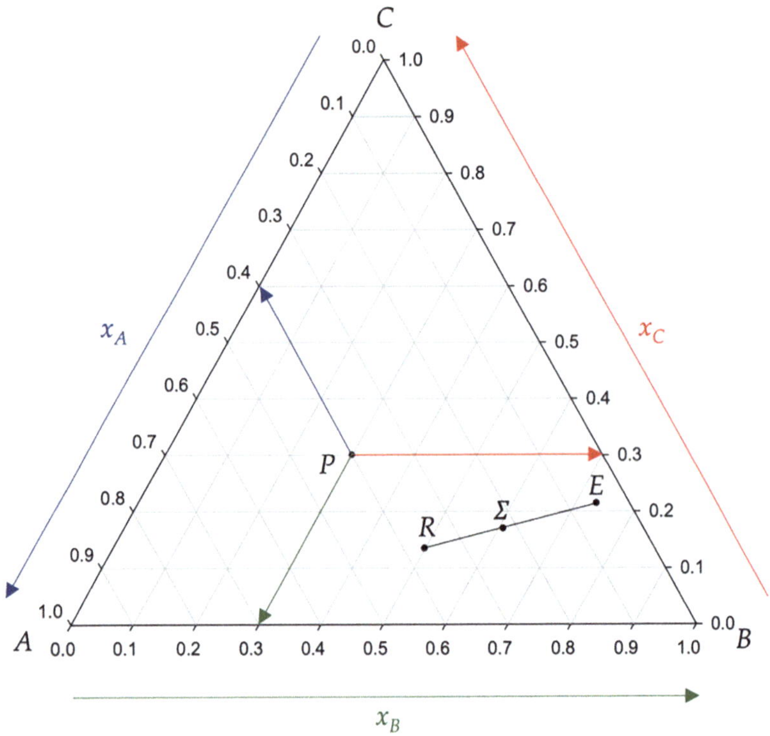

Fig. 2.1 Equilateral triangle diagram for a ternary system. A, B and C represent the carrier solvent, the extracting solvent and the solute, respectively

The use of triangular diagrams is very convenient for graphically solving material balance problems in liquid-liquid extraction. Indeed, these diagrams satisfy the alignment condition for streams resulting from the mixing or splitting of mixtures. Let, Fig. 2.1, R and E (the reason for these two symbols will be clear later) be two liquid phases mixed to give a resulting stream Σ (symbols also denote the streams mass flow rates for a continuous process): the point representing Σ lies on the segment connecting the points representing the original mixtures R and E. Moreover, the position of Σ can be determined by the already-presented *lever rule*:

$$\boxed{\frac{R}{E} = \frac{\overline{\Sigma E}}{\overline{R\Sigma}}} \tag{2.1}$$

Alternative types of representation of ternary systems, other than the equilateral triangle, involve the use of right-angled triangles, that can be employed to expand one composition scale relative to another. For instance, the mass fraction of C can be plotted on the vertical axis as a function of the mass fraction of B on the horizontal axis (Fig. 2.2). Once the mass fractions of B and C are identified as Cartesian coordinates, the fraction of the third component (A) can be calculated from the

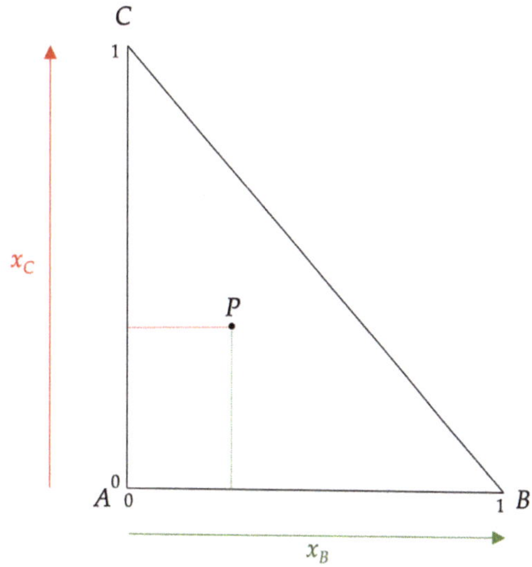

Fig. 2.2 Right-angled triangular diagram for a ternary system

mass fraction congruence relation $x_A + x_B + x_C = 1$. The lever rule for mixtures is also applicable to this coordinate system.

2.2.2 Liquid-Liquid Equilibria

The equilibrium conditions for liquid phases in a ternary system, at fixed temperature and pressure, can conveniently be represented using triangular diagrams. Most systems are such that one pair of components (the carrier solvent and the extracting solvent) are only *partially miscible*, while the other two pairs are totally miscible. In the example shown in Fig. 2.3, component C (the solute to be extracted) is completely soluble in both A (the carrier solvent from which C has to be extracted) and B (the solvent that extracts C), whereas A and B are only partially miscible (this helps B in selectively extracting C from the A-C mixture).

A binary system of A and B with composition represented by point J on the diagram separates into two immiscible equilibrium phases, whose compositions are given by points L and M. The curve that separates the region in which a homogeneous solution is thermodynamically stable from that in which two liquid phases coexist at equilibrium is called the *binodal solubility curve*, and the region beneath it is called the *miscibility gap*. For example, Fig. 2.3, point Z located outside the miscibility gap represents a single-phase system. A point Σ inside the miscibility gap results in the formation of two separate homogeneous solutions, R and E, in equilibrium, located on the solubility curve. The phases R and E are richer respectively in carrier solvent A and extracting solvent B (as it can be seen by applying the procedure illustrated in Fig. 2.1), and they are called *conjugate solutions*. These phases are also referred, in LLE, to as *raffinate* (R) and extract (E), respectively, and

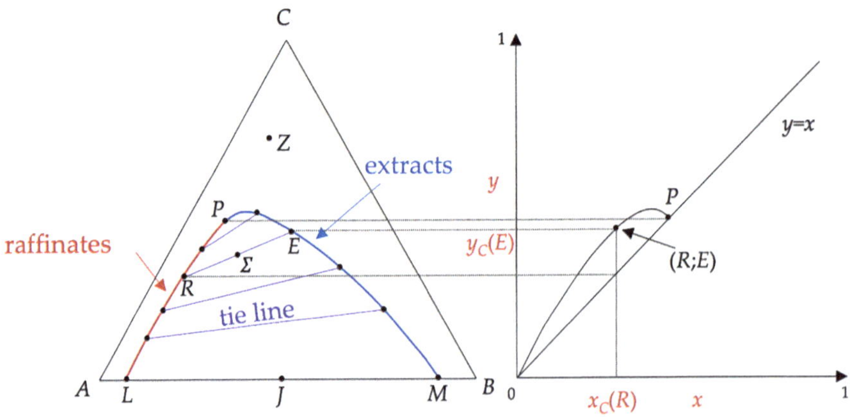

Fig. 2.3 On the left: triangular diagram for a ternary system with a single pair of partially miscible components. On the right: distribution curve of the solute ($D_C > 1$)

the segment connecting them is called the *conjugate line* or *tie line*. In the system under scrutiny (Fig. 2.3), there also exists a *critical point* on the miscibility curve, called the *plait point P*, at which a single homogeneous solution exists. The plait point separates the *raffinate branch* (curve PL) from the *extract branch* (curve PM).

The partitioning of the solute C between raffinate and extract at equilibrium can be expressed by the *distribution coefficient* of the solute D_C, which represents the ratio between the mass fraction of the solute in the extract $y_C(E)$ and that in the raffinate $x_C(R)$:

$$\boxed{D_C = \frac{y_C(E)}{x_C(R)}} \tag{2.2}$$

By plotting, on a Cartesian diagram, all pairs of compositions of C at the ends of a tie line, one obtains the *distribution curve* (Fig. 2.3). The plait point P lies on the diagonal $y = x$ of this diagram. Constructing the distribution curve is very simple if the lengths of the x- and y-axes are set equal to the height of the equilateral triangle. Considering, for example, the tie line $\overline{RE}$ in Fig. 2.3:

- the abscissa $x_C(R)$ on the x-y diagram is obtained by drawing a horizontal line from point R to the diagonal, and projecting the value to the x-axis;
- the ordinate $y_C(E)$ is obtained by drawing a horizontal line from point E to the y-axis.

The point with coordinates ($x_C(R)$;$y_C(E)$) represents the distribution of C in the raffinate R and in the extract E at equilibrium. The position of the distribution curve with respect to the diagonal depends on the slope of the tie lines. In the case examined, we have $D_C > 1$. Since usually only some tie lines are determined experimentally, interpolation of the known points of the distribution curve can be

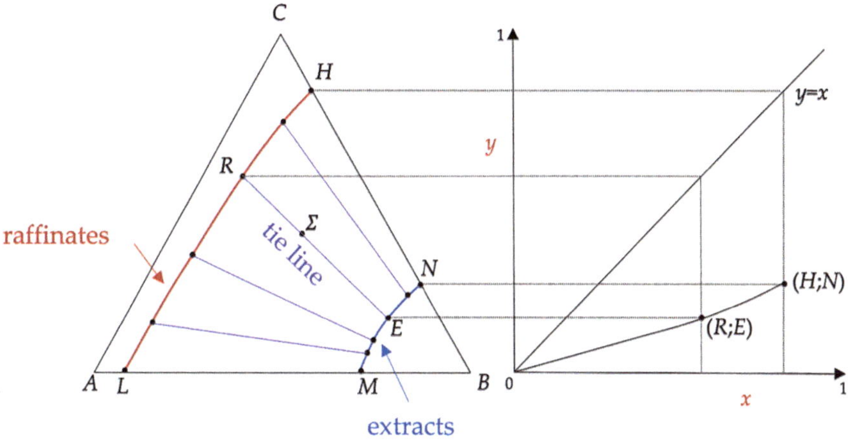

Fig. 2.4 On the left: triangular diagram for a ternary system with two pairs of partially miscible components. On the right: distribution curve of the solute ($D_C < 1$)

used to identify the unknown tie lines, using a procedure inverse to the one described above.

Some ternary systems, such as aniline/*n*-heptane/methylcyclohexane (McCabe et al. 2001), exhibit two partially miscible pairs. With reference to Fig. 2.4, A and C are completely soluble (as usual), whereas pairs A-B (as usual) and B-C (as not usual) are only partially soluble. The curves LH and MN represent here the raffinate and extract branches, respectively. The region of the triangle between the two curves identifies the miscibility gap (two-phase system), while points outside the miscibility gap represent homogeneous ternary solutions. In the considered case, the distribution curve lies below the diagonal, and therefore $D_C < 1$. *Solutropic systems*, instead, exhibit tie lines whose slope changes direction, passing through a horizontal tie line (QS in Fig. 2.5, whose characteristic point on the Cartesian diagram must lie on the diagonal). As a result, the distribution curve intersects the diagonal.

For ternary systems that exhibit partial miscibility of at least one pair of components, the *selectivity* α (analogous to the relative volatility in distillation) can be defined as the ratio between the distribution coefficient of the solute D_C and that of the carrier solvent D_A:

$$\boxed{\alpha = \frac{D_C}{D_A} = \frac{\frac{y_C(E)}{x_C(R)}}{\frac{y_A(E)}{x_A(R)}} = \frac{y_C(E)x_A(R)}{x_C(R)y_A(E)}} \tag{2.3}$$

where $x_A(R)$ and $y_A(E)$ represent the mass fraction of the component A in the raffinate R and extract E, respectively.

The selection of an extracting solvent to be adopted for a liquid-liquid extraction process depends on several factors, some of which are listed below (for a more detailed analysis cf., for example, Green and Perry 2008):

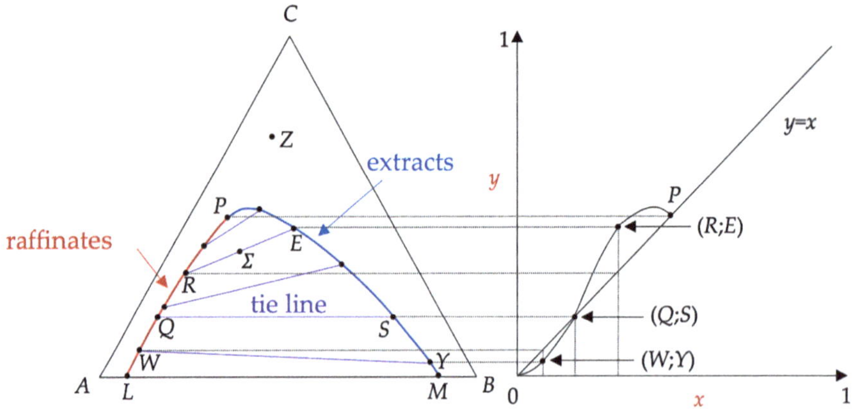

Fig. 2.5 On the left: triangular diagram for a ternary system exhibiting solutropic behaviour. On the right: distribution curve of the solute (D_C assumes values >1 and < 1)

- a value of the distribution coefficient greater than unity reduces solvent consumption and increases solute concentration in the extract, Eq. 2.2. Separation is still feasible when $D_C < 1$, although with high solvent requirements;
- the value of selectivity should be as large as possible, greater than unity, to ensure efficient separation. For $\alpha = 1$, extraction is not feasible. The higher the selectivity, the higher the efficiency of the extracting solvent B in embodying C rather than A upon LLE;
- the carrier A and the extracting B solvents should have low mutual solubility and different densities, to ease the separation;
- the extracting solvent must be chemically inert with respect to the other components of the system and to the construction materials of the equipment;
- the extracting solvent should be readily available at low cost and with minimum environmental impact.

2.3 Single-Stage Process

2.3.1 Mass Balances

The *single-stage* LLE process involves the use of a *single contact unit between two liquid phases*. In the following, the stage is assumed to be *ideal*, implying *that the two liquid phases exiting the unit are at thermodynamic equilibrium* (a similar concept has been encountered in distillation). A more detailed analysis of the stage efficiency concept, types of equipment and their design based on mass transfer rates are covered in specialised textbooks (Green and Perry 2008; Treybal 1981).

A scheme of a single-stage continuous LLE process is reported in Fig. 2.6. Let F be the mass flow rate of a binary mixture feed with mass fractions $x_C(F)$ and $x_A(F)$ for C and A, respectively, to be separated by means of a mass flow rate B of

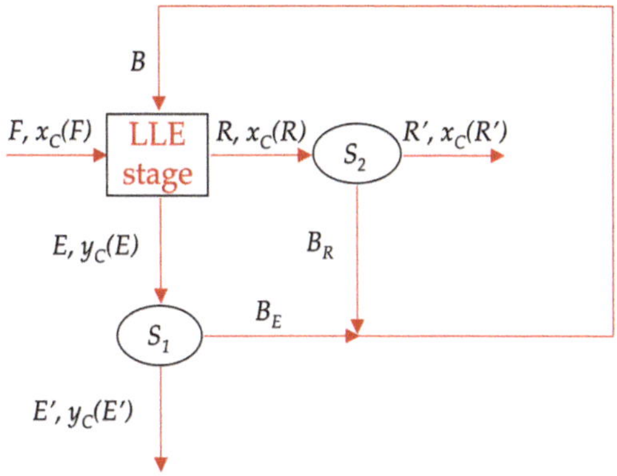

Fig. 2.6 Scheme of a single-stage continuous LLE process with extracting solvent separation units (denoted with S_1 and S_2). After separation, the streams of extracting solvent (B_R and B_E) are re-sent as stream B to the process

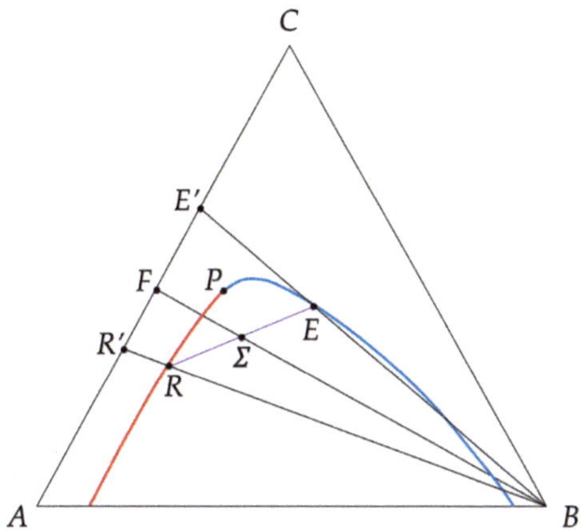

Fig. 2.7 Triangular diagram for single-stage LLE

extracting solvent. We assume, for simplicity, that the solvent is pure (fresh solvent, namely the stream B is made up of 100% B-component). In the industrial practice, the extracting solvent may contain non-negligible fractions of the carrier solvent A and solute C, if it is obtained as a recycle stream from a previous operation. The contact between the phases F and B, if mixed in appropriate proportions (see below), gives rise to the raffinate R and the extract E exiting the stage at equilibrium, with solute C composition of $x_C(R)$ and $y_C(E)$, respectively. To fix ideas, in the following discussion it is assumed that only the A-B pair exhibits partial solubility, and that $D_C > 1$ (Fig. 2.7).

The overall mass balance for the LLE stage in Fig. 2.6 can be expressed as:

$$\boxed{F + B = R + E} \tag{2.4}$$

The mass balances for B (fed as pure) and C components around the LLE stage can be respectively written as:

$$\boxed{B = Rx_B(R) + Ey_B(E)} \tag{2.5}$$

$$\boxed{Fx_C(F) = Rx_C(R) + Ey_C(E)} \tag{2.6}$$

where $x_B(R)$ and $y_B(E)$ denote the mass fractions of B in the raffinate and extract, respectively.

The mass fraction of A in raffinate $x_A(R)$ and extract $y_A(E)$ can be computed once known the mass fractions of B and C components in the raffinate R and extract E, by applying the congruence condition.

The recovery of C in the extract ($Rec\%$) is:

$$\boxed{Rec\% = 100\frac{Ey_C(E)}{Fx_C(F)} = 100\frac{Fx_C(F) - Rx_C(R)}{Fx_C(F)}} \tag{2.7}$$

The determination of the flow rates and compositions of the streams exiting the stage can be obtained using the Hunter-Nash's method, which applies the lever rule (being the graphical interpretation of the mass balances) and equilibrium conditions (represented by tie lines) to the ternary diagram.

Given the composition of the binary mixture F, it can be located on the ternary diagram along the side $\overline{AC}$ in Fig. 2.7. Let Σ be the stream resulting from mixing the B and F streams. It is located along the segment $\overline{BF}$, and its position can be identified using the lever rule once the mass flow rate value for B and F is known:

$$\frac{B}{F} = \frac{\overline{F\Sigma}}{\overline{\Sigma B}} \tag{2.8}$$

The formation of the two phases E and R, and consequently the separation process, is only possible if the mass flow rates of B and F mixed in the stage produce a mixing point Σ pertaining to the miscibility gap. Otherwise, meeting in the stage, B and F would produce a ternary system that cannot be separated in a phase richer in C (the extract) and a raffinate poorer in C.

The overall mass balance (Eq. 2.4) can therefore be written as:

$$\boxed{F + B = R + E = \Sigma} \tag{2.9}$$

Take the tie line passing through Σ (conceptually, there must exist): the point Σ also lies on this tie line $\overline{RE}$, and the lever rule says:

$$\frac{R}{E} = \frac{\overline{\Sigma E}}{\overline{R\Sigma}} \tag{2.10}$$

Coupling the overall mass balance Eq. 2.9 with the lever rule Eq. 2.10 applied graphically (i.e., knowing the length of the segments on the diagram), the flow rates R and E can be determined, as they are the two unknowns of the system of Eqs. 2.9 and 2.10. Their compositions can be read directly from the ternary diagram.

The extract and raffinate streams exiting the extraction stage can be sent to separation units (S_1 and S_2 in Fig. 2.6), typically distillation columns, to remove the solvent B and obtain two streams, which we will call *R'* and *E'*, respectively richer and poorer in component A with respect to the feed. In the case of complete solvent separation, the streams *R'* and *E'* are two binary mixtures of A and C, with mass fractions of C equal to $x_C(R')$ and $y_C(E')$, respectively (raffinate and extract compositions on a B-free basis). The overall mass balances on S_1 and S_2 are, respectively:

$$E = E' + B_E \tag{2.11}$$

$$R = R' + B_R \tag{2.12}$$

where B_E and B_R represent (assuming complete B solvent separation) the mass flow rates of extracting solvent (100% B) recovered from units S_1 and S_2, respectively.

The following mass balances also hold:

$$B = B_R + B_E \tag{2.13}$$

$$F = R' + E' \tag{2.14}$$

From Eqs. 2.11 and 2.12, the positions of the streams *R'* and *E'* can be easily located on the ternary diagram (Fig. 2.7). The streams *E*, *E'* and B_E (the latter assumed as pure extracting solvent, therefore positioned at the B-vertex) are aligned, with *E* being the mixing point; thus, the position of stream *E'* is obtained by extending the segment $\overline{BE}$ along the side $\overline{AC}$ (since *E'* is a binary A-C mixture). Similarly, the streams *R*, *R'* and B_R (the latter assumed as pure extracting solvent, therefore positioned at the B-vertex) are aligned, with *R* being the mixing point; thus, the position of stream *R'* is obtained by extending the segment $\overline{BR}$ along the side $\overline{AC}$ (since *R'* is a binary A-C mixture). Graphically, Fig. 2.7, the passionate reader can verify Eq. 2.14 as well.

2.3.2 Operating Limits of Extracting Solvent Flow Rate

The flow rates and compositions of the extract and raffinate exiting the stage depend not only on the equilibrium conditions of the system but also on the ratio *B*/*F*. Indeed, the latter determines the position of the mixing point Σ and, consequently, the tie line passing through it (that, in turn, determines *R* and *E*, Fig. 2.7). In particular, there are two limiting values within which the extracting solvent flow rate can vary in order to ensure the separation process. These limiting solvent flow rates are identified from the intersection of the segment $\overline{FB}$ with the binodal solubility curve.

Refer to Fig. 2.8. The *maximum extracting solvent flow rate* B_{max} is obtained when the mixing stream Σ_{max} (corresponding to the intersection of the segment $\overline{FB}$ with the extract branch) is considered. In fact, applying the lever rule (see Eq. 2.9):

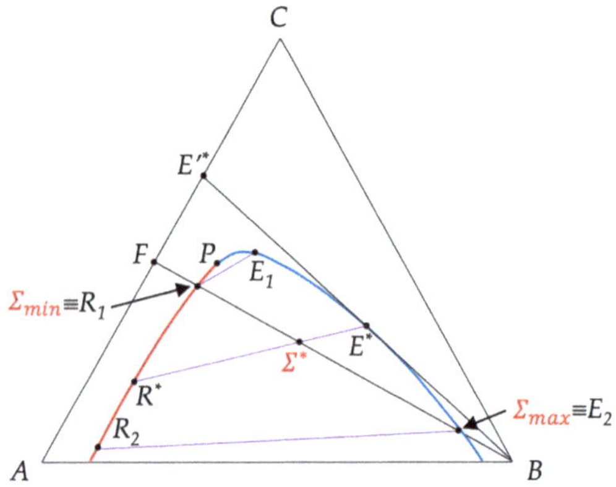

Fig. 2.8 Operating limits of the extracting solvent flow rate for single-stage LLE

$$\boxed{\frac{B_{max}}{F} = \frac{\overline{F\Sigma_{max}}}{\overline{\Sigma_{max}B}}} \tag{2.15}$$

the segment $\overline{F\Sigma_{max}}$ is the longest possible. Equation 2.15 allows to determine the value for B_{max}. Once known Σ_{max}, we can consider the tie line passing through it. We will call this tie line, for the purposes of Fig. 2.8, $\overline{R_2E_2}$. At this condition, the obtained raffinate (R_2) has the lowest possible concentration of component C. But, at the same time, the flow rate of R_2 is infinitesimal, as evident from the application of the lever rule (the point Σ_{max} coincides with E_2; see Eq. 2.9):

$$\frac{R_2}{E_2} = \frac{\overline{\Sigma_{max}E_2}}{\overline{R_2\Sigma_{max}}} \to 0 \tag{2.16}$$

The *minimum extracting solvent flow rate* B_{min} is instead obtained when the mixing stream Σ_{min} (corresponding to the intersection of the segment $\overline{FB}$ with the raffinate branch) is considered. In fact, applying the lever rule (see Eq. 2.9):

$$\boxed{\frac{B_{min}}{F} = \frac{\overline{F\Sigma_{min}}}{\overline{\Sigma_{min}B}}} \tag{2.17}$$

the segment $\overline{F\Sigma_{min}}$ is the shortest possible. Equation 2.17 allows to determine the value for B_{min}. Once known Σ_{min}, we can consider the tie line passing through it. We will call this tie line, for the purposes of Fig. 2.8, $\overline{R_1E_1}$. In this case, an infinitesimal extract flow rate E_1 is obtained (the point Σ_{min} coincides with R_1; see Eq. 2.9):

$$\frac{R_1}{E_1} = \frac{\overline{\Sigma_{min}E_1}}{\overline{R_1\Sigma_{min}}} \to \infty \tag{2.18}$$

Finally, the maximum purity of C in the extract (see Fig. 2.6; we will call this stream E'^{*}) is achieved for an extracting solvent flow rate such that the application of the lever rule between streams B and F gives a stream Σ^* which individuates the tie line

$\overline{R^*E^*}$, where the segment $\overline{BE^*}$ is tangent to the binodal solubility curve (Fig. 2.8). On the $\overline{AC}$ side of the triangle, no point of the "E' type" can be placed "above" (i.e., richer in C) E'^*.

The previous analysis shows that the compositions of the extract and raffinate vary within well-defined limits, as the extracting solvent flow rate B changes between B_{min} and B_{max}. To overcome the separation limitations imposed by a single extraction stage, multistage systems are used, as we will see in the following.

Example: Design of a Single-Stage LLE Process

An aqueous solution (water representing the A component) with mass flow rate $F = 5500$ kg/h and containing a mass fraction of formic acid (C component) $x_C(F) = 0.614$ must be processed in a single-stage LLE to recover the acid. The extracting solvent (B component) is a mixture tributyl phosphate/dodecane (termed TBD). Table 2.1 reports equilibrium data for the ternary system A/B/C at 25 °C. We wish to determine the recovery of formic acid obtainable using an extracting solvent mass flow rate equal to 50% of the maximum operating value ($B = 0.5B_{max}$).

As first step, we represent the feed and equilibrium data on the triangular diagram with the corresponding tie lines (Fig. 2.9). Of course, the binary mixture F will be placed on the $\overline{AC}$ side of the triangle, corresponding to a fraction of water (A) equal to $x_A(F) = 1–0.614 = 0.386$. The construction of the tie lines is in accordance with Table 2.1 caption. Data show that $y_C(E) < x_C(R)$ throughout the whole range, thus $D_C < 1$. After having

(continued)

Table 2.1 Equilibrium data for the ternary system formic acid/water/TBD at 25 °C (Malmary et al. 1997). For each row, the raffinate and the extract phases are in equilibrium and, therefore, data on the same row represent points connected by a tie line in Fig. 2.9

Tie line data (mass fractions)					
Raffinate phase			Extract phase		
Formic acid $x_C(R)$	Water $x_A(R)$	TBD $x_B(R)$	Formic acid $y_C(E)$	Water $y_A(E)$	TBD $y_B(E)$
0.000	0.996	0.004	0.000	0.027	0.973
0.065	0.931	0.004	0.050	0.029	0.921
0.126	0.870	0.004	0.068	0.030	0.902
0.179	0.817	0.004	0.078	0.030	0.892
0.234	0.761	0.005	0.090	0.031	0.879
0.286	0.709	0.005	0.097	0.031	0.872
0.324	0.669	0.007	0.104	0.033	0.863
0.370	0.619	0.011	0.111	0.033	0.856
0.409	0.577	0.014	0.118	0.033	0.849
0.471	0.479	0.050	0.131	0.039	0.830
0.490	0.391	0.119	0.133	0.040	0.827

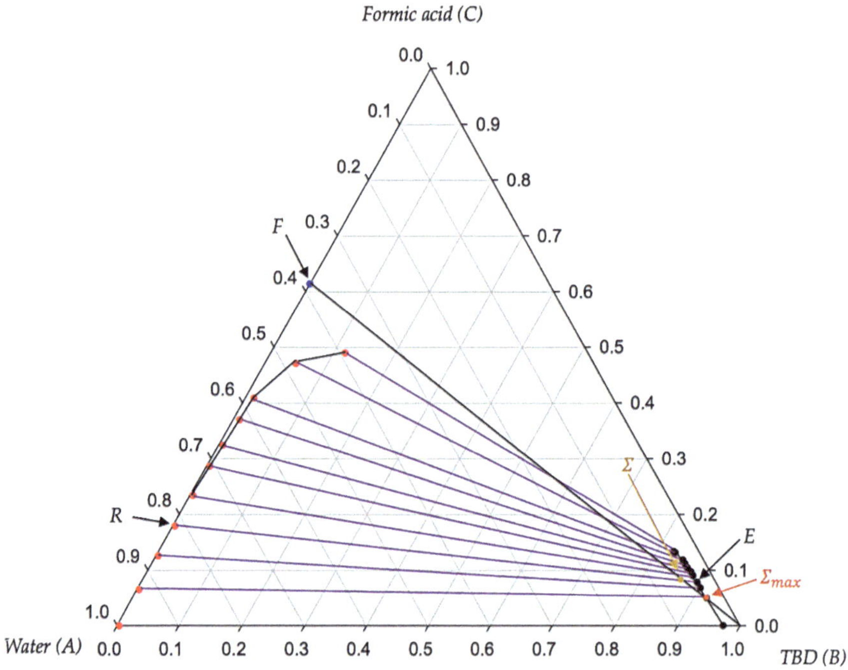

Fig. 2.9 Hunter-Nash's method applied for the design of a single-stage LLE process

determined the tie lines, we can also plot on Fig. 2.9 the raffinates and extracts branches, by linear interpolation of data.

The position of Σ_{max}, the point representing the mixing between F and B_{max}, is obtained as previously done in Fig. 2.8, i.e. by the intersection of the line $\overline{BF}$ with the binodal solubility curve at the point closer to B-vertex (that is, on the extract branch). The maximum extracting solvent mass flow rate can be determined from Eq. 2.15: graphically, we find $\overline{F\Sigma_{max}}\,/\overline{\Sigma_{max}B} = 13.55$, and we derive $B_{max} = 74{,}525$ kg/h and, therefore, the operating mass flow rate $B = 0.5B_{max} = 37{,}262.5$ kg/h. We can now determine the position of the actual mixing point Σ by applying the lever rule $B/F = (\overline{F\Sigma}/\overline{\Sigma B}) = (\overline{FB} - \overline{\Sigma B})/\overline{\Sigma B} \Rightarrow \overline{\Sigma B} = \overline{FB}/(1 + B/F)$ has a known length. It should be noted that the graphical length of segments depends on the scale adopted for drawing the diagram, but their ratio must obviously obey the lever rule. Of course, we end up with a mixing point Σ that, resulting from the sum of F and B (the latter, lower than B_{max}), is farther from the B-vertex than Σ_{max}. The obtained Σ lies on a given tie line (Fig. 2.9), so its endpoints directly identify the raffinate R and extract E exiting the

(continued)

equilibrium stage. It is worth recalling that if unknown tie lines are needed (i.e., not directly retrievable from the annexed table as Table 2.1 in our case), the distribution curve must be constructed following the approach in Figs. 2.3, 2.4 and 2.5. Now, we read on the graph $y_C(E) = 0.078$ and $x_C(R) = 0.179$. The mass flow rates of R and E can be computed by solving the system consisting of the overall material balance Eq. 2.9 and the lever rule Eq. 2.10:

$$\begin{cases} F + B = R + E = \Sigma = 42{,}762.5 \text{ kg/h} \\ \dfrac{R}{E} = \dfrac{\overline{\Sigma E}}{\overline{R\Sigma}} = 0.033 \text{ (graphically)} \end{cases} \Rightarrow \begin{cases} R = 1366.1 \text{ kg/h} \\ E = 41{,}396.4 \text{ kg/h} \end{cases} \tag{2.19}$$

The recovery of the solute in E is obtained as (Eq. 2.7) $Rec\% = 100Ey_C(E)/(Fx_C(F)) = 95.6\%$.

2.4 Multistage Crossflow Process

In the following, we will analyse cascade processes, maintaining the assumption of ideal (equilibrium) stage. As in the case of fractional distillation, for multistage extraction systems there are design and verification problems. In design problems, given the feed specifications, it is usually required to determine the number of stages N needed to achieve a certain purity of the raffinate (or a given solute recovery) at a fixed extracting solvent (B) flow rate; alternatively, it can be necessary to evaluate the number of stages and the extracting solvent flow rate required to achieve a desired separation in terms of extract and raffinate compositions. In verification problems, it is typically requested to assess whether an existing unit (fixed N) can accomplish a given separation process.

In a *multistage crossflow LLE*, the feed F is introduced into the first stage of the cascade. In each stage, we assume that pure extracting solvent (B) is added, and the raffinate leaving each stage is fed as the input to the next stage (Fig. 2.10). The

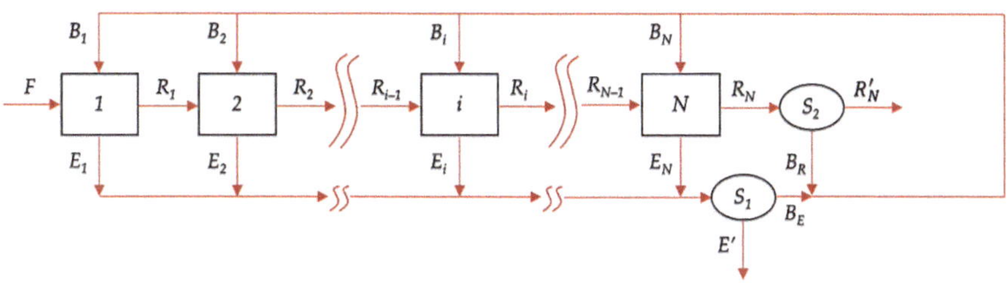

Fig. 2.10 Scheme of a multistage crossflow LLE process (separation units for recovery of the extracting solvent are denoted with S_1 and S_2). After separation, the streams of extracting solvent (B_R and B_E) are re-sent as streams B_1 ... B_N to the process

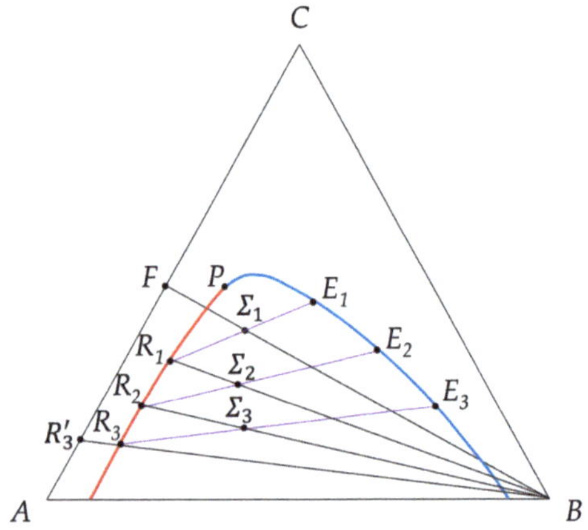

Fig. 2.11 Triangular diagram for multistage crossflow LLE (3 stages are considered in the example)

raffinate leaving the last stage can be sent to a separator for recovery of the extracting solvent, as can the extract stream obtained by mixing all the extracts leaving the stages of the extractor train. The extracting solvent streams (and their mass flow rates) are indexed according to the stage in which they are fed, as B_1, B_2 ... B_N, while the extract and raffinate streams are denoted as usual, i.e. with an index representing the stage from which they exit, E_1, E_2 ... E_N, and R_1, R_2 ... R_N, respectively.

For a more detailed analysis of the process, we consider a theoretical design example for which the cascade consists of 3 equilibrium stages, and that the solvent flow rates B_1, B_2 and B_3 are known (on a triangular graph, these three streams are represented by the B-vertex as they are assumed, as usual, composed of 100% B). The approach can then be extended to any number N of extraction stages. It is required to determine the purity of the raffinate R_3' (on a solvent-free basis) exiting the cascade. Let the ternary system be, for instance, characterised by a single A-B pair partially soluble and that $D_C > 1$. The example is depicted with the aid of Fig. 2.11.

The overall material balances for stages 1, 2, and 3 are given by, respectively (cf. Figure 2.10 with $N = 3$):

$$\boxed{F + B_1 = R_1 + E_1 = \Sigma_1} \tag{2.20}$$

$$\boxed{R_1 + B_2 = R_2 + E_2 = \Sigma_2} \tag{2.21}$$

$$\boxed{R_2 + B_3 = R_3 + E_3 = \Sigma_3} \tag{2.22}$$

The mixing of streams F and B_1 yields the mixing point Σ_1, whose location (known the values for F and B_1) is determined by the lever rule:

$$\frac{F}{B_1} = \frac{\overline{\Sigma_1 B}}{\overline{F\Sigma_1}} \tag{2.23}$$

The streams R_1 and E_1, leaving the first stage, are located at the ends of the tie line passing through Σ_1. Their flow rates can be determined by combining Eq. 2.20 with the lever rule:

$$\frac{R_1}{E_1} = \frac{\overline{\Sigma_1 E_1}}{\overline{R_1 \Sigma_1}} \tag{2.24}$$

Now, the mixing of streams R_1 and B_2 in the second stage gives the mixing point Σ_2, whose position on the segment $\overline{R_1 B}$ is again defined by the lever rule:

$$\frac{R_1}{B_2} = \frac{\overline{\Sigma_2 B}}{\overline{R_1 \Sigma_2}} \tag{2.25}$$

The streams R_2 and E_2, leaving the second stage, are located at the ends of the tie line passing through Σ_2, with mass flow rates determined by Eq. 2.21 combined with the lever rule:

$$\frac{R_2}{E_2} = \frac{\overline{\Sigma_2 E_2}}{\overline{R_2 \Sigma_2}} \tag{2.26}$$

Similar considerations apply to the third stage, where the lever rule allows the determination of the mixing stream Σ_3, resulting from the mixing of R_2 and B_3:

$$\frac{R_2}{B_3} = \frac{\overline{\Sigma_3 B}}{\overline{R_2 \Sigma_3}} \tag{2.27}$$

Finally, the extract and raffinate streams leaving the third stage are positioned at the endpoints of the tie line through Σ_3, and their flow rates are obtained by combining Eq. 2.22 with the lever rule:

$$\frac{R_3}{E_3} = \frac{\overline{\Sigma_3 E_3}}{\overline{R_3 \Sigma_3}} \tag{2.28}$$

Once the position of R_3 is known, we can observe, Fig. 2.10:

$$R'_3 + B_R = R_3 \tag{2.29}$$

where B_R is represented by the B-vertex. Thus, prolonging the line $\overline{BR_3}$ with the side $\overline{AC}$ (where R'_3 is located, as this is a stream free of B), it is possible to get the composition of R'_3. Application of the lever rule would allow to evaluate its flow rate, as well.

In general, it is observed that mixing a raffinate with fresh solvent generates a new raffinate that is poorer in component C than the one entering the stage. Therefore, the use of multiple extractors makes it possible to overcome the lower limit of C content in the raffinate observed for a single stage (by increasing the number of stages, a raffinate of desired purity in A can be obtained once the extracting solvent is removed).

Example: Verification of a Multistage Crossflow LLE Process

A liquid stream with a mass flow rate $F = 150$ kg/h and an equal-by-weight composition of methanol/benzene (methanol is the C component, benzene is the A component) must be treated in a two-stage LLE process operated in crossflow at 20 °C, in order to obtain a raffinate with benzene purity greater than 99%. Water (B component) is used as the extracting solvent, and its mass flow rate fed to the second stage is twice that fed to the first one. Table 2.2 reports equilibrium data for the ternary system methanol/benzene/water at 20 °

(continued)

Table 2.2 Equilibrium data for the ternary system methanol/benzene/water at 20 °C (Triday 1984). For each row, the raffinate and the extract phases are in equilibrium and, therefore, data on the same row represent points connected by a tie line in Fig. 2.12

Tie line data (mass fractions)					
Raffinate phase			Extract phase		
Methanol $x_C(R)$	Benzene $x_A(R)$	Water $x_B(R)$	Methanol $y_C(E)$	Benzene $y_A(E)$	Water $y_B(E)$
0.0007	0.9967	0.0026	0.0914	0.0020	0.9066
0.0055	0.9915	0.0030	0.1603	0.0026	0.8371
0.0114	0.9852	0.0034	0.2734	0.0036	0.7230
0.0199	0.9760	0.0041	0.4322	0.0073	0.5605
0.0371	0.9574	0.0055	0.5033	0.0245	0.4722
0.0667	0.9249	0.0084	0.5547	0.0400	0.4053
0.1339	0.8500	0.0161	0.6182	0.1279	0.2539
0.1708	0.8080	0.0212	0.6187	0.1496	0.2317
0.1789	0.7986	0.0225	0.6161	0.1698	0.2141

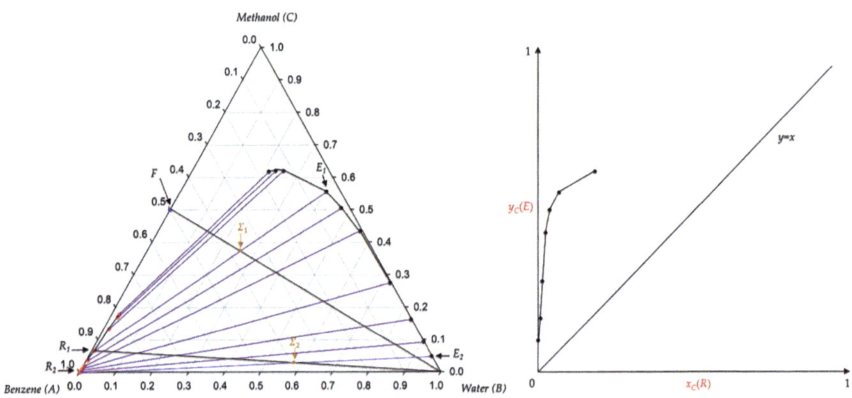

Fig. 2.12 Hunter-Nash's method applied for the verification of a multistage crossflow LLE process

C. We wish to determine the total solvent flow rate required to carry out the separation.

With given data, we plot (Fig. 2.12) the tie lines, distribution curve and feed composition ($x_C(F) = 0.5$). This problem can be solved by trial-and-error, fixing e.g. the flow rate for B_1 and verifying that $x_C(R_2) < 0.01$. The solution is verified with $B_1 = 51.7$ kg/h. In fact, first locate the position of the mixing point at the first stage, Σ_1, by the lever rule $F/B_1 = (\overline{\Sigma_1 B}/\overline{F\Sigma_1}) = 2.90$ and knowing (graphically) the sum of the two segments, $\overline{F\Sigma_1}+\overline{\Sigma_1 B} = \overline{FB}$. The tie line passing through Σ_1 identifies the streams exiting the first stage, E_1 and R_1. Their mass flow rates can be computed by solving the system:

$$\begin{cases} F + B_1 = R_1 + E_1 = \Sigma_1 = 201.7\ \text{kg/h} \\ \dfrac{R_1}{E_1} = \dfrac{\overline{\Sigma_1 E_1}}{\overline{R_1 \Sigma_1}} = 0.6\ \text{(graphically)} \end{cases} \Rightarrow \begin{cases} R_1 = 75.6\ \text{kg/h} \\ E_1 = 126.1\ \text{kg/h} \end{cases} \quad (2.30)$$

Now we go to the stage $N = 2$. Since we have been instructed that $B_2 = 2B_1 = 103.4$ kg/h, the position of the mixing point at the second stage, Σ_2, is obtained by the lever rule $R_1/B_2 = (\overline{\Sigma_2 B}/\overline{R_1 \Sigma_2}) = 0.73$ and knowing (graphically) the sum of the two segments, $\overline{R_1\Sigma_2}+\overline{\Sigma_2 B} = \overline{R_1 B}$. The tie line passing through Σ_2 demonstrates the correctness of the value chosen for B_1, as we obtain a raffinate R_2 of the requested purity. Finally, the total solvent flow rate is $B = B_1 + B_2 = 155.1$ kg/h.

2.5 Multistage Countercurrent Flow Process

2.5.1 Mass Balances and Operating Lines

In a *multistage countercurrent flow* configuration, shown in Fig. 2.13, the feed F is introduced at one end of the cascade (stage 1), while the extracting solvent B is fed from the opposite end (stage N). Along the sequence of stages, the feed gives rise to

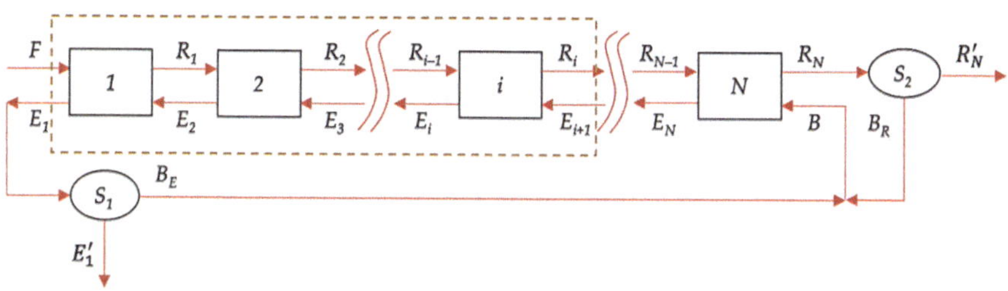

Fig. 2.13 Scheme of a multistage countercurrent flow LLE process (separation units for recovery of the extracting solvent are denoted with S_1 and S_2). After separation, the streams of extracting solvent (B_R and B_E) are re-sent as stream B to the process

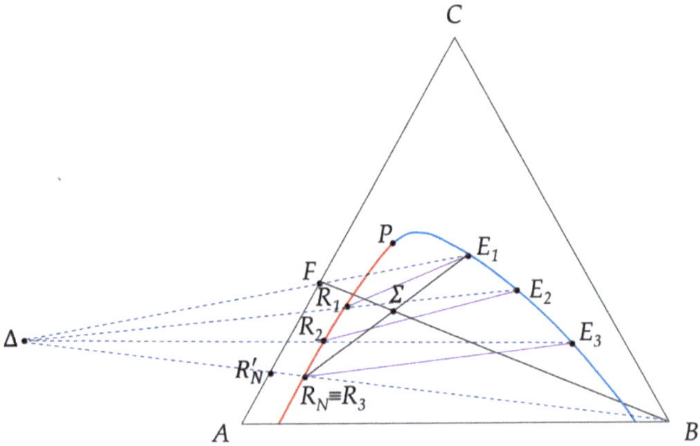

Fig. 2.14 Triangular diagram for multistage countercurrent flow LLE (3 stages are considered in the example)

raffinates $R_1 \dots R_i \dots R_N$, which become progressively depleted in C from stage 1 to N, while the solvent gives rise to extracts $E_N \dots E_i \dots E_1$, which become progressively enriched in C from stage N to 1. This contact scheme is more efficient than the crossflow one. For example, for a fixed raffinate purity and number of stages, countercurrent extraction requires a lower extracting solvent flow rate than the crossflow process. More complex schemes, involving e.g. reflux, are treated in reference textbooks, such as Treybal 1981.

To analyse in detail the process, let us consider a design problem for a battery of extractors operating in countercurrent mode, where the goal is to purify a feed stream F of known composition and flow rate with a fixed flow rate B of extracting solvent, in order to obtain a raffinate R_N of specified composition. So, we need to determine N. Assume that the ternary system is such that components A and B are partially miscible and that the distribution coefficient $D_C > 1$.

The overall material balance on the entire cascade of N stages is:

$$\boxed{F + B = E_1 + R_N = \Sigma} \tag{2.31}$$

It is observed that the point Σ does not result from an actual mixing of streams and, furthermore, the streams E_1 and R_N are not in equilibrium. It is possible to locate Σ on the triangular diagram by means of the lever rule (Fig. 2.14):

$$\frac{B}{F} = \frac{\overline{F\Sigma}}{\overline{\Sigma B}} \tag{2.32}$$

Once the position of R_N (fixed by the problem) and Σ is known, the position of E_1 can be easily found from the extension of $\overline{R_N\Sigma}$ to the extract branch (E_1, R_N and Σ are aligned as per Eq. 2.31). The flow rates of extract E_1 and raffinate R_N can be evaluated by combining Eq. 2.31 with the lever rule:

$$\frac{R_N}{E_1} = \frac{\overline{\Sigma E_1}}{\overline{R_N \Sigma}} \tag{2.33}$$

The determination of the number of equilibrium stages requires, as discussed for fractional distillation, the alternate application of operating and equilibrium conditions (the latter represented, on the triangular diagram, by tie lines). To define the operating conditions, we firstly write the mass balance considering as control surface the portion of the cascade enclosing the stage 1 and the generic *i*-th stage (cf. the dashed region in the scheme of Fig. 2.13):

$$F + E_{i+1} = E_1 + R_i \tag{2.34}$$

i.e.:

$$E_{i+1} - R_i = E_1 - F = \Delta \tag{2.35}$$

where Δ is defined as the *difference* or *operating point*, and *represents the difference between the extract and raffinate flow rates crossing at the interstage* (or net mass flow).

Equation 2.31 can be thus rearranged in the form:

$$B - R_N = E_1 - F = \Delta \tag{2.36}$$

From the mass balance applied to each stage (Eq. 2.34), the following equalities hold:

$$\boxed{E_1 - F = E_2 - R_1 = \ldots = E_{i+1} - R_i = B - R_N = \Delta} \tag{2.37}$$

Thus, the net mass flow is constant throughout the battery of extractors (in the absence of addition or withdrawal of intermediate streams). The equalities expressed in Eq. 2.37 constitute *operating conditions*, as they link the streams crossing between two successive stages. In particular, on the triangular diagram (Fig. 2.14), the interstage streams are aligned with the point Δ defining the *operating line*. It is emphasised that the definition of the stream Δ as the difference between extract and raffinate flow rates is purely arbitrary (it could also be defined as the opposite difference). Furthermore, being a fictitious stream, its value may be positive or negative.

From Eq. 2.36, it is possible to locate the point Δ, obtained as the intersection of the lines $\overline{FE_1}$ and $\overline{BR_N}$ (Fig. 2.14). The determination of the number of stages is now obtained by the graphical construction of Hunter-Nash, starting, for example, from the first stage of the battery. The streams E_1 and R_1 are in equilibrium, so the tie line through E_1 (the position of the latter has been previously determined) individuates, on the raffinate branch, the position of R_1. The two streams R_1 and E_2, which cross between the first and second stage, are linked by the operating condition expressed in Eq. 2.37, so the position of E_2 is obtained from the intersection of the line $\overline{\Delta R_1}$ with the extract branch. From Eq. 2.36, the value of Δ can be calculated, and the value of E_2 is then determined applying the lever rule:

$$\frac{|\Delta|}{E_2} = \frac{\overline{R_1 E_2}}{\overline{\Delta R_1}} \tag{2.38}$$

where the absolute value of Δ is used since line segments are obviously positive quantities (whereas Δ may be also negative). The value of R_1 is then calculated as:

$$R_1 = E_2 - \Delta \tag{2.39}$$

The construction proceeds by alternating equilibrium and operating conditions, and, from the material balances coupled with the lever rule, all intermediate extract and raffinate flow rates can be determined. In the graphical construction shown as an example, it is obtained that $R_N \equiv R_3$, and therefore three equilibrium stages are required to carry out the operation.

2.5.2 Effect of the Extracting Solvent Flow Rate

Consider the effect of varying the flow rate of the extracting solvent B, keeping R_N fixed, as illustrated in Fig. 2.15. By increasing B, point Σ will of course move toward the vertex of the pure solvent, and the operating point (the net mass flow) Δ moves away from R_N toward the left in the triangular diagram. In fact, consider an increase in B: the point Σ will shift to Σ', then E_1 will move to E_1' as it will result from the extension of the segment $\overline{R_N \Sigma'}$ on the extract branch, and the operating point, obtained now as the intersection of the lines $\overline{FE_1'}$ and $\overline{BR_N}$, will shift from Δ to Δ'. Continuing this reasoning, the increase in the solvent flow rate will eventually lead to a condition (mixing point Σ_∞) in which the line passing through F and the extract

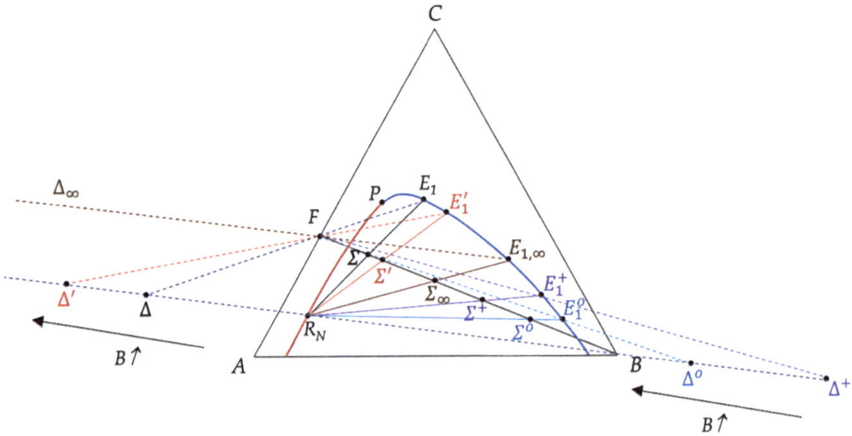

Fig. 2.15 Effect of the extracting solvent flow rate for multistage countercurrent flow LLE

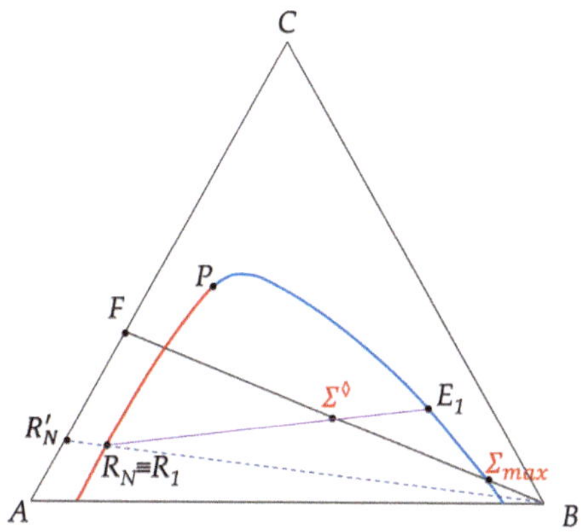

Fig. 2.16 Identification of the mixing points corresponding to $N = 1$ ($\Sigma^{\diamond}$) and B_{max} (Σ_{max}) for countercurrent flow LLE

leaving the cascade ($E_{1,\infty}$ in Fig. 2.15) becomes parallel to $\overline{BR_N}$, and the operating point is represented by the intersection of two parallel lines, i.e. a point (Δ_∞) individuating their direction. Under this condition, all the operating lines are parallel to $\overline{BR_N}$. Further increases in solvent flow (see Σ^+ and Σ^o or, respectively, E_1^+ and E_1^o in Fig. 2.15) will place the operating point (Δ^+ and Δ^o, respectively) to the right of the triangular diagram and, the higher B, the further to the left the operating point Δ is.

The effect of increasing the B results in a reduction of the number of required stages, because the driving force for separation increases in each stage. There exists a solvent flow rate $B^{\diamond}$ that allows achieving the desired separation in a single equilibrium stage ($N = 1$). Obviously, under this condition, E_1 and $R_N \equiv R_1$ are at equilibrium. Therefore, $B^{\diamond}$ is obtained by considering the tie line for R_N and by locating the mixing point $\Sigma^{\diamond}$ at the intersection between this tie line and the segment $\overline{FB}$ (Fig. 2.16). The lever rule will then allow for the evaluation of $B^{\diamond}$:

$$\frac{B^{\diamond}}{F} = \frac{\overline{F\Sigma^{\diamond}}}{\overline{\Sigma^{\diamond}B}} \tag{2.40}$$

For $B > B^{\diamond}$, the purity of the A raffinate would exceed the required process specification, leading to excessive solvent consumption and higher total process costs. The upper operating limit for the flow rate, B_{max}, is the one determined for single-stage extraction, i.e. locating the mixing point (Σ_{max}) at the intersection of the line $\overline{FB}$ with the extract branch (cf. Fig. 2.16), following the approach illustrated in Fig. 2.8.

The minimum solvent mass flow rate, B_{min}, corresponds to the situation where a theoretically infinite number of stages would be needed to carry out the operation. This pinch condition (a concept already seen in distillation) occurs when an operating line passing through Δ coincides with a tie line. Determining the operating point corresponding to B_{min}, denoted as Δ_{min}, and the operational working conditions requires specific considerations based on the solute distribution coefficient, D_C.

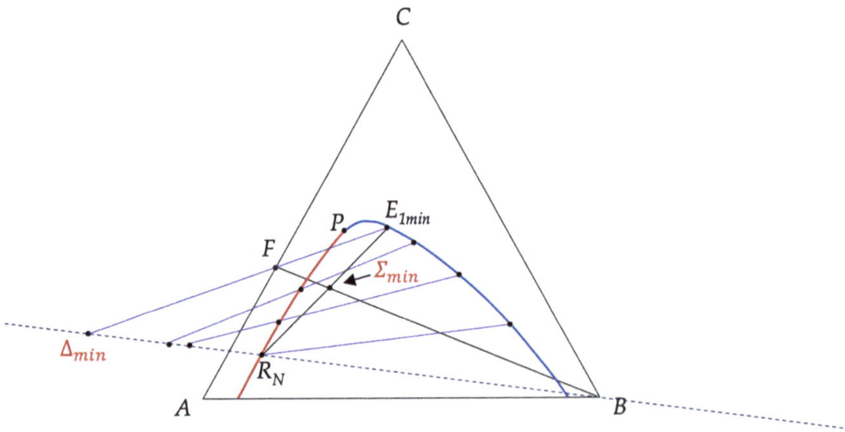

Fig. 2.17 Determination of the minimum extracting solvent mass flow rate for multistage countercurrent flow LLE (case $D_C > 1$)

For $D_C > 1$, the mass fraction of solute in the extract is greater than in the raffinate at equilibrium. This case is shown in Fig. 2.17. To determine B_{min}, the conjugate lines between the one passing through the feed and that through R_N are extended along the line $\overline{BR_N}$. This allows determining the locus of points where operating conditions coincide with equilibrium. This segment, indicated $\overline{\Delta_{min}R_N}$ in Fig. 2.17, defines the non-operating conditions, and the upper end Δ_{min} is determined by the tie line that intersects the line $\overline{BR_N}$ at the farthest point from vertex B on the left of the triangular diagram. For $B > B_{min}$, Δ is located outside the segment $\overline{\Delta_{min}R_N}$, and the operation is feasible with a finite number of stages (please refer to Fig. 2.15: the higher B, the further to the left the operating point Δ is). Generally, the position of Δ_{min} is obtained from the tie line extended through F, but this must be verified case by case. Once the position of Δ_{min} is determined, E_{1min} (e.g., Fig. 2.17, the extract exiting the first stage that, in pinch conditions, is in equilibrium with F) is positioned on the extract branch corresponding to this limiting tie line (i.e., in the example of Fig. 2.17, the one passing per F). Thus, the position of Σ_{min} is immediately obtained from the intersection of $\overline{R_NE_{1\,min}}$ with $\overline{FB}$, and the minimum flow rate B_{min} is obtained by applying the lever rule:

$$\frac{B_{min}}{F} = \frac{\overline{F\Sigma_{min}}}{\overline{\Sigma_{min}B}} \tag{2.41}$$

For $D_C < 1$, the procedure for determining Δ_{min} is the same as for $D_C > 1$. However, in this case, the pinch points obtained from the intersections with $\overline{BR_N}$ of the conjugate lines between the feed and R_N lie on the right side of the triangular diagram (Fig. 2.18). Here, the most limiting condition for the minimum extracting solvent flow rate corresponds to Δ_{min} being identified by the pinch point closest to the vertex representing B. All operating conditions are those for which Δ is located within the segment $\overline{B\Delta_{min}}$. Also in this case, usually, the position of Δ_{min} is obtained

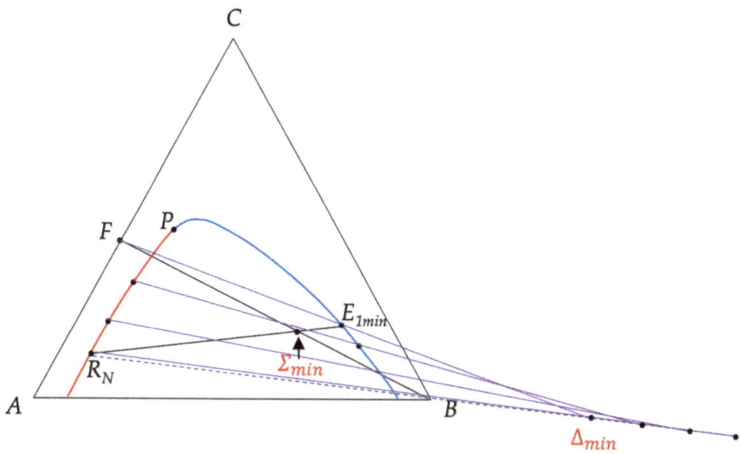

Fig. 2.18 Determination of the minimum extracting solvent mass flow rate for multistage countercurrent flow LLE (case $D_C < 1$)

from the tie line extended through F (this is the case depicted in Fig. 2.18), although not being a general rule. Once the position of Δ_{min} is determined, E_{1min} is positioned in the same way as above. Thus, the position of Σ_{min} is obtained, and the minimum flow rate B_{min} is known by Eq. 2.41.

For solutropic systems (which show inversion of the inequality between $y_C(E)$ and $x_C(R)$, cf. Fig. 2.5), the pinch points are located on both sides of the triangular diagram. In this case, the minimum solvent flow rate is determined by considering the pinch conditions along the line $\overline{BR_N}$ to the right of the point representing the extracting solvent.

Concluding, as a general indication, since B_{min} is non-operative (infinite number of stages), and since the increase in B makes N decrease but operating costs (i.e., for the use of solvent) increase, the optimal extracting solvent flow rate is about 1.5–2.5 times the minimum value (although it will depend on the specific case under scrutiny and its economic parameters).

Example: Design of a Multistage Countercurrent Flow LLE Process

It is required to design a cascade of countercurrent extraction stages operating at 22 °C with castor oil (derived from *Ricinus communis*) as extracting solvent (B component). The system must process an aqueous stream containing 50% ethanol (water is the A component, ethanol is the C component), with a feed flow rate $F = 5000$ kg/h, in order to obtain a raffinate phase containing 3% ethanol (percentages are in mass, as usual). Table 2.3 reports tie line data for the ternary system ethanol/water/castor oil at 22 °C. We wish to determine the

(continued)

Table 2.3 Equilibrium data for the ternary system ethanol/water/castor oil at 22 °C (Hasan et al. 2023). For each row, the raffinate and the extract phases are in equilibrium and, therefore, data on the same row represent points connected by a tie line in Fig. 2.19

Tie line data (mass fractions)					
Raffinate phase			Extract phase		
Ethanol $x_C(R)$	Water $x_A(R)$	Castor oil $x_B(R)$	Ethanol $y_C(E)$	Water $y_A(E)$	Castor oil $y_B(E)$
0.0273	0.9727	0.0000	0.0095	0.0267	0.9638
0.1159	0.8840	0.0001	0.0196	0.0327	0.9477
0.1935	0.8065	0.0000	0.0414	0.0366	0.9220
0.2410	0.7590	0.0000	0.0551	0.0431	0.9018
0.2839	0.7161	0.0000	0.0677	0.0482	0.8841
0.3787	0.6212	0.0001	0.1010	0.0517	0.8473
0.4240	0.5760	0.0000	0.1182	0.0554	0.8264
0.4513	0.5486	0.0001	0.1313	0.0531	0.8156
0.4720	0.5270	0.0010	0.1321	0.0579	0.8100
0.5580	0.4410	0.0010	0.1822	0.0596	0.7582

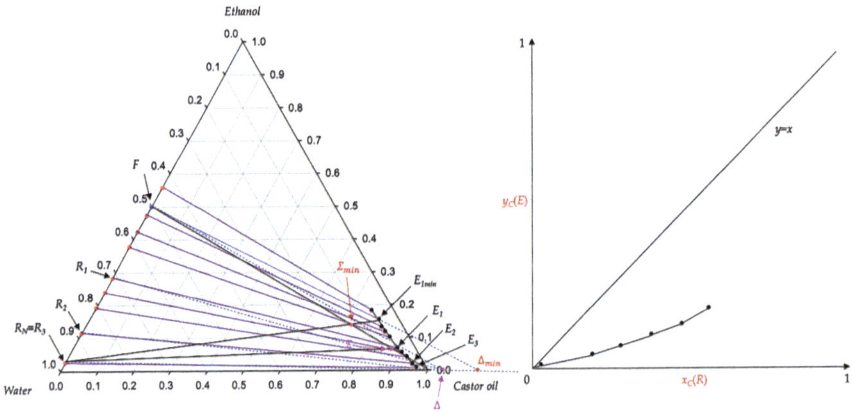

Fig. 2.19 Hunter-Nash's method applied for the design of a multistage countercurrent flow LLE process

number of equilibrium stages required to carry out the separation, operating with $B = 2.5B_{min}$.

With the available data, we can represent on the triangular diagram (Fig. 2.19) the position of the feed ($x_C(F) = 0.5$), the position of the raffinate exiting the cascade ($x_C(R_N) = 0.03$), the tie lines and the corresponding equilibrium curve. We recognize that it is $D_C < 1$ (the case illustrated in Fig. 2.18).

(continued)

For this system, we graphically verify that Δ_{min} derives from the tie line extended through F on the line $\overline{BR_N}$, identifying E_{1min} on the extract branch. Given the global mass balance, $F + B_{min} = E_{1min} + R_N = \Sigma_{min}$, we graphically find Σ_{min} from the intersection of $\overline{FB}$ with $\overline{R_N E_{1\,min}}$. Then, we determine B_{min} from the lever rule Eq. 2.41, $B_{min} = F(\overline{F\Sigma_{min}}/\overline{\Sigma_{min}B}) = 5000 \cdot 2.61 = 13{,}050$ kg/h. Consequently, $B = 32{,}625$ kg/h.

Applying again the lever rule, we locate Σ: $B/F = (\overline{F\Sigma}/\overline{\Sigma B}) = 6.52$ and knowing (graphically) the sum of the two segments, $\overline{F\Sigma} + \overline{\Sigma B} = \overline{FB}$. Needless to say, Σ is closest to the B vertex than Σ_{min}. The global material balance enables finding E_1 by prolonging $\overline{R_N\Sigma}$ on the extract branch, and the tie line through E_1 determines R_1 on the raffinate branch.

The operating line can be now drawn, by recalling Eq. 2.37, $E_1 - F = E_2 - R_1 = \cdots = E_{i+1} - R_i = \cdots = B - R_N = \Delta$. Hence, we find Δ (the net mass flow) from the intersection of $\overline{BR_N}$ and $\overline{FE_1}$. Please take your time to note that Δ is closer to vertex B than point Δ_{min}. Using the operating point Δ, we locate E_2 from the intersection of $\overline{\Delta R_1}$ with the extract branch, and the tie line through E_2 allows identifying R_2. We repeat the alternation of operating and equilibrium conditions, finding that $R_3 \equiv R_N$, thus $N = 3$.

2.6 Background Information

- Ternary systems can be represented on triangular diagrams (Fig. 2.1). Mixture compositions are determined graphically, and the lever rule (Eq. 2.1) allows the evaluation of the relative amounts of phases upon mixing.
- Liquid-liquid equilibria are described by the binodal curve and tie lines (Figs. 2.3, 2.4 and 2.5). The solute distribution in extract and raffinate is quantified by the distribution coefficient (Eq. 2.2).
- Single-stage extraction (Fig. 2.6) is analysed through overall and component material balances (Eqs. 2.4–2.6). The Hunter-Nash's method combines equilibrium tie lines and the lever rule for determining raffinate and extract compositions and flow rates (Fig. 2.7).
- Multistage crossflow extraction is modelled by recursive mass balances (Eqs. 2.20–2.22), with each stage receiving fresh solvent (Fig. 2.10). Tie line constructions and the application of the lever rule allow the analysis of the process (Fig. 2.11).
- For multistage countercurrent extraction, the overall mass balance (Eq. 2.31) defines the mixing point, while the operating conditions are expressed through the difference point (net mass flow in Eq. 2.37). The Hunter-Nash's graphical construction is used to determine the number of equilibrium stages (Fig. 2.14). The effect of solvent flow rate and its operating limits are graphically determined (Figs. 2.15, 2.16, 2.17, 2.18).

References

Blackadder DA, Nedderman RM (1971) A handbook of unit operations. Academic, New York

Green DW, Perry RH (2008) Perry's chemical engineers' handbook. McGraw-Hill, New York

Hasan MJ, Yeganeh F, Ciric A, Chen P, Vasquez ES, Ureña-Benavides EE (2023) Liquid-liquid equilibria of water + ethanol + castor oil and the effect of cellulose nanocrystal/Fe_3O_4 and lignin/Fe_3O_4 nanoparticles. J Chem Thermodyn 180:107007. https://doi.org/10.1016/j.jct.2023.107007

Hunter TG, Nash AW (1934) The application of physico-chemical principles to the design of liquid-liquid contact equipment. Part II: application of phase-rule graphical methods. J Soc Chem Ind 53:95T–106T. https://doi.org/10.1002/jctb.5000531407

Malmary G, Faizal M, Albet J, Molinier J (1997) Liquid-liquid equilibria of acetic, formic, and oxalic acids between water and tributyl phosphate + dodecane. J Chem Eng Data 42:985–987. https://doi.org/10.1021/je9700939

McCabe WL, Smith JC, Harriott P (2001) Unit operations of chemical engineering. McGraw-Hill, Singapore

Seader JD, Henley EJ (2006) Separation process principles. John Wiley & Sons, Hoboken

Treybal RE (1981) Mass-transfer operations. McGraw-Hill, Singapore

Triday JO (1984) Liquid-liquid equilibria for the system benzene-water-methanol. J Chem Eng Data 29:321–324. https://doi.org/10.1021/je00037a029

Momentum Transport

3

Abstract

After having analysed two types of unit operations (distillation and extraction) in a way that has not required a specific analysis of transport phenomena, this chapter begins a trilogy concerning such mechanisms, with reference to systems of interest for particular unit operations and chemical reactors. The first episode of the trilogy is dedicated to the transport of momentum, which is presented by focusing on the flow of a fluid inside a conduit (cylindrical, horizontal, with steady and one-dimensional flow). The concepts of laminar and turbulent motion are illustrated, also with the fundamental aid of the Reynolds number, and the general law of molecular transport is presented, which, applied to the momentum, becomes the Newton's law on viscosity (for Newtonian fluids) and relates the viscosity to the shear stress and the radial gradient of axial velocity. For laminar motion, the radial field of both shear stress and velocity is obtained (this latter profile is then also described for turbulent motion). Finally, the definition of friction factor (and its relation to the Reynolds number) is used to show how, even in absence of information on the mean radial velocity, it is possible to argue the prevailing flow regime.

Nomenclature

A	Area [length^2]
A_W	Wetted surface area [length^2]
D	Diameter [length]
F	Force [mass·length/time^2]
F_S	Shear force [mass·length/time^2]
f	Fanning friction factor [−]
k	Specific kinetic energy [mass/(length·time^2)]
L	Length [length]

F. Montagnaro et al., *Elements of Unit Operations and Chemical Reactors*,
https://doi.org/10.1007/978-3-032-18046-9_3

m	Molar mass [mass/moles]
$\mathcal{P}$	Pressure [mass/(length·time^2)]
$\mathcal{P}_0$	Inlet pressure [mass/(length·time^2)]
$\mathcal{P}_L$	Outlet pressure [mass/(length·time^2)]
Q	Volumetric flow rate [volume/time]
R	Radius [length]
$\mathfrak{R}$	Gas constant [energy/(moles·temperature)]
r	Radial coordinate [length]
Re	Reynolds number [−]
S	Cross-section area [length^2]
T	Temperature [temperature]
v	Fluid velocity [length/time]
$\bar{v}$	Mean fluid velocity [length/time]
v^+	Dimensionless velocity [−]
v^*	Friction velocity [length/time]
v_{max}	Maximum fluid velocity [length/time]
W	Mass flow rate [mass/time]
y	Radial coordinate [length]
y^+	Dimensionless radial coordinate [−]
z	Axial coordinate [length]
$\lvert \Delta\mathcal{P} \rvert$	Pressure drop [mass/(length·time^2)]
μ	Dynamic viscosity [mass/(length·time)]
μ_{GAS}	Dynamic viscosity for gas [mass/(length·time)]
μ_{LIQ}	Dynamic viscosity for liquid [mass/(length·time)]
ν_k	Kinematic viscosity [length^2/time]
ρ	Density [mass/volume]
τ	Shear stress [mass/(length·time^2)]
τ_w	Shear stress at system wall [mass/(length·time^2)]

3.1 General Characteristics

This textbook has been given, in the title, the term “elements”, with the intention of focusing on basic concepts that can then be explored in depth by the reader, with subsequent studies. With reference to *transport phenomena*, they will therefore be treated in a simplified way (the simplifications will be, when needed, made explicit), and by choosing some case studies that can be useful in developing the topics of this book (which mostly deals with basic unit operations and chemical reactors). For any further reading concerning transport phenomena, the scholar is referred to specific texts such as, for example, Bird et al. 2007.

In classical physics, the *momentum* of an “object” depends on its velocity, v (properly speaking, v is meant as a local velocity, at a certain position in space and time). Referring specifically to *fluids* (liquids, gases), if two fluid elements have

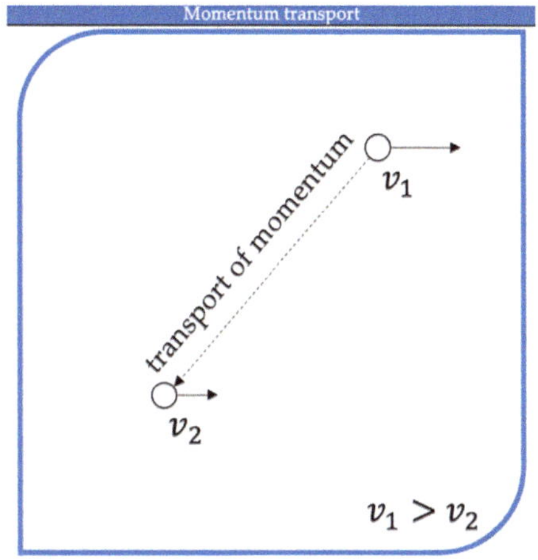

Fig. 3.1 Transport of momentum: schematic concept

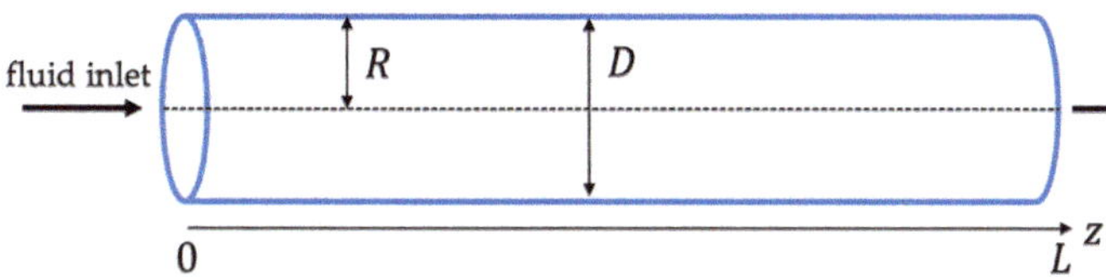

Fig. 3.2 Reference system: cylindrical tube, horizontal, stationary and one-dimensional flow

different velocity values, there will be a *transport of momentum* from the fluid element that has higher velocity to the one that has lower velocity (Fig. 3.1). The difference in v (the *spatial gradient* of velocity) represents the *driving force* that regulates this transport phenomenon.

3.2 Flow in Conduits

Very often, for the purposes of the subsequent discussion of some unit operations and chemical reactors, we will refer to the flow in a system defined by solid walls and, more specifically, consisting of a *cylindrical tube* (we define, from now on, its length L, diameter D, and radius R). In this conduit, *horizontal* (meaning, by horizontal direction, that perpendicular to the direction of the gravitational force), a fluid with *stationary motion, isothermal and with constant composition*, flows. We will analyse the system in a simplified (two-dimensional) way, considering the axial coordinate (z) and the radial coordinate (r). The system is represented in Fig. 3.2.

Fluid dynamics arises because, from the outside, we feed the system at $z = 0$ (e.g., by means of a pump or a compressor) with our fluid. This determines a dominant motion along the z-axis, the main direction of flow that we will examine (*one-dimensional flow*). Therefore, when we talk about velocity v, we will refer more specifically to the *axial component* of the velocity vector.

Flowing in the system, the fluid adhering to the solid walls has zero velocity, due to the interfacial tension between the fluid and the solid. However, since the fluid elements far from the wall have their own non-zero velocity, the existence of a *velocity field* can be intuited, with variations of v that determine a transport of momentum (Fig. 3.1). In particular, we will examine the variations of v possessed by the fluid elements as a function of their radial coordinate ($v(r)$).

3.3 Laminar and Turbulent Flow

If, overall, the fluid moves with *low values of* v, the fluid elements flow *without noticeable lateral mixing*: we have layers ("*laminae*") of fluid sliding over each other, the *laminar flow*. Laminar flow is a *molecular transport*, and depends on the motion of the individual molecules (generally, each molecule has a certain amount of momentum, energy and mass associated with it). Moreover, in laminar flow, we do not expect to observe eddies/swirls of fluid. In a famous experiment published in the year 1883 (Fig. 3.3), Reynolds [Osborne Reynolds (1842–1912), British scientist, Professor at Owens College; see Reynolds 1883] immersed a coloured tracer inside a main fluid (water) flowing in a glass tube at a certain velocity. More in particular, the small flow of water-soluble dye was introduced, through a capillary (a "trumpet mouthpiece", in the Reynolds' words), with a velocity equal to that of water. For low values of velocity (both of the carrier fluid and of the dye), Reynolds observed that the filaments of coloured fluid flowed intact, without cross-mixing (hence, there was laminar flow, i.e. without eddies).

At *high values of* v, there will be lateral mixing, eddies (sometimes so large that they can be perceived by the naked eye) and *turbulent flow* (more generally, turbulent transport depends on the disorderly motion of sets of molecules). In the Reynolds experiment, for high velocity values, the dye dispersed uniformly in water (due to the presence of lateral mixing and eddies, Fig. 3.3): turbulent flow was,

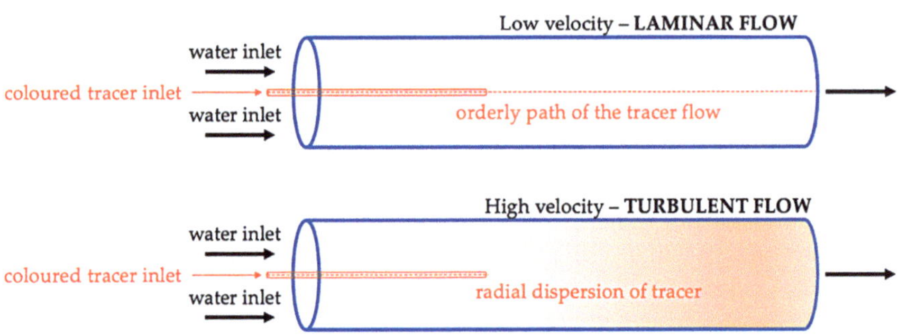

Fig. 3.3 Schematisation of the Reynolds experiment (the reader is kindly redirected to Reynolds 1883, for the original description and graphical illustration; in the experiment, the glass tube was just over 1 m long, and diameters from about 0.5 to 2.5 cm were used). For clarity, only the flow of the tracer is depicted

therefore, observed, which superimposes (but does not eliminate) the molecular/ laminar flow.

3.4 The Viscosity Law for Newtonian Fluids

3.4.1 Shear

Generally speaking, by *shear* we mean the lateral displacement of a layer of fluid with respect to another layer, a phenomenon related to an external force (the *shear force* F_S). The ratio between the shear force and the area of the layer over which this force is exerted, is called *shear stress* (indicated with τ). The shear force and, therefore, the shear stress, are induced by the circumstance that (in our reference system) layers of fluid at different radial coordinate possess different velocity: the force is exerted by the layers at higher velocity on the layers at lower velocity, Fig. 3.4.

3.4.2 General Law of Molecular Transport

As already met in distillation and extraction, the *flow rate* $\mathbb{F}$ of a given physical quantity is dimensionally expressed as *the ratio between this quantity and time*. With *flux* $\dot{\mathbb{F}}$ of a given flow rate, we instead indicate *the ratio between the flow rate and a characteristic area A*:

$$\dot{\mathbb{F}} = \frac{\mathbb{F}}{A} \tag{3.1}$$

Once introduced a characteristic quantity $\mathbb{Q}$ (the one that, showing a gradient, causes the transport phenomenon along the direction x), and a generic constant $\mathbb{K}$, we propose the following *general law of molecular transport*:

$$\boxed{\dot{\mathbb{F}} = -\mathbb{K}\frac{\mathrm{d}\mathbb{Q}}{\mathrm{d}x}} \tag{3.2}$$

This law indicates that the flux is proportional, via the constant $\mathbb{K}$, to the gradient of the quantity $\mathbb{Q}$ that determines the transport phenomenon (the higher the gradient,

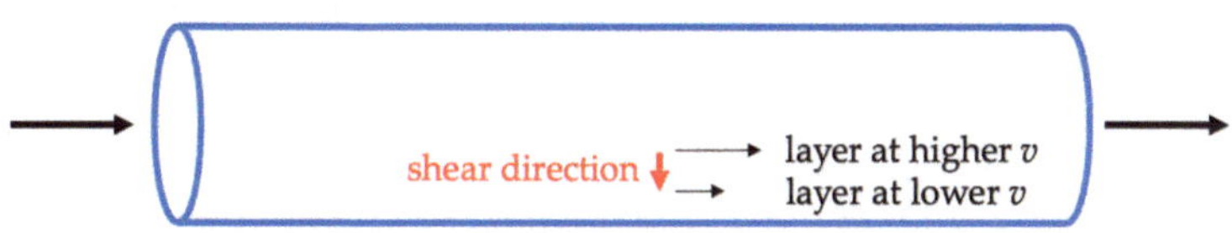

Fig. 3.4 Fluid flow in a circular tube: scheme of the onset of shear exerting from the layer at higher, to the layer at lower, axial velocity

the higher the flux, and vice versa). The minus sign in Eq. 3.2 indicates that the *flux is positive in the opposite direction to that for which the gradient is positive* (Fig. 3.1).

3.4.3 Newtonian Fluids

Momentum is the product between mass and velocity. The flow rate of momentum is, dimensionally, expressed as (the importance of *dimensional analysis* has been enlightened in the inspiring work by Astarita 1997):

$$\frac{\text{momentum}}{\text{time}} = \frac{[\text{mass} \cdot \text{velocity}]}{[\text{time}]}$$
$$= \left[\frac{\text{mass} \cdot \frac{\text{length}}{\text{time}}}{\text{time}} = \frac{\text{mass} \cdot \text{length}}{\text{time}^2} = \text{mass} \cdot \text{acceleration} = \text{force}\right] \tag{3.3}$$

Thus, the flow rate of momentum is a force: in our case, it is the shear force ($\mathbb{F} = F_S$). Then, with flux in Eqs. 3.1 and 3.2 we mean the ratio between the shear force and the characteristic area, i.e. the flux of momentum is the shear stress ($\dot{\mathbb{F}} = F_S/A = \tau$). We deal with *Newtonian fluids*, defined as those for which the shear stress τ can be expressed (regardless of process time) according to Eq. 3.2. Examples of these fluids are gases, and many liquids (molecular weight < 5000, approximately).

The *Newton's law of viscosity* translates Eq. 3.2 for momentum transfer: in this case, the quantity that determines the momentum transport is (Fig. 3.1) the velocity ($\mathbb{Q} = v$) and, for our reference system, x is the radial coordinate r:

$$\boxed{\tau = -\mu \frac{\mathrm{d}v}{\mathrm{d}r}} \tag{3.4}$$

For Newtonian fluids, the shear stress is proportional to the velocity radial gradient, via a constant $\mathbb{K} = \mu$, termed *(dynamic) viscosity*, that depends on the nature of the fluid (it is a property of the material, regardless the fluid dynamics of the system). This parameter characterises the resistance to flow of a fluid (at fixed velocity radial gradient, high viscosity means high shear stress). The higher the viscosity, the higher the resistance to flow. The minus sign reminds us that the shear stress acts in the opposite direction to that of the velocity increase (cf. Fig. 3.4). In other words, the shear stress is exerted from layers with higher velocity towards layers with lower velocity.

What we generally call viscosity is referred to as "dynamic", as there is another viscosity (*kinematic viscosity*) that we will sometimes use (where ρ is the fluid density, considered constant along the flow):

$$\nu_k = \frac{\mu}{\rho} \tag{3.5}$$

Although not considered here, we recall that there are fluids with non-Newtonian behaviour, e.g. sludges, rubbery fluids, emulsions, polymers, suspensions, pastes, slurries. Other examples of non-Newtonian fluids are thixotropic fluids (the term derives from Greek words indicating "touch" and "transformation", giving the idea of a fluid that is transformed when touched), whose viscosity can decrease (their flowability increases) if a shear is applied over time while keeping the system in agitation (example: ketchup, some paints, some clays); and rheopectic fluids, which have the opposite behaviour (here, the parent Greek words indicate "flow" and "condensed", giving the idea of a fluid that, by flowing, becomes more viscous and with reduced flowability; example: gypsum pastes, some lubricants, some polymers).

3.4.4 Viscosity

To find the dimensions for μ from the equation that defines it (Eq. 3.4), we observe that the dimensions for τ are those of a stress:

$$\tau = \frac{\text{force}}{\text{area}} = \frac{[\text{mass} \cdot \text{acceleration}]}{[\text{area}]} = \left[\frac{\text{mass} \cdot \frac{\text{length}}{\text{time}^2}}{\text{length}^2} = \frac{\text{mass}}{\text{length} \cdot \text{time}^2} \right] \tag{3.6}$$

and those for a velocity spatial gradient are:

$$\frac{\mathrm{d}v}{\mathrm{d}r} = \frac{\text{velocity}}{\text{length}} = \frac{\left[\frac{\text{length}}{\text{time}}\right]}{[\text{length}]} = \left[\frac{1}{\text{time}}\right] \tag{3.7}$$

Isolating μ in Eq. 3.3, the dimensional analysis for the viscosity reads:

$$\mu = -\frac{\tau}{\frac{\mathrm{d}v}{\mathrm{d}r}} = \frac{\text{stress}}{\frac{\text{velocity}}{\text{length}}} = \frac{\left[\frac{\text{force}}{\text{area}}\right]}{\left[\frac{1}{\text{time}}\right]} = \left[\frac{\frac{\text{mass} \cdot \frac{\text{length}}{\text{time}^2}}{\text{length}^2}}{\frac{1}{\text{time}}} = \frac{\text{mass}}{\text{length} \cdot \text{time}} \right] \tag{3.8}$$

For instance, we can find μ expressed in measure units of [kg/(m·s) = Pa·s] (since [Pa = kg/(m·s^2)]) or, being the value of viscosity for room temperature water equal to about 10^{-3} Pa·s, a measure unit called centipoise (1 cP = 10^{-3} Pa·s) can be found [named after Jean Léonard Marie Poiseuille (1797–1869), French physicist/ physiologist, *alma mater* at Polytechnic School of Paris; see Brillouin 1930; Sutera and Skalak 1993]. Often, the viscosity of room temperature water is approximated at 1 cP. Finally, the dimensions for the kinematic viscosity (Eq. 3.5) are:

$$\nu_k = \frac{\text{viscosity}}{\text{density}} = \frac{\left[\frac{\text{mass}}{\text{length} \cdot \text{time}}\right]}{\left[\frac{\text{mass}}{\text{volume}}\right]} = \left[\frac{\text{length}^2}{\text{time}}\right] \tag{3.9}$$

The dynamic viscosity depends on temperature T, much less on pressure $\mathcal{P}$ (unless the pressure is very high). Therefore, along an isothermal flow, we consider μ constant (this is a characteristic for Newtonian fluids). The viscosity of liquids is generally (much) greater than that of gases and, contrary to what happens for gases (where, according to the kinetic theory of gases, μ increases with T), for liquids μ decreases as T increases. Using the formulae provided in the Perry's handbook (Green and Perry 2008), it is (with appropriate constants listed as a function of the species), for liquids (at either 1 atm or the vapour pressure, whichever is higher) and gases (at either 1 atm or the vapour pressure, whichever is lower), respectively:

$$\begin{cases} \mu_{LIQ} = \exp\left[C_1 + \frac{C_2}{T} + C_3 \ln(T) + C_4 T^{C_5}\right] \\ \mu_{GAS} = \frac{C_1 T^{C_2}}{1 + \frac{C_3}{T} + \frac{C_4}{T^2}} \end{cases} \tag{3.10}$$

where T is expressed in [K] and μ is obtained in [Pa·s].

Example: Viscosity for Water and Air

Application of Eq. 3.10 allows to calculate, for a reference liquid (water) and gas (air), the values of viscosity as T increases (in our example, from 20 °C to 80 °C, properly expressed in [K]), at $\mathcal{P} = 1$ atm. Table 3.1 lists the values for the constants (Green and Perry 2008), from which we can obtain the dynamic viscosity as function of temperature (values in Pa·s have been multiplied per 10^3, so to express the viscosity in cP). We can observe that, increasing T from 20 °C to 80 °C, the dynamic viscosity of water decreases by about 65%, while that of air increases by about 15%. The liquid water viscosity is, actually, (much) greater than that of the reference gas. Moreover, once known the values for density, it is also possible to list the values for the kinematic viscosity. For liquid water, ρ is taken from Green and Perry 2008; for air, ρ is obtained by application of the ideal gas law:

(continued)

Table 3.1 Evaluation of viscosity for liquid water and air (atmospheric pressure)

Species	C_1	C_2	C_3	C_4	C_5
Liquid water	−52.843	3703.6	5.866	$-5.879 \cdot 10^{-29}$	10
Air	$1.425 \cdot 10^{-6}$	0.5039	108.3	0	
	μ [cP]	ρ [kg/m^3]	ν_k [m^2/s]		
Liquid water (20 °C)	1.021	1001.9	$1.02 \cdot 10^{-6}$		
Liquid water (40 °C)	0.671	996.4	$6.74 \cdot 10^{-7}$		
Liquid water (60 °C)	0.474	987.9	$4.80 \cdot 10^{-7}$		
Liquid water (80 °C)	0.355	978.2	$3.63 \cdot 10^{-7}$		
Air (20 °C)	0.018	1.205	$1.49 \cdot 10^{-5}$		
Air (40 °C)	0.019	1.128	$1.68 \cdot 10^{-5}$		
Air (60 °C)	0.020	1.060	$1.89 \cdot 10^{-5}$		
Air (80 °C)	0.021	1.000	$2.10 \cdot 10^{-5}$		

$$\rho = \frac{\mathcal{P}}{\Re T} m \tag{3.11}$$

where m is the molar mass of the gas (assume 29 g/mol), and $\Re$ the ideal gas constant (0.0821 atm·L/(mol·K)).

3.5 The Reynolds Number

Let us focus on the ratio between inertial and viscous forces, that is, between the momentum transported by turbulent mechanism (governed by inertial forces) and the momentum transported by molecular/laminar mechanism (governed by viscous forces). Inertial and viscous forces can be dimensionally expressed as, respectively:

$$\begin{cases} \text{inertial forces} = [\text{mass} \cdot \text{acceleration} = \text{volume} \cdot \text{density} \cdot \text{acceleration}] \\ \text{viscous forces} = [\text{surface} \cdot \text{shear stress} = \text{surface} \cdot \text{viscosity} \cdot \text{velocity spatial gradient}] \end{cases} \tag{3.12}$$

where, in expressing the shear stress, we have used Eq. 3.4, that depicts the shear stress as the product between the dynamic viscosity and the velocity spatial gradient. The ratio between the two is:

$$\frac{\text{inertial forces}}{\text{viscous forces}} = \frac{[\text{volume} \cdot \text{density} \cdot \text{acceleration}]}{[\text{surface} \cdot \text{viscosity} \cdot \text{velocity spatial gradient}]} = \left[\frac{\frac{\text{volume}}{\text{surface}} \cdot \text{density} \cdot \frac{\text{acceleration}}{\text{velocity spatial gradient}}}{\text{viscosity}} \right] \tag{3.13}$$

where [volume/surface] is a characteristic length (in our case, the duct diameter D). Moreover, we observe that (see Eq. 3.7):

$$\left[\frac{\text{acceleration}}{\text{velocity spatial gradient}} \right] = \left[\frac{\frac{\text{length}}{\text{time}^2}}{\frac{1}{\text{time}}} = \frac{\text{length}}{\text{time}} = \text{velocity} \right] \tag{3.14}$$

Coupling Eqs. 3.13 and 3.14, we define the *Reynolds number* as the *ratio between the turbulent and the laminar transport of momentum*:

$$\boxed{Re = \frac{\text{turbulent transport of momentum}}{\text{laminar transport of momentum}} = \frac{\text{inertial forces}}{\text{viscous forces}} = \left[\frac{\text{length} \cdot \text{velocity} \cdot \text{density}}{\text{viscosity}} \right] = \frac{Dv\rho}{\mu}} \tag{3.15}$$

Reynolds observed that the dimensionless group just presented is able to determine a discrimination between the types of motion. More specifically, as Re increases, there is a transition *from laminar flow* (prevalence of viscous forces, i.e. of momentum transported by molecular mechanism) *to turbulent flow* (prevalence of inertial forces, i.e. of momentum transported by turbulent mechanism). The flow of fluids within conduits can be considered laminar for $Re < 2100$. Beyond that, there is a transition region and then the turbulent flow begins to dominate (starting from about $Re = 4000$–10,000). For $Re > 10^4$, typically the turbulent flow can be said to be fully developed.

3.6 Mean Velocity

In our cylindrical pipe of cross-section area $S = \pi D^2/4$, we have a field of axial velocity along the radial coordinate, $v(r)$, and we want to determine the mean value of this velocity ($\overline{v}$), Fig. 3.5. Given a certain position, the *volumetric flow rate* [volume/time] is:

$$Q = \overline{v}S \tag{3.16}$$

that is, the higher the mean velocity, the higher Q. The volumetric flow rate, thus, represents how much fluid (in terms of volume) flows in a certain position per unit time. An important consequence of Eq. 3.16 is that, at constant Q (stationary system) and S (cylindrical duct), *the mean velocity along the radial coordinate is constant along the axis* (that is, v varies along the radial coordinate, but its axial mean value remains the same, whatever the axial position). The *mass flow rate* [mass/time] is:

$$W = \rho Q = \rho \overline{v} S \tag{3.17}$$

which must be constant for the law of conservation of mass. In a differential area, dS, a differential mass flow rate circulates (where ρ is considered constant; *incompressible flow*):

$$\mathrm{d}W = \rho \mathrm{d}Q \tag{3.18}$$

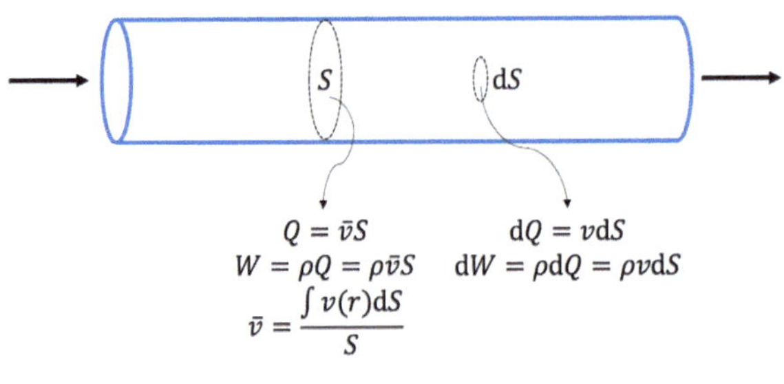

Fig. 3.5 Main quantities for flow inside a circular tube (steady, isothermal, constant density and composition, no effect of gravity). With v we indicate the local axial velocity, with $v(r)$ the field of axial velocity along the radial coordinate, while with $\overline{v}$ we indicate its mean value. The axial position of S and dS is irrelevant

But, locally, the differential volumetric flow rate is the product between the local velocity v and the differential area:

$$dQ = v dS \tag{3.19}$$

Then, joining Eqs. 3.18 and 3.19:

$$dW = \rho v dS \tag{3.20}$$

Integrating along the whole passage area of the conduit:

$$W = \rho \int v(r) dS \tag{3.21}$$

where we have highlighted that, along this area, v is a function of the radial position r. Comparing Eqs. 3.17 and 3.21, we get the *mean velocity along the radial coordinate*:

$$\boxed{\overline{v} = \frac{\int v(r) dS}{S}} \tag{3.22}$$

which is (Lagrange's mean value theorem) the ratio between the surface integral of the $v(r)$ field, and the extension of the surface itself.

3.7 Laminar Flow

3.7.1 Balance of Forces

Consider our reference system where a fluid flows enlivened by *laminar motion* (predominance of viscous forces, $Re < 2100$). The fluid flows from $z = 0$ (pressure $\mathcal{P}_0$) to $z = L$ (pressure $\mathcal{P}_L$), thus it must be $\mathcal{P}_0 > \mathcal{P}_L$ (the overall *pressure drop* is $|\Delta\mathcal{P}| = \mathcal{P}_0 - \mathcal{P}_L$; it is generally expressed in absolute term, as the symbol Δ would require to write "outlet–inlet", but we prefer to have here a positive quantity). The system is *axisymmetric*. Moreover, we can define the radial coordinate r starting from $r = 0$ at the axis and ending at $r = R$ at one (any) of the walls (Fig. 3.6). Take a control volume of thickness dz, with z the axial coordinate. Be this control volume not extended from wall to wall, but referred to a generic radial coordinate r (and axisymmetric). The control region has differential volume $\pi r^2 dz$.

Let us write a *balance of forces* around our control volume, stating that the forces acting in the direction of the flow are equal to the forces acting against the direction of the flow. In the direction of the flow, a force related to the local pressure $\mathcal{P}(z)$ acts on the control volume. We will write this force as: $\mathcal{P}(z)\pi r^2$, as πr^2 is the surface of the control volume on which the pressure $\mathcal{P}(z)$ is exerted. Against the direction of the flow, two forces are active on the control volume: the first is $\mathcal{P}(z + dz)\pi r^2$ (of course, $\mathcal{P}(z + dz) < \mathcal{P}(z)$, as the fluid must flow in the direction indicated by z); the second is

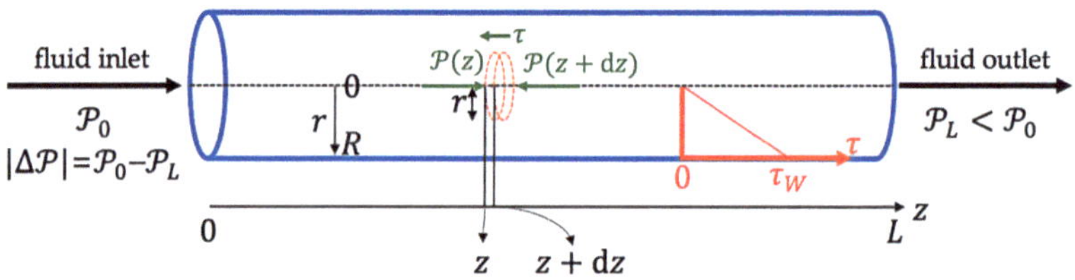

Fig. 3.6 Reference system with indication of pressures, axial and radial coordinates, differential control volume with acting forces (expressed in terms of force per unit surface), and radial profile of the shear stress, which is zero on the axis and increases linearly up to the value assumed on the wall

the shear force F_S. The shear force can be expressed as the product between the shear stress τ and the surface $2\pi r\mathrm{d}z$: in fact, the shear is exerted from the outside of the control volume on the volume itself, in all directions; if the fluid contained in the control volume moves in the z direction, the remaining fluid external to the control volume will oppose a viscous stress in the opposite direction, acting on the surface $2\pi r\mathrm{d}z$ (which can be seen as the lateral surface of the "coin" representing our control volume). Therefore, the force balance reads:

$$\mathcal{P}(z)\pi r^2 = \mathcal{P}(z+\mathrm{d}z)\pi r^2 + F_S \tag{3.23}$$

But:

$$F_S = \tau 2\pi r\mathrm{d}z \tag{3.24}$$

Then:

$$\mathcal{P}(z)\pi r^2 = \mathcal{P}(z+\mathrm{d}z)\pi r^2 + \tau 2\pi r\mathrm{d}z \tag{3.25}$$

$$\frac{\mathcal{P}(z+\mathrm{d}z)-\mathcal{P}(z)}{\mathrm{d}z} = \frac{\mathrm{d}\mathcal{P}}{\mathrm{d}z} = -\frac{2\tau}{r} \tag{3.26}$$

Equation 3.26 confirms that the axial pressure gradient is negative: in fact, if the motion occurs along the positive direction of z, clearly the pressure must decrease along z; the only force in the direction of the flow, related to the pressure in z, must be able to counteract both the force related to the pressure in z + dz and the shear force.

3.7.2 Radial Profile for the Shear Stress

Equation 3.26 has been obtained for a generic value of the radial coordinate r. Thus, it can be written also for $r = R$ (i.e., at system wall), where the shear stress will be denoted as $\tau(r = R) = \tau_w$:

$$\frac{\mathrm{d}\mathcal{P}}{\mathrm{d}z} = -\frac{2\tau_w}{R} \tag{3.27}$$

Comparing Eqs. 3.26 and 3.27:

$$\frac{\tau}{r} = \frac{\tau_w}{R} \tag{3.28}$$

that is, the expression of the radial profile for the shear stress (Fig. 3.6):

$$\tau(r) = \frac{r}{R}\tau_w \tag{3.29}$$

This fundamental result, when discussed in the light of the Newton Eq. 3.4, allows us to grasp that:

- when $r = 0$ (i.e., at the axis), τ is zero, i.e., near the axis we do not expect radial variations in the fluid velocity;
- when r increases (i.e., moving from the axis towards the wall), τ linearly increases (as we move away from the axis, the differences in velocity between adjacent fluid layers are increasingly marked);
- when we reach the wall ($r = R$) the value for the shear stress is maximum (τ_w), therefore the maximum radial gradient of the axial velocity value is observed at the wall.

Equation 3.27 can be also integrated by the Fourier method (separation of variables), between $z = 0$ ($\mathcal{P} = \mathcal{P}_0$) and $z = L$ ($\mathcal{P} = \mathcal{P}_L$), i.e. over the whole system, with the aim of obtaining the value for τ_w:

$$\int_{\mathcal{P}_0}^{\mathcal{P}_L} \mathrm{d}\mathcal{P} = -\frac{2\tau_w}{R}\int_0^L \mathrm{d}z \tag{3.30}$$

$$|\Delta\mathcal{P}| = \frac{2\tau_w}{R}L \Rightarrow \tau_w = \frac{|\Delta\mathcal{P}|R}{2L} \tag{3.31}$$

Equation 3.31 directly relates the shear stress at wall to a measurable quantity (the overall pressure drop) and to the geometrical features of the system.

3.7.3 Radial Profile for the Axial Velocity

To find the $v(r)$ profile for laminar flow, start back from the Newton's law of viscosity:

$$\frac{\mathrm{d}v}{\mathrm{d}r} = -\frac{\tau}{\mu} \tag{3.32}$$

Substitute the $\tau(r)$ profile, Eq. 3.29, and then integrate by separation of variables between $r = R$ (where $v = 0$ as the fluid elements adhering to the solid walls are not in movement) and a generic radial coordinate r, where the fluid elements have a generic velocity v:

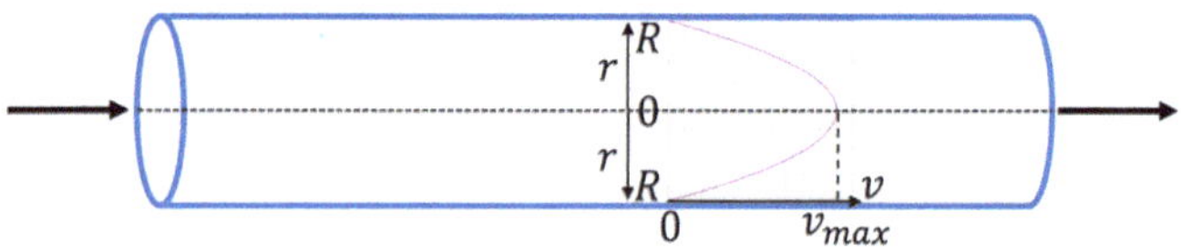

Fig. 3.7 Parabolic profile along the radial coordinate for the axial velocity in laminar flow: at the axis, the velocity is maximum, and the velocity gradient is minimum (as it is the shear stress); moving away from the axis towards the walls, the velocity decreases and the velocity gradient increases (as the shear stress). The axial position is irrelevant

$$\frac{\mathrm{d}v}{\mathrm{d}r} = -\frac{\tau_W}{\mu R} r \tag{3.33}$$

$$\int_0^v \mathrm{d}v = -\frac{\tau_W}{\mu R} \int_R^r r \mathrm{d}r \tag{3.34}$$

Maths follows:

$$v(r) = -\frac{\tau_W}{2\mu R}\left[r^2\right]_R^r = \frac{\tau_W}{2\mu R}\left(R^2 - r^2\right) = \frac{\tau_W R^2}{2\mu R}\left[1 - \left(\frac{r}{R}\right)^2\right] = \frac{\tau_W R}{2\mu}\left[1 - \left(\frac{r}{R}\right)^2\right] \tag{3.35}$$

Substitute with the expression for the wall shear stress obtained in Eq. 3.31:

$$\boxed{v(r) = \frac{|\Delta \mathcal{P}| R^2}{4\mu L}\left[1 - \left(\frac{r}{R}\right)^2\right]} \tag{3.36}$$

i.e., the fundamental equation representing a *parabolic radial profile for the velocity in laminar flow*, Fig. 3.7. The axial velocity is zero when $r = R$ (walls), increases following a parabolic path when r decreases (i.e., moving from wall to axis), and values the following maximum when $r = 0$ (axis):

$$v_{max} = \frac{|\Delta \mathcal{P}| R^2}{4\mu L} \tag{3.37}$$

The axial velocity, Eq. 3.36, is proportional to the pressure drop, and inversely proportional to the fluid viscosity. Moreover, we remind that the velocity field along the radial coordinate is the same, and it is expressed by the field in Eq. 3.36, whatever the axial position ($v(r)$ does not depend on z).

Therefore, we are realising the following: in laminar flow, the fastest fluid is that travelling on the axis, and the velocity gradually decreases from the axis to the walls. To find the mean radial value for the velocity profile, we can use Eq. 3.22:

$$\overline{v} = \frac{\int v(r)\mathrm{d}S}{S} = \frac{\frac{|\Delta\mathcal{P}|R^2}{4\mu L}\int_0^R \left[1-\left(\frac{r}{R}\right)^2\right]\mathrm{d}(\pi r^2)}{\pi R^2} = \frac{\frac{|\Delta\mathcal{P}|R^2}{4\mu L}\int_0^R \left[1-\left(\frac{r}{R}\right)^2\right]2\pi r\mathrm{d}r}{\pi R^2} \tag{3.38}$$

Simplify recalling Eq. 3.37:

$$\overline{v} = \frac{2v_{max}}{R^2}\int_0^R \left[1-\left(\frac{r}{R}\right)^2\right] r\mathrm{d}r = \frac{2v_{max}}{R^2}\left[\int_0^R r\mathrm{d}r - \int_0^R \frac{r^3}{R^2}\mathrm{d}r\right] \tag{3.39}$$

$$\overline{v} = \frac{2v_{max}}{R^2}\left\{\left[\frac{r^2}{2}\right]_0^R - \left[\frac{r^4}{4R^2}\right]_0^R\right\} = \frac{2v_{max}}{R^2}\left(\frac{R^2}{2} - \frac{R^2}{4}\right) \tag{3.40}$$

That is, the obvious result illustrating, for a parabolic profile, a *mean value equal to 50% of the maximum value*:

$$\boxed{\overline{v} = \frac{v_{max}}{2} = \frac{|\Delta\mathcal{P}|R^2}{8\mu L} = \frac{|\Delta\mathcal{P}|D^2}{32\mu L}} \tag{3.41}$$

Example: Analysis of Laminar Flow in a Cylindrical Pipe

Consider a cylindrical pipe ($D = 0.05$ m; $L = 1$ m) where water flows at 20 °C (from Table 3.1, $\mu = 1.021 \cdot 10^{-3}$ kg/(m·s) and $\rho = 1001.9$ kg/m^3) enlivened by laminar motion, a hypothesis that we will verify at the end of the example. Be the pressure drop between fluid inlet and outlet, evaluated by a pressure gauge, $|\Delta\mathcal{P}| = 0.27$ kg/(m·s^2). This pressure drop permits the fluid flowing in the system. Application of Eqs. 3.31 and 3.37 allows to calculate, respectively, $\tau_w = 3.38 \cdot 10^{-3}$ kg/(m·s^2) and $v_{max} = 0.041$ m/s. That is, water flows with a mean velocity $\overline{v}$, Eq. 3.41, of 2 cm/s. Insertion of $\overline{v}$ into the Reynolds number, Eq. 3.15, returns $Re = 1000$ (approximately), which is far lower than the upper limit ($Re = 2100$) for which laminar flow is effective. Please note that this would have been valid also using v_{max}, instead of $\overline{v}$, in the expression for Re, still <2100. Finally, Fig. 3.8 reports the solution of the velocity radial profile for the system at hand.

3.8 Turbulent Flow

3.8.1 Conceptual Presentation

As already highlighted, *turbulent flow* is associated with the presence of lateral mixing and eddies/vortexes (also visible ones), Fig. 3.9. Even the smallest vortexes can contain up to 10^{12} molecules (McCabe et al. 2001): therefore, turbulence is to be

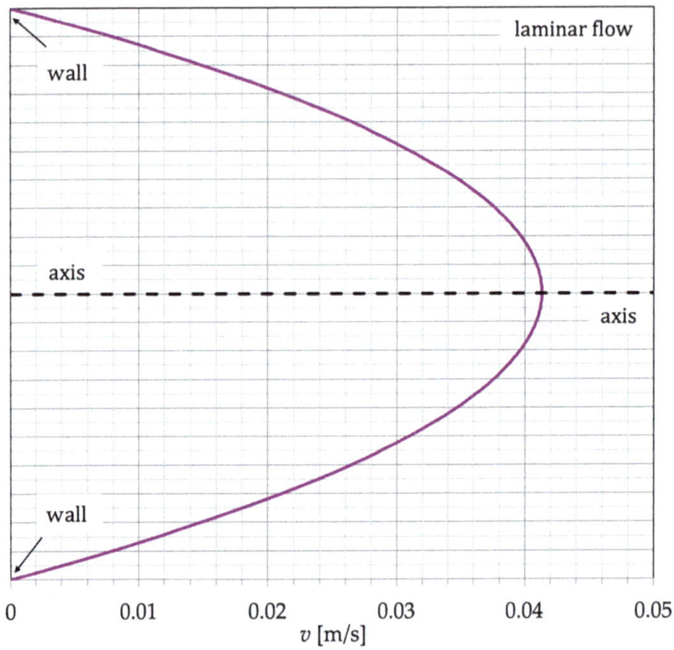

Fig. 3.8 Solution of the velocity field (radial profile of the axial velocity, Eq. 3.36) for a pure laminar flow in a cylindrical conduit (water at 20 °C, data in the example description)

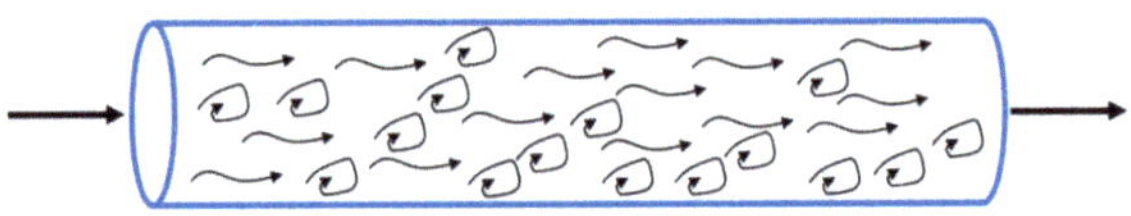

Fig. 3.9 Schematic representation of a turbulent flow in a cylindrical pipe, with presence of fluid in motion following disorderly path and in presence of eddies/vortexes

understood as a *macroscopic phenomenon*. Given the turbulent nature of the motion here considered, when we talk here about the velocity of the fluid element, we will refer to its *average value over time*, since there are temporal fluctuations of v around an average value. The predominance of inertial forces makes Re increase respect to the case of laminar flow (cf. Eq. 3.15). We will refer to turbulent flow, in this scenario, starting from about $Re = 4000–10{,}000$.

Typically, the effect of the wall in turbulent flow is confined to a zone adjacent to it, also called *boundary layer* or *viscous sublayer* (Fig. 3.10). We can define the momentum boundary layer as a zone in which the motion is influenced by the presence of a solid boundary (in our case, the wall of the cylindrical duct). In this zone, as we will see later, the velocity of the fluid increases, starting from its value (obviously, zero) at the wall. The motion in the boundary layer is, essentially, still laminar. This zone is not free from eddies (occasionally, they can migrate from the

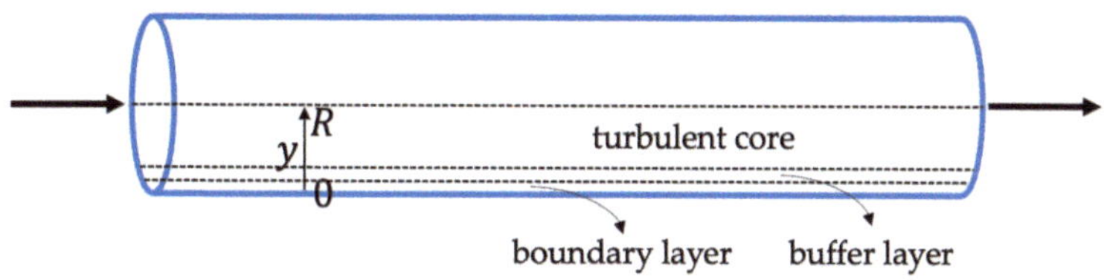

Fig. 3.10 Schematisation of the three zones dividing our system (turbulent flow in a cylindrical conduit). The system is axisymmetric. The extension of the three zones is arbitrary (not to scale), for the sake of clarity. Please observe that now we prefer to define a radial coordinate y that is opposite to r: $y = 0$ at the wall, and increases up to $y = R$ at the axis

turbulent zone towards the boundary layer), even if the turbulent effects can be usually neglected in the viscous sublayer.

Following the boundary layer, moving away from the wall and towards the axis, we encounter a transition zone (*buffer layer*), in which we find both viscous and inertial phenomena. While the axial velocity keeps increasing, we reach the *turbulent core*, where the viscous shear is negligible compared to the eddies. Turbulent vortexes are established starting from a certain value of *Re*, and the relevance of the turbulent core against the other two zones (the buffer layer and the boundary layer) gradually becomes more extensive as *Re* increases. An extreme of this concept, which will not be considered here, but becomes very important when we will deal with chemical reactors, is represented by the flow in a system with very high *Re*, up to a point where the whole radial coordinate is characterised by turbulence (no presence of the other two zones), i.e. the plug flow.

3.8.2 Quantitative Analysis

First, recognising the new way we have redefined the radial coordinate in Fig. 3.10, the three zones can be subdivided in an elegant way, via a dimensionless radial coordinate:

$$y^+ = \frac{yv^*\rho}{\mu} \tag{3.42}$$

that, in its general appearance, resembles the Reynolds number in Eq. 3.15. The velocity appearing in Eq. 3.42 is named friction velocity and it is defined as:

$$v^* = \sqrt{\frac{|\tau_w|}{\rho}} \tag{3.43}$$

Now the radial coordinate is directed from the wall to the axis (Fig. 3.10), while the shear stress is directed from zones where the velocity is higher to those where the velocity is lower (Fig. 3.4). As the velocity at wall is nil, the shear stress is positive in the direction of the wall, i.e. opposite to the direction of y^+. To avoid confusion, in Eq. 3.43, the shear stress at the wall is defined as absolute value. The reader is left to

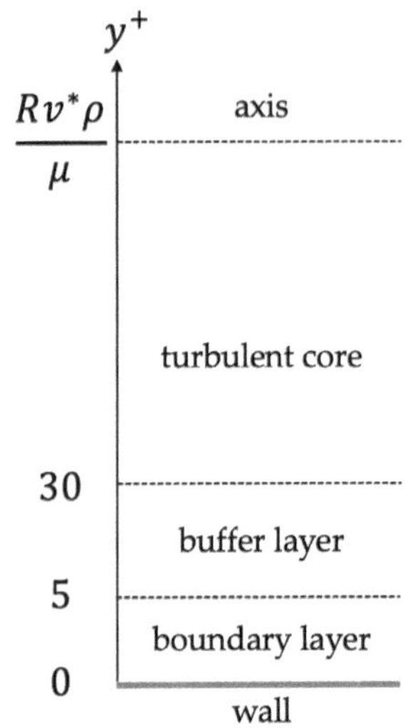

Fig. 3.11 Subdivision of the three zones for turbulent flow as a function of the non-dimensional radial coordinate, from a wall up to the system axis. The extension of the three zones is arbitrary (not to scale)

verify the dimensional correctness of Eq. 3.43, once given the dimensions for τ_w [mass/(length·time^2)] and ρ [mass/length^3]. From the theory reported in devoted books (Bird et al. 2007; McCabe et al. 2001), and on the basis of the works, among others, of von Kármán [Theodore von Kármán (1881–1963), Hungarian-American scientist, Director at California Institute of Technology; see von Kármán 1921] and Prandtl [Ludwig Prandtl (1875–1953), German physicist, Professor at Technical University of Hannover; see Prandtl 1933], the extension of the three zones is (Fig. 3.11), under the assumption of a fully developed turbulent flow:

$$\begin{cases} 0 \le y^+ \le 5 \Leftrightarrow 0 \le y \le \frac{5\mu}{v^*\rho} \Rightarrow \text{boundary layer} \\ 5 \le y^+ \le 30 \Leftrightarrow \frac{5\mu}{v^*\rho} \le y \le \frac{30\mu}{v^*\rho} \Rightarrow \text{buffer layer} \\ 30 \le y^+ \le \frac{Rv^*\rho}{\mu} \Leftrightarrow \frac{30\mu}{v^*\rho} \le y \le R \Rightarrow \text{turbulent core} \end{cases} \tag{3.44}$$

That is, the boundary layer extends from the wall ($y^+ = 0$ since $y = 0$) up to a radial coordinate y corresponding to $y^+ = 5$; the buffer layer starts from this point and extends up to a radial coordinate y corresponding to $y^+ = 30$; the turbulent core starts from this point and extends up to a maximum value for y^+, corresponding to $y = R$. The advantage of using y^+, instead of y, is that we can generalise the approach, without making reference to specific values for the radial coordinate depending on the system at hand from time to time. As we will demonstrate in the following example, although conceptually relevant, *the extension of the boundary and buffer layers is quantitatively negligible* when compared with the zone occupied by the turbulent core.

Example: Extension of the Three Zones for Turbulent Flow in a Cylindrical Pipe

In the same cylindrical pipe as above ($D = 0.05$ m; $L = 1$ m), air flows at 80 °C (from Table 3.1, $\mu = 2.1 \cdot 10^{-5}$ kg/(m·s) and $\rho = 1$ kg/m^3) enlivened by turbulent flow (later on in this chapter, we will verify this hypothesis). Be the pressure drop between fluid inlet and outlet, evaluated by a pressure gauge, $|\Delta\mathcal{P}| = 19.24$ kg/(m·s^2). Application of Eq. 3.31 allows to calculate $|\tau_w| = 0.2405$ kg/(m·s^2). Thus, Eq. 3.43, the friction velocity is $v^* = 0.4904$ m/s. This means that, in axis, the (maximum) value for y^+ is, last of Eq. 3.44 and Fig. 3.11, equal to 583.8. So, in a scale from 0 to 583.8, the boundary layer extends for 5/583.8 = 0.86% (0.215 mm over a radius $R = 2.5$ cm), the buffer layer extends for (30–5)/583.8 = 4.28% (1.07 mm), and the remainder 94.86% (2.3715 cm) represents the turbulent core. This is a general feature, representing the noticeable predominance of the turbulent zone vs. the "non-turbulent" layers.

To make this quantitative analysis fully dimensionless, express not only the radial coordinate, but also the velocity, in non-dimensional form:

$$v^+ = \frac{v}{v^*} \tag{3.45}$$

Thus, the radial profile for the axial velocity in turbulent flow will be obtained as $v^+(y^+)$. Start from the boundary layer. As we have demonstrated in the previous example, the viscous sublayer is located very close to the wall, then we can consider $\tau \cong \tau_w$. Rewrite the Newton equation (that rules the momentum transport in the laminar layer) as:

$$\frac{\mathrm{d}v}{\mathrm{d}y} = \frac{|\tau_w|}{\mu} \tag{3.46}$$

From Eqs. 3.45 and 3.42, it is, respectively:

$$\begin{cases} v = v^+ v^* \\ y = \dfrac{y^+ \mu}{v^* \rho} \end{cases} \tag{3.47}$$

Substituting in Eq. 3.46, with all parameters constant but v^+ and y^+:

$$\frac{(v^*)^2 \rho}{\mu} \frac{\mathrm{d}v^+}{\mathrm{d}y^+} = \frac{|\tau_w|}{\mu} \tag{3.48}$$

Rewrite Eq. 3.43 as:

$$(v^*)^2 \rho = |\tau_w| \tag{3.49}$$

Equation 3.48 becomes:

$$\frac{|\tau_w|}{\mu}\frac{dv^+}{dy^+}=\frac{|\tau_w|}{\mu} \tag{3.50}$$

$$\frac{dv^+}{dy^+}=1 \Rightarrow dv^+=dy^+ \tag{3.51}$$

At tube wall ($y^+ = 0$), it is $v = 0$ (and, Eq. 3.45, $v^+ = 0$). Thus:

$$\int_0^{v^+} dv^+ = \int_0^{y^+} dy^+ \Rightarrow v^+ = y^+ \tag{3.52}$$

Very interestingly, in the viscous sublayer, the non-dimensional velocity is equal to the non-dimensional coordinate. While a solid theoretical background has brought us to Eq. 3.52 for the boundary layer, empirical expressions are available for $v^+(y^+)$ relationships in the other two zones. Assuming those reported in McCabe et al. 2001 (further discussion in Bird et al. 2007), we land at:

$$\boxed{\begin{cases} 0 \le y^+ \le 5 \Rightarrow v^+ = y^+ \text{(for boundary layer)} \\ 5 \le y^+ \le 30 \Rightarrow v^+ = 5\ln(y^+) - 3.05 \text{ (for buffer layer)} \\ 30 \le y^+ \le \dfrac{Rv^*\rho}{\mu} \Rightarrow v^+ = 2.5\ln(y^+) + 5.5 \text{ (for turbulent core)} \end{cases}} \tag{3.53}$$

The relationships presented are reasonably valid only if the turbulent flow is well developed (*Re* at least 10^4). Furthermore, it has to be noted that the distinction into three zones is for schematisation purposes, but is not clear-cut. The $v^+(y^+)$ expressions in Eq. 3.53 allow to fix a distance y from the wall, calculate y^+ (linearly dependent on y, Eq. 3.42), then calculate v^+ (Eq. 3.53) and, finally, v (linear dependent on v^+, Eq. 3.45) in that position. It can be verified that the relationships are correctly connected between the various zones. At the boundary between the viscous sublayer and the buffer zone ($y^+ = 5$), it is $v^+ = 5$ both by applying the first of Eq. 3.53 for the boundary layer (where $y^+ = 5$ represents the upper limit) and the second of Eq. 3.53 for the buffer zone (where $y^+ = 5$ represents the lower limit). At the boundary between the buffer layer and the turbulent core ($y^+ = 30$), it is $v^+ = 14$ both by applying the second of Eq. 3.53 for the buffer zone (where $y^+ = 30$ represents the upper limit) and the third of Eq. 3.53 for the turbulent core (where $y^+ = 30$ represents the upper limit). A general graphic representing Eq. 3.53 is illustrated in Fig. 3.12.

This analysis shows, for turbulent flow, an increasing profile of v^+ (and, therefore, of v, see Eq. 3.45; v and v^+ are linearly dependent each other) with respect to y^+ (and, therefore, with respect to y, see Eq. 3.42; y and y^+ are linearly dependent each other). The dependence of the velocity on the distance from the wall is more marked in the boundary layer (first of Eq. 3.53, which is linear), compared to the milder trends in the other two zones (equations characterised by the presence of the logarithm): this is in line with the conceptual outline of a velocity which, although increasing as one moves away from the wall, tends to vary less as the axis is approached.

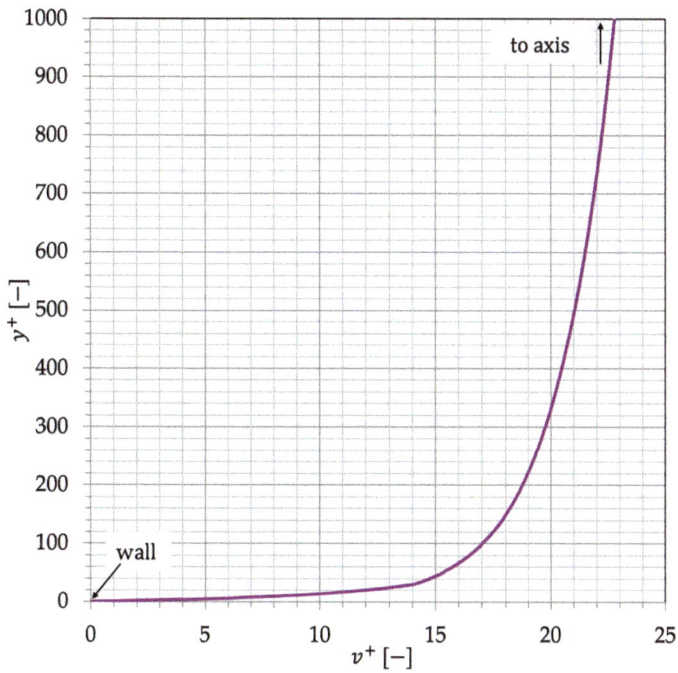

Fig. 3.12 Radial profile of velocity for turbulent flow in pipes (universal trend). The non-dimensional velocity is plotted, in abscissa, against the non-dimensional distance from the wall (in ordinate), arbitrarily reported up to $y^+ = 1000$

For laminar motion, taking the maximum value for the velocity (i.e., the one that occurs on the axis), we have seen that the average velocity value is 50% of this maximum (Eq. 3.41). A similar approach is adopted in turbulent motion. Neglecting, as seen, the relevance of the boundary layer and the intermediate zone, and having noted that also in this case the maximum velocity is obtained on the axis (Fig. 3.12), the average velocity ($\overline{v}$) is estimated as equal to a certain % with respect to this maximum. Typically, $\overline{v}$ is 75–80%, 80–85%, 85–90% of v_{max} for, respectively, $Re = 10^4, 10^5, 10^6$. This is in line with the conceptual outline according to which, *as Re increases, the motion becomes increasingly more turbulent*, therefore *the average velocity value gets closer and closer to the maximum value attained on the axis*. In other words, as *Re* increases, the velocity profile becomes flatter and flatter. At the theoretical limit of plug flow (dealt in this textbook when discussing ideal chemical reactors), the mean and maximum velocity values would coincide. Of course, more refined calculations are possible to determine $\overline{v}$, using Eq. 3.22 and integrating the velocity radial profile starting from Eq. 3.53.

Example: Analysis of Turbulent Flow in a Cylindrical Conduit

May we extend the previous example, to solve the velocity field. First, at axis ($y^+ = 583.8$), the maximum non-dimensional velocity is $v^+ = 2.5\ln(y^+) + 5.5 = 21.4$ (Eq. 3.53) that, translated in m/s, equals $v_{max} = 21.4 \cdot 0.4904 = 10.5$ m/s (Eq. 3.45). Using this value as order of magnitude to individuate *Re*, it is, Eq. 3.15, *Re* approx. 25,000. This means a fully developed turbulent flow, and we can assume a mean velocity $\overline{v}$ of 80% v_{max} (circa 8.4 m/s, that would return, more properly, *Re* of around 20,000).

Application of Eq. 3.53 allows to draw Fig. 3.13, where we can observe that, respect to the profile in laminar flow (e.g., Fig. 3.8), the radial profile for the axial velocity is markedly flatter. In Fig. 3.13, the values on the abscissa have been expressed as actual velocity, obtained in terms of $v = v^+v^*$ (Eq. 3.45), where v^+ has been calculated as a function of y^+ (from $y^+ = 0$, walls, to $y^+ = 583.8$, axis; the system is axisymmetric) according to Eq. 3.53.

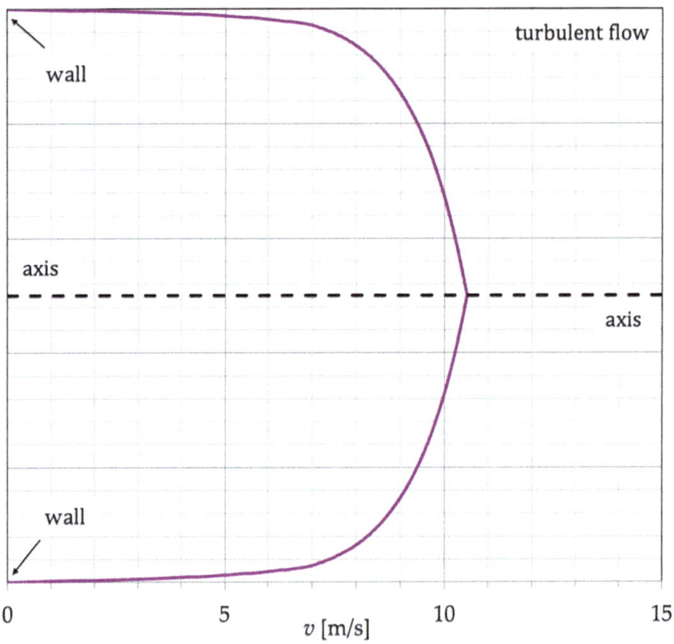

Fig. 3.13 Solution of the velocity field (radial profile of the axial velocity, Eq. 3.53) for a fully developed turbulent flow in a cylindrical duct (air at 80 °C, data in the examples description). Please note that the field in Eq. 3.53 fails in representing a nil velocity gradient at axis, a minor issue commonly accepted, given the empirical nature of the expressions in Eq. 3.53

3.9 Friction Factor

The last topic, with which we conclude this chapter dedicated to the ideal rudiments of momentum transport in systems of interest for this book, is the friction factor. Let us observe the motion of the fluid inside the cylindrical conduit at a macroscopic level. The overall force F associated with the motion of the fluid can be simply expressed as the product between the overall pressure drop (between fluid entry and exit) and the duct cross-section area:

$$F = |\Delta \mathcal{P}|\pi R^2 \tag{3.54}$$

Define now the wetted surface area (i.e., the internal lateral surface of the system, which is "wetted" by the passage of the fluid):

$$A_W = 2\pi RL \tag{3.55}$$

and the specific kinetic energy (i.e., the fluid kinetic energy per unit volume):

$$k = \frac{\rho \overline{v}^2}{2} \tag{3.56}$$

The force F is proportional to the product $A_W k$ (as a matter of fact, the higher the wetted surface, and the higher the kinetic energy, the higher F; moreover, the product between [surface] and [energy/volume] returns [energy/length], i.e., a force). The equation that states this proportionality is, actually, the definition for the dimensionless *friction factor* f (Fanning 1877):

$$F = f A_W k \tag{3.57}$$

i.e., the friction factor is the constant relating F and the product $A_W k$. Equivalently:

$$|\Delta \mathcal{P}|\pi R^2 = f 2\pi RL\frac{\rho \overline{v}^2}{2} \Rightarrow \boxed{f = \frac{R}{L}\frac{|\Delta \mathcal{P}|}{\rho \overline{v}^2} = \frac{D}{2L}\frac{|\Delta \mathcal{P}|}{\rho \overline{v}^2} = \frac{\text{total (laminar + turbulent) transport of momentum}}{\text{turbulent transport of momentum}}} \tag{3.58}$$

The Fanning friction factor can be seen as the ratio between the total (laminar +turbulent) momentum transport (the numerator shows the overall pressure drop, defining the overall force F as per Eq. 3.54) and the transport of momentum by turbulent mechanism (the denominator shows the mean velocity illustrating the inertial mechanisms, see Eq. 3.15 for Re, but it does not present the viscosity, which would instead have taken into account the laminar viscous mechanisms).

We are now looking for $f(Re)$ relations for laminar and turbulent flow. First, note that, comparing Eqs. 3.58 and 3.15, we expect an *inverse dependence* of f on Re, of the type (with a and b proper constants):

$$f = \frac{a}{Re^b} \tag{3.59}$$

since, as the relevance of turbulent transport mechanisms (and the mean fluid velocity) increases, the number of Reynolds increases while the friction factor decreases.

For laminar motion ($Re < 2100$), Eq. 3.41 holds for the mean velocity. Rewrite to express the pressure drop, obtaining the *Hagen–Poiseuille equation* [named, in addition to Poiseuille, after Gotthilf Heinrich Ludwig Hagen (1797–1884), German engineer, Professor at Building Academy in Berlin; see Hagen 1839; Poiseuille 1844]:

$$|\Delta \mathcal{P}| = \frac{32\mu L\bar{v}}{D^2} \tag{3.60}$$

that is, the longer the system-the more viscous the fluid-the higher the velocity-the lower the diameter, the higher the pressure drop, as intuitive. Substitute Eq. 3.60 into Eq. 3.58:

$$f = \frac{D}{2L}\frac{\frac{32\mu L\bar{v}}{D^2}}{\rho\bar{v}^2} = \frac{16\mu}{D\bar{v}\rho} = \frac{16}{Re} \tag{3.61}$$

which, *q.e.d.*, is a relationship of the type exposed in Eq. 3.59 (where $a = 16$ and $b = 1$). Couples of values for a and b for turbulent motion are empirically reported in literature (Bird et al. 2007; McCabe et al. 2001), giving rise to the following scheme:

$$\boxed{\begin{cases} f = \dfrac{16}{Re} \text{ for laminar motion } (Re < 2100) \\ f = \dfrac{0.046}{Re^{0.2}}, \text{ or } f = \dfrac{0.0791}{Re^{0.25}}, \text{ for turbulent motion} \end{cases}} \tag{3.62}$$

Equation 3.59 can be graphically illustrated on a chart (the *Moody representation*, Moody 1944) where both the abscissa and the ordinate are reported in logarithmic scale. In this case, Eq. 3.59 becomes $\ln(f) = \ln(a/Re^b) = \ln(a) - b\ln(Re)$: on a plane where f is plotted vs. Re on a log-log basis, Eq. 3.59 is a straight line having slope $-b$ (i.e., -1 for laminar flow, and -0.2 for turbulent flow, choosing the first of the two options in Eq. 3.62). The Moody chart is illustrated in Fig. 3.14.

The relationship reported for turbulent motion, Eq. 3.62, is valid for hydraulically smooth pipes. However, pipes can have *roughness*, indicated by a parameter that measures the average protuberances of the roughness of the internal walls. These protuberances, which are not able to influence the laminar motion (and, therefore, the value of f in the laminar case) unless they are very significant, can instead affect f in the turbulent case: the higher the roughness, the higher the friction factor (the other parameters being fixed, this means that the pressure drop is greater in rough pipes, see Eq. 3.58). Therefore, Moody plots can be accompanied (for turbulent flow) by parametric $f(Re)$ curves in the ratio between the height of the protuberances and the diameter of the tube, a ratio typically in the range 10^{-5}–10^{-3} (e.g., for a diameter of 10 cm, we may have wall roughness with height of the protuberances of 1 μm–0.1 mm). Obviously, f is an increasing function (fixed Re) of this ratio (termed

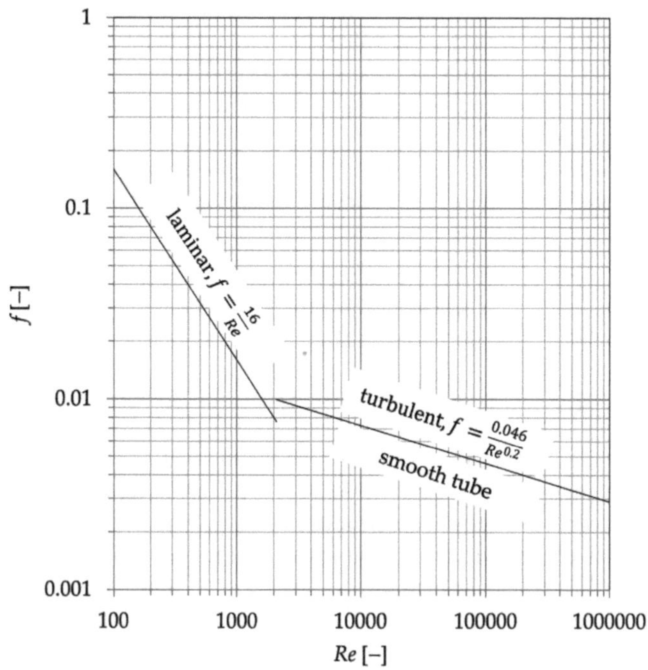

Fig. 3.14 Moody's plot reporting the Fanning friction factor for tube flow vs. Reynolds number. For simplicity, the turbulent zone starts at $Re = 2100$, and hydraulically smooth tubes are considered. Log-log plot

"relative roughness" in the original work by Moody 1944), as the interested reader may check in literature (Bird et al. 2007; McCabe et al. 2001).

In previous examples, we have seen that the type of motion (laminar or turbulent) had to be hypothesised and, then, a check on Re was performed, once the average velocity was known. Actually, this can be avoided by means of a more elegant approach. Start from the relation $f = 16/Re$ (Eq. 3.61) valid for laminar motion. Recalling the general expression for the friction factor, Eq. 3.58, if we multiply side by side $f = 16/Re$ by Re^2, we will obtain, on the left-hand side of the equation highlighted in the box, an expression that no longer depends on the (unknown) average velocity $\bar{v}$:

$$\begin{cases} f\,Re^2 = \dfrac{D}{2L}\dfrac{|\Delta\mathcal{P}|}{\rho\bar{v}^2}\left(\dfrac{D\bar{v}\rho}{\mu}\right)^2 = \boxed{\dfrac{D^3|\Delta\mathcal{P}|\rho}{2L\mu^2} = 16\,Re} \\ \text{for } Re \le 2100, \text{ i.e. for } \dfrac{D^3|\Delta\mathcal{P}|\rho}{2L\mu^2} \le 33{,}600 \end{cases} \tag{3.63}$$

This expression, as indicated, is valid only in the range of existence of laminar flow, i.e. up to $Re = 2100$. Therefore, the left-hand side of Eq. 3.63, which does not depend on the mean velocity $\bar{v}$ and is therefore known if the other parameters are known, cannot be greater than the value that the right-hand side ($16Re$) assumes when $Re = 2100$, i.e. it cannot be greater than 33,600. If this happens, the flow can be considered laminar, Re can be calculated as the left-hand side divided by 16, and $\bar{v}$ can be obtained from Eq. 3.15.

If Eq. 3.63 is not verified, restart from Eq. 3.62 for turbulent regime ($f = 0.046/Re^{0.2}$). Adopt the same approach to make $\bar{v}$ disappear, multiplying both sides per Re^2:

$$\begin{cases} f\,Re^2 = \boxed{\dfrac{D^3|\Delta\mathcal{P}|\rho}{2L\mu^2}} = 0.046\,Re^{1.8} \\ \text{for } Re > 2100, \text{ i.e. for } \dfrac{D^3|\Delta\mathcal{P}|\rho}{2L\mu^2} > 44{,}000 \text{ (ca.)} \end{cases} \tag{3.64}$$

This expression, as indicated, is valid only in the interval of existence of turbulent flow, i.e. (for simplicity) starting from $Re = 2100$. Therefore, the left-hand side of Eq. 3.64, which does not depend on the mean velocity $\bar{v}$ and is therefore known if the other parameters are known, must be at least equal to the value that the right-hand side ($0.046Re^{1.8}$) assumes when $Re = 2100$, i.e. it must be at least equal to ca. 44,000. If this happens, the flow can be considered turbulent, Re can be calculated by isolating it on the right-hand side, and finally $\bar{v}$ can be obtained from Eq. 3.15. The concluding examples to follow will deal with these approaches.

Example: Finding the Flow Regime Via Friction Factor

Take our reference system from an example above: cylindrical pipe ($D = 0.05$ m; $L = 1$ m) where water flows at 20 °C ($\mu = 1.021 \cdot 10^{-3}$ kg/(m·s) and $\rho = 1001.9$ kg/m^3), $|\Delta\mathcal{P}| = 0.27$ kg/(m·s^2). Calculate:

$$f\,Re^2 = \frac{D^3|\Delta\mathcal{P}|\rho}{2L\mu^2} = 16{,}219 < 33{,}600 \tag{3.65}$$

Thus, the flow regime is laminar, and we can use Eq. 3.63:

$$Re = \frac{f\,Re^2}{16} = \frac{16{,}219}{16} = 1014 < 2100 \tag{3.66}$$

and, finally (Eq. 3.15):

$$\bar{v} = \frac{\mu Re}{D\rho} = 2 \text{ cm/s} \tag{3.67}$$

that confirms what previously seen for $\bar{v}$.

In another example, same geometry, we have air flowing at 80 °C ($\mu = 2.1 \cdot 10^{-5}$ kg/(m·s) and $\rho = 1$ kg/m^3), $|\Delta\mathcal{P}| = 19.24$ kg/(m·s^2). Calculate:

(continued)

$$f\,Re^2 = \frac{D^3|\Delta\mathcal{P}|\rho}{2L\mu^2} = 2.724 \cdot 10^6 > 44{,}000 \tag{3.68}$$

We are in a well developed turbulent regime. Use Eq. 3.59:

$$Re = \left(\frac{fRe^2}{0.046}\right)^{\frac{1}{1.8}} = \left(\frac{2.724 \cdot 10^6}{0.046}\right)^{\frac{1}{1.8}} = 20{,}798 > 2100 \tag{3.69}$$

and, finally (Eq. 3.15):

$$\bar{v} = \frac{\mu Re}{D\rho} = 8.7\ \text{m/s} \tag{3.70}$$

that, also in this case, confirms what previously seen (the value of the average velocity is close to that, 8.4 m/s, estimated starting from v_{max} in the above example, remembering that the equations for turbulent flow are empirical).

3.10 Background Information

- The transport of momentum is driven by a velocity gradient (Fig. 3.1). The Newton's law of viscosity (Eq. 3.4) relates the shear stress to the velocity spatial gradient via a proportionality constant depending on the fluid nature: the dynamic viscosity. This is valid for Newtonian fluids.
- Laminar flow is a molecular transport (orderly fluid path), while turbulent flow is a macroscopic phenomenon (disorderly fluid path with eddies/vortexes). Transition from laminar to turbulent regime is indicated by the increase in the Reynolds number (Eq. 3.15), that represents the ratio between turbulent and laminar transport of momentum. The higher the Reynolds number, the more turbulent the fluid flow.
- Our reference system is a horizontal cylindrical conduit, axisymmetric. The flow is stationary, one-dimensional, and incompressible (constant density).
- For laminar flow, the axial velocity is nil at walls, and increases moving towards the axis following a parabolic trend (Fig. 3.7). The mean velocity is 50% the maximum velocity (Eq. 3.41).
- For turbulent flow, three zones are individuated (Fig. 3.10): the boundary layer, a buffer zone, and the turbulent core. The turbulent core accounts for most of the system. The radial profile for velocity is given in Eq. 3.53, and the mean velocity is a percentage (always >50%, and increasing as the Reynolds number increases) of the maximum velocity. The velocity profile is flatter than the parabolic one for laminar flow (Figs. 3.12 and 3.13).
- The friction factor represents the ratio between total and turbulent transport of momentum (Eq. 3.58). The friction factor is inversely proportional to the

Reynolds number (Eq. 3.62), with expressions according to the flow regime (Fig. 3.14). The combination of the friction factor and Reynolds number allows to individuate the flow regime without knowing in advance the average velocity (Eqs. 3.63 and 3.64).

References

Astarita G (1997) Dimensional analysis, scaling, and orders of magnitude. Chem Eng Sci 52:4681–4698. https://doi.org/10.1016/S0009-2509(97)85420-6

Bird RB, Stewart WE, Lightfoot EN (2007) Transport phenomena. John Wiley & Sons, New York

Brillouin M (1930) Jean Leonard Marie Poiseuille. J Rheol 1:345–348. https://doi.org/10.1122/1.2116329

Fanning JT (1877) A practical treatise on hydraulic and water-supply engineering: relating to the hydrology, hydrodynamics, and practical construction of water-works, in North America. Van Nostrand, New York

Green DW, Perry RH (2008) Perry's chemical engineers' handbook. McGraw-Hill, New York

Hagen G (1839) Ueber die Bewegung des Wassers in engen cylindrischen Röhren. Ann Phys (Berlin) 122:423–442. https://doi.org/10.1002/andp.18391220304

McCabe WL, Smith JC, Harriott P (2001) Unit operations of chemical engineering. McGraw-Hill, Singapore

Moody LF (1944) Friction factors for pipe flow. Trans ASME 66:671–684

Poiseuille JLM (1844) Recherches expérimentales sur Ie mouvement des liquides dans les tubes de très-petits diamètres. Académie des Sciences, Paris

Prandtl L (1933) Über Tragflügel kleinsten induzierten Widerstandes. ZFM 24:305–306

Reynolds O (1883) An experimental investigation of the circumstances which determine whether the motion of water shall be direct or sinuous, and of the law of resistance in parallel channels. Proc R Soc Lond 35:84–99. https://doi.org/10.1098/rspl.1883.0018

Sutera SP, Skalak R (1993) The history of Poiseuille's law. Annu Rev Fluid Mech 25:1–19. https://doi.org/10.1146/annurev.fl.25.010193.000245

von Kármán T (1921) Über laminare und turbulente Reibung. Z Angew Math Mech 1:233–252. https://doi.org/10.1002/zamm.19210010401

4 Energy Transport

Abstract

The second argument concerning transport phenomena relates to energy. We begin with the treatment of thermal conduction, whereby the general law of molecular transport is transformed into the Fourier's equation. The introduction of the concept of thermal conduction allows us to define the Prandtl number, seen as the ratio between the molecular transport of momentum and that of energy. We then continue with thermal convection, which we apply to the case of a fluid animated by fully developed turbulent motion. We use the film theory to concentrate the resistances to energy transfer in a boundary layer adjacent to the wall, and we define the Nusselt number as the ratio between the total transport and the molecular transport of energy. The analysis of some cases (energy transport through flat or cylindrical walls) allows us to highlight how the heat flux can be expressed in terms of the ratio between driving force and resistance, or of the product between driving force and heat transfer coefficient (the latter being the reciprocal of the resistance), and how the overall resistance to energy transport can be expressed as the sum of the resistances to the individual transport phenomena placed in series.

Nomenclature

$\mathbb{C}$	Specific heat [energy/(mass·temperature)]
D	Diameter [length]
h	Convective heat transfer coefficient [energy/(length^2·time·temperature)]
h_c	Convective heat transfer coefficient (cold fluid) [energy/(length^2·time·temperature)]
h_h	Convective heat transfer coefficient (hot fluid) [energy/(length^2·time·temperature)]
k_T	Thermal conductivity [energy/(length·time·temperature)]

F. Montagnaro et al., *Elements of Unit Operations and Chemical Reactors*,
https://doi.org/10.1007/978-3-032-18046-9_4

$k_{T,GAS}$	Thermal conductivity for gas [energy/(length·time·temperature)]
$k_{T,LIQ}$	Thermal conductivity for liquid [energy/(length·time·temperature)]
L	Length [length]
Nu	Nusselt number [–]
Pr	Prandtl number [–]
q	Energy flow rate [energy/time]
$\dot{q}$	Energy flux [energy/(length^2·time)]
$\dot{q}_e$	Energy flux at external side [energy/(length^2·time)]
$\dot{q}_i$	Energy flux at internal side [energy/(length^2·time)]
$\mathcal{R}$	Overall resistance [length^2·time·temperature/energy]
$\mathcal{R}_{cond}$	Conductive resistance [length^2·time·temperature/energy]
$\mathcal{R}_{cond,e}$	External conductive resistance [length^2·time·temperature/energy]
$\mathcal{R}_{cond,i}$	Internal conductive resistance [length^2·time·temperature/energy]
$\mathcal{R}_{conv}$	Convective resistance [length^2·time·temperature/energy]
$\mathcal{R}_{conv,c}$	Convective resistance (cold fluid) [length^2·time·temperature/energy]
$\mathcal{R}_{conv,h}$	Convective resistance (hot fluid) [length^2·time·temperature/energy]
$\mathcal{R}_e$	Overall resistance at external side [length^2·time·temperature/energy]
$\mathcal{R}_i$	Overall resistance at internal side [length^2·time·temperature/energy]
r	Radial coordinate [length]
r_e	External radial coordinate [length]
r_i	Internal radial coordinate [length]
Re	Reynolds number [–]
T	Temperature [temperature]
T_c	Temperature of the cold fluid [temperature]
T_e	Temperature at external side [temperature]
T_f	Temperature of the fluid [temperature]
T_h	Temperature of the hot fluid [temperature]
T_i	Temperature at internal side [temperature]
T_k	Temperature at generic k position [temperature]
T_w	Temperature of the wall [temperature]
U	Overall heat transfer coefficient [energy/(length^2·time·temperature)]
U_e	Overall heat transfer coefficient at external side [energy/(length^2·time·temperature)]
U_i	Overall heat transfer coefficient at internal side [energy/(length^2·time·temperature)]
x	Direction of transport [length]
α_T	Thermal diffusivity [length^2/time]
$\|\Delta T\|$	Thermal driving force (absolute value) [temperature]
δ	Thickness [length]
μ	Dynamic viscosity [mass/(length·time)]
ν_k	Kinematic viscosity [length^2/time]
ρ	Density [mass/volume]

4.1 Mechanisms of Energy Transport

If in thermal contact (i.e. in a condition in which energy can be exchanged), two phases (or, e.g., two fluid elements) at different temperature T will show *transport of energy* from the phase with higher to the one with lower temperature (Fig. 4.1). The difference in T (the gradient of temperature) represents the *driving force* that regulates this transport phenomenon. Apart from transport by radiation (not addressed in this book; cf. Bird et al. 2007 for details), the two main mechanisms of energy transport are: *conduction* (a molecular phenomenon such as the *laminar flow* addressed in momentum transport) and *convection* (a macroscopic phenomenon associated with the fluid motion). We will refer to *stationary systems* of interest for the purposes of this textbook.

4.2 Thermal Conduction

4.2.1 General Characteristics

Let us start with *thermal conduction*, just as, in the chapter dedicated to the transport of momentum, we began with the discussion of laminar motion. In the case of conduction (*a molecular transport of energy*), there is *no observable motion of matter* with reference to the phases that exchange energy. For example, if a gradient of T is established on the two sides of a solid wall, there will be a conductive heat flow through the wall. This happens due to the movement of electrons/molecules/atoms (depending on the system). For simplicity, here reference is made to solids, but it is recalled that conduction is also associated with other states of aggregation (fluids). The reference is initially to solids because in this case the convective mechanism (which we will refer to in the following) is assumed to be irrelevant,

Fig. 4.1 Transport of energy: schematic concept

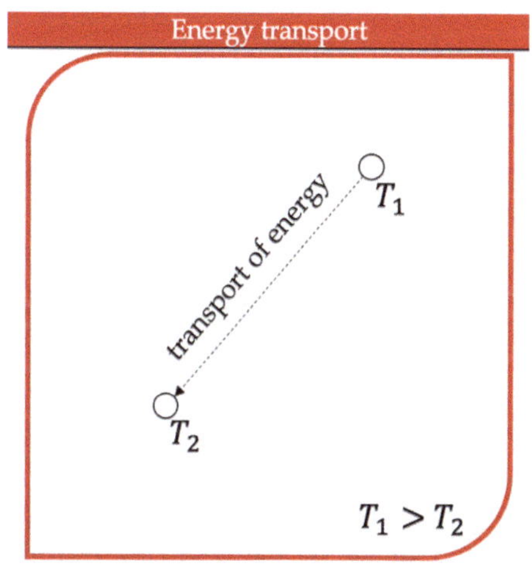

and therefore it is easier to separate the two phenomena (and to illustrate, for the moment, the topic in terms of conduction only).

4.2.2 The Fourier's Law on Conduction

Recall the general law of molecular transport presented in Eq. 3.2:

$$\boxed{\dot{\mathbb{F}} = -\mathbb{K}\frac{\mathrm{d}\mathbb{Q}}{\mathrm{d}x}} \tag{4.1}$$

Here, once defined the energy flow rate as q [energy/time], the energy flux $\dot{q}$ [energy/(length^2·time)] takes on the role of $\dot{\mathbb{F}}$. The quantity that determines the transport phenomenon is, Fig. 4.1, $\mathbb{Q} = T$, while x is the generic direction of transport. Let us remind again that the minus sign in Eq. 4.1 indicates that the flux is positive in the opposite direction to that for which the gradient is positive. We can therefore write the *Fourier's law on conduction* [named after Jean-Baptiste Joseph Fourier (1768–1830), French scientist, Professor at Polytechnic School of Paris; see Fourier 1822]:

$$\boxed{\dot{q} = -k_T\frac{\mathrm{d}T}{\mathrm{d}x}} \tag{4.2}$$

In other words, the heat flux by thermal conduction is proportional to the temperature gradient, via a constant $\mathbb{K} = k_T$, the *thermal conductivity*, that depends on the nature of the material (regardless the fluid dynamics of the system, as for the viscosity). The higher the thermal conductivity, the higher the heat flux exchanged by conduction. At this point, the reader has realised the symmetry between the Newton's law governing laminar momentum transport (Eq. 3.4) and the Fourier's law governing molecular energy transport (Eq. 4.2), both obeying the general law of molecular transport.

4.2.3 Thermal Conductivity

To find the dimensions for k_T from the equation that defines it (Eq. 4.2), we observe that:

$$k_T = -\frac{\dot{q}}{\frac{\mathrm{d}T}{\mathrm{d}x}} = \frac{\left[\frac{\text{energy}}{\text{length}^2\cdot\text{time}}\right]}{\left[\frac{\text{temperature}}{\text{length}}\right]} = \left[\frac{\text{energy}}{\text{length}\cdot\text{time}\cdot\text{temperature}}\right] \tag{4.3}$$

For instance, we can find k_T expressed in measure units of [J/(m·s·K)]. The thermal conductivity depends on temperature T, much less on pressure (unless the pressure is, e.g., higher than 10 bar for a gaseous phase). The thermal conductivity can assume values on the order of 10–100 J/(m·s·K) for conductive (e.g., metallic) solids, or on the order of 1 J/(m·s·K) for solids with low conductivity (insulating materials) such as glass, bricks, concrete. The value for wood is even lower. The thermal conductivity for liquids is generally lower than for solids: e.g., for water at room temperature it is $k_T = 0.6$ J/(m·s·K). For gases, k_T is even lower: e.g., for air at room temperature it is 0.026 J/(m·s·K). More in detail, using the formulae provided in the Perry's handbook (Green and Perry 2008), it is (with appropriate constants listed as a function of the species), for liquids (at either 1 atm or the vapour pressure, whichever is higher) and gases (at either 1 atm or the vapour pressure, whichever is lower), respectively:

$$\begin{cases} k_{T,LIQ} = C_1 + C_2 T + C_3 T^2 + C_4 T^3 + C_5 T^4 \\ k_{T,GAS} = \dfrac{C_1 T^{C_2}}{1 + \dfrac{C_3}{T} + \dfrac{C_4}{T^2}} \end{cases} \tag{4.4}$$

where T is expressed in [K] and k_T is obtained in [J/(m·s·K)].

Example: Thermal Conductivity for Water and Air

Application of Eq. 4.4 allows to calculate, for a reference liquid (water) and gas (air), the values of thermal conductivity as T increases (in our example, from 20 °C to 80 °C, properly expressed in [K]). Table 4.1 lists the values for the appropriate constants (Green and Perry 2008). We can observe that, increasing T from 20 °C to 80 °C, the thermal conductivity of water increases by about 11%, while that of air increases by about 15%. The liquid water conductivity is, actually, (much) greater than that of the reference gas.

Table 4.1 Evaluation of the thermal conductivity for liquid water and air (atmospheric pressure)

Species	C_1	C_2	C_3	C_4	C_5
Liquid water	−0.432	$5.7255 \cdot 10^{-3}$	$-8.078 \cdot 10^{-6}$	$1.861 \cdot 10^{-9}$	0
Air	$3.1417 \cdot 10^{-4}$	0.7786	−0.7116	2121.7	
	k_T [J/(m·s·K)]				
Liquid water (20 °C)	0.599				
Liquid water (40 °C)	0.626				
Liquid water (60 °C)	0.648				
Liquid water (80 °C)	0.664				
Air (20 °C)	0.026				
Air (40 °C)	0.027				
Air (60 °C)	0.028				
Air (80 °C)	0.030				

4.2.4 Thermal Diffusivity and Prandtl Number

Define $\mathbb{C}$ the specific heat of a substance, expressed here on mass basis [energy/(mass·temperature)]. We can express the *thermal diffusivity* as:

$$\alpha_T = \frac{k_T}{\rho \mathbb{C}} \tag{4.5}$$

The thermal diffusivity is the analogous of the kinematic viscosity ν_k in Eq. 3.5; they have also the same dimensions [length^2/time] (for such reason, we could define ν_k also as *momentum diffusivity*). Please observe that, as like as ν_k is a function of μ, the analogous of ν_k (the thermal diffusivity α_T) is a function of the analogous of μ (that is, the transport property relevant for the molecular phenomenon, in this case the thermal conductivity k_T that takes on the role of $\mathbb{K}$ in the general equation for molecular transport).

The ratio between the momentum diffusivity and the thermal diffusivity is the (obviously, non-dimensional) *Prandtl number*:

$$\boxed{Pr = \frac{\text{molecular (laminar) transport of momentum}}{\text{molecular (conductive) transport of energy}} = \frac{\nu_k}{\alpha_T} = \frac{\mu \mathbb{C}}{k_T}} \tag{4.6}$$

As like as for ν_k and α_T (or, equivalently, for μ, $\mathbb{C}$ and k_T), also Pr is a sole function of the material we are considering (of course, at given conditions). The fluid dynamics does not influence Pr, that represents the *ratio between the transport of momentum and the transport of energy, both following molecular (laminar/conductive) mechanisms*.

The analogies between quantities, and phenomena, observed so far are illustrated in Fig. 4.2 where, for simplicity, x indicates a generic direction of transport also for momentum.

Example: Prandtl Number for Water and Air

Table 4.2 synoptically illustrates the relevant quantities for water and air as function of T in the selected range. The values for dynamic viscosity, density and kinematic viscosity are taken from Table 3.1, while those for thermal conductivity from Table 4.1. Start from liquid water, for which we can safely assume $\mathbb{C}$ constant and equal to 4.18 kJ/(kg·K). The values for the Prandtl number vary from about 7 (room temperature) to about 2 (at $T = 80$ °C), decreasing with T as, Eq. 4.6, μ decreases and k_T increases. For air, we assume $\mathbb{C}$ constant and equal to 1 kJ/(kg·K), observing that Pr remains substantially anchored at 0.7 (as both μ and k_T increase as T increases).

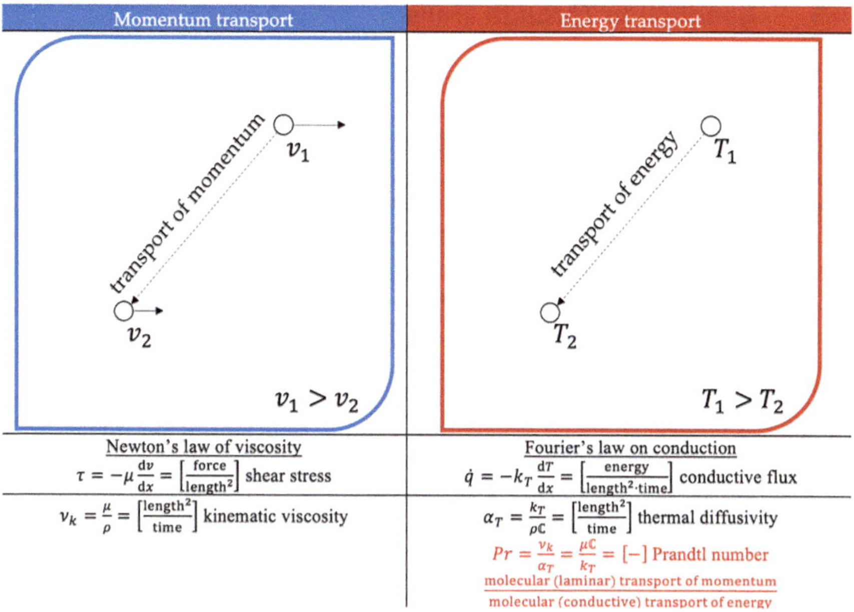

Fig. 4.2 An abecedary of transport phenomena analogies—Part A

Table 4.2 Evaluation of the Prandtl number for liquid water and air (atmospheric pressure)

Species	μ [kg/(m·s)]	ρ [kg/m^3]	ν_k [m^2/s]	k_T [kJ/(m·s·K)]	α_T [m^2/s]	Pr [–]
Liquid water (20 °C)	$1.021 \cdot 10^{-3}$	1001.9	$1.02 \cdot 10^{-6}$	$0.599 \cdot 10^{-3}$	$1.43 \cdot 10^{-7}$	7.1
Liquid water (40 °C)	$0.671 \cdot 10^{-3}$	996.4	$6.74 \cdot 10^{-7}$	$0.626 \cdot 10^{-3}$	$1.50 \cdot 10^{-7}$	4.5
Liquid water (60 °C)	$0.474 \cdot 10^{-3}$	987.9	$4.80 \cdot 10^{-7}$	$0.648 \cdot 10^{-3}$	$1.57 \cdot 10^{-7}$	3.1
Liquid water (80 °C)	$0.355 \cdot 10^{-3}$	978.2	$3.63 \cdot 10^{-7}$	$0.664 \cdot 10^{-3}$	$1.62 \cdot 10^{-7}$	2.2
Air (20 °C)	$0.018 \cdot 10^{-3}$	1.205	$1.49 \cdot 10^{-5}$	$0.026 \cdot 10^{-3}$	$2.16 \cdot 10^{-5}$	0.7
Air (40 °C)	$0.019 \cdot 10^{-3}$	1.128	$1.68 \cdot 10^{-5}$	$0.027 \cdot 10^{-3}$	$2.39 \cdot 10^{-5}$	0.7
Air (60 °C)	$0.020 \cdot 10^{-3}$	1.060	$1.89 \cdot 10^{-5}$	$0.028 \cdot 10^{-3}$	$2.64 \cdot 10^{-5}$	0.7
Air (80 °C)	$0.021 \cdot 10^{-3}$	1.000	$2.10 \cdot 10^{-5}$	$0.030 \cdot 10^{-3}$	$3.00 \cdot 10^{-5}$	0.7

4.3 Conduction Through Walls

4.3.1 Flat Wall

Consider the system in Fig. 4.3: a flat wall (*constant cross-section*) of a solid material (having thickness δ) exposed, at its two sides, to two different temperatures (e.g., hot T_1 on one side and cold T_2 on the other; how this is realised, for the moment is irrelevant, while it will be addressed further on in this chapter). This exposure determines the onset of a conductive thermal flux, $\dot{q}$ in Eq. 4.2, from the side at higher temperature towards the side at lower temperature. The flux develops along the constant cross-section perpendicular to the direction of transfer, i.e. the area of the lateral surface (height of the wall multiplied by its depth, as it were a parallelepiped; the scheme of Fig. 4.3 is in 2D for the sake of the simplicity). Let x be defined, as usual, as the direction of the thermal flow, positive from the side at T_1 to that at T_2. Finally, let $|\Delta T| = T_1 - T_2$ be the *thermal driving force* in absolute value.

First, find a relation between driving force and flux, by integration of the Fourier's equation via the method (separation of variables) named after the same scientist. To do this, we adopt a simplification: as k_T is a function of T, we should insert this dependence under integral, but we consider the variation of k_T (for the material of the wall) between T_1 and T_2 to be of minor relevance, so to keep an average (and constant) value for the thermal conductivity. This assumption will be also valid in similar parts of this chapter. It is, from Eq. 4.2:

$$\int_{T_1}^{T_2} \mathrm{d}T = -\frac{\dot{q}}{k_T}\int_0^{\delta} \mathrm{d}x \Rightarrow T_2 - T_1 = -\frac{\dot{q}}{k_T}\delta \Rightarrow |\Delta T| = \frac{\dot{q}}{k_T}\delta \tag{4.7}$$

The integration in Eq. 4.7 has been realised also noting that the *thermal flux is constant* as it is the ratio between a flow rate (constant under the hypothesis of a stationary system) and an area of transfer, that is constant for this geometry. Equation 4.7 is rewritten as:

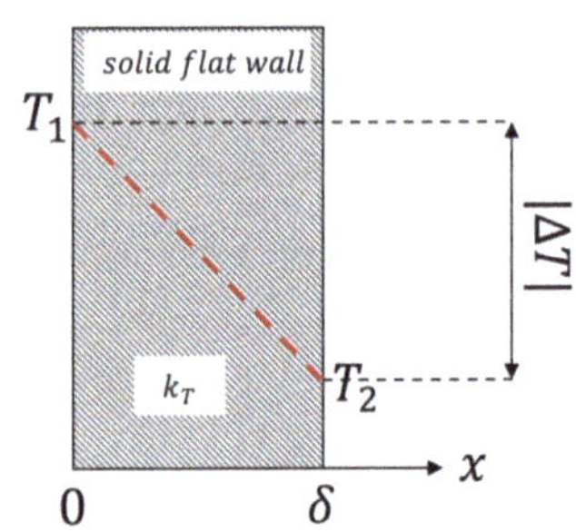

Fig. 4.3 Energy conduction through a flat wall, following a driving force ($T_1 > T_2$). The conductive flux is directed from the side kept at T_1 to the side kept at T_2. The temperature profile along x is linear

$$\dot{q} = \frac{k_T}{\delta} |\Delta T| \tag{4.8}$$

where the *resistance* to energy conduction is:

$$\mathcal{R}_{cond} = \frac{\delta}{k_T} \tag{4.9}$$

since the higher the wall thickness, and the lower the material conductivity, the higher the resistance opposed to the conductive energy flux. A thermal resistance is expressed, cf. Eq. 4.9, in [length^2·time·temperature/energy]. Coupling Eqs. 4.8 and 4.9, we can therefore recognise the following, general, result:

$$\boxed{\text{flux} = \frac{\text{driving force}}{\text{resistance}}} \tag{4.10}$$

which can be considered a universal condition for transport phenomena. *The higher the driving force, and the lower the resistance, the higher the flux.* Of course, driving force and resistance will assume specific expressions as a function of the system at hand.

As graphically anticipated in Fig. 4.3, the temperature profile is linear. In fact, rewriting Eq. 4.7 by integration between $x = 0$ ($T = T_1$) and a generic value for x (and for T), it is:

$$\int_{T_1}^{T} \mathrm{d}T = -\frac{\dot{q}}{k_T}\int_0^x \mathrm{d}x \Rightarrow T(x) - T_1 = -\frac{\dot{q}}{k_T}x \Rightarrow T(x) = T_1 - \frac{\dot{q}}{k_T}x \tag{4.11}$$

The reciprocal of a thermal resistance is called *heat transfer coefficient* (in this case, related to conductive phenomena), expressed as [energy/(length^2·time·temperature)]. The higher the resistance, the lower the heat transfer coefficient. Then, Eq. 4.10 can be extended:

$$\boxed{\begin{array}{c} \text{flux} = \dfrac{\text{driving force}}{\text{resistance}} \\ \text{heat transfer coefficient} = \dfrac{1}{\text{resistance}} \\ \text{flux} = \text{heat transfer coefficient} \times \text{driving force} \end{array}} \tag{4.12}$$

We realise here, as relevant result, that *the energy flux can be expressed as the product between the heat transfer coefficient and the driving force* (the higher the coefficient and/or the driving force, the higher the flux, and vice versa; think, for example, of a system with high thickness, low conductivity, low driving force; all these aspects will lower the heat transfer coefficient and the driving force, determining a low conductive heat flux, and vice versa). Simple extensions to a series of flat walls, as like as to a series of non-flat walls with geometry addressed in the following, can be found in, e.g., Bird et al. 2007.

4.3.2 Cylindrical Wall

The planar geometry (flat wall) has allowed us to identify, in particular, the general relations in Eq. 4.12, *which are valid for any type of geometry*. However, of greater interest for many chemical plants involving energy transfer is the cylindrical geometry, as we will see in the next chapter.

Consider here the system in Fig. 4.4: a cylindrical wall (*variable cross-section*) of a solid material exposed, at its two sides, to two different temperatures (e.g., hot T_i on the internal side and cold T_e on the external side). This exposure determines the onset of a conductive thermal flux, $\dot{q}$ in Eq. 4.2, from the side at higher temperature towards the side at lower temperature. The cylindrical wall extends from an internal radial coordinate r_i to an external radial coordinate r_e, setting the origin of the radial coordinate ($r = 0$) at the centre of symmetry of the system. Let r be defined as the direction of the thermal flow, positive from the side at T_i to that at T_e. The material has mean thermal conductivity k_T, and the driving force is $|\Delta T| = T_i - T_e$.

The area, through which the conductive heat flux is transferred, is now variable and equal, at a given value for r, to $2\pi rL$, where L is the length of the cylindrical wall (not reported in the 2D picture of Fig. 4.4). Therefore, the *thermal flux is variable* as it corresponds to the ratio between a flow rate q (constant under the hypothesis of a stationary system) and an area of transfer, that is variable for this geometry:

$$\dot{q} = \frac{q}{2\pi rL} \tag{4.13}$$

Being q constant, let us express the conductive flow rate by Eq. 4.13 and inserting the Fourier's law (where we substitute x with r, to indicate the cylindrical geometry):

$$q = \dot{q}2\pi rL = -k_T \frac{\mathrm{d}T}{\mathrm{d}r} 2\pi rL \tag{4.14}$$

Integrate by separation of variables (with constant q, k_T and L):

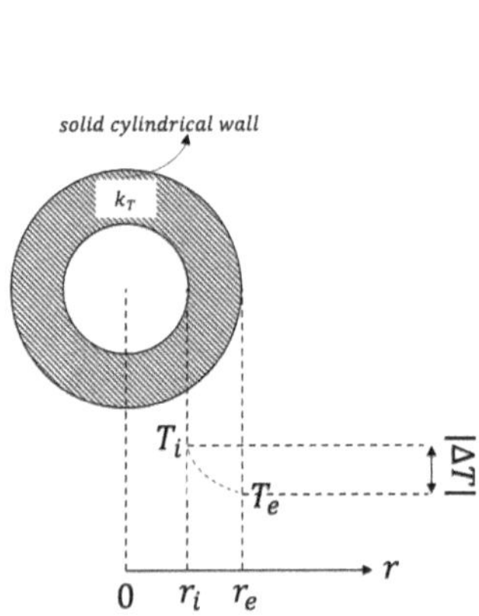

Fig. 4.4 Energy conduction through a cylindrical wall, following a driving force ($T_i > T_e$). The conductive flux is directed from the side kept at T_i to the side kept at T_e. The temperature profile along r is concave upward (i.e., convex)

$$\int_{T_i}^{T_e} \mathrm{d}T = -\frac{q}{k_T 2\pi L}\int_{r_i}^{r_e}\frac{1}{r}\mathrm{d}r \Rightarrow T_e - T_i = -\frac{q}{k_T 2\pi L}\ln\left(\frac{r_e}{r_i}\right) \Rightarrow |\Delta T| = \frac{q}{k_T 2\pi L}\ln\left(\frac{r_e}{r_i}\right) \tag{4.15}$$

Note how, for the logarithm, the argument is >1 (and, therefore, the logarithm itself assumes values > 0). Equation 4.15 is rewritten as:

$$q = \frac{k_T 2\pi L}{\ln\left(\frac{r_e}{r_i}\right)}|\Delta T| \tag{4.16}$$

Aiming at an expression for the conductive flux, since it is not constant, we must specify which transfer area we are referring to. For the transfer areas at radial coordinate r_i and r_e, respectively, it is (from Eq. 4.13):

$$\begin{cases} \dot{q}_i = \dfrac{q}{2\pi r_i L} = \dfrac{k_T}{r_i \ln\left(\dfrac{r_e}{r_i}\right)}|\Delta T| \\ \dot{q}_e = \dfrac{q}{2\pi r_e L} = \dfrac{k_T}{r_e \ln\left(\dfrac{r_e}{r_i}\right)}|\Delta T| \end{cases} \tag{4.17}$$

with thermal flux not depending on L. We encounter again the phenomenological picture of Eq. 4.12: the flux is the ratio between the thermal driving force and the conductive resistance that now equals, respectively:

$$\begin{cases} \mathcal{R}_{cond,i} = \dfrac{r_i \ln\left(\dfrac{r_e}{r_i}\right)}{k_T} \\ \mathcal{R}_{cond,e} = \dfrac{r_e \ln\left(\dfrac{r_e}{r_i}\right)}{k_T} \end{cases} \tag{4.18}$$

Again (cf. Eq. 4.9), the conductive resistance takes into account, at the numerator, the wall thickness (here represented by the ratio r_e/r_i) and, at the denominator, the thermal conductivity of the material. Please observe that the dimensions of the resistance, [length^2·time·temperature/energy], are the same for both geometries (flat and cylindrical). Back to Eq. 4.12, we can say that the reciprocal of the resistance is the heat transfer coefficient and, also in this case, we can see the thermal flux as the product between the transfer coefficient and the driving force.

As graphically anticipated in Fig. 4.4, the temperature profile $T(r)$ is not linear, but decreases following a trend with upward concavity. Let us demonstrate this. Rewrite Eq. 4.15 by integration between $r = r_i$ ($T = T_i$) and a generic value for r (and for T):

$$\int_{T_i}^{T} \mathrm{d}T = -\frac{q}{k_T 2\pi L}\int_{r_i}^{r}\frac{1}{r}\mathrm{d}r \Rightarrow T(r) - T_i = -\frac{q}{k_T 2\pi L}\ln\left(\frac{r}{r_i}\right) \Rightarrow T(r) = T_i - \frac{q}{k_T 2\pi L}\ln\left(\frac{r}{r_i}\right) \tag{4.19}$$

Now, derive two times:

$$\frac{\mathrm{d}T}{\mathrm{d}r} = -\frac{q}{k_T 2\pi L r} \tag{4.20}$$

$$\frac{\mathrm{d}^2T}{\mathrm{d}r^2} = \frac{q}{k_T 2\pi L r^2} > 0$$

q.e.d.

4.4 Thermal Convection

4.4.1 General Characteristics

Now we present the phenomenon of *energy convection*, i.e. an energy flow *associated with the motion of a fluid*. For the purposes of this textbook, we will directly refer to the case study of a fluid animated by a *fully developed turbulent motion*, often thought of as being inside a cylindrical duct. Under this hypothesis, the so-called *film theory* is assumed to be valid, according to which, in a simplifying approach, the temperature of the fluid is variable only in the thermal *boundary layer (film)* in proximity to the wall (whatever that may be), while the temperature is assumed constant in the turbulent bulk (core) of the fluid. This, in other words, is equivalent to saying that the convective resistance to energy transport is considered to be concentrated *only inside the film* in which the temperature gradient occurs. What we are talking about here is the *forced convection*, since there is also another convective phenomenon, *natural (free) convection* (e.g., when thermal currents are the result of forces generated by differences in density in the fluid, in turn caused by temperature gradients in the mass of fluid). Natural convection, although not strictly excluded in our study, will be neglected since it has a lesser effect.

In the case of the conductive phenomenon, we have previously seen how the energy flux can be expressed by the product of a transfer coefficient and a driving force (Eq. 4.12). This concept can also be replicated in the case of the convective mechanism. Take, for example (Fig. 4.5), a fluid enlivened by turbulent motion, with bulk average temperature termed T_f, and in presence of a wall that is at a different temperature, termed T_w. Just as an example, let us imagine that the wall is hotter than the fluid ($T_w > T_f$). This driving force, $|\Delta T| = T_w - T_f$, determines the onset of a convective flow of energy (that can embody conductive phenomena inside the fluid which are of minor relevance vs. convective ones, as like as, in the turbulent

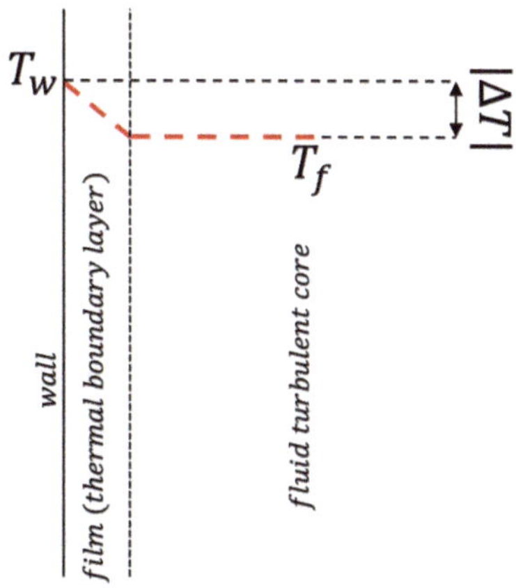

Fig. 4.5 Scheme for convective energy flux in presence of a fluid in turbulent motion (at average constant bulk temperature T_f) confined by a wall (at different temperature T_w, in the example $T_w > T_f$). The temperature is assumed to vary only in the close proximity of the wall (film), where the convective resistance is concentrated. The convective flux is here directed from the wall to the fluid core

transport of momentum, we have neglected laminar phenomena in the turbulent fluid bulk):

$$\dot{q} = h|\Delta T| \tag{4.21}$$

where, with h, we indicate the *heat transfer coefficient* (local, for convective mechanism).

Unlike thermal conductivity, h is *not an intrinsic property* of the fluid, but also depends on the thermo-fluid dynamic conditions of the system. The coefficient h, of course, has the usual dimensions of [energy/(length^2·time·temperature)]. Similarly, here too we can define a *convective resistance* which, according to Eq. 4.12, can be now expressed as:

$$\mathcal{R}_{conv} = \frac{1}{h} \tag{4.22}$$

4.4.2 Nusselt Number

Now we discuss how the convective heat transfer coefficient, h, can be determined. First, we can express the ratio between the *total heat flux* in Eq. 4.21 (the relation was proposed imagining that it can also embody the less relevant conductive phenomena within the fluid), and the *conductive heat flux* that, by reworking Fourier's Eq. 4.2 and expressing it in absolute value, can be approximated to $k_T|\Delta T|/D$, with D being the characteristic dimension of the system (e.g., the diameter of the duct in which the fluid flows). This ratio, naturally dimensionless, is called the *Nusselt number* [named after Ernst Kraft Wilhelm Nusselt (1882–1957), German engineer, Professor at University of Karlsruhe; see Nusselt and Schmidt 1982]:

$$\boxed{Nu = \frac{\text{total transport of energy}}{\text{molecular (conductive) transport of energy}} = \frac{h|\Delta T|}{\frac{k_T|\Delta T|}{D}} = \frac{hD}{k_T}} \tag{4.23}$$

We can observe that, under the hypotheses of fully developed turbulent flow, we expect Nu much greater than unity. There are numerous relations, in the literature (Bird et al. 2007; McCabe et al. 2001), very often empirical, that link Nu to other dimensionless numbers capable of characterising the properties of the fluid (e.g., Pr) and the thermo-fluid dynamic conditions of the system (e.g., Re). Leaving the reader to apply these relations case by case (we will present some of them in the chapter dedicated to unit operations based on energy transport), here it is sufficient to remember that, once Nu is known, we will obtain, as desired:

$$h = \frac{k_T}{D} Nu \tag{4.24}$$

4.5 Convection and Conduction Through Walls

4.5.1 Flat Wall

Let us return to the case study illustrated in Fig. 4.3, extending it as in Fig. 4.6. It is a solid flat wall where, on one side, a hot fluid flows in turbulent motion (at the bulk temperature equal to T_h) and, on the other, a cold fluid flows in turbulent motion (at the bulk temperature equal to T_c). The resulting overall driving force, $|\Delta T| = T_h-$

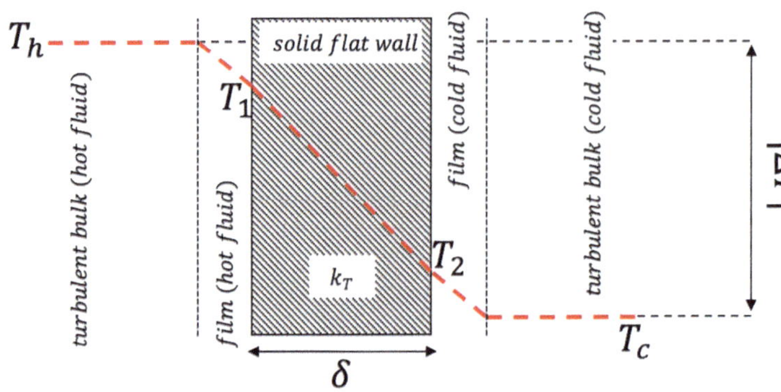

Fig. 4.6 Energy convection/conduction in a system composed by a flat wall, where at two sides there are two fluids animated by fully developed turbulent motion. For both hot and cold fluid, the average bulk temperature is constant. Convective resistances are concentrated in the two films

T_c, determines a thermal flux $\dot{q}$ (constant, since the section crossed by the flux is constant) that takes into account three phenomena:

- convective energy transport from the bulk of the hot fluid to the wall, whose temperature is called T_1;
- conductive energy transport through the wall;
- convective energy transport from the wall, whose temperature is called T_2, to the bulk of the cold fluid.

In this scheme, the convective energy transport in relation to the hot fluid is governed by a local heat transfer coefficient h_h, and the flux is expressed as (cf. Eq. 4.21):

$$\dot{q} = h_h(T_h - T_1) \tag{4.25}$$

considering, according to the film theory, that the resistance to energy transport is concentrated in the boundary layer and is equal to:

$$\mathcal{R}_{conv,h} = \frac{1}{h_h} \tag{4.26}$$

The conductive energy transport through the wall is described by the previous Eq. 4.8:

$$\dot{q} = \frac{k_T}{\delta}(T_1 - T_2) \tag{4.27}$$

with resistance, Eq. 4.9:

$$\mathcal{R}_{cond} = \frac{\delta}{k_T} \tag{4.28}$$

Finally, the convective energy transport in relation to the cold fluid is governed by a local heat transfer coefficient h_c, and the flux is expressed as (cf. Eq. 4.21):

$$\dot{q} = h_c(T_2 - T_c) \tag{4.29}$$

considering, according to the film theory, that the resistance to energy transport is concentrated in the boundary layer and is equal to:

$$\mathcal{R}_{conv,c} = \frac{1}{h_c} \tag{4.30}$$

Let us now express the local driving forces from the previous expressions:

$$\begin{cases} T_h - T_1 = \dfrac{\dot{q}}{h_h} \\ T_1 - T_2 = \dfrac{\dot{q}\delta}{k_T} \\ T_2 - T_c = \dfrac{\dot{q}}{h_c} \end{cases} \tag{4.31}$$

Summing up the three in Eq. 4.30, we obtain the overall driving force:

$$|\Delta T| = T_h - T_c = \dot{q}\left(\frac{1}{h_h} + \frac{\delta}{k_T} + \frac{1}{h_c}\right) \tag{4.32}$$

In terms of heat flux, it is:

$$\dot{q} = \left(\frac{1}{h_h} + \frac{\delta}{k_T} + \frac{1}{h_c}\right)^{-1} |\Delta T| \tag{4.33}$$

In light of the general laws in Eq. 4.12, and calling U the *overall heat transfer coefficient*:

$$U = \left(\frac{1}{h_h} + \frac{\delta}{k_T} + \frac{1}{h_c}\right)^{-1} \tag{4.34}$$

This allows to define an *overall resistance*:

$$\mathcal{R} = \frac{1}{U} = \frac{1}{h_h} + \frac{\delta}{k_T} + \frac{1}{h_c} = \mathcal{R}_{conv,h} + \mathcal{R}_{cond} + \mathcal{R}_{conv,c} \tag{4.35}$$

The expert reader will not have missed the analogy with the laws of classical physics (e.g., in the case of electrical resistances placed in series), according to which the overall resistance to transport is equal, for resistances in series (as is the case in Fig. 4.6), *to the sum of the individual resistances*. We will therefore say, in conclusion, that:

$$\boxed{\dot{q} = U|\Delta T|} \tag{4.36}$$

Example: Analysis of Convection/Conduction for the Case of a Flat Wall
Be the system the one represented in Fig. 4.6. Assume the following data:

- fluids at bulk temperature $T_h = 120$ °C (steam) and $T_c = 20$ °C (water), thus $|\Delta T| = 100$ °C $= 100$ K;

(continued)

- convective heat transfer coefficients equal to $h_h = 10^4$ J/(m^2·s·K) and $h_c = 10^3$ J/(m^2·s·K). We will see, in another chapter, how to calculate h via the Nusselt number and Eq. 4.24;
- wall made of brick, $\delta = 10$ cm, $k_T = 1$ J/(m·s·K).

Resistances are, therefore:

$$\begin{cases} \mathcal{R}_{conv,h} = \dfrac{1}{h_h} = 10^{-4}\ \dfrac{\text{m}^2\cdot\text{s}\cdot\text{K}}{\text{J}} \\ \mathcal{R}_{cond} = \dfrac{\delta}{k_T} = 10^{-1}\ \dfrac{\text{m}^2\cdot\text{s}\cdot\text{K}}{\text{J}} \\ \mathcal{R}_{conv,c} = \dfrac{1}{h_c} = 10^{-3}\ \dfrac{\text{m}^2\cdot\text{s}\cdot\text{K}}{\text{J}} \end{cases} \tag{4.37}$$

and, thus, the global resistance (Eq. 4.35) is:

$$\mathcal{R} = \mathcal{R}_{conv,h} + \mathcal{R}_{cond} + \mathcal{R}_{conv,c} = 0.1011\ \frac{\text{m}^2\cdot\text{s}\cdot\text{K}}{\text{J}} \tag{4.38}$$

The overall heat transfer coefficient is (Eq. 4.35):

$$U = \frac{1}{\mathcal{R}} = 9.891\ \frac{\text{J}}{\text{m}^2\cdot\text{s}\cdot\text{K}} \tag{4.39}$$

and the overall convective/conductive thermal flux is, Eq. 4.36:

$$\dot{q} = U|\Delta T| = 989.1\ \frac{\text{J}}{\text{m}^2\cdot\text{s}} \tag{4.40}$$

We might want to calculate the intermediate temperatures, T_1 and T_2 in Fig. 4.6. Well, from Eq. 4.31:

$$\begin{cases} T_h - T_1 = \dfrac{\dot{q}}{h_h} = 0.10\,^\circ\text{C} \Rightarrow T_1 = 119.9\,^\circ\text{C} \\ T_1 - T_2 = \dfrac{\dot{q}\delta}{k_T} = 98.91\,^\circ\text{C} \Rightarrow T_2 = 20.99\,^\circ\text{C} \\ T_2 - T_c = \dfrac{\dot{q}}{h_c} = 0.99\,^\circ\text{C} \Rightarrow T_c = 20\,^\circ\text{C (verified)} \end{cases} \tag{4.41}$$

In this example, T_1 is very close to T_h (and T_2 is very close to T_c), as the convective resistances are very low, compared to the conductive resistance (and, in fact, the most relevant driving force is the conductive one through the wall, where most of the resistance is concentrated). This can be appreciated from Eq. 4.37 and relies on the properties of the wall (having relatively high thickness and being made of brick, a common insulating material).

4.5.2 Cylindrical Wall

Now we return to the case study illustrated in Fig. 4.4, extending it as in Fig. 4.7. It is a solid cylindrical wall where, on one side, a hot fluid flows in turbulent motion (at the bulk temperature equal to T_h) and, on the other, a cold fluid flows in turbulent motion (at the bulk temperature equal to T_c). The resulting overall driving force, $|\Delta T| = T_h - T_c$, determines a thermal flux $\dot{q}$ (variable, since the section crossed by the flux is variable) that takes into account three phenomena:

- convective energy transport from the bulk of the hot fluid to the wall, whose temperature is called T_i;
- conductive energy transport through the wall;
- convective energy transport from the wall, whose temperature is called T_e, to the bulk of the cold fluid.

In this scheme, while the flux is not constant, the energy flow rate q remains constant (stationary system), and it can be expressed, as usual (Eq. 4.13), via the product between the flux and the characteristic area. By the aid of Eq. 4.21, the convective energy transport in relation to the hot fluid, governed by a local heat transfer coefficient h_h, is characterised by the following expression for the flow rate:

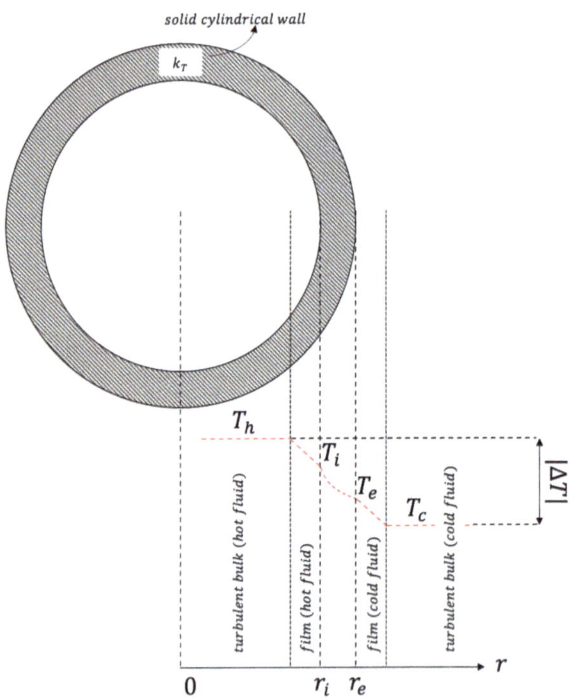

Fig. 4.7 Energy convection/conduction in a system composed by a cylindrical wall, where at two sides there are two fluids animated by fully developed turbulent motion. For both hot and cold fluid, the average bulk temperature is constant. Convective resistances are concentrated in the two films

$$q = h_h (T_h - T_i) 2\pi r_i L \tag{4.42}$$

The conductive energy transport through the wall is described by the previous Eq. 4.16:

$$q = \frac{k_T 2\pi L}{\ln\left(\frac{r_e}{r_i}\right)} (T_i - T_e) \tag{4.43}$$

Finally, the convective energy transport in relation to the cold fluid, governed by a local heat transfer coefficient h_c, is characterised by the following expression for the flow rate:

$$q = h_c (T_e - T_c) 2\pi r_e L \tag{4.44}$$

Let us now express the local driving forces from the previous expressions:

$$\begin{cases} T_h - T_i = \dfrac{q}{h_h 2\pi r_i L} \\ T_i - T_e = \dfrac{q \ln\left(\dfrac{r_e}{r_i}\right)}{k_T 2\pi L} \\ T_e - T_c = \dfrac{q}{h_c 2\pi r_e L} \end{cases} \tag{4.45}$$

Summing up the three in Eq. 4.45, we obtain the overall driving force:

$$|\Delta T| = T_h - T_c = \frac{q}{2\pi L} \left[\frac{1}{r_i h_h} + \frac{\ln\left(\frac{r_e}{r_i}\right)}{k_T} + \frac{1}{r_e h_c} \right] \tag{4.46}$$

or, equivalently:

$$q = 2\pi L \left[\frac{1}{r_i h_h} + \frac{\ln\left(\frac{r_e}{r_i}\right)}{k_T} + \frac{1}{r_e h_c} \right]^{-1} |\Delta T| \tag{4.47}$$

Aiming at an expression for the thermal flux, since it is not constant, we must specify which transfer area we are referring to. For the transfer areas at radial coordinate r_i and r_e, respectively, it is:

$$\begin{cases} \dot{q}_i = \dfrac{q}{2\pi r_i L} = \left[\dfrac{1}{h_h} + \dfrac{r_i \ln\left(\frac{r_e}{r_i}\right)}{k_T} + \dfrac{r_i}{r_e h_c}\right]^{-1} |\Delta T| \\ \dot{q}_e = \dfrac{q}{2\pi r_e L} = \left[\dfrac{r_e}{r_i h_h} + \dfrac{r_e \ln\left(\frac{r_e}{r_i}\right)}{k_T} + \dfrac{1}{h_c}\right]^{-1} |\Delta T| \end{cases} \tag{4.48}$$

with thermal flux not depending on the system length, L.

In light of the general laws in Eq. 4.12, the overall heat transfer coefficient referring to transfer areas at radial coordinate r_i and r_e, respectively, is:

$$\begin{cases} U_i = \left[\dfrac{1}{h_h} + \dfrac{r_i \ln\left(\frac{r_e}{r_i}\right)}{k_T} + \dfrac{r_i}{r_e h_c}\right]^{-1} \\ U_e = \left[\dfrac{r_e}{r_i h_h} + \dfrac{r_e \ln\left(\frac{r_e}{r_i}\right)}{k_T} + \dfrac{1}{h_c}\right]^{-1} \end{cases} \tag{4.49}$$

Equation 4.49 allows to define an overall resistance, expressing as usual the sum of the individual resistances. The overall resistance referring to transfer areas at radial coordinate r_i and r_e, respectively, is:

$$\begin{cases} \mathcal{R}_i = \dfrac{1}{U_i} = \dfrac{1}{h_h} + \dfrac{r_i \ln\left(\dfrac{r_e}{r_i}\right)}{k_T} + \dfrac{r_i}{r_e h_c} \\ \mathcal{R}_e = \dfrac{1}{U_e} = \dfrac{r_e}{r_i h_h} + \dfrac{r_e \ln\left(\dfrac{r_e}{r_i}\right)}{k_T} + \dfrac{1}{h_c} \end{cases} \tag{4.50}$$

We will therefore say, in conclusion, that:

$$\boxed{\begin{cases} \dot{q}_i = U_i |\Delta T| \\ \dot{q}_e = U_e |\Delta T| \end{cases}} \tag{4.51}$$

Example: Analysis of Convection/Conduction for the Case of a Cylindrical Wall

Be the system the one represented in Fig. 4.7. Assume the following data:

- fluids at bulk temperature $T_h = 120$ °C (steam) and $T_c = 20$ °C (water), thus $|\Delta T| = 100$ °C $= 100$ K;
- convective heat transfer coefficients equal to $h_h = 10^4$ J/(m^2·s·K) and $h_c = 10^3$ J/(m^2·s·K);

(continued)

- wall made of brick, $r_i = 10$ cm, $r_e = 20$ cm, $L = 1$ m (the length of the cylindrical wall, not illustrated in Fig. 4.7), $k_T = 1$ J/(m·s·K).

The overall resistance referring to transfer areas at radial coordinate r_i and r_e, respectively, is (Eq. 4.49):

$$\begin{cases} \mathcal{R}_i = 10^{-4} + 0.0693 + 5 \cdot 10^{-4} = 0.0699 \dfrac{\text{m}^2 \cdot \text{s} \cdot \text{K}}{\text{J}} \\ \mathcal{R}_e = 2 \cdot 10^{-4} + 0.1386 + 10^{-3} = 0.1398 \dfrac{\text{m}^2 \cdot \text{s} \cdot \text{K}}{\text{J}} \end{cases} \tag{4.52}$$

Again, the most relevant resistance is the conductive one, through the solid wall. The overall heat transfer coefficient referring to transfer areas at radial coordinate r_i and r_e, respectively, is (Eq. 4.49):

$$\begin{cases} U_i = \dfrac{1}{\mathcal{R}_i} = 14.306 \dfrac{\text{J}}{\text{m}^2 \cdot \text{s} \cdot \text{K}} \\ U_e = \dfrac{1}{\mathcal{R}_e} = 7.153 \dfrac{\text{J}}{\text{m}^2 \cdot \text{s} \cdot \text{K}} \end{cases} \tag{4.53}$$

The overall convective/conductive thermal flux referring to transfer areas at radial coordinate r_i and r_e, respectively, is (Eq. 4.50):

$$\begin{cases} \dot{q}_i = U_i |\Delta T| = 1430.6 \dfrac{\text{J}}{\text{m}^2 \cdot \text{s}} \\ \dot{q}_e = U_e |\Delta T| = 715.3 \dfrac{\text{J}}{\text{m}^2 \cdot \text{s}} \end{cases} \tag{4.54}$$

Although the energy flux is variable (due to the system geometry, it is smaller when the crossed area is larger, i.e. at $r = r_e$, cf. Eq. 4.47), the energy flow rate is constant, as can be verified by the application of Eq. 4.47: $q = \dot{q}_i 2\pi r_i L = \dot{q}_e 2\pi r_e L = 898.42$ J/s.

Now, known the flow rate, we can calculate the intermediate temperatures, T_i and T_e in Fig. 4.7. From Eq. 4.44:

$$\begin{cases} T_h - T_i = \dfrac{q}{h_h 2\pi r_i L} = 0.14\,^\circ\text{C} \Rightarrow T_i = 119.86\,^\circ\text{C} \\ T_i - T_e = \dfrac{q \ln\left(\dfrac{r_e}{r_i}\right)}{k_T 2\pi L} = 99.16\,^\circ\text{C} \Rightarrow T_e = 20.7\,^\circ\text{C} \\ T_e - T_c = \dfrac{q}{h_c 2\pi r_e L} = 0.7\,^\circ\text{C (verified)} \end{cases} \tag{4.55}$$

Again, T_i is very close to T_h (and T_e is very close to T_c), as the convective resistances are very low, compared to the conductive resistance.

4.6 Background Information

- The transport of energy is driven by a temperature gradient (Fig. 4.1). The two main mechanisms of energy transport addressed here are conduction (a molecular phenomenon) and forced convection (a macroscopic phenomenon that we associate to the motion of a fluid in turbulent regime).
- The Fourier's law on conduction (Eq. 4.2) relates the conductive heat flux to the temperature spatial gradient via a proportionality constant depending on the fluid nature: the thermal conductivity.
- The analogies between momentum and energy transport are illustrated in Fig. 4.2. The ratio between molecular transport of momentum and molecular transport of energy is the Prandtl number (Eq. 4.6).
- When convection is present, the ratio between total transport of energy and molecular transport of energy is the Nusselt number (Eq. 4.23). The knowledge of the Nusselt number allows, Eq. 4.24, to calculate the convective heat transfer coefficient.
- In Eq. 4.12, we find the general relationships stating that the energy flux is the ratio between driving force and resistance, or the product between driving force and heat transfer coefficient (the latter being the reciprocal of the resistance).
- The overall resistance to energy transport can be expressed as the sum of the resistances to the individual energy transport phenomena placed in series.

References

Bird RB, Stewart WE, Lightfoot EN (2007) Transport phenomena. John Wiley & Sons, New York
Fourier JBJ (1822) Théorie analytique de la chaleur. Didot, Paris
Green DW, Perry RH (2008) Perry's chemical engineers' handbook. McGraw-Hill, New York
McCabe WL, Smith JC, Harriott P (2001) Unit operations of chemical engineering. McGraw-Hill, Singapore
Nusselt W, Schmidt E (1982) Selected publications. Hemisphere, Washington

Heat Exchangers

5

Abstract

The concepts of energy transport phenomena are now applied to the design of heat exchangers with particular reference to those with double pipe, whose geometry recalls a previously seen system (i.e., heat exchange through cylindrical walls). The heat exchangers are analysed in their geometric characteristics, and in terms of flow in the internal and external tube (with counterflow feeding mode, generally preferred to the equicurrent case). We continue with the energy balances, and the presentation of the operating lines method to determine flow rate and outlet temperature of the service fluid. The local and overall transfer coefficients are then presented in detail. For the local ones, it is necessary to study the relationships between dimensionless numbers. Therefore, it is possible to develop and obtain the design equation that allows, for example, to calculate the heat transfer area (and length) for a double pipe heat exchanger.

Nomenclature

A_T	Energy transfer area [length^2]
A_e	Energy external transfer area of the internal tube [length^2]
A_i	Energy internal transfer area of the internal tube [length^2]
$\mathbb{C}_C$	Specific heat of the cold fluid [energy/(mass·temperature)]
$\mathbb{C}_H$	Specific heat of the hot fluid [energy/(mass·temperature)]
$\mathbb{C}_e$	Specific heat of the external fluid [energy/(mass·temperature)]
$\mathbb{C}_i$	Specific heat of the internal fluid [energy/(mass·temperature)]
D_{eq}	Equivalent diameter for the external tube (annulus) [length]
D_i	Internal diameter for the external tube (annulus) [length]
d_e	External diameter for the internal tube [length]
d_i	Internal diameter for the internal tube [length]

F. Montagnaro et al., *Elements of Unit Operations and Chemical Reactors*,
https://doi.org/10.1007/978-3-032-18046-9_5

h_e Local convective heat transfer coefficient (external annulus) [energy/(length^2·time·temperature)]
h_i Local convective heat transfer coefficient (internal tube) [energy/(length^2·time·temperature)]
k_T Thermal conductivity for wall material [energy/(length·time·temperature)]
$k_{T,e}$ Thermal conductivity of the external fluid [energy/(length·time·temperature)]
$k_{T,i}$ Thermal conductivity of the internal fluid [energy/(length·time·temperature)]
L Length [length]
Nu Nusselt number [−]
Nu_e Nusselt number for the external fluid [−]
Nu_i Nusselt number for the internal fluid [−]
Pr Prandtl number [−]
Pr_e Prandtl number for the external fluid [−]
Pr_i Prandtl number for the internal fluid [−]
q Power of the heat exchanger [energy/time]
$\dot{q}$ Energy flux [energy/(length^2·time)]
$\mathcal{R}_d$ Fouling resistance [length^2·time·temperature/energy]
$\mathcal{R}_e$ Overall resistance at external wall of the internal tube [length^2·time·temperature/energy]
$\mathcal{R}_i$ Overall resistance at internal wall of the internal tube [length^2·time·temperature/energy]
r_e External radius for the internal tube [length]
r_i Internal radius for the internal tube [length]
Re Reynolds number [−]
Re_e Reynolds number for the external fluid [−]
Re_i Reynolds number for the internal fluid [−]
S Cross-section area [length^2]
T Temperature of the hot fluid [temperature]
T_c Critical temperature [temperature]
T_{in} Inlet temperature of the hot fluid [temperature]
T_{out} Outlet temperature of the hot fluid [temperature]
T_r Reduced temperature [−]
T_{ref} Reference temperature [temperature]
T_{out}^{min} Minimum outlet temperature of the hot fluid [temperature]
t Temperature of the cold fluid [temperature]
t_{in} Inlet temperature of the cold fluid [temperature]
t_{out} Outlet temperature of the cold fluid [temperature]
t_{out}^{max} Maximum outlet temperature of the cold fluid [temperature]
U Overall heat transfer coefficient [energy/(length^2·time·temperature)]
U_e Overall heat transfer coefficient at external wall of the internal tube [energy/(length^2·time·temperature)]

U_i	Overall heat transfer coefficient at at internal wall of the internal tube [energy/(length^2·time·temperature)]
$\bar{v}$	Mean fluid velocity (radial) [length/time]
W	Mass flow rate [mass/time]
W_C	Mass flow rate of the cold fluid [mass/time]
W_H	Mass flow rate of the hot fluid [mass/time]
W_e	Mass flow rate of the external fluid [mass/time]
W_i	Mass flow rate of the internal fluid [mass/time]
W_C^{min}	Minimum mass flow rate of the cold fluid [mass/time]
W_H^{min}	Minimum mass flow rate of the hot fluid [mass/time]
z	Exchanger axial coordinate [length]
α	Semi-empirical exponent for Prandtl number [−]
ΔT	Thermal driving force [temperature]
ΔT_A	Thermal driving force at "A" section [temperature]
ΔT_B	Thermal driving force at "B" section [temperature]
ΔT_{lm}	Logarithmic mean driving force [temperature]
$\lvert\lambda_C\rvert$	Latent heat for the cold fluid (liquid-vapour phase transition) [energy/mass]
$\lvert\lambda_H\rvert$	Latent heat for the hot fluid (liquid-vapour phase transition) [energy/mass]
μ_e	Dynamic viscosity of the external fluid [mass/(length·time)]
μ_i	Dynamic viscosity of the internal fluid [mass/(length·time)]
ρ	Density [mass/volume]

5.1 Double Pipe Exchangers

The concepts we have discussed so far on transport phenomena (in particular on energy transport) are applied to an industrial unit operation of interest, namely the *heat exchangers* (for further information, please refer to more specialised books such as Kern 1983; McCabe et al. 2001). These systems allow heat exchange between a "cold" and a "hot" fluid through a solid wall (in cylindrical geometry), by means of convective and conductive phenomena (see, e.g., the diagram in Fig. 4.7). We will consider stationary systems. Heat exchange can occur by means of *latent or sensible heat*. For simplicity, the sensible case is considered as the base case, therefore with fluids that release and acquire heat by changing their temperature. Reference to the latent case will be made when appropriate. Of course, one of the two fluids (the cold one to be heated or the hot one to be cooled) is to be considered as the *process fluid* (i.e. for which the exchanger must be designed), while the other is the *service fluid* (i.e. the "sacrificial" one).

The heat exchangers referred to here are based on two coaxial tubes, and they are named *double pipe exchangers* (for other configurations, such as shell-and-tube exchangers, refer to the literature such as Kern 1983). In structures often arranged in the form of *hairpins* (a simplified scheme is illustrated in Fig. 5.1, further elaborated in "linearised" form for the sake of simplicity, Fig. 5.2), for example,

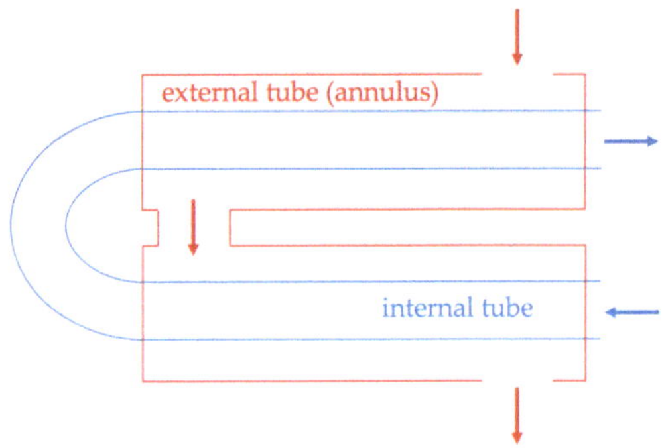

Fig. 5.1 Schematic representation of a double pipe exchanger in the form of hairpin

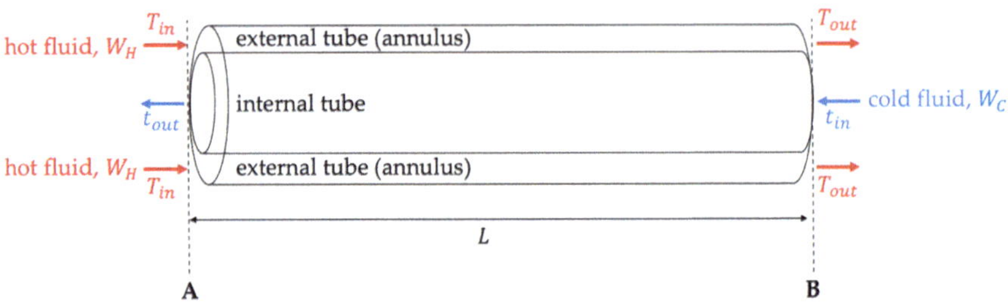

Fig. 5.2 "Linearised" representation of a double pipe exchanger (coaxial tubes). The choice of the fluid to be sent into the internal tube, and that to be sent into the annulus, as well as the direction of feeding for the two fluids, does not affect the generality of the discussion

there will be a fluid (e.g. cold) circulating in the *internal tube* and a fluid (e.g. hot) circulating in the *annulus* (the annular region otherwise called the external tube).

Referring to the illustration in Fig. 5.2 (the choice of the fluid to be sent into the internal tube, and the one to be sent into the annulus, depends on aspects such as, e.g., their fouling and corrosiveness potential, and the material the tubes are made of), let t_{in} and t_{out} be the inlet and outlet temperature, respectively, of the cold fluid whose mass flow rate is W_C. Similarly, T_{in} and T_{out} for the hot fluid, whose mass flow rate is W_H.

From a geometric point of view, the system has length L. The internal tube has internal and external diameters d_i and d_e respectively, while the annulus (external tube) has internal diameter D_i, cf. the end view illustrated in Fig. 5.3. The heat exchange phenomena between the exchanger and the external environment are neglected (the system is well insulated with respect to the outside, i.e. it is *globally adiabatic*; therefore, it is not interesting to refer to the external diameter of the external tube).

The heat exchange occurs, from the hot fluid (with generic temperature T) to the cold fluid (with generic temperature t), under the influence of a *thermal driving force* $\Delta T = T - t$, with $T > t$. It is evident that, between section "A" and section "B" of the exchanger (Fig. 5.2, the two far ends of the system), there will be infinite values for

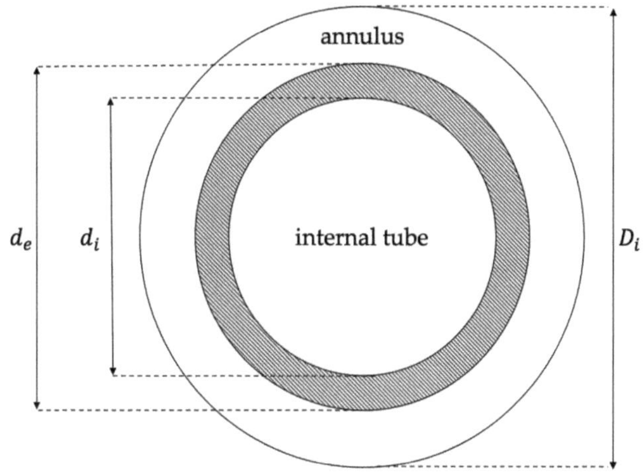

Fig. 5.3 End view for a double pipe exchanger

the driving force, depending on the section we are considering. In particular, let the two driving forces at the two ends be:

$$\begin{cases} \Delta T_A = T_{in} - t_{out} \\ \Delta T_B = T_{out} - t_{in} \end{cases} \tag{5.1}$$

As we will see later, it is useful to define the *logarithmic mean driving force*, that is, to express the logarithmic mean between ΔT_A and ΔT_B:

$$\Delta T_{lm} = \frac{\Delta T_A - \Delta T_B}{\ln\left(\frac{\Delta T_A}{\Delta T_B}\right)} = \frac{\Delta T_B - \Delta T_A}{\ln\left(\frac{\Delta T_B}{\Delta T_A}\right)} \tag{5.2}$$

5.2 Counterflow Operation

The reader will not have missed the choice, made in Fig. 5.2, to feed the two fluids in *counterflow* (actually, this is valid also in Fig. 5.1, where, in the upper part of the hairpin, the hot fluid flows from right to left while the cold fluid flows from left to right; and vice versa in the lower part of the hairpin). This choice is now discussed and motivated. It is, however, recalled that this choice, as well as that regarding which fluid to send into the internal tube, and which into the annulus, does not influence the generality of the results we will arrive at.

Countercurrent flow allows thermal profiles of the type in Fig. 5.4. That is, along the axis of the double pipe exchanger, we observe driving forces $\Delta T = T - t$ approximately constant and certainly much more uniform than it would occur in the case of *equicurrent flow*. In other words, an equicurrent flow would determine a non-homogeneous exchanger, with high driving forces at the inlet of the two fluids

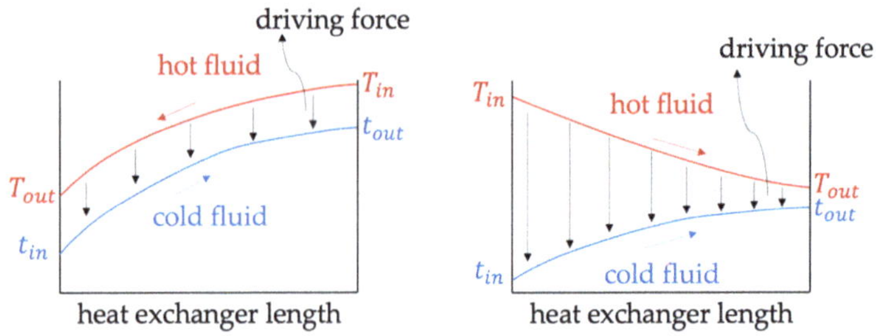

Fig. 5.4 Typical temperature profiles (arbitrary trends), along the heat exchanger axis, for the hot fluid and the cold fluid. On the left: countercurrent flow; on the right: equicurrent flow

(high thermal flux as per scheme in Eq. 4.12) and vice versa at the outlet (low thermal flux). A counterflow exchanger therefore presents much more homogeneous thermal fluxes along its axis (while, for example, it is noted that, for an equicurrent exchanger, further increasing the length would only lead to residual advantages, see Fig. 5.4 in which the two thermal profiles tend to meet, vanishing the driving force).

In the countercurrent case, moreover, an outlet temperature of the cold fluid higher than the outlet temperature of the hot fluid could also be favourably determined ($t_{out} > T_{out}$ as proposed in Fig. 5.4; this could never happen for equicurrent feeding, for which T_{out} and t_{out} are relative to the same section and therefore the temperature of the cold fluid could not be higher than the temperature of the hot fluid), further enhancing the system efficacy.

Cases where an equicurrent flow may be preferred are those where, for example, there is a need to immediately heat the cold fluid (liquid) because it is very viscous (e.g. applications in the food industry). In fact, we know that, for a liquid, the viscosity decreases as the temperature increases (cf. Eq. 3.10). Or, vice versa, when it is necessary to immediately cool a hot fluid, perhaps because it is coming out of a chemical reactor and for which one wants, as soon as possible, to stop the course of reactions. In such situations, the steep temperature variations possible in the inlet section for each of the two fluids may suggest operating in equicurrent flow.

A special case is that of a fluid that changes its state of matter (heat exchange with latent heat, with constant temperature for the fluid that changes state): consider the simple case of using condensing water vapour as hot fluid. Here, as it can also be seen in Fig. 5.5, feeding in counter- or equi-current mode is equivalent (the equicurrent case has been drawn in the figure).

5.3 Energy Balance

Since, in this system, the variation in total energy can be confused with the variation in enthalpy (i.e. avoiding to consider the secondary variation of other forms of energy, such as potential/kinetic/mechanical), the *energy balance* (stationary and

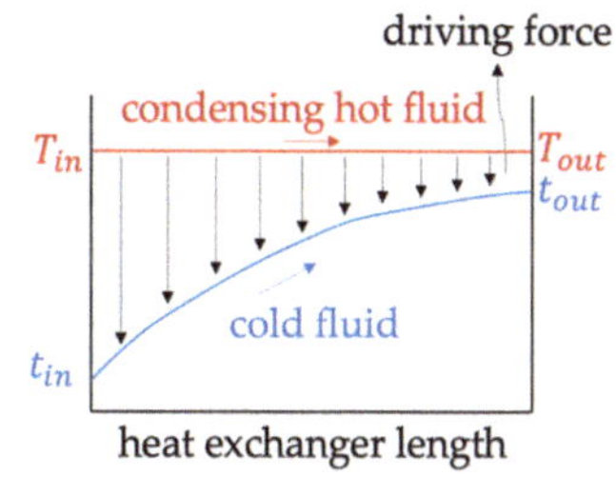

Fig. 5.5 Typical temperature profiles (arbitrary trends), along the heat exchanger axis, for condensing hot fluid and a cold fluid. Countercurrent and equicurrent flows are equivalent

adiabatic system, absence of chemical reactions) is written by equating the enthalpic flow rate [energy/time] entering through the two streams with the enthalpic flow rate exiting through the two streams:

$$W_C \mathbb{C}_C (t_{in} - T_{ref}) + W_H \mathbb{C}_H (T_{in} - T_{ref}) = W_C \mathbb{C}_C (t_{out} - T_{ref}) + W_H \mathbb{C}_H (T_{out} - T_{ref}) \tag{5.3}$$

with specific heat of the two fluids, $\mathbb{C}_C$ (for the cold fluid) and $\mathbb{C}_H$ (for the hot fluid) to be considered, for simplicity, averaged in the thermal range experienced by each fluid, and where T_{ref} is a reference temperature for the enthalpy. Rearranging, we obtain the thermal flow rate (or *power*) [energy/time] associated with the operation of the heat exchanger:

$$\boxed{q = W_C \mathbb{C}_C (t_{out} - t_{in}) = W_H \mathbb{C}_H (T_{in} - T_{out})} \tag{5.4}$$

that is, the enthalpic flow rate given off by the hot fluid is acquired by the cold fluid, in the hypotheses formulated above.

In the particular case of a pure fluid that changes its state of matter (for example, condensing vapour as hot fluid, which enters saturated and not superheated, and which exits not subcooled), the energy balance equation will change to (with mass latent heat of condensation for the hot fluid, $|\lambda_H|$, clearly calculated at the conditions of the state change):

$$\boxed{q = W_C \mathbb{C}_C (t_{out} - t_{in}) = W_H \,|\, \lambda_H \,|} \tag{5.5}$$

Vice versa, if we have an evaporating cold fluid that subtracts heat to a hot fluid, calling $|\lambda_C|$ the mass heat of vaporisation of the cold fluid:

$$\boxed{q = W_C |\lambda_C| = W_H \mathbb{C}_H (T_{in} - T_{out})} \tag{5.6}$$

Example: Analysis of a System with Sensible Heat Exchange

Consider a heat exchanger (counterflow) according to the scheme in Fig. 5.2. The process fluid is a hot organic liquid with flow rate W_H = 10,000 kg/h,

(continued)

which must be cooled from $T_{in} = 60$ °C to $T_{out} = 30$ °C. Assume properties similar to benzene, so $\mathbb{C}_H = 1.81$ kJ/(kg·K). This information is sufficient to determine the heat power of the exchanger from Eq. 5.4:

$$q = W_H \mathbb{C}_H (T_{in} - T_{out}) = 543{,}000 \text{ kJ/h} \quad (5.7)$$

The operation is carried out using, as service fluid, cold liquid water at a temperature of $t_{in} = 20$ °C. It is required that water will heat up to a temperature of $t_{out} = 40$ °C. Assuming $\mathbb{C}_C = 4.18$ kJ/(kg·K), reworking Eq. 5.4 allows us to obtain the flow rate of water to be fed:

$$W_C = \frac{q}{\mathbb{C}_C (t_{out} - t_{in})} = 6495.2 \text{ kg/h} \quad (5.8)$$

Example: Analysis of a System with Latent Heat Exchange

Consider a heat exchanger according to the scheme in Fig. 5.2. The process fluid is a cold organic liquid with flow rate $W_C = 10{,}000$ kg/h, which must be heated from $t_{in} = 15$ °C to $t_{out} = 55$ °C. Assume properties similar to benzene, so $\mathbb{C}_C = 1.81$ kJ/(kg·K). This information is sufficient to determine the heat power of the exchanger from Eq. 5.4:

$$q = W_C \mathbb{C}_C (t_{out} - t_{in}) = 724{,}000 \text{ kJ/h} \quad (5.9)$$

The operation is carried out using, as service fluid, condensing steam at $T_{in} = T_{out} = 100$ °C (cf. Fig. 5.5). Assuming $|\lambda_H| = 2260$ kJ/kg, reworking Eq. 5.5 allows us to obtain the flow rate of steam to be fed:

$$W_H = \frac{q}{|\lambda_H|} = 320.4 \text{ kg/h} \quad (5.10)$$

5.4 Operating Lines

Let us consider the base case scenario of exchange with sensible heat for both fluids. In Eq. 5.4 we note that, for the process fluid (we assume the hot fluid as the process and the cold fluid as the service fluid, and counterflow operation, but the treatise would be conceptually the same even with the roles of the fluids reversed and in different cases), the mass flow rate (W_H), the inlet temperature (T_{in}) and the outlet temperature (T_{out}) must be known, as well as the average value of the specific heat ($\mathbb{C}_H$). This allows us to calculate, as seen, q (the thermal power of the exchanger). While, for the service fluid, we assume that the average value of the specific heat ($\mathbb{C}_C$) and the inlet temperature (t_{in}) are known. Therefore, we have two unknown

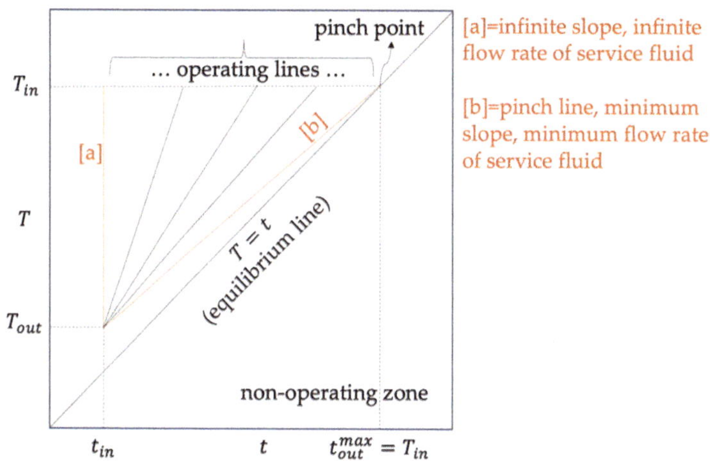

Fig. 5.6 Operating chart for countercurrent heat exchanger (sensible heat exchange), with hot process fluid and cold service fluid

quantities for the service fluid (the mass flow rate W_C and the outlet temperature t_{out}). These two unknown quantities are linked to each other. Determining the value of the mass flow rate of the service fluid (W_C) will allow us, via Eq. 5.4, to calculate its outlet temperature (t_{out}):

$$t_{out} = t_{in} + \frac{W_H \mathbb{C}_H (T_{in} - T_{out})}{W_C \mathbb{C}_C} \quad (5.11)$$

In more detail, given $q = W_H\mathbb{C}_H(T_{in}-T_{out})$, if the mass flow rate of the service fluid (W_C) is increased, its outlet temperature (t_{out}) will decrease (the inlet/outlet temperature difference for the service fluid will be less marked, if a greater flow rate absorbs the same thermal power q), and vice versa.

To address the problem of choosing the mass flow rate value of the service fluid (W_C), consider the plane of Fig. 5.6 which reports the temperature of the hot fluid T against the temperature of the cold fluid t. Draw the diagonal ($T = t$) on the plane, which represents the *thermal equilibrium conditions*. Below the diagonal (the temperature of the hot fluid would be lower than the temperature of the cold fluid), there is the non-working zone. First of all, it can be sensed that *the greater the distance between the operating conditions and the equilibrium conditions, the smaller the exchange area necessary for the operation*.

Now rewrite Eq. 5.4 as:

$$\frac{T_{in} - T_{out}}{t_{out} - t_{in}} = \frac{W_C \mathbb{C}_C}{W_H \mathbb{C}_H} \quad (5.12)$$

which represents the *operating line of the heat exchanger*, passing through the points (t_{in};T_{out}), with known coordinates, and (t_{out};T_{in}), where we only know the value of the ordinate T_{in}. This operating line has slope $W_C\mathbb{C}_C/W_H\mathbb{C}_H$, and it is the locus of points having coordinates t (the temperature of the cold fluid) and T (the temperature of the hot fluid), where t and T are the temperature values of the two fluids, respectively, that exchange energy in a given (generic) section of the heat exchanger. The operating line will therefore pass through the point of known coordinates (t_{in}; T_{out}), which in our example (Fig. 5.2) represents the section of inlet of the cold fluid and outlet of the hot fluid (both temperatures are in fact known). Its slope will instead depend on the value chosen for W_C (all other values being known), in the sense that the greater this flow rate, the greater the slope.

For infinite mass flow rate of service fluid (W_C), a vertical line will be obtained (Fig. 5.6, line [a]). This means that the operation will occur without temperature variation for the service fluid (obviously $t_{out} = t_{in}$ is the only possible condition, the product between an infinite flow rate and a zero thermal difference will give a finite thermal power q; in other words, if an infinite flow rate of the service fluid absorbs the thermal power released by the process fluid, the service fluid will not see its temperature change, cf. Eq. 5.11) and the distance between operating and equilibrium conditions will be the maximum possible. Therefore, *with maximum driving force, this would allow a minimum exchange area*, but this cannot be considered an operating condition for obvious reasons (this reminds us of what we saw, for example, in the case of distillation: infinite reflux ratio, maximum distance from equilibrium, maximum driving force, minimum number of stages).

As the mass flow rate of the service fluid (W_C) decreases (Fig. 5.6), the slope will decrease and, clearly, t_{out} will increase (if the thermal power released by the process fluid is absorbed by a progressively lower flow rate of the service fluid, the latter will progressively see its outlet temperature increase), with savings on the flow rate of the service fluid but an increase in the exchange area (since the distance between operating and equilibrium conditions decreases).

At the other conceptual extreme, by imposing the *pinch condition* $t_{out}^{max} = T_{in}$ (operating point belonging to the equilibrium diagonal), an operating line with minimum slope (Fig. 5.6, line [b]) will be obtained (i.e., the minimum flow rate of service fluid, W_C, to be used). However, this is not an operating condition either, as an infinite exchange area must be foreseen (a characteristic of operations in pinch conditions, as already seen, for example, in the case of distillation: minimum reflux ratio, infinite number of stages).

Thus, the actual operating line will be comprised between [a] and [b] in Fig. 5.6. It is necessary to mediate between the effects, first of all by finding *the minimum value for the flow rate of the service fluid*, which in our example is, from Eq. 5.4 for $t_{out}^{max} = T_{in}$:

$$W_C^{min} = \frac{W_H \mathbb{C}_H (T_{in} - T_{out})}{\mathbb{C}_C \left(t_{out}^{max} - t_{in}\right)} \tag{5.13}$$

The value of W_C^{min} must then be appropriately increased, to obtain a value that results in not too high exchange area, but without having to work with a value that is too high for the flow rate of the service fluid. The determination of this flow rate allows, by entering the energy balance equation (Eq. 5.11), to calculate the operating value

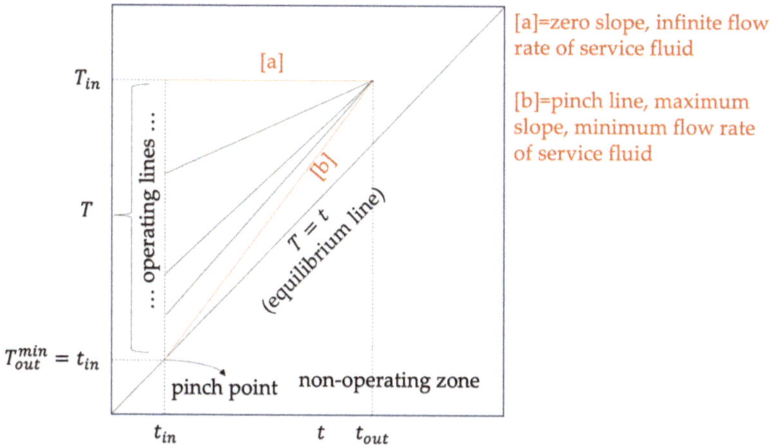

Fig. 5.7 Operating chart for countercurrent heat exchanger (sensible heat exchange), with cold process fluid and hot service fluid

for t_{out}, therefore to determine the operating line on the chart in Fig. 5.6, and to know the logarithmic mean driving force, ΔT_{lm}, by application of Eqs. 5.1 and 5.2.

Example: Optimal Operation of a Heat Exchanger
Please rework a previous example: the process fluid is a hot organic liquid ($W_H = 10{,}000$ kg/h, $T_{in} = 60$ °C, $T_{out} = 30$ °C, $\mathbb{C}_H = 1.81$ kJ/(kg·K)) and the service fluid is cold liquid water ($t_{in} = 20$ °C, $\mathbb{C}_C = 4.18$ kJ/(kg·K)). Its minimum (pinch) mass flow rate is, Eq. 5.13, $W_C^{min} = 3247.6$ kg/h. Under an economic analysis that suggests operating with $W_C = 2W_C^{min} = 6495.2$ kg/h, Eq. 5.11 gives $t_{out} = 40$ °C.

If the role of process and service fluid is reversed, the situation is conceptually identical. In Eq. 5.4 we now note that, for the process fluid, the mass flow rate (W_C), the inlet temperature (t_{in}) and the outlet temperature (t_{out}) must be known. This allows us to calculate q. While, for the service fluid, the inlet temperature (T_{in}) is known. Therefore, we have two unknown quantities for the service fluid (the mass flow rate W_H and the outlet temperature T_{out}). These two unknown quantities are linked to each other. Determining the value of the mass flow rate of the service fluid (W_H) will allow us, via Eq. 5.4, to calculate its outlet temperature (T_{out}):

$$T_{out} = T_{in} - \frac{W_C \mathbb{C}_C (t_{out} - t_{in})}{W_H \mathbb{C}_H} \tag{5.14}$$

To address the problem of choosing the mass flow rate value of the service fluid (W_H), consider the plane of Fig. 5.7. The operating line of the heat exchanger is in Eq. 5.12, passing through the points (t_{in};T_{out}), where we only know the value of the abscissa t_{in}, and (t_{out},T_{in}), with known coordinates. This latter point represents, in our example (Fig. 5.2), the section of outlet of the cold fluid and inlet of the hot fluid (both temperatures are in fact known). Its slope will instead depend on the value chosen for W_H (all other values being known), in the sense that the greater this flow rate, the smaller the slope.

For infinite mass flow rate of service fluid (W_H), a horizontal line will be obtained (Fig. 5.7, line [a]). This means that the operation will occur without temperature variation for the service fluid (obviously $T_{out} = T_{in}$ is the only possible condition, cf. Eq. 5.14) and the distance between operating and equilibrium conditions will be the maximum possible. Therefore, with maximum driving force, this would allow a minimum exchange area, but this cannot be considered an operating condition. As the mass flow rate of the service fluid (W_H) decreases (Fig. 5.7), the slope will increase and, clearly, T_{out} will decrease, with savings on the flow rate of the service fluid but an increase in the exchange area. At the other conceptual extreme, by imposing the pinch condition $T_{out}^{min} = t_{in}$ (operating point belonging to the equilibrium diagonal), an operating line with maximum slope (Fig. 5.7, line [b]) will be obtained (i.e., the minimum flow rate of service fluid, W_H, to be used). However, this is not an operating condition either.

Thus, the actual operating line will be comprised between [a] and [b] in Fig. 5.7. It is necessary to mediate between the effects, first of all by finding the minimum value for the flow rate of the service fluid, which in our example is, from Eq. 5.4 for $T_{out}^{min} = t_{in}$:

$$W_H^{min} = \frac{W_C \mathbb{C}_C (t_{out} - t_{in})}{\mathbb{C}_H \left(T_{in} - T_{out}^{min}\right)} \tag{5.15}$$

The value of W_H^{min} must then be appropriately increased. The determination of this flow rate allows, by entering the energy balance equation (Eq. 5.14), to calculate the operating value for T_{out}, and therefore to determine the operating line, and to know the logarithmic mean driving force.

5.5 Overall Heat Transfer Coefficients

Actually, the case previously presented in Fig. 4.7 (heat exchange between two fluids through a separating wall in cylindrical geometry) represents that of interest for our heat exchanger, with characteristic diameters defined in Fig. 5.3.

The relations in Eq. 4.50 can be rewritten here:

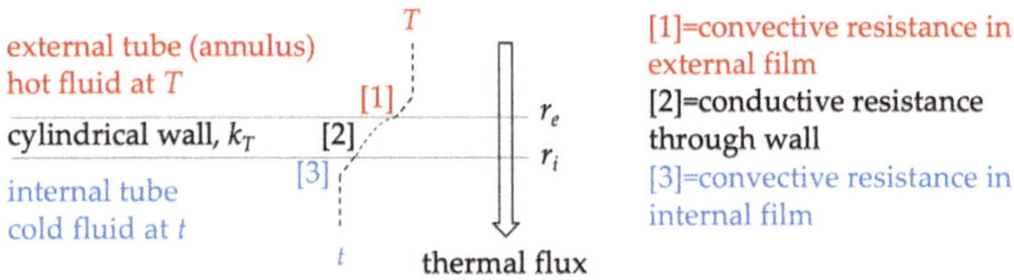

Fig. 5.8 Resistances in series for heat transfer from a hot fluid circulating in the annulus (at temperature T) to a cold fluid circulating in the internal tube (at temperature t) through a cylindrical wall (cf. Fig. 4.7). The concept does not change if the position of the two fluids is reversed

$$\begin{cases} \mathcal{R}_i = \dfrac{1}{U_i} = \dfrac{1}{h_i} + \dfrac{r_i \ln\left(\dfrac{r_e}{r_i}\right)}{k_T} + \dfrac{r_i}{r_e h_e} + \sum \mathcal{R}_d \\ \mathcal{R}_e = \dfrac{1}{U_e} = \dfrac{r_e}{r_i h_i} + \dfrac{r_e \ln\left(\dfrac{r_e}{r_i}\right)}{k_T} + \dfrac{1}{h_e} + \sum \mathcal{R}_d \end{cases} \tag{5.16}$$

which remind us that three resistances in series are active, connected to (we refer to Fig. 5.8, the basic case for which the hot fluid circulates in the annulus, and the cold fluid in the tube, but the relations have general validity), respectively:

- convective energy transport (governed by a local convective heat transfer coefficient, h_e) from the bulk of the fluid in the external annulus towards the external wall of the internal tube, described by radius $r_e = d_e/2$ (Fig. 5.3; [1] in Fig. 5.8);
- conductive energy transport (governed by the thermal conductivity of the material, k_T) through the wall of the internal tube ([2] in Fig. 5.8);
- convective energy transport (governed by a local convective heat transfer coefficient, h_i) from the internal wall of the internal tube, described by radius $r_i = d_i/2$ (Fig. 5.3; [3] in Fig. 5.8), to the bulk of the fluid in the internal tube.

In Eqs. 5.16, $\mathcal{R}_i$ and $\mathcal{R}_e$ are the overall resistances, while U_i and U_e:

$$\boxed{\begin{cases} U_i = \left[\frac{1}{h_i} + \frac{r_i \ln\left(\frac{r_e}{r_i}\right)}{k_T} + \frac{r_i}{r_e h_e} + \sum \mathcal{R}_d\right]^{-1} \\ U_e = \left[\frac{r_e}{r_i h_i} + \frac{r_e \ln\left(\frac{r_e}{r_i}\right)}{k_T} + \frac{1}{h_e} + \sum \mathcal{R}_d\right]^{-1} \end{cases}} \tag{5.17}$$

are the *overall heat transfer coefficients*, with reference to the internal wall and the external wall of the internal tube, respectively.

The reader will have noticed, with reference to Eq. 4.50, that in Eqs. 5.16 and 5.17 there is a fourth term. This is the sum of the global *fouling ("dirt") resistances*, $\mathcal{R}_d$. It is considered, in the operation of an exchanger, that the fluids, in their motion, can for example leave sediments on the walls, or cause phenomena of corrosion, crystallisation, decomposition, polymerisation, or biological phenomena if the fluid contains microorganisms, algae, bacteria, determining fouling and therefore offering a greater resistance to heat transfer. The value of $\mathcal{R}_d$ is usually assigned semi-empirically, and it can refer to either, of both, fluids. Namely, $\sum \mathcal{R}_d$ is a global resistance taking into account, if needed, fouling phenomena on both sides of the internal tube (the sum of the internal and external fluid contributions). In order of magnitude, $\mathcal{R}_d$ varies e.g. between 10^{-3} and 10^{-4} $m^2 \cdot s \cdot K/J$ in canonical cases and considering average service times between cleaning (McCabe et al. 2001).

5.6 Local Heat Transfer Coefficients

5.6.1 General Considerations

It is therefore evident, from Eq. 5.17, that the knowledge of the global heat transfer coefficients, U_i and U_e, must pass through the knowledge of the local heat transfer coefficients, h_i and h_e, once the radial geometry of the system (the diameters of Fig. 5.3), the thermal conductivity of the material that constitutes the wall and, if necessary, the fouling resistances have been fixed.

In this discussion, we will refer to properties that are functions of temperature. Even assuming constant specific heats (as already said), it remains true that properties such as viscosity (Eq. 3.10) and thermal conductivity (Eq. 4.4) of fluids are dependent on temperature. We will therefore consider an average value of $T = (T_{in} + T_{out})/2$ (and $t = (t_{in} + t_{out})/2$) between inlet and outlet for the hot fluid (and the cold fluid, respectively), at which to calculate the viscosity and thermal conductivity. *This approach is considered valid for fully developed turbulent flow*, i.e. considering that the two fluids circulate in their respective tubes with a Reynolds number greater than 10,000. For intermediate situations of turbulent flow (i.e., Reynolds number greater than 2100 but less than 10,000), an error of about 10% should be considered in the design of the heat exchanger. For still different situations, or for more refined approaches which are outside the scope of this textbook, please refer to the literature (Kern 1983; McCabe et al. 2001).

As a further consideration for cognitive purposes, it is recalled that the Reynolds number depends on the viscosity, which in turn is a function, as mentioned, of the temperature. Therefore, strictly speaking, for each of the two fluids, two Reynolds numbers should be calculated, one at the inlet conditions, the other at the outlet conditions, and it should be verified that the two fluids are in turbulent regime at the two far ends. In fact, temperature variations could, in some cases, cause the viscosity to vary (which decreases with temperature for liquids, and increases with temperature for gases, cf. Eq. 3.10) to the point that a fluid could change its fluid dynamic regime (from turbulent to laminar, for example) during its motion in the heat exchanger. Further on, it will be presented how to calculate *Re* for the system at hand.

5.6.2 Equivalent Diameter for Heat Exchangers Annulus

The fluid circulating in the internal tube is characterised by a relevant dimension which is the diameter of the tube itself, d_i (Fig. 5.9). For the external tube, instead (Fig. 5.9), the fluid does not occupy the entire section. In the design of heat exchangers, when necessary, reference is made to the *equivalent diameter for the annulus*, defined as 4 times the passage section divided by the wetted perimeter for the purposes of heat exchange (i.e., Fig. 5.9, *only the perimeter offered by the external surface of the internal tube*; the perimeter offered by the internal surface of the annulus, in fact, although "wet", is not involved in the heat exchange since the system is globally adiabatic and thermally insulated from the external environment):

$$D_{eq} = 4\frac{\frac{\pi}{4}\left(D_i^2 - d_e^2\right)}{\pi d_e} = \frac{D_i^2 - d_e^2}{d_e} \tag{5.18}$$

5.6.3 Non-Dimensional Numbers

First, from Eq. 3.17, we observe that:

$$\rho\bar{v} = \frac{W}{S} \tag{5.19}$$

that is, the product of density and mean velocity of a fluid can be expressed as the ratio of the mass flow rate W to the cross-sectional passage area S (i.e., this ratio is the mass flux). This is useful in heat exchangers, where reference is made to mass flow rates.

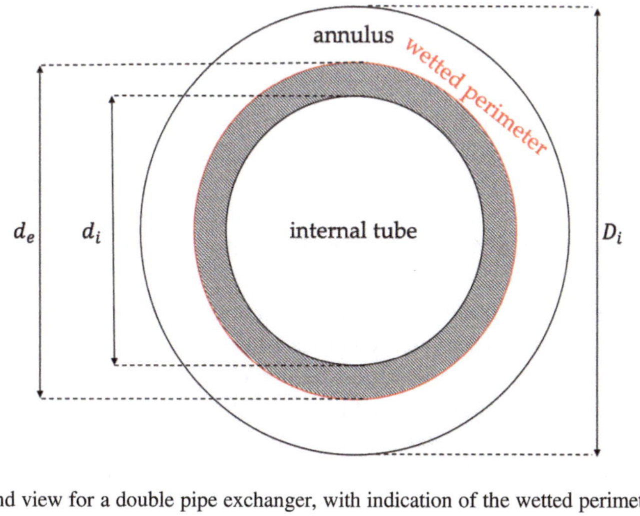

Fig. 5.9 End view for a double pipe exchanger, with indication of the wetted perimeter

At this point, the characteristic Reynolds numbers for the internal fluid and the external fluid can be defined:

$$\begin{cases} Re_i = \dfrac{W_i d_i}{\frac{\pi}{4} d_i^2 \mu_i} \\ Re_e = \dfrac{W_e D_{eq}}{\frac{\pi}{4}(D_i^2 - d_e^2)\mu_e} \end{cases} \tag{5.20}$$

where Eq. 5.19 has been recalled. The subscripts "i" and "e", applied to the mass flow rate W and the viscosity (and, further on, to the specific heat and the fluid thermal conductivity) refer, respectively, to the fluid circulating in the internal tube and in the external tube. We also observe the passage cross-sectional area for the fluid in the internal tube ($\pi d_i^2/4$) and in the external tube ($\pi(D_i^2 - d_e^2)/4$), and that the characteristic dimension for the Reynolds number is d_i for the internal tube, and D_{eq} (Eq. 5.18) for the external tube.

Let us now define the corresponding Prandtl numbers (Eq. 4.6):

$$\begin{cases} Pr_i = \dfrac{\mu_i \mathbb{C}_i}{k_{T,i}} \\ Pr_e = \dfrac{\mu_e \mathbb{C}_e}{k_{T,e}} \end{cases} \tag{5.21}$$

For the properties of the fluids, remember what was said above, that is, that they must be calculated at their respective average temperatures.

Finally, the corresponding Nusselt numbers are defined as (Eq. 4.23):

$$\begin{cases} Nu_i = \dfrac{h_i d_i}{k_{T,i}} \\ Nu_e = \dfrac{h_e D_{eq}}{k_{T,e}} \end{cases} \tag{5.22}$$

where it is observed, again, that the characteristic dimension for the external side Nusselt number is the equivalent diameter (Eq. 5.18).

5.6.4 Evaluation of the Local Coefficients

As already highlighted in the chapter devoted to energy transport, there are numerous literature relations (very often empirical) that link Nu to other dimensionless numbers capable of characterising the properties of the fluid (e.g., Pr) and the thermo-fluid dynamic conditions of the system (e.g., Re). For fluids exchanging energy in a heat exchanger, both animated by fully developed turbulent flow, the following relations $Nu(Re,Pr)$ can be considered (for this and other approaches, refer to the literature, e.g. Bird et al. 2007; Kern 1983; McCabe et al. 2001):

$$Nu = 0.023Re^{0.8}Pr^{\alpha}\begin{cases} \alpha = 0.4 \text{ for cold fluids that heat up} \\ \alpha = 0.3 \text{ for hot fluids that cool down} \end{cases} \quad (5.23)$$

Therefore, given a fluid, first consider whether it circulates in the internal tube or in the annulus, for the correct definition of its *Nu* (Eq. 5.22), *Re* (Eq. 5.20) and *Pr* (Eq. 5.21) numbers. Then, examine whether it heats up or cools down, and apply the corresponding relation in Eq. 5.23. Finally, once *Nu* is known, isolate the local heat transfer coefficient in Eq. 5.22 and get it (for a first, rough, estimate, typically these coefficients have values [J/(m^2·s·K)] in the range 1–50 for air, 50–1500 for oils, 300–20,000 for water, 1500–50,000 for boiling water, 5000–100,000 for condensing water vapour (Bird et al. 2007; McCabe et al. 2001). Once obtained h_i and h_e, it is possible to return to Eq. 5.17 and calculate U_i and U_e. At this point, we are ready to develop the design equation for a heat exchanger.

5.7 Design Equations

Figure 5.10 re-elaborates Fig. 5.2, indicating a control volume of extension dz which we will refer to, with z the main direction of motion (arbitrarily chosen from section "A" towards section "B"). This control volume is associated with a differential energy transfer area, which we will generically call dA_T. In this control volume, we assume that the hot fluid has a bulk temperature T, and the cold fluid a bulk temperature t (for fully developed turbulent flow, we consider these temperatures radially constant). The choice of sending the hot fluid into the annulus, and the cold fluid into the internal tube, i.e. the direction of the heat flow (in this case, from the annulus to the internal tube), does not influence the validity of what we are about to obtain.

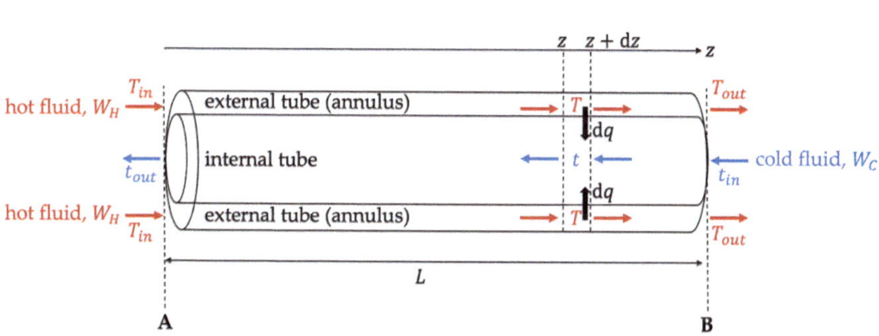

Fig. 5.10 Representation of a double pipe exchanger. The choice of the fluid to be sent into the internal tube, and that to be sent into the annulus, as well as the direction of feeding for the two fluids, does not affect the generality of the discussion. The figure indicates the differential control volume, with bulk temperatures for both fluids, and the direction of the differential thermal power, which is transferred from the hot to the cold fluid

In general, the scheme of Eq. 4.12 holds: the heat flux is the product of a transfer coefficient (U) and a driving force $\Delta T = T - t$. We can therefore write, with reference to our control volume:

$$d\dot{q} = \frac{dq}{dA_T} = U\Delta T = U(T - t) \tag{5.24}$$

Note, in Eq. 5.24, the already-known concept that the thermal flux is the ratio between thermal power q and transfer area.

Let us write the energy balance on the hot fluid, a balance that we can express in terms of enthalpy as already mentioned. The balance on the differential control volume is written in dimensions [energy/time], and must respond to the concept that the difference in enthalpy between the input (at z) and the output (at z + dz), for a stationary (and, globally adiabatic) system, corresponds to the differential thermal power (dq) transferred to the cold fluid (cf. Eq. 5.24):

$$W_H \mathbb{C}_H [T(z) - T(z + dz)] = dq = U[T(z) - t(z)]dA_T \tag{5.25}$$

$$-W_H \mathbb{C}_H dT(z) = U[T(z) - t(z)]dA_T \tag{5.26}$$

$$\frac{dT(z)}{T(z) - t(z)} = -\frac{U}{W_H \mathbb{C}_H} dA_T \tag{5.27}$$

Similarly, the energy balance on the differential control volume for the cold fluid responds to the concept that the difference in enthalpy between the exit (at level z) and the entrance (at level z + dz), for a stationary system, corresponds to the differential thermal power dq acquired from the hot fluid (cf. Eq. 5.24):

$$W_C \mathbb{C}_C [t(z) - t(z + dz)] = dq = U[T(z) - t(z)]dA_T \tag{5.28}$$

$$-W_C \mathbb{C}_C dt(z) = U[T(z) - t(z)]dA_T \tag{5.29}$$

$$\frac{dt(z)}{T(z) - t(z)} = -\frac{U}{W_C \mathbb{C}_C} dA_T \tag{5.30}$$

We subtract Eqs. 5.27 and 5.30:

$$\frac{d[T(z) - t(z)]}{T(z) - t(z)} = \frac{d\Delta T}{\Delta T} = U\left(\frac{1}{W_C \mathbb{C}_C} - \frac{1}{W_H \mathbb{C}_H}\right) dA_T \tag{5.31}$$

Let us now focus on the term in parentheses in the right-hand side of Eq. 5.31. From Eq. 5.4, the following holds:

$$\begin{cases} \dfrac{1}{W_C \mathbb{C}_C} = \dfrac{t_{out} - t_{in}}{q} \\ \dfrac{1}{W_H \mathbb{C}_H} = \dfrac{T_{in} - T_{out}}{q} \end{cases} \tag{5.32}$$

$$\frac{1}{W_C \mathbb{C}_C} - \frac{1}{W_H \mathbb{C}_H} = \frac{1}{q}(t_{out} - t_{in} - T_{in} + T_{out}) = \frac{1}{q}$$
$$\times [(T_{out} - t_{in}) - (T_{in} - t_{out})] = \frac{1}{q}(\Delta T_B - \Delta T_A) \tag{5.33}$$

where we recalled, Eq. 5.1, the definitions for the driving force at the ends "A" and "B" of the exchanger (Fig. 5.10). We substitute Eq. 5.33 into Eq. 5.31:

$$\frac{\mathrm{d}\Delta T}{\Delta T} = \frac{U}{q}(\Delta T_B - \Delta T_A)\mathrm{d}A_T \tag{5.34}$$

We integrate by separation of variables, between section "A" and section "B", considering the coefficient U constant and equal to an average value, as stated above (U is therefore put outside the integral, an approach valid for both fluids having fully developed turbulent flow):

$$\int_{\mathrm{A}}^{\mathrm{B}} \frac{\mathrm{d}\Delta T}{\Delta T} = \frac{U}{q}(\Delta T_B - \Delta T_A)\int_{\mathrm{A}}^{\mathrm{B}} \mathrm{d}A_T \tag{5.35}$$

$$[\ln(\Delta T)]_{\mathrm{A}}^{\mathrm{B}} = \ln\left(\frac{\Delta T_B}{\Delta T_A}\right) = \frac{U}{q}(\Delta T_B - \Delta T_A)A_T \tag{5.36}$$

where, with A_T, we mean the overall heat transfer area of the double pipe exchanger, developed between section "A" and section "B". Let us rewrite Eq. 5.36:

$$q = UA_T \frac{\Delta T_B - \Delta T_A}{\ln\left(\frac{\Delta T_B}{\Delta T_A}\right)} \tag{5.37}$$

where the logarithmic mean driving force of Eq. 5.2 appears, to give the *design equation of the double pipe heat exchanger in its general formulation*:

$$q = UA_T \Delta T_{lm} \tag{5.38}$$

This equation, obtained for the case of countercurrent flow, with the hot fluid in the annulus and the cold fluid in the internal tube, for heat exchange with sensible heat for both fluids, *is generally valid also in all other situations* (of course, assuming that U can be considered constant as average value). The Eq. 5.38 relates the heat exchanger thermal power to: (1) the overall heat transfer coefficient; (2) the overall heat transfer area; (3) the logarithmic mean driving force. *When each of these three quantities is greater, the greater is the heat exchanger thermal power.*

At this point, let us discuss the expression for the transfer area. In fact, it is possible to refer to the internal transfer area of the inner tube which, for internal diameter of the inner tube equal to d_i (Fig. 5.9) in a heat exchanger of length L, is:

$$A_T = A_i = \pi d_i L \tag{5.39}$$

Or, equivalently, it is possible to refer to the external transfer area of the internal tube which, for external diameter of the internal tube equal to d_e (Fig. 5.9) in a heat exchanger of length L, is equal to:

$$A_T = A_e = \pi d_e L \tag{5.40}$$

Each of the values, A_i or A_e, can be used in Eq. 5.38, coupling the related overall heat transfer coefficient, U_i or U_e respectively (Eq. 5.17). Therefore, the final formulation of the design equation for a double pipe heat exchanger is:

$$\boxed{\begin{cases} q = U_i A_i \Delta T_{lm} \\ q = U_e A_e \Delta T_{lm} \end{cases}} \tag{5.41}$$

Altogether, in sequence:

- the knowledge of all the parameters for the process fluid allows to calculate the thermal power of the heat exchanger, q;
- the approach with the operating lines method allows to determine the mass flow rate and outlet temperature for the service fluid. This, in particular, makes known the 4 extreme (section "A" and section "B") temperatures, therefore the logarithmic mean driving force, ΔT_{lm} (Eq. 5.2);
- the Eqs. 5.17 determine the global heat transfer coefficient (U_i or U_e), with the approaches presented for the evaluation of the local coefficients.

All this allows us to identify, as an unknown, the transfer area in Eq. 5.41:

$$\boxed{\begin{cases} A_i = \dfrac{q}{U_i \Delta T_{lm}} \\ A_e = \dfrac{q}{U_e \Delta T_{lm}} \end{cases}} \tag{5.42}$$

It is observed, for example, that for heat exchangers of fixed power q, the transfer area can be smaller if the transfer coefficient, and/or the average driving force, are higher (as expected). For systems with geometrically defined diameters, the real unknown is the *length needed for the heat exchange*, which can be calculated using one of the two expressions:

$$\boxed{\begin{cases} L = \dfrac{A_i}{\pi d_i} \\ L = \dfrac{A_e}{\pi d_e} \end{cases}} \tag{5.43}$$

Example: Design of a Double Pipe Exchanger

A pharmaceutical company uses, for the production of isopropyl acetate, liquid acetic acid that must be previously heated from $t_{in} = 25$ °C to $t_{out} = 80$ °C. For this purpose, a stainless steel countercurrent heat exchanger is used. A mass flow rate $W_C = 4500$ kg/h of acetic acid is sent into the internal tube, while liquid benzonitrile at $T_{in} = 160$ °C is used as service fluid in the annulus.

We wish to determine the heat exchanger length, assuming that:

- the specific heat capacity of the process and service fluids can be taken as $\mathbb{C}_C = 2.17$ kJ/(kg·K) and $\mathbb{C}_H = 1.85$ kJ/(kg·K), respectively;
- the flow rate of the service fluid is double the minimum value ($W_H = 2W_H^{min}$);
- the thermal conductivity k_T of the steel which the exchanger is made of, is 22.3 J/(m·s·K);
- the inner and outer diameters of the internal tube are $d_i = 0.154$ m and $d_e = 0.168$ m, respectively, while the inner diameter of the annulus is $D_i = 0.173$ m;
- the sum of the global dirt resistances is $\sum \mathcal{R}_d = 2.0 \cdot 10^{-4}$ m^2·s·K/J.

The application of the energy balance in Eq. 5.4, referred to the process fluid (acetic acid), allows us to determine the heat power of the exchanger:

$$q = W_C \mathbb{C}_C (t_{out} - t_{in}) = 537{,}075 \text{ kJ/h} \tag{5.44}$$

while Eq. 5.15 helps in obtaining the minimum mass flow rate for the service fluid (benzonitrile):

$$W_H^{min} = \frac{W_C \mathbb{C}_C (t_{out} - t_{in})}{\mathbb{C}_H \left(T_{in} - T_{out}^{min}\right)} = 2150.45 \text{ kg/h} \tag{5.45}$$

where $T_{out}^{min} = t_{in}$. According to the instructions received, it is $W_H = 2W_H^{min} =$ 4300.90 kg/h.

Once determined the operating flow rate for the service fluid, application of the energy balance in Eq. 5.4 allows to obtain its exit temperature:

$$T_{out} = T_{in} - \frac{q}{W_H \mathbb{C}_H} = 92.50\,°\text{C} \tag{5.46}$$

We are now ready to calculate the logarithmic mean driving force from Eq. 5.2:

(continued)

$$\Delta T_{lm} = \frac{\Delta T_A - \Delta T_B}{\ln\left(\frac{\Delta T_A}{\Delta T_B}\right)} = 73.57\,°\text{C} = 73.57\ \text{K} \tag{5.47}$$

where, Eq. 5.1:

$$\begin{cases} \Delta T_A = T_{in} - t_{out} = 80\,°\text{C} \\ \Delta T_B = T_{out} - t_{in} = 67.50\,°\text{C} \end{cases} \tag{5.48}$$

<u>*Process fluid (cold, internal tube)*</u>

<u>Step *i1*: evaluate viscosity and thermal conductivity</u>

The general expression for the dynamic viscosity (μ_i) is, Eq. 3.10:

$$\mu_i = \exp\left[C_1 + \frac{C_2}{t} + C_3 \ln(t) + C_4 t^{C_5}\right] \tag{5.49}$$

with the following set of constants for acetic acid (Green and Perry 2008):

$$\begin{cases} C_1 = -9.03 \\ C_2 = 1212.3 \\ C_3 = -0.322 \\ C_4 = C_5 = 0 \end{cases} \tag{5.50}$$

In Eq. 5.49, which gives back viscosity in [kg/(m·s)], please recall that we are using, as already discussed, $t = (t_{in} + t_{out})/2 = 325.65$ K (the measure unit required in Eq. 5.49). We obtain $\mu_i = 7.69 \cdot 10^{-4}$ kg/(m·s).

The general expression for the thermal conductivity ($k_{T,i}$) is, Eq. 4.4:

$$k_{T,i} = C_1 + C_2 t + C_3 t^2 + C_4 t^3 + C_5 t^4 \tag{5.51}$$

with the following set of constants for acetic acid (Green and Perry 2008):

$$\begin{cases} C_1 = 0.214 \\ C_2 = -0.0001834 \\ C_3 = C_4 = C_5 = 0 \end{cases} \tag{5.52}$$

From Eq. 5.51, which gives back conductivity in [J/(m·s·K)], we obtain $k_{T,i} = 0.154$ J/(m·s·K).

<u>Step *i2*: evaluate Reynolds and Prandtl number</u>

In being so kind to express $W_i = W_C$ in [kg/s] and $\mathbb{C}_i = \mathbb{C}_C$ in [J/(kg·K)], from Eqs. 5.20 and 5.21 we get, respectively:

(continued)

$$\begin{cases} Re_i = \dfrac{W_i d_i}{\frac{\pi}{4} d_i^2 \mu_i} = 13{,}446.00 \text{ (fully developed turbulent flow)} \\ Pr_i = \dfrac{\mu_i \mathbb{C}_i}{k_{T,i}} = 10.84 \end{cases} \quad (5.53)$$

<u>Step *i3*: evaluate the local heat transfer coefficient via Nusselt number</u>

The Nusselt number for a cold fluid that heats up is calculated through Eq. 5.23:

$$Nu_i = 0.023\, Re_i^{0.8} Pr_i^{0.4} = 119.84 \quad (5.54)$$

To conclude, application of Eq. 5.22 gives the local heat transfer coefficient:

$$\boxed{h_i = \frac{Nu_i k_{T,i}}{d_i} = 119.84\ \frac{\text{J}}{\text{m}^2 \cdot \text{s} \cdot \text{K}}} \quad (5.55)$$

<u>*Service fluid (hot, external tube)*</u>

<u>Step *e1*: evaluate viscosity and thermal conductivity</u>

Start with the viscosity (μ_e):

$$\mu_e = \exp\left[C_1 + \frac{C_2}{T} + C_3 \ln(T) + C_4 T^{C_5}\right] \quad (5.56)$$

with the following set of constants for benzonitrile (Green and Perry 2008):

$$\begin{cases} C_1 = -20.236 \\ C_2 = 1737.4 \\ C_3 = 1.3531 \\ C_4 = C_5 = 0 \end{cases} \quad (5.57)$$

With $T = (T_{in} + T_{out})/2 = 399.40$ K, we obtain $\mu_e = 4.18 \cdot 10^{-4}$ kg/(m·s).

We continue with the thermal conductivity ($k_{T,e}$):

$$k_{T,e} = C_1 + C_2 T + C_3 T^2 + C_4 T^3 + C_5 T^4 \quad (5.58)$$

with the following set of constants for benzonitrile (Green and Perry 2008):

$$\begin{cases} C_1 = 0.21284 \\ C_2 = -0.00021587 \\ C_3 = C_4 = C_5 = 0 \end{cases} \quad (5.59)$$

We obtain $k_{T,e} = 0.127$ J/(m·s·K).

<u>Step *e2*: evaluate Reynolds and Prandtl number</u>

First, we need to calculate the annulus equivalent diameter from Eq. 5.18:

(continued)

$$D_{eq} = \frac{D_i^2 - d_e^2}{d_e} = 0.010 \text{ m} \tag{5.60}$$

Then, expressing $W_e = W_H$ in [kg/s] and $\mathbb{C}_e = \mathbb{C}_H$ in [J/(kg·K)], from Eqs. 5.20 and 5.21 we get, respectively:

$$\begin{cases} Re_e = \dfrac{W_e D_{eq}}{\frac{\pi}{4}\left(D_i^2 - d_e^2\right)\mu_e} = 21{,}354.36 \text{ (fully developed turbulent flow)} \\ Pr_e = \dfrac{\mu_e \mathbb{C}_e}{k_{T,e}} = 6.09 \end{cases} \tag{5.61}$$

Step *e3*: evaluate the local heat transfer coefficient via Nusselt number

The Nusselt number for a hot fluid that cools down is calculated through Eq. 5.23:

$$Nu_e = 0.023\, Re_e^{0.8} Pr_e^{0.3} = 115.00 \tag{5.62}$$

To conclude, application of Eq. 5.22 gives the local heat transfer coefficient:

$$\boxed{h_e = \frac{Nu_e k_{T,e}}{D_{eq}} = 1460.50 \ \frac{\text{J}}{\text{m}^2 \cdot \text{s} \cdot \text{K}}} \tag{5.63}$$

Overall heat transfer coefficient

We will refer, Eq. 5.17, to U_i, but the procedure is equivalent if one wants to use U_e instead. The internal, wall, external and fouling resistances are, respectively (where $r_i = d_i/2$ and $r_e = d_e/2$):

$$\begin{cases} \dfrac{1}{h_i} = 8.34 \cdot 10^{-3} \ \dfrac{\text{m}^2 \cdot \text{s} \cdot \text{K}}{\text{J}} \\ \dfrac{r_i \ln\left(\dfrac{r_e}{r_i}\right)}{k_T} = 3.0 \cdot 10^{-4} \ \dfrac{\text{m}^2 \cdot \text{s} \cdot \text{K}}{\text{J}} \\ \dfrac{r_i}{r_e h_e} = 6.28 \cdot 10^{-4} \ \dfrac{\text{m}^2 \cdot \text{s} \cdot \text{K}}{\text{J}} \\ \sum \mathcal{R}_d = 2.0 \cdot 10^{-4} \ \dfrac{\text{m}^2 \cdot \text{s} \cdot \text{K}}{\text{J}} \end{cases} \tag{5.64}$$

Thus, it is:

(continued)

$$U_i = \left[\frac{1}{h_i} + \frac{r_i \ln\left(\frac{r_e}{r_i}\right)}{k_T} + \frac{r_i}{r_e h_e} + \sum \mathcal{R}_d \right]^{-1} = 105.62 \ \frac{\mathrm{J}}{\mathrm{m}^2 \cdot \mathrm{s} \cdot \mathrm{K}} \tag{5.65}$$

Transfer area and length of the heat exchanger

From Eq. 5.42, it is (please remember to convert q in [J/s]):

$$A_i = \frac{q}{U_i \Delta T_{lm}} = 19.20 \ \mathrm{m}^2 \tag{5.66}$$

Given the value for d_i, application of Eq. 5.43 allows us to determine the overall length as:

$$L = \frac{A_i}{\pi d_i} = 39.71 \ \mathrm{m} \tag{5.67}$$

which can be operationally obtained, cf. Fig. 5.1, by arranging several hairpins in series.

Example: Verification of a Double Pipe Exchanger

Starting from the previous example, to reduce the risks associated with benzonitrile handling, it is decided to switch to condensing steam at given pressure ($T_{in} = T_{out} = 110$ °C) as service (hot) fluid. We want to highlight how the previously designed heat exchanger must be modified, by keeping, for the process fluid, the same flow rate and inlet/outlet temperatures (that is, Eq. 5.44, q is left unchanged). For the sake of the simplicity, the value for $h_e = 4000$ J/(m^2·s·K) is given (the local heat transfer coefficient steam-side, which is higher than the value obtained in Eq. 5.63 for benzonitrile). As steam brings about much less fouling resistance than an organic liquid, assume $\sum \mathcal{R}_d = 10^{-4}$ m^2·s·K/J (i.e., 50% vs. the previous case).

To apply Eq. 5.5 (the energy balance in the case of a condensing pure fluid), we need to evaluate (Green and Perry 2008) the mass latent heat of condensation for steam. The approach has been presented in Eq. 1.48:

$$|\lambda_H| = C_1 (1 - T_r)^{C_2 + C_3 T_r + C_4 T_r^2} \tag{5.68}$$

(continued)

$$\begin{cases} C_1 = 5.2053 \cdot 10^7 \\ C_2 = 0.3199 \\ C_3 = -0.212 \\ C_4 = 0.25795 \end{cases} \tag{5.69}$$

Equation 5.68 gives $|\lambda_H|$ in [J/kmol], where $T_r = T/T_c$ is the non-dimensional reduced temperature. Putting values in [K], $T = T_{in} = T_{out} = 383.15$ K and the critical temperature for water is $T_c = 647.10$ K. Thus, it is $T_r = 0.59$ and $|\lambda_H| = 40{,}386{,}782$ J/kmol, easily converted into 2241.84 kJ/kg for water vapour. Well, we are now ready to use Eq. 5.5:

$$q = W_H |\lambda_H| \tag{5.70}$$

from which we get the flow rate of service fluid, $W_H = 239.57$ kg/h (please observe that this flow rate is far lower than the previous one, as the exchange by latent heat is generally more efficient than that by sensible heat, i.e., a heat of state change is far higher than the product between a specific heat and a difference in temperature).

Let us enter into the core of the problem. The extreme temperatures for the service fluid are different from the previous case, thus the logarithmic mean driving force is different:

$$\begin{cases} \Delta T_A = T_{in} - t_{out} = 30\,^\circ\mathrm{C} \\ \Delta T_B = T_{out} - t_{in} = 85\,^\circ\mathrm{C} \end{cases} \tag{5.71}$$

$$\Delta T_{lm} = \frac{\Delta T_A - \Delta T_B}{\ln\left(\frac{\Delta T_A}{\Delta T_B}\right)} = 52.81\ \mathrm{K} \tag{5.72}$$

The decrease in the driving force (cf. Eq. 5.47) acts against the efficiency of the exchanger. Since the properties of the process fluid have not changed, h_i is the same as the one calculated in Eq. 5.55, and:

$$\begin{cases} \dfrac{1}{h_i} = 8.34 \cdot 10^{-3}\ \dfrac{\mathrm{m^2 \cdot s \cdot K}}{\mathrm{J}} \\ \dfrac{r_i \ln\left(\dfrac{r_e}{r_i}\right)}{k_T} = 3.0 \cdot 10^{-4}\ \dfrac{\mathrm{m^2 \cdot s \cdot K}}{\mathrm{J}} \\ \dfrac{r_i}{r_e h_e} = 2.29 \cdot 10^{-4}\ \dfrac{\mathrm{m^2 \cdot s \cdot K}}{\mathrm{J}} \\ \sum \mathcal{R}_d = 10^{-4}\ \dfrac{\mathrm{m^2 \cdot s \cdot K}}{\mathrm{J}} \end{cases} \tag{5.73}$$

(continued)

$$U_i = \left[\frac{1}{h_i} + \frac{r_i \ln\left(\frac{r_e}{r_i}\right)}{k_T} + \frac{r_i}{r_e h_e} + \sum \mathcal{R}_d \right]^{-1} = 111.50 \ \frac{\mathrm{J}}{\mathrm{m}^2 \cdot \mathrm{s} \cdot \mathrm{K}} \tag{5.74}$$

The value for U_i slightly increases vs. the previous case, as fouling resistances decreased and the heat transfer coefficient for the service fluid increased.

The transfer area results:

$$A_i = \frac{q}{U_i \Delta T_{lm}} = 25.34 \ \mathrm{m}^2 \tag{5.75}$$

Namely:

$$L = \frac{A_i}{\pi d_i} = 52.40 \ \mathrm{m} \tag{5.76}$$

That is, we verify that the counteracting effects of the decrease in ΔT_{lm} and the increase in U_i determine an overall worsening of the exchanger performances. The previously designed exchanger is not suitable for the present application, while we may consider the possibility to increase its length (by ca. 32%) via additional hairpins placed in series.

5.8 Background Information

- Double pipe heat exchangers, schematised as in Fig. 5.2, typically operate with countercurrent flow to allow greater efficiency.
- Referring to heat exchange operated with sensible heat load, and therefore with temperature variation for both the process fluid and the service fluid, the energy balance equation, which defines the exchanger power, is in Eq. 5.4.
- The approach presented in Figs. 5.6 and 5.7 illustrate how to determine the optimal flow rate, and therefore the outlet temperature, for the service fluid.
- The scheme in Table 5.1 presents a synopsis of the operating procedure to design a heat exchanger, through the evaluation of the local and global energy transfer coefficients.

Table 5.1 Design of a heat exchanger (approach generally valid for fully developed turbulent flow for both fluids, either in countercurrent and equicurrent mode, for either purely sensible and purely latent heat flow, no matter which fluid is sent in the annulus and which in the internal tube): synopsis

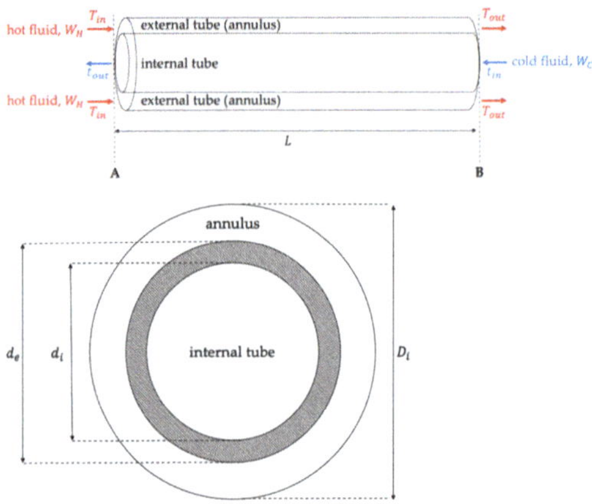	Two flow rates and four extreme temperatures known by the approach in Figs. 5.6 and 5.7 $q = W_C\mathbb{C}_C(t_{out} - t_{in}) = W_H\mathbb{C}_H(T_{in} - T_{out})$ *(or, examples in Eqs. 5.5 and 5.6 for latent heat case)* $\Delta T_{lm} = \dfrac{\Delta T_A - \Delta T_B}{\ln\left(\frac{\Delta T_A}{\Delta T_B}\right)}$ $D_{eq} = \dfrac{D_i^2 - d_e^2}{d_e}$ Fluids properties calculated at their respective average temperatures $T = (T_{in} + T_{out})/2$ and $t = (t_{in} + t_{out})/2$
Internal tube-side	Annulus-side
$Re_i = \dfrac{W_i d_i}{\frac{\pi}{4} d_i^2 \mu_i}$	$Re_e = \dfrac{W_e D_{eq}}{\frac{\pi}{4}\left(D_i^2 - d_e^2\right)\mu_e}$
$Pr_i = \dfrac{\mu_i \mathbb{C}_i}{k_{T,i}}$	$Pr_e = \dfrac{\mu_e \mathbb{C}_e}{k_{T,e}}$
$Nu = 0.023 Re^{0.8} Pr^{\alpha} \begin{cases} \alpha = 0.4 \text{ for cold fluids that heat up} \\ \alpha = 0.3 \text{ for hot fluids that cool down} \end{cases}$	
$h_i = \dfrac{k_{T,i} Nu_i}{d_i}$	$h_e = \dfrac{k_{T,e} Nu_e}{D_{eq}}$
$\begin{cases} U_i = \left[\frac{1}{h_i} + \frac{r_i \ln\left(\frac{r_e}{r_i}\right)}{k_T} + \frac{r_i}{r_e h_e} + \sum \mathcal{R}_d\right]^{-1} \Rightarrow \text{use in } q = U_i A_i \Delta T_{lm} \Rightarrow A_i = \pi d_i L = \dfrac{q}{U_i \Delta T_{lm}} \Rightarrow L \\ U_e = \left[\frac{r_e}{r_i h_i} + \frac{r_e \ln\left(\frac{r_e}{r_i}\right)}{k_T} + \frac{1}{h_e} + \sum \mathcal{R}_d\right]^{-1} \Rightarrow \text{use in } q = U_e A_e \Delta T_{lm} \Rightarrow A_e = \pi d_e L = \dfrac{q}{U_e \Delta T_{lm}} \Rightarrow L \end{cases}$	*(one of the two)*

References

Bird RB, Stewart WE, Lightfoot EN (2007) Transport phenomena. John Wiley & Sons, New York

Green DW, Perry RH (2008) Perry's chemical engineers' handbook. McGraw-Hill, New York

Kern DQ (1983) Process heat transfer. McGraw-Hill, Tokyo

McCabe WL, Smith JC, Harriott P (2001) Unit operations of chemical engineering. McGraw-Hill, Singapore

Evaporators

6

Abstract

This chapter deals with the operation of evaporation for concentrating solutions containing non-volatile solutes. After introducing the process objectives and main industrial applications, the effect of solute presence on the boiling temperature is analysed through the concept of boiling-point elevation. The design principles of the single-effect evaporator are then presented by means of mass and enthalpy balances, together with the definition of the steam economy. Finally, the discussion extends to multiple-effect systems.

Nomenclature

A_T	Energy transfer area [length^2]
E	Steam economy [−]
F	Mass flow rate of feed [mass/time]
F_s	Mass flow rate of steam at the heat exchanger [mass/time]
H_F	Specific enthalpy of feed [energy/mass]
H_L	Specific enthalpy of concentrated solution [energy/mass]
H_V	Specific enthalpy of vapour [energy/mass]
L	Mass flow rate of concentrated solution [mass/time]
m	Molar mass [mass/moles]
P_s	Steam pressure at the heat exchanger [mass/(length·time^2)]
P_w^0	Vapour pressure of pure water [mass/(length·time^2)]
p_w	Partial pressure of water [mass/(length·time^2)]
q	Power of the heat exchanger [energy/time]
T_F	Feed temperature [temperature]
T_L	Concentrated solution temperature [temperature]
T_V	Vapour temperature [temperature]
T_{bs}	Boiling point of the solution in the evaporator [temperature]

F. Montagnaro et al., *Elements of Unit Operations and Chemical Reactors*,
https://doi.org/10.1007/978-3-032-18046-9_6

T_{bw}	Boiling point of pure water [temperature]
T_s	Steam temperature at the heat exchanger [temperature]
U	Overall heat transfer coefficient [energy/(length^2·time·temperature)]
V	Mass flow rate of vapour [mass/time]
x	Mass fraction of solute in the concentrated solution [−]
x'_w	Mole fraction of water in the solution [−]
y	Mass fraction of solute in the vapour phase [−]
z_F	Mass fraction of solute in the feed solution [−]
γ_w	Activity coefficient of water [−]
$\lvert\lambda_V\rvert$	Latent heat of liquid-vapour phase transition [energy/mass]
$\lvert\lambda_s\rvert$	Latent heat of condensation of steam [energy/mass]
π	Pressure in the evaporator vapour space [mass/(length·time^2)]

6.1 Generalities

The objective of *evaporation* is to concentrate a solution consisting of one or more non-volatile solutes and a volatile solvent (commonly, water) by applying heat, generally provided by condensing steam. The process is carried out on a solution (*thin liquor*) by vaporising a portion of the solvent to produce a concentrated stream of dense liquid (*thick liquor*). Usually, the thick liquor is the valuable product, while the vapour is condensed and discarded.

The main industrial applications of evaporation include:

- concentration of solutions in the food, pharmaceutical, and chemical industries (e.g., transforming fruit juice into concentrate or reducing the volume of liquid extracts);
- recovery of products such as salts, sugars, or active ingredients in concentrated liquid solutions;
- reduction of wastewater volume in order to lower disposal costs and environmental impact;
- desalination of seawater for the production of drinking water;
- energy recovery; in some systems (such as black liquor evaporation in pulp and paper mills), the concentrated liquor can in fact be burned for energy and chemical recovery.

The process is carried out in heat exchange equipment defined *evaporators,* and the most widely adopted ones comprise those based on horizontal tube, short and long vertical tube, falling and agitated film, and forced circulation systems. The main factors ruling the choice of the most appropriate evaporator are liquid viscosity, heat sensitivity, occurring of foaming and scales formation and deposition. A detailed analysis of the different evaporators configurations and selection criteria can be found in specific texts (Coulson and Richardson 2023; Green and Perry 2008; McCabe et al. 2001; Seader and Henley 2006).

6.2 Boiling Point Elevation

The design of an evaporator requires the knowledge of the solution boiling point T_{bs}. In the presence of a solute in aqueous solution, the boiling temperature of the solution (at a fixed pressure) is higher than that of pure water, T_{bw}. This temperature difference is called the *boiling point elevation*. In the common case of a solute with negligible vapour pressure, the pressure in the vapour space of the evaporator, π, equals the partial pressure of H_2O in the solution, p_w. For non-ideal systems, the modified Raoult's law can be applied:

$$\boxed{\pi = p_w = \gamma_w P_w^0 x_w'} \qquad (6.1)$$

where γ_w and x_w' represent, respectively, the activity coefficient and the mole fraction of water in the solution ($\gamma_w \rightarrow 1$ for dilute solutions). P_w^0 is the vapour pressure of the pure water at the same temperature of the solution. Correlations and data for the evaluation of the activity coefficients can be found, e.g., in Green and Perry 2008.

Alternatively to Eq. 6.1, the boiling temperature of a solution can be determined using the *Dühring's rule* stating that T_{bs} linearly increases with the boiling point of pure water as the pressure rises (Dühring 1878). The linear relationship holds over a moderate pressure range. Its graphical representation is commonly referred to as *Dühring-line chart*, where lines are reported parametric of the solute concentration as depicted in Fig. 6.1. It is observed that the boiling point elevation (i.e., T_{bs}–T_{bw}) increases with the solute mass fraction in solution, z_F, at fixed pressure (i.e., at fixed T_{bw}, the higher z_F, the higher T_{bs}). Obviously, in the case of zero solute concentration it is $T_{bs} = T_{bw}$, and our line corresponds to the diagonal. Dühring-line charts for aqueous solutions of NaOH and NaCl solutes are reported in Coulson and Richardson 2023; McCabe et al. 2001. Nomographs for aqueous solutions of other solutes can be also adopted to evaluate the boiling point elevation (cf., e.g., Green and Perry 2008).

Example: Boiling Point Elevation of an Aqueous NaCl Solution

An aqueous solution containing a mass fraction of NaCl $z_F = 0.37$ has to be vaporised at pressure $\pi = 0.27$ bar. Firstly, we can evaluate the boiling point of pure water at the evaporation pressure either by steams tables (cf., e.g., Coulson and Richardson 2023) or via the Antoine's equation shown in a previous chapter (see Eq. 1.4). In the present case, we obtain $T_{bw}(\pi = 0.27 \text{ bar}) = 340$ K. From the Dühring-line chart of aqueous NaCl solutions (retrievable in Coulson and Richardson 2023), we read, at $z_F = 0.37$

(continued)

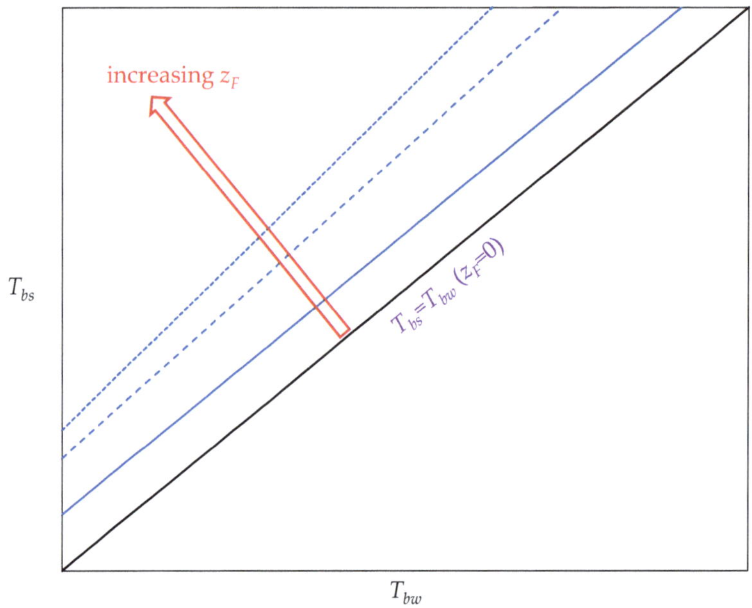

Fig. 6.1 Dühring-line chart (plots are parametric of the solute mass fraction in the solution, z_F)

(parametric line), the solution boiling point $T_{bs} = 348$ K (the abscissa corresponding to our T_{bw}). Thus, the boiling point elevation is 8 K.

The vapour pressure of pure water at 348 K is $P_w^0 = 0.38$ bar. Moreover, the datum $z_F = 0.37$ allows the direct calculation of the solvent (water) molar fraction in the solution:

$$x'_w = \frac{\frac{1-z_F}{m_{H_2O}}}{\frac{1-z_F}{m_{H_2O}} + \frac{z_F}{m_{NaCl}}} = 0.85 \tag{6.2}$$

where the molar mass m of water and NaCl is taken at 18 and 58.4 g/mol, respectively. This allows to compute the activity coefficient of water in solution via Eq. 6.1: $\gamma_w = \pi/(P_w^0 x'_w) = 0.84$.

6.3 Single-Effect Evaporator

The *single-effect evaporator* is based on a single stage of evaporation, where the obtained vapour is generally condensed and discarded. Figure 6.2 shows a schematic representation of a continuous equipment with relevant quantities:

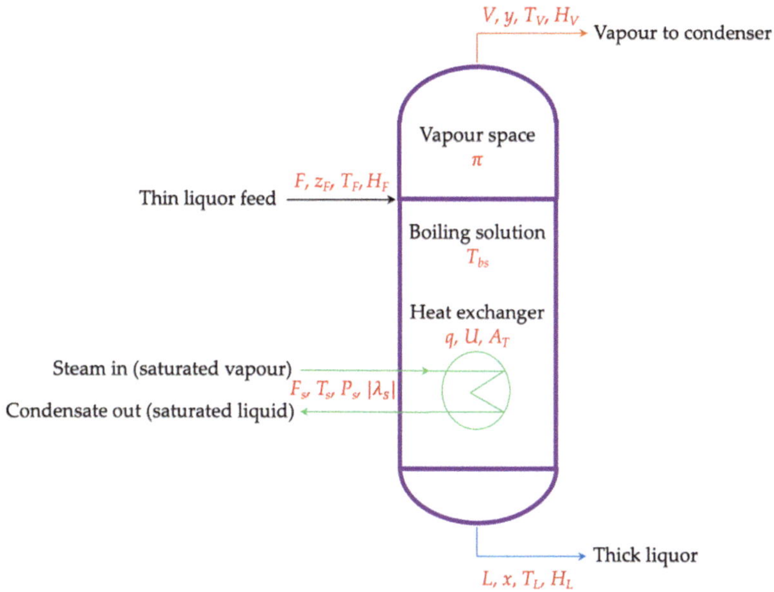

Fig. 6.2 Scheme of a single-effect evaporator

- feed (thin liquor) with mass flow rate F, solute mass fraction z_F, temperature T_F and specific enthalpy H_F;
- vapour with mass flow rate V, solute mass fraction y, temperature T_V and specific enthalpy H_V;
- concentrated solution (thick liquor) with mass flow rate L, solute mass fraction x, temperature T_L and specific enthalpy H_L;
- the heating medium (service fluid) in the heat exchanger is condensing steam with mass flow rate F_s, temperature T_s, pressure P_s, latent heat of condensation $|\lambda_s|$ (properties equal to those of the outlet condensate);
- in the heat exchanger, the thermal power q is transferred to the solution at temperature T_{bs} in the evaporator. It is characterised by energy transfer area A_T and overall heat transfer coefficient U (cf. the concepts in Chap. 5);
- the pressure in the vapour space is π.

The following hypotheses are commonly applied for the design of a single-effect evaporator (McCabe et al. 2001; Seader and Henley 2006):

- non-volatile solute ($y = 0$);
- the system is perfectly mixed. Therefore, it is $T_{bs} = T_L = T_V$;
- the driving force for heat transfer is T_s–T_{bs};
- negligible heat losses from the equipment.

The equations governing the system translate, respectively, the following aspects: global mass balance, Eq. 6.3; mass balance on solute, Eq. 6.4, for $y = 0$; the thermal energy entering the system via the heat exchanger, plus that related to the thin liquor feed, must equal the contributions associated to the two exiting streams, i.e. V and L, Eq. 6.5 written as [energy/time]; energy balance on the heat exchanger (see Eq. 5.5), Eq. 6.6; heat exchanger design equation (see Eq. 5.38), Eq. 6.7:

$$\boxed{F = L + V} \tag{6.3}$$

$$\boxed{F z_F = Lx} \tag{6.4}$$

$$\boxed{q = VH_V + LH_L - FH_F} \tag{6.5}$$

$$\boxed{q = F_s \mid \lambda_s \mid} \tag{6.6}$$

$$\boxed{q = UA_T(T_s - T_{bs})} \tag{6.7}$$

It is highlighted that the enthalpies of liquid solutions can be evaluated, e.g., from enthalpy-composition diagrams available in the literature (cf., e.g., Green and Perry 2008; McCabe et al. 2001). An important parameter quantifying the performance of an evaporator is the *steam economy* E defined as (the higher this parameter, the more efficient the system):

$$\boxed{E = \frac{V}{F_s}} \tag{6.8}$$

Example: Design of a Single-Effect Evaporator

An aqueous solution at $T_F = 316$ K, with mass flow rate $F = 1000$ kg/h and containing a mass fraction of NaOH $z_F = 0.3$, has to be concentrated at $x = 0.6$ in a single-effect evaporator operated at $\pi = 0.46$ bar. The heat exchanger uses condensing steam at $P_s = 4$ bar (we can assume $T_s = 417$ K) as service fluid, and it is characterised by $U = 2$ kJ/(m^2·s·K).

Determine: (i) the thick liquor, vapour and steam mass flow rates; (ii) the energy transfer area; (iii) the evaporator steam economy.

Firstly, we compute L and V. From Eq. 6.4, it is $L = 500$ kg/h, that substituted in Eq. 6.3 gives $V = 500$ kg/h.

The pure water boiling point at 0.46 bar is (from Antoine's Eq. 1.4) $T_{bw} = 353$ K. Thus, the solution boiling point at the same pressure can be derived from Dühring-line chart for the system H_2O-NaOH at 60% (by mass) concentration (McCabe et al. 2001) as $T_{bs} = 408$ K. The specific enthalpy of the superheated vapour V ($\pi = 0.46$ bar; $T_V = T_{bs} = 408$ K) is (from steam tables) $H_V = 2751.0$ kJ/kg; the specific enthalpy of feed F ($z_F = 0.3$;

(continued)

$T_F = 316$ K) and thick liquor L ($x = 0.6$; $T_L = T_{bs} = 408$ K) can be determined from the enthalpy-composition diagram of the system H_2O-NaOH (McCabe et al. 2001) as $H_F = 162.8$ kJ/kg and $H_L = 744.3$ kJ/kg.

Now, we get the power of the heat exchanger from Eq. 6.5 as $q = 1584.85 \cdot 10^3$ kJ/h; since $|\lambda_s| = 2132.9$ kJ/kg at $P_s = 4$ bar (from steam tables), we derive from Eq. 6.6 the steam flow rate as $F_S = 743.0$ kg/h. The energy transfer area is (Eq. 6.7) $A_T = 24.5$ m^2. Finally we derive the steam economy (Eq. 6.8) $E = V/F_s = 500/743 = 0.67$.

6.4 Multiple-Effect Evaporator

A single-effect evaporator, although simple from a design standpoint, uses steam inefficiently. In fact, about 1–1.3 kg of steam are required to evaporate 1 kg of water from the solution to be concentrated (McCabe et al. 2001). If the vapour generated in one evaporator is used as heat source in the heat exchanger of a second stage, a double-effect system is obtained. The heat of the initial steam is reused in the second effect, and the amount of water evaporated per unit mass of steam supplied to the first effect is almost doubled (higher energy efficiency). *Multi-effect evaporator* adopts different stages between the steam inlet and the vapour condenser, to increase the process steam economy.

Different configurations are available according to the feeding pattern, including *forward feed*, *backward feed*, *mixed feed* and *parallel feed* (details can be found, e.g., in Coulson and Richardson 2023; McCabe et al. 2001). As an example, Fig. 6.3 shows the most common forward feed scheme where both feed and fresh steam enter the first effect, the concentrated product exiting one stage is fed to the subsequent one and the pressure monotonically decreases from the first to the last effect.

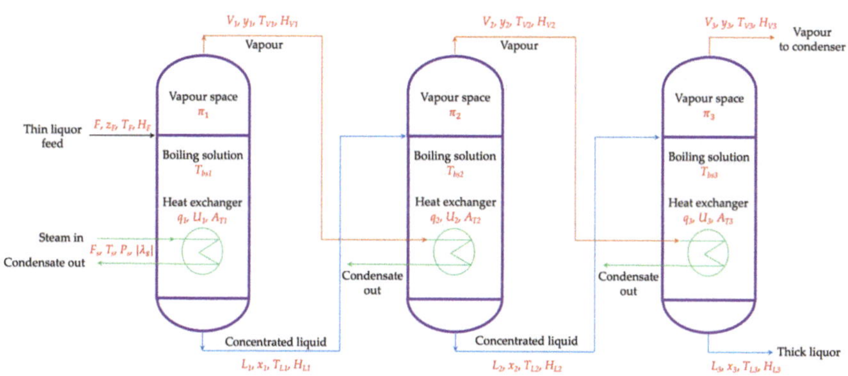

Fig. 6.3 Scheme of a triple-effect evaporator with forward feed

Example: Design of a Double-Effect Evaporator

An aqueous solution of a sugar, with mass flow rate $F = 5000$ kg/h and $z_F = 0.1$, has to be concentrated in a double-effect evaporator with forward feed, to obtain a thick liquor with a mass fraction of sugar $x_2 = 0.25$. The solution enters at boiling temperature of the first effect, and boiling point elevation effects are negligible. Assume, as latent heat of vaporisation in the first effect, $|\lambda_{V_1}| = 2300$ kJ/kg and equal to $|\lambda_s|$; for the second effect, it is $|\lambda_{V_2}| = 2400$ kJ/kg.

Determine: (i) the mass flow rate of the product thick liquor L_2; (ii) the mass flow rates of water evaporated from the solution (V_1 and V_2, respectively); (iii) the steam flow rate F_s and the steam economy E.

The non-volatile solute mass balance on the entire equipment allows to calculate the thick liquor mass flow rate as $L_2 = Fz_F/x_2 = 2000$ kg/h. Consequently, the global mass balance gives $V_1 + V_2 = F - L_2 = 3000$ kg/h.

The energy balance in the second effect is $V_1|\lambda_{V_1}| = V_2|\lambda_{V_2}|$, therefore $V_1/V_2 = 1.04$. Being $V_1 + V_2 = 3000$ kg/h, it is $V_1 = 1529.4$ kg/h and $V_2 = 1470.6$ kg/h. The energy balance for the first effect is $V_1|\lambda_{V_1}| = F_s|\lambda_s|$, from which we obtain $F_s = V_1 = 1529.4$ kg/h. Finally, the steam economy is $E = (V_1 + V_2)/F_s = 1.96$. The obtained value for E testifies the good efficiency of a double-effect evaporator.

6.5 Background Information

- The boiling-point elevation of liquid solutions is discussed through either the modified Raoult's law for non-ideal systems (Eq. 6.1) or the Dühring-line chart (Fig. 6.1).
- The single-effect evaporator is analysed via mass and energy balances (Eqs. 6.3–6.7), and defining the process steam economy as the vapour mass flow rate generated per unit mass flow rate of steam fed to the heat exchanger (Eq. 6.8).
- Multiple-effect configurations (Fig. 6.3) are presented, highlighting the increase in thermal efficiency through vapour reuse.

References

Coulson JM, Richardson JF (2023) Coulson and Richardson's chemical engineering: separation processes (vol. 2B). Butterworth-Heinemann, Oxford

Dühring E (1878) Neue Grundgesetze zur rationellen Physik und Chemie. Fues's Verlag (R. Reisland), Leipzig

Green DW, Perry RH (2008) Perry's chemical engineers' handbook. McGraw-Hill, New York

McCabe WL, Smith JC, Harriott P (2001) Unit operations of chemical engineering. McGraw-Hill, Singapore

Seader JD, Henley EJ (2006) Separation process principles. John Wiley & Sons, Hoboken

Mass Transport

7

Abstract

The trilogy on transport phenomena ends with the chapter dedicated to the mass transport. We will begin with molecular transfer, that is, diffusion, presenting Fick's law and the diffusion coefficient, with the associated Schmidt number that represents the ratio between the molecular transport of momentum and mass. We will then address the problem of transfer in the presence of convective motions, appropriately defining the transport coefficients and the Sherwood number, seen as the ratio between the total transport and the molecular transport of mass. In particular, our attention will focus on the transfer of mass between phases, in the hypotheses of validity of the film theory. We will identify the expressions for flux, resistances, local and global transfer coefficients.

Nomenclature

C_A	Concentration of A species [moles/volume]
C_{TOT}	Total concentration [moles/volume]
D	Characteristic dimension [length]
$\mathcal{D}$	Diffusion coefficient [length^2/time]
h	Convective heat transfer coefficient [energy/(length^2·time·temperature)]
h_e	Convective heat transfer coefficient at external side [energy/(length^2·time·temperature)]
h_i	Convective heat transfer coefficient at internal side [energy/(length^2·time·temperature)]
K_x	Liquid phase global mass transfer coefficient [moles/(length^2·time)]
K_y	Gas phase global mass transfer coefficient [moles/(length^2·time)]
k_T	Thermal conductivity [energy/(length·time·temperature)]
k_m	Convective mass transfer coefficient [length/time]
k_x	Liquid phase mass transfer coefficient [moles/(length^2·time)]

F. Montagnaro et al., *Elements of Unit Operations and Chemical Reactors*,
https://doi.org/10.1007/978-3-032-18046-9_7

k_y	Gas phase mass transfer coefficient [moles/(length^2·time)]
L	Length [length]
m	Henry's equilibrium constant [–]
N_A	Molar flow rate of A species [moles/time]
$\dot{N}_A$	Molar flux of A species [moles/(length^2·time)]
Nu	Nusselt number [–]
$\mathcal{P}$	Pressure [mass/(length·time^2)]
Pr	Prandtl number [–]
$\mathfrak{R}$	Gas constant [energy/(moles·temperature)]
r_e	External radial coordinate [length]
r_i	Internal radial coordinate [length]
Re	Reynolds number [–]
Sc	Schmidt number [–]
Sh	Sherwood number [–]
T	Temperature [temperature]
U_e	Overall heat transfer coefficient at external side [energy/(length^2·time·temperature)]
U_i	Overall heat transfer coefficient at internal side [energy/(length^2·time·temperature)]
x	Direction of transport [length]
x_A	Molar fraction of A in liquid phase [–]
x_A^i	Molar fraction of A in liquid phase at interface [–]
x_A^*	Virtual molar fraction of A in liquid phase [–]
y_A	Molar fraction of A in gas phase [–]
y_A^i	Molar fraction of A in gas phase at interface [–]
y_A^*	Virtual molar fraction of A in gas phase [–]
α_T	Thermal diffusivity [length^2/time]
$\lvert\Delta C_A\rvert$	Concentration driving force (absolute value) [moles/volume]
$\lvert\Delta x_A\rvert$	Driving force in A molar fraction (liquid phase, absolute value) [–]
$\lvert\Delta y_A\rvert$	Driving force in A molar fraction (gas phase, absolute value) [–]
μ	Dynamic viscosity [mass/(length·time)]
ν_k	Kinematic viscosity [length^2/time]
ρ	Density [mass/volume]

7.1 Generalities

Just as differences in velocity or temperature are the basis of, respectively, momentum or energy transfer phenomena, *mass transfer phenomena* are mostly caused by concentration gradients. For example, given a fluid solvent containing the chemical species A, and two regions of the fluid where a gradient in C_A (the concentration of the species A) occurs, this difference in concentration will represent the *driving force* of the mass transport from the region with higher concentration of A to the region

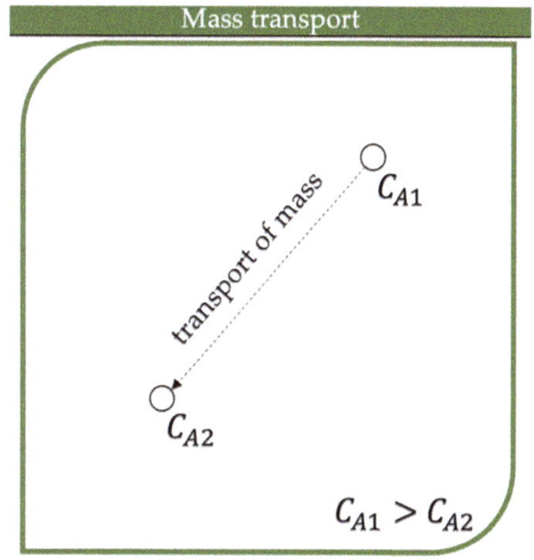

Fig. 7.1 Transport of mass: schematic concept

with lower concentration of A (Fig. 7.1). As done for the other two transport phenomena, also here we will refer to *stationary systems* of interest for the purposes of this textbook. Moreover, we will consider *binary mixtures* (e.g., a component A in a fluid solvent). Systems are taken as isothermal. Extensions on mass transport will be found, as usual, in devoted literature (Bird et al. 2007; McCabe et al. 2001).

7.2 Mass Diffusion

7.2.1 Definition

The motion of A we are referring to is named *diffusion*. Mass diffusion refers to the movement of a given component within a mixture. As we have just seen, it essentially occurs as a result of the existence of a concentration gradient, with mass diffusing from areas at higher concentration to areas at lower concentration. Diffusion can therefore be seen as the *molecular transport* of mass, although it should not be thought of as a phenomenon that occurs only in stagnant situations (similarly to what was said about its molecular analogues for momentum transport, i.e. laminar flow, and energy transport, i.e. conduction). In fact, even in the presence of eddies and turbulent flow, molecular diffusion is active inside the smaller vortices (just as laminar components are active, although usually neglected, in turbulent flow, and thermal conduction is present, although usually neglected, during heat convection).

7.2.2 The Fick's Law on Diffusion

Recall again the general law of molecular transport presented in Eq. 3.1:

$$\boxed{\dot{\mathbb{F}} = -\mathbb{K}\frac{\mathrm{d}\mathbb{Q}}{\mathrm{d}x}} \tag{7.1}$$

Here, once defined the molar flow rate of A as N_A [moles/time] (of course, the molar flow rate can be easily transformed into a mass flow rate, if needed), the molar flux of A, $\dot{N}_A$ [moles/(length^2·time)], takes on the role of $\dot{\mathbb{F}}$. The quantity that determines the transport phenomenon is, Fig. 7.1, $\mathbb{Q} = C_A$, while x is the generic direction of transport (as usual). Let us remind again that the minus sign in Eq. 7.1 indicates that the flux is positive in the opposite direction to that for which the gradient is positive. We can therefore write the *Fick's law on diffusion* [named after Adolf Eugen Fick (1829–1901), German physiologist, Professor at University of Würzburg; see Fick 1855]:

$$\boxed{\dot{N}_A = -\mathcal{D}\frac{\mathrm{d}C_A}{\mathrm{d}x}} \tag{7.2}$$

In other words, the molar flux by diffusion is proportional to the gradient in A concentration, via a constant $\mathbb{K} = \mathcal{D}$, the *diffusion coefficient*, that depends on the nature of the diffusing species and of the solvent in which it diffuses (regardless the fluid dynamics of the system, as for the viscosity/thermal conductivity). The higher the diffusion coefficient, the higher the molar flux exchanged by diffusion. At this point, the analogy between the Newton's law governing laminar momentum transport (Eq. 3.4), the Fourier's law governing molecular energy transport (Eq. 4.2), and the Fick's law governing diffusion mass transport (Eq. 7.2) has been completely unveiled. All of these obey the general law of molecular transport, as illustrated in Fig. 7.2.

7.2.3 Diffusion Coefficient

To find the dimensions for $\mathcal{D}$ from the equation that defines it (Eq. 7.2), we observe that:

$$\mathcal{D} = -\frac{\dot{N}_A}{\frac{\mathrm{d}C_A}{\mathrm{d}x}} = \frac{\left[\frac{\text{moles}}{\text{length}^2 \cdot \text{time}}\right]}{\left[\frac{\text{moles}}{\text{length}^4}\right]} = \left[\frac{\text{length}^2}{\text{time}}\right] \tag{7.3}$$

As an alternative interpretation, let us look at the dimensions just obtained for diffusivity (the diffusion coefficient). It can be seen as a "molecular velocity". In the convective case, a fluid element with its own velocity [length/time] will take, to travel a distance L, a time equal to the ratio between L and its velocity. In the diffusive case, however, the concept of velocity is replaced by that of diffusivity. Here, *the motion is stochastic*, and the fluid element is seen as a *random walker* that, to travel the distance L, will actually take a distance having order of magnitude L^2

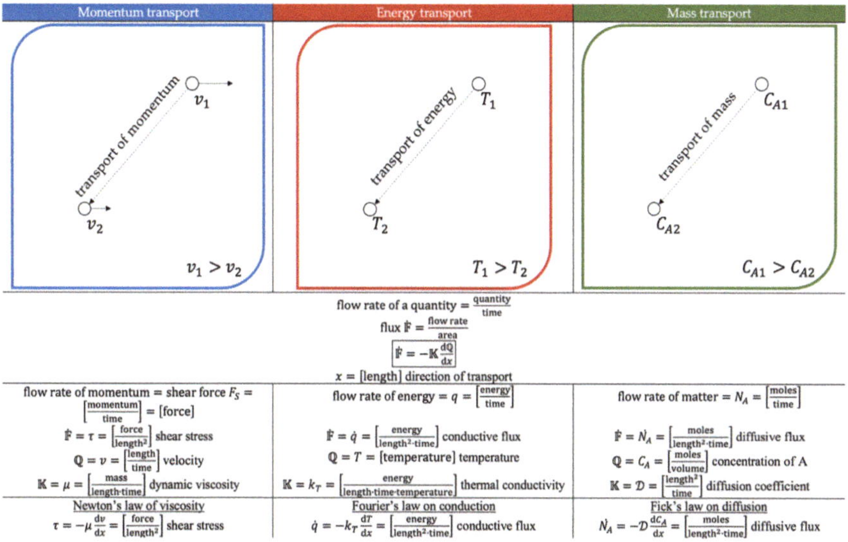

Fig. 7.2 An abecedary of transport phenomena analogies—Part B

(since the motion is random). Therefore, the molecular velocity will be expressed in [length^2/time], i.e. the diffusion coefficient, and the time necessary to travel the distance L will result from the ratio between L^2 and diffusivity.

Diffusivity depends on temperature (increasingly, according to, e.g., power laws) and pressure. The diffusivity of a gas in another gas can be on the order of 10^{-5} m^2/s, but in liquids it can drop to values of 10^{-9} m^2/s; even lower is the diffusivity in the solid state. While the constant $\mathbb{K}$ in the case of the transport of momentum (dynamic viscosity μ, Eq. 3.10) and energy (thermal conductivity k_T, Eq. 4.4) could be expressed by means of general relations present in the literature, it is not possible to give, for the case of the mass transport, analogous relations for $\mathcal{D}$, whose values will in fact depend not only on the species that diffuses (the compound A), but also on the nature of the substance (e.g., the solvent) in which the species diffuses. Tables with reference data can be found in the literature (Green and Perry 2008). For further details on diffusion, it is also possible to refer to other works of ours (Balsamo and Montagnaro 2015; Montagnaro 2024a), apart from the extensive theory on the diffusion coefficient offered in specialised textbooks (Bird et al. 2007; McCabe et al. 2001).

7.2.4 Schmidt Number

In Fig. 7.3 it is highlighted how the mass diffusivity $\mathcal{D}$ is (see the dimensions) the analogue of the kinematic viscosity (the momentum diffusivity ν_k) and of the thermal diffusivity α_T. While, in Eq. 4.6, the ratio between the momentum diffusivity and the thermal diffusivity allowed us to define the number of Prandtl, similarly, here, the

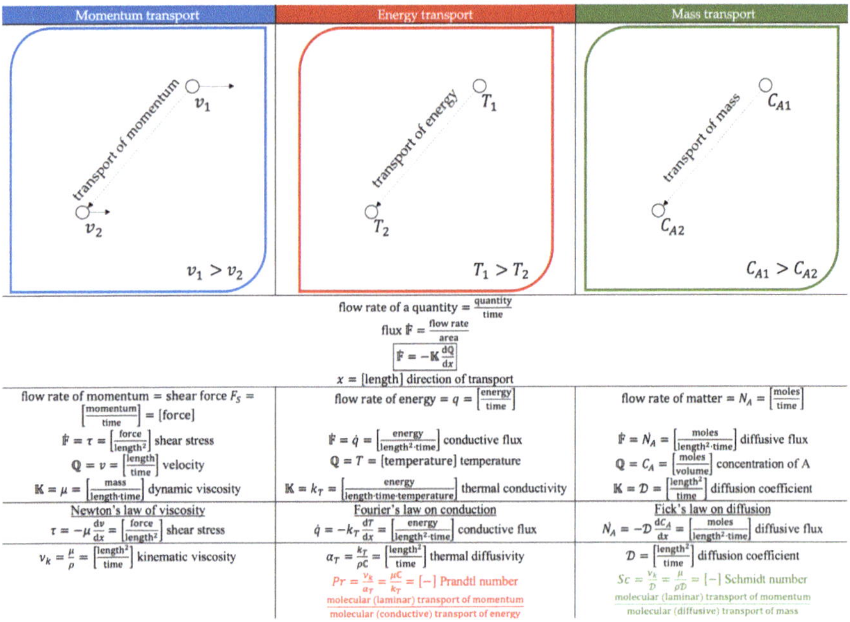

Fig. 7.3 An abecedary of transport phenomena analogies—Part C

ratio between the momentum diffusivity and the mass diffusivity leads us to the definition of the analogue ("mass side") of *Pr*, namely the Schmidt number [named after Ernst Heinrich Wilhelm Schmidt (1892–1975), German physicist, Professor at Technical University of Munich; see Schmidt 1926]:

$$\boxed{Sc = \frac{\text{molecular (laminar) transport of momentum}}{\text{molecular (diffusive) transport of mass}} = \frac{\nu_k}{\mathcal{D}} = \frac{\mu}{\rho\mathcal{D}}} \tag{7.4}$$

As like as for ν_k and $\mathcal{D}$ (or, equivalently, for μ, ρ and $\mathcal{D}$), also *Sc* is a sole function of the phase we are considering (of course, at given conditions). The fluid dynamics does not influence *Sc*, that represents the *ratio between the transport of momentum and the transport of mass, both following molecular (laminar/diffusive) mechanisms*. For diffusion in the gas phase, for example, $Sc = 0.5$ approximately (a typical order of magnitude), but in condensed phases *Sc* can become much higher (much lower diffusivity): in liquid phases, *Sc* can vary e.g. between 10^2 and 10^5.

7.3 Mass Transfer by Convection

7.3.1 Mass Transfer Coefficients

In the case of fluids animated by turbulent motion, in the presence of a driving force in terms of concentration of species A ($|\Delta C_A|$), we will write the convective mass flux (of A, from higher to lower concentration regions) in full analogy to what was done in the case of energy transfer (Eq. 4.21):

$$\dot{N}_A = k_m |\Delta C_A| \tag{7.5}$$

We then define the analogue of h (the convective energy transfer coefficient), namely k_m, the *convective mass transfer coefficient.* The dimensional analysis of Eq. 7.5 returns k_m in [length/time]. Please observe that, as like as h has dimensions of [thermal conductivity/length], k_m has dimensions of [diffusivity/length], being the mass diffusivity the analogous of the thermal conductivity.

It is often useful to adopt, for the driving force, the molar fraction of the species involved in the transfer, rather than its concentration. Start defining y_A or x_A, respectively, the molar fraction of A if in gas or liquid phase. If C_{TOT} is the total concentration of the phase, it is:

$$\begin{cases} |\Delta C_A| = C_{TOT}|\Delta y_A| = \dfrac{\mathcal{P}}{\Re T}|\Delta y_A| & \text{gas phase} \\ |\Delta C_A| = C_{TOT}|\Delta x_A| & \text{liquid phase} \end{cases} \tag{7.6}$$

For the gaseous phase, in particular, we have equivalently expressed the total gas concentration as a function of pressure $\mathcal{P}$ and temperature T, according to the ideal gas law. Therefore, the convective mass flux reads:

$$\begin{cases} \dot{N}_A = k_m C_{TOT}|\Delta y_A| = k_m \dfrac{\mathcal{P}}{\Re T}|\Delta y_A| & \text{gas phase} \\ \dot{N}_A = k_m C_{TOT}|\Delta x_A| & \text{liquid phase} \end{cases} \tag{7.7}$$

Grouping constant parameters, we can define a mass transfer coefficient for the gas phase and another for the liquid phase, respectively:

$$\begin{cases} k_y = k_m C_{TOT} = k_m \dfrac{\mathcal{P}}{\Re T} & \text{gas phase} \\ k_x = k_m C_{TOT} & \text{liquid phase} \end{cases} \tag{7.8}$$

Both k_y and k_x are expressed as the product between k_m and a concentration, thus they have the same dimensions of a molar flux [moles/(length^2·time)]. Concluding (for more insights into mass transfer coefficients, the reader could refer to another our work, Inglezakis et al. 2020), we can write the molar flux, elegantly, as:

$$\begin{cases} \dot{N}_A = k_y|\Delta y_A| & \text{gas phase} \\ \dot{N}_A = k_x|\Delta x_A| & \text{liquid phase} \end{cases} \tag{7.9}$$

7.3.2 Sherwood Number

Now we illustrate how the convective mass transfer coefficient, k_m, can be determined. We will develop an analogous discussion as the one that led us to the definition of the Nusselt number Nu (cf. the analogies in Fig. 7.4). First, we can express the ratio between the *total flux* in Eq. 7.5 (again, the presence of convective motion does not exclude the less relevant molecular-diffusive phenomena within the fluid), and the *diffusive flux* that, by reworking Fick's Eq. 7.2 and expressing it in absolute value, can be approximated to $\mathcal{D}|\Delta C_A|/D$, with D being the characteristic dimension of the system. This ratio, naturally dimensionless, is called the *Sherwood number* [named after Thomas Kilgore Sherwood (1903–1976), U.-S. chemical engineer, Professor at Massachusetts Institute of Technology; see Sherwood et al. 1975]:

Momentum transport	Energy transport	Mass transport

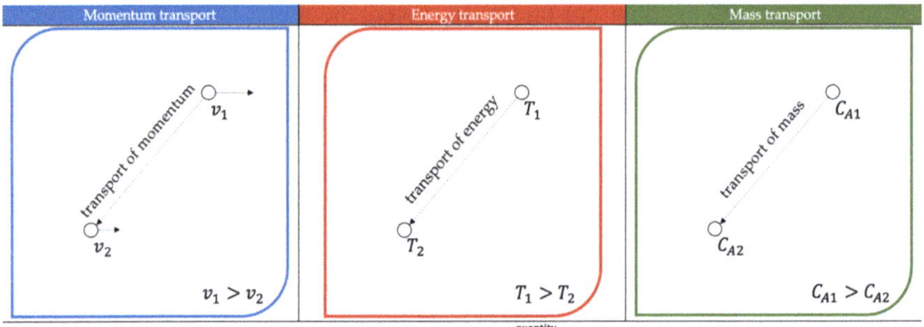

Momentum transport	Energy transport	Mass transport
	flow rate of a quantity = $\frac{\text{quantity}}{\text{time}}$ flux $\dot{\mathbb{F}} = \frac{\text{flow rate}}{\text{area}}$ $\boxed{\dot{\mathbb{F}} = -\mathbb{K}\frac{d\mathbb{Q}}{dx}}$ x = [length] direction of transport	
flow rate of momentum = shear force $F_S = \left[\frac{\text{momentum}}{\text{time}}\right]$ = [force] $\dot{\mathbb{F}} = \tau = \left[\frac{\text{force}}{\text{length}^2}\right]$ shear stress $\mathbb{Q} = v = \left[\frac{\text{length}}{\text{time}}\right]$ velocity $\mathbb{K} = \mu = \left[\frac{\text{mass}}{\text{length}\cdot\text{time}}\right]$ dynamic viscosity	flow rate of energy = $q = \left[\frac{\text{energy}}{\text{time}}\right]$ $\dot{\mathbb{F}} = \dot{q} = \left[\frac{\text{energy}}{\text{length}^2\cdot\text{time}}\right]$ conductive flux $\mathbb{Q} = T$ = [temperature] temperature $\mathbb{K} = k_T = \left[\frac{\text{energy}}{\text{length}\cdot\text{time}\cdot\text{temperature}}\right]$ thermal conductivity	flow rate of matter = $N_A = \left[\frac{\text{moles}}{\text{time}}\right]$ $\dot{\mathbb{F}} = \dot{N}_A = \left[\frac{\text{moles}}{\text{length}^2\cdot\text{time}}\right]$ diffusive flux $\mathbb{Q} = C_A = \left[\frac{\text{moles}}{\text{volume}}\right]$ concentration of A $\mathbb{K} = \mathcal{D} = \left[\frac{\text{length}^2}{\text{time}}\right]$ diffusion coefficient
Newton's law of viscosity $\tau = -\mu\frac{dv}{dx} = \left[\frac{\text{force}}{\text{length}^2}\right]$ shear stress	Fourier's law on conduction $\dot{q} = -k_T\frac{dT}{dx} = \left[\frac{\text{energy}}{\text{length}^2\cdot\text{time}}\right]$ conductive flux	Fick's law on diffusion $\dot{N}_A = -\mathcal{D}\frac{dC_A}{dx} = \left[\frac{\text{moles}}{\text{length}^2\cdot\text{time}}\right]$ diffusive flux
$\nu_k = \frac{\mu}{\rho} = \left[\frac{\text{length}^2}{\text{time}}\right]$ kinematic viscosity	$\alpha_T = \frac{k_T}{\rho C} = \left[\frac{\text{length}^2}{\text{time}}\right]$ thermal diffusivity $Pr = \frac{\nu_k}{\alpha_T} = \frac{\mu C}{k_T} = [-]$ Prandtl number $\frac{\text{molecular (laminar) transport of momentum}}{\text{molecular (conductive) transport of energy}}$	$\mathcal{D} = \left[\frac{\text{length}^2}{\text{time}}\right]$ diffusion coefficient $Sc = \frac{\nu_k}{\mathcal{D}} = \frac{\mu}{\rho\mathcal{D}} = [-]$ Schmidt number $\frac{\text{molecular (laminar) transport of momentum}}{\text{molecular (diffusive) transport of mass}}$
	$\dot{q} = h\lvert\Delta T\rvert = \left[\frac{\text{energy}}{\text{length}^2\cdot\text{time}}\right]$ convective flux $h = \left[\frac{\text{energy}}{\text{length}^2\cdot\text{time}\cdot\text{temperature}}\right]$ convective heat transfer coefficient	$\dot{N}_A = k_m\lvert\Delta C_A\rvert = \left[\frac{\text{moles}}{\text{length}^2\cdot\text{time}}\right]$ convective flux $k_m = \left[\frac{\text{length}}{\text{time}}\right]$ convective mass transfer coefficient
	$Nu = \frac{hD}{k_T} = [-]$ Nusselt number $\frac{\text{total transport of energy}}{\text{molecular (conductive) transport of energy}}$ e.g., $Nu(Re, Pr)$	$Sh = \frac{k_m D}{\mathcal{D}} = [-]$ Sherwood number $\frac{\text{total transport of mass}}{\text{molecular (diffusive) transport of mass}}$ e.g., $Sh(Re, Sc)$
	$Re = \frac{Dv\rho}{\mu} = [-]$ Reynolds number $\frac{\text{turbulent transport of momentum}}{\text{molecular (laminar) transport of momentum}}$	

Fig. 7.4 An abecedary of transport phenomena analogies—Part D

$$\boxed{Sh = \frac{\text{total transport of mass}}{\text{molecular (diffusive) transport of mass}} = \frac{k_m|\Delta C_A|}{\frac{\mathcal{D}|\Delta C_A|}{D}} = \frac{k_m D}{\mathcal{D}}} \tag{7.10}$$

In literature (e.g., our review paper on the subject, Montagnaro 2024b) we can find relations that link *Sh* to other dimensionless numbers capable of characterising the properties of the fluid (e.g., *Sc*) and the thermo-fluid dynamic conditions of the system (e.g., *Re*). For example, we have proposed an almost general relationship having appearance:

$$Sh = \alpha + \beta Sc^{\sigma} Re^{\lambda} \tag{7.11}$$

with α, β, σ and λ proper constants depending on the system. Leaving all the details to the analysis of the literature, here we report a chart (Fig. 7.5) where, once *Re* is

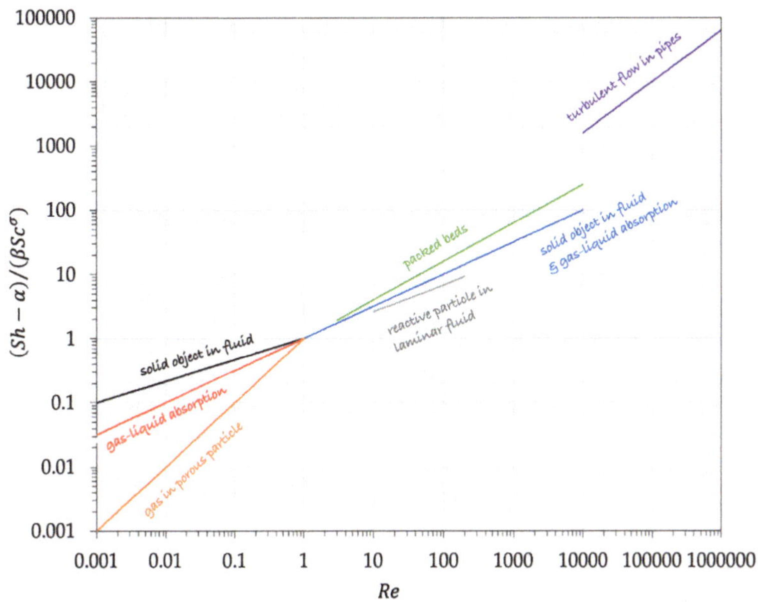

Study case	α	β	σ	D
Solid object in fluid	2	0.991	1/3	Sphere diameter
Solid object in fluid	2	0.6	1/3	Sphere diameter
	2	0.69	1/3	Sphere diameter
	0	0.61	1/3	Cylinder diameter
Reactive particle in laminar fluid (under mass transfer control)	0.922	1	1/3	Particle diameter
Turbulent flow mass transfer to pipe walls	0	0.023	1/3	Pipe diameter
Packed beds	0	1.1	1/3	Particle diameter
Gas-liquid absorption (gas bubble, liquid in Stokes flow)	2	0.6415	0.5	Bubble diameter
Gas-liquid absorption (in a liquid falling film)	0	1.128	0.5	Film length
Gas-liquid absorption (liquid drops falling in gas)	2	0.6	1/3	Drop diameter
Trapped gas in porous particle	0	0.084	0.728	Particle diameter

Fig. 7.5 Synoptical chart (valid for selected study cases) for the evaluation of the Sherwood number according to Eq. 7.11 (adapted from Montagnaro, 2024b). The accompanying table gives the required details (constants and characteristic dimension D to be used in *Re* and *Sh*; the colour of the text in the table corresponds to the colour of the straight line on the log-log plot for which the equation is valid in the indicated range of Reynolds number)

known, it is possible to obtain the ordinate (the group $(Sh–\alpha)/(\beta Sc^{\sigma})$) and, known Sc, one can calculate Sh by application of the constants as reported in Fig. 7.5 for selected study cases. Being these cases different each other, the chart also indicates what is the characteristic dimension (D) to be used in Re and Sh. Once Sh is known, we will obtain, as desired:

$$k_m = \frac{\mathcal{D}}{D} Sh \tag{7.12}$$

and, finally, k_y and k_x from Eq. 7.8.

7.4 Convective Mass Transfer Between Phases

7.4.1 General Presentation

As a study case of interest for the purposes of this book, we will refer specifically to the *convective transport of mass between different phases*. Let us consider two phases (a gaseous one and a liquid one), in both of which the compound A is dissolved in a solvent. The two phases are separated by a solid interface, as will be detailed in the next chapter dedicated to gas-liquid absorption. For the moment, let us point out that (Fig. 7.6):

- in the turbulent bulk of the gaseous phase, we express the concentration of A in terms of its molar fraction, y_A;
- in the turbulent bulk of the liquid phase, we will have A present with molar fraction x_A;
- according to the "two-film" theory (Whitman 1923), y_A and x_A remain constant in the bulk of the gaseous and liquid phases, respectively, while the resistances to the mass transfer are concentrated in the two films (gas-side and liquid-side boundary layers) near the inert solid separation interface (which represents the place where the mass transfer occurs);
- at the interface, the molar fraction of A in the gas takes the value y_A^i and the molar fraction of A in the liquid takes the value x_A^i. *These two values are in equilibrium* each other;
- we are in a condition in which the transfer of A from the gaseous phase to the liquid phase is favoured (absorption of A), and not vice versa (more details will be provided in the next chapter), while both the absorption of the gaseous solvent in the liquid and the desorption of the liquid solvent in the gas are neglected (the two solvents, therefore, can be considered inert).

The reader will have noticed the analogies of the scheme in Fig. 7.6 with, for example, the case of heat exchange through a wall (temperature is on average constant within the bulk of the two fluids, and variable only in the two films, see

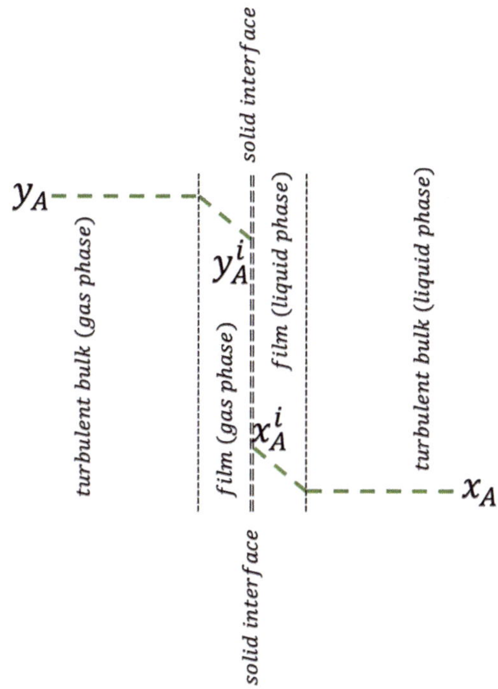

Fig. 7.6 Two-film theory for mass transfer of A from a gaseous to a liquid phase. For both phases, the average bulk molar fraction of A is constant. Convective resistances are concentrated in the two films

e.g. Figs. 4.6 and 4.7). Clearly, the separation between the boundary layer (film) and the turbulent core, for each of the two phases, is not clear-cut, but it is rendered as in Fig. 7.6 for simplicity of exposition.

7.4.2 Main Equations

The equilibrium relationship holding at the interface, expressed in terms of molar fractions and in the hypothesis of compound A *sufficiently diluted in both the gaseous and liquid phases*, can be written according to the Henry's law:

$$y_A^i = m x_A^i \tag{7.13}$$

where m is the dimensionless equilibrium constant, known from the literature. Considering that, in this case, the local driving force on the gas side and the liquid side is, respectively (Fig. 7.6):

$$\begin{cases} |\Delta y_A| = y_A - y_A^i & \text{gas phase} \\ |\Delta x_A| = x_A^i - x_A & \text{liquid phase} \end{cases} \tag{7.14}$$

the A molar flux, see Eq. 7.9 (constant for a stationary system), will be expressed respectively as:

$$\begin{cases} \dot{N}_A = k_y\left(y_A - y_A^i\right) & \text{gas phase} \\ \dot{N}_A = k_x\left(x_A^i - x_A\right) & \text{liquid phase} \end{cases} \tag{7.15}$$

7.4.3 Local and Global Mass Transfer Coefficients

The mass transfer coefficients in Eq. 7.15 can be seen as *local*. In fact, k_y refers to a local gas-side driving force (y_A–y_A^i), and k_x to a local liquid-side driving force (x_A^i–x_A). However, we can also refer to *global* driving forces. To do this, we need to find expressions for the global driving force which, see Fig. 7.6, cannot be simply y_A–x_A, since it is formally incorrect (we cannot subtract a molar fraction in the gas phase with another one in the liquid phase). The problem can be circumvented as follows. Assume:

- $y_A^* = mx_A$ as the *virtual* molar fraction of A in the gas phase that would stay in equilibrium with x_A (in the field of validity of the Henry's law). This allows us to define a global *virtual* driving force (as it does not, actually, exist) in terms of y_A–y_A^*; the coefficient that multiplies this driving force will be a global (overall) gas-side mass transfer coefficient, which we will denote by K_y;
- $x_A^* = y_A/m$ as the *virtual* molar fraction of A in the liquid phase that would stay in equilibrium with y_A (in the field of validity of the Henry's law). This allows us to define a global *virtual* driving force (as it does not, actually, exist) in terms of x_A^*–x_A; the coefficient that multiplies this driving force will be a global (overall) liquid-side mass transfer coefficient, which we will denote by K_x.

At this point, we obtain the following expressions for the A molar flux:

$$\begin{cases} \dot{N}_A = K_y\left(y_A - y_A^*\right) & \text{gas phase} \\ \dot{N}_A = K_x\left(x_A^* - x_A\right) & \text{liquid phase} \end{cases} \tag{7.16}$$

7.4.4 Relation Between Local and Global Mass Transfer Coefficients

Aim is to see how the overall mass transfer coefficients (K_y and K_x) can be calculated starting from the local mass transfer coefficients (k_y and k_x) which, in turn, depend on k_m (Eq. 7.8), the mass transfer coefficient obtained from the Sherwood number (e.g., by Eq. 7.12). Start with the gas phase. Isolating y_A^i from Eq. 7.15 (gas phase), it is:

$$y_A^i = y_A - \frac{\dot{N}_A}{k_y} \tag{7.17}$$

At the same time, Eq. 7.15 (liquid phase) can be rewritten in terms of y_A^i by the equilibrium Eq. 7.13:

$$\dot{N}_A = k_x\left(\frac{y_A^i}{m} - x_A\right) \tag{7.18}$$

Isolating y_A^i, it is:

$$y_A^i = \frac{m\dot{N}_A}{k_x} + mx_A = \frac{m\dot{N}_A}{k_x} + y_A^* \tag{7.19}$$

By comparison of Eqs. 7.17 and 7.19, we end up into a relation independent of y_A^i (the value at interface, which is not easy to verify, while the bulk values for the molar fractions are much more easily obtainable):

$$y_A - \frac{\dot{N}_A}{k_y} = \frac{m\dot{N}_A}{k_x} + y_A^* \tag{7.20}$$

Isolating the molar flux:

$$\dot{N}_A = \frac{y_A - y_A^*}{\frac{1}{k_y} + \frac{m}{k_x}} \tag{7.21}$$

By comparison of Eq. 7.21 with Eq. 7.16 (gas phase), we obtain the desired expression for the overall gas phase mass transfer coefficient as a function of the two local mass transfer coefficients:

$$K_y = \left(\frac{1}{k_y} + \frac{m}{k_x}\right)^{-1} \tag{7.22}$$

Symmetrically, for the liquid phase. Isolating x_A^i from Eq. 7.14 (liquid phase), it is:

$$x_A^i = x_A + \frac{\dot{N}_A}{k_x} \tag{7.23}$$

At the same time, Eq. 7.14 (gas phase) can be rewritten in terms of x_A^i by the equilibrium Eq. 7.12:

$$\dot{N}_A = k_y\left(y_A - mx_A^i\right) \tag{7.24}$$

Isolating x_A^i, it is:

$$x_A^i = -\frac{\dot{N}_A}{mk_y} + \frac{y_A}{m} = -\frac{\dot{N}_A}{mk_y} + x_A^* \tag{7.25}$$

By comparison of Eqs. 7.22 and 7.24, we end up into a relation independent of x_A^i (a value at interface difficult to obtain):

$$x_A + \frac{\dot{N}_A}{k_x} = -\frac{\dot{N}_A}{mk_y} + x_A^* \tag{7.26}$$

Isolating the molar flux:

$$\dot{N}_A = \frac{x_A^* - x_A}{\frac{1}{mk_y} + \frac{1}{k_x}} \tag{7.27}$$

By comparison of Eq. 7.26 with Eq. 7.15 (liquid phase), we obtain the desired expression for the overall liquid phase mass transfer coefficient as a function of the two local mass transfer coefficients:

$$K_x = \left(\frac{1}{mk_y} + \frac{1}{k_x} \right)^{-1} \tag{7.28}$$

7.4.5 Resistances and Overall Transfer Coefficients

When dealing with the transport of energy, we have presented Eq. 4.12 that, due to the strict analogies between transport phenomena, is valid also in this case (by substituting "heat" with "mass"):

$$\boxed{\begin{aligned} \text{flux} &= \frac{\text{driving force}}{\text{resistance}} \\ \text{mass transfer coefficient} &= \frac{1}{\text{resistance}} \\ \text{flux} &= \text{mass transfer coefficient} \times \text{driving force} \end{aligned}} \tag{7.29}$$

We keep referring, as a study case, to the two-film theory of Fig. 7.6 that, as observed, is an analogous of the heat transfer from a hot, to a cold, fluid through a wall. Considering for example the two relations in Eq. 7.16, we highlight that the flux is the product between a (overall) mass transfer coefficient and a (overall) driving force. Therefore, Eq. 7.29, the two global resistances are, from Eqs. 7.22 and 7.28, respectively:

$$\boxed{\begin{cases} \dfrac{1}{K_y} = \dfrac{1}{k_y} + \dfrac{m}{k_x} \\ \dfrac{1}{K_x} = \dfrac{1}{mk_y} + \dfrac{1}{k_x} \end{cases}} \tag{7.30}$$

As like as for the reference case in energy transport, also in this case we have *processes in series*, and the overall resistance results from *the sum of the individual resistances*. For the first relation in Eq. 7.30, the overall mass transfer resistance (referred to gas), $1/K_y$, is the sum of the (two) individual resistances in series during the transfer of A from the gas to the liquid phase (Fig. 7.6), i.e. the local resistance

Gas side	Liquid side
$k_m = \left[\frac{\text{length}}{\text{time}}\right]$ in $N_A = k_m \lvert \Delta C_A \rvert = \left[\frac{\text{moles}}{\text{length}^2 \cdot \text{time}}\right]$	
Local mass transfer coefficients	
$k_y = k_m C_{TOT} = k_m \frac{\mathcal{P}}{\Re T} = \left[\frac{\text{moles}}{\text{length}^2 \cdot \text{time}}\right]$ in $N_A = k_y (y_A - y_A^i)$	$k_x = k_m C_{TOT} = \left[\frac{\text{moles}}{\text{length}^2 \cdot \text{time}}\right]$ in $N_A = k_x (x_A^i - x_A)$
Global mass transfer coefficients	
$K_y = \left[\frac{\text{moles}}{\text{length}^2 \cdot \text{time}}\right]$ in $N_A = K_y (y_A - y_A^*)$ where $y_A^* = m x_A$	$K_x = \left[\frac{\text{moles}}{\text{length}^2 \cdot \text{time}}\right]$ in $N_A = K_x (x_A^* - x_A)$ where $x_A^* = y_A / m$
Relations between global and local mass transfer coefficients	
$\frac{1}{K_y} = \frac{1}{k_y} + \frac{m}{k_x}$	$\frac{1}{K_x} = \frac{1}{m k_y} + \frac{1}{k_x}$

Fig. 7.7 Synoptic illustration for local and global mass transfer coefficients, study case of the two-film theory in Fig. 7.6

referred to gas ($1/k_y$) and the local resistance referred to liquid ($1/k_x$ corrected by multiplying per m). For the second relation in Eq. 7.30, the overall mass transfer resistance (referred to liquid), $1/K_x$, is the sum of the (two) individual resistances in series during the transfer of A from the gas to the liquid phase (Fig. 7.6), i.e. the local resistance referred to gas ($1/k_y$ corrected by dividing per m) and the local resistance referred to liquid ($1/k_x$).

In other words, Eqs. 7.22 and 7.28 (or, equivalently, Eq. 7.30) can be seen as the analogous of the two relations in Eq. 4.49 for heat transfer, noting in particular that the correction parameter here is m, while, in the energy case we presented, it is r_i/r_e. As like as, in the energy case, for the overall internal transfer coefficient (U_i), the correction parameter is applied to the resistance on the external side, and for the overall external transfer coefficient (U_e) the correction parameter is applied to the resistance on the internal side, in our case, for the overall gas-side transfer coefficient (K_y), the correction parameter is applied to the liquid-side resistance, and for the overall liquid-side transfer coefficient (K_x), the correction parameter is applied to the gas-side resistance. The two cases are vis-à-vis compared (writing Eq. 4.49 in a more general form as presented in Eq. 5.17, excluding the fouling resistances:

$$\left\{\begin{aligned} U_i &= \left[\frac{1}{h_i} + \frac{r_i \ln\left(\frac{r_e}{r_i}\right)}{k_T} + \frac{r_i}{r_e h_e}\right]^{-1} \\ U_e &= \left[\frac{r_e}{r_i h_i} + \frac{r_e \ln\left(\frac{r_e}{r_i}\right)}{k_T} + \frac{1}{h_e}\right]^{-1} \end{aligned}\right. \Leftrightarrow \left\{\begin{aligned} K_y &= \left(\frac{1}{k_y} + \frac{m}{k_x}\right)^{-1} \\ K_x &= \left(\frac{1}{m k_y} + \frac{1}{k_x}\right)^{-1} \end{aligned}\right. \tag{7.31}$$

with the peculiar energy transfer resistance through the wall that, of course, can not apply to the case of mass transfer.

Figure 7.7 offers, in conclusion, a synoptic scheme for the expressions of fluxes and mass transfer coefficients encountered here.

Example: Absorption of Ammonia from Air to Water Under Turbulent Conditions

Under the hypotheses of validity of the two-film theory, Fig. 7.6, we consider the mass transfer by absorption of NH_3 from air to water. Assume the following data:

- ammonia is sufficiently diluted in both air and water (both streams are animated by turbulent motion), with $y_A = 0.03$ and $x_A = 0.01$. Therefore, we can safely confuse gas with air, and liquid with water;
- the operation is carried out at room conditions, for which the density of air and water can be taken as, cf. Table 3.1, 1.2 kg/m^3 and 1000 kg/m^3, respectively. Considering 29 kg/kmoles and 18 kg/kmoles as the molar mass for air and water, respectively, it is $C_{TOT} = 0.04$ kmol/m^3 for air and $C_{TOT} = 55.6$ kmol/m^3 for water;
- under the operating conditions adopted, literature reports $m = 1.2$ and $k_m = 0.9$ m/h.

Refer to the scheme in Fig. 7.7. First, we can calculate the local mass transfer coefficients:

$$\begin{cases} k_y = k_m C_{TOT} = 0.036 \dfrac{\text{kmol}}{\text{m}^2 \cdot \text{h}} \\ k_x = k_m C_{TOT} = 50.04 \dfrac{\text{kmol}}{\text{m}^2 \cdot \text{h}} \end{cases} \tag{7.32}$$

where, in particular, we observe that the mass transfer is much more favoured on the liquid side (the limiting stage is the mass transfer on the gas side). Now we can calculate the overall mass transfer coefficients:

$$\begin{cases} K_y = \left(\dfrac{1}{k_y} + \dfrac{m}{k_x} \right)^{-1} = 0.036 \dfrac{\text{kmol}}{\text{m}^2 \cdot \text{h}} \\ K_x = \left(\dfrac{1}{mk_y} + \dfrac{1}{k_x} \right)^{-1} = 0.043 \dfrac{\text{kmol}}{\text{m}^2 \cdot \text{h}} \end{cases} \tag{7.33}$$

We note that the order of magnitude for the overall coefficients is the same as for k_y (the lowest coefficient, i.e. that representing the limiting stage, for processes in series, regulates the value of the overall coefficient, as usual; or, equivalently, the global resistance is dictated by the highest resistance, that for the limiting stage). To use Eq. 7.16 for the determination of the ammonia molar flux, we calculate the virtual molar fractions:

(continued)

$$\begin{cases} y_A^* = mx_A = 0.012 \\ x_A^* = \dfrac{y_A}{m} = 0.025 \end{cases} \tag{7.34}$$

We are now ready to calculate the flux, that must (of course) coincide using either of the two relations in Eq. 7.16:

$$\begin{cases} \dot{N}_A = K_y\left(y_A - y_A^*\right) = 6.5 \cdot 10^{-4}\ \dfrac{\text{kmol}}{\text{m}^2 \cdot \text{h}} \\ \dot{N}_A = K_x\left(x_A^* - x_A\right) = 6.5 \cdot 10^{-4}\ \dfrac{\text{kmol}}{\text{m}^2 \cdot \text{h}} \end{cases} \tag{7.35}$$

7.5 Background Information

- The transport of mass is driven by a concentration gradient (Fig. 7.1). The molecular phenomenon is named diffusion.
- The Fick's law on diffusion (Eq. 7.2) relates the diffusive flux to the concentration spatial gradient via a proportionality constant depending on the species we are considering: the diffusion coefficient.
- The ratio between molecular transport of momentum and molecular transport of mass is the Schmidt number (Eq. 7.4).
- When mass transfer by convection is present, the ratio between total transport of mass and molecular transport of mass is the Sherwood number (Eq. 7.10). The knowledge of the Sherwood number allows, Eq. 7.12, to calculate the convective mass transfer coefficient.
- The two-film theory is applied (Fig. 7.6) to illustrate mass transfer between phases. Figure 7.7 shows the expressions for local and overall mass transfer coefficients.
- In Eq. 7.29, we find the general relationships stating that the flux is the ratio between driving force and resistance, or the product between driving force and mass transfer coefficient (the latter being the reciprocal of the resistance).
- The overall resistance to mass transport can be expressed as the sum of the resistances to the individual transport phenomena placed in series.
- Finally, Fig. 7.8 illustrates an abecedary on the analogies encountered for the three (momentum, energy, mass) transport phenomena.

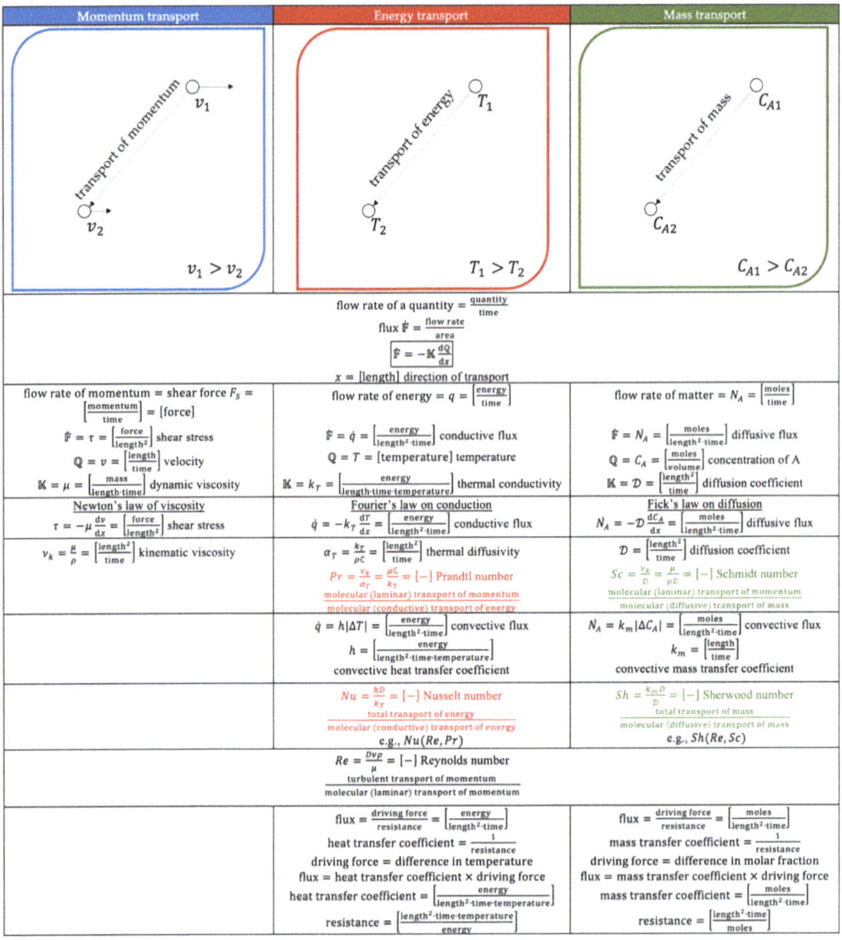

Fig. 7.8 An abecedary of transport phenomena analogies—Part E (concluded). Note that, in the last row, the driving force for mass transfer has been thought as a difference in molar fraction (and not concentration), that determines, in turn, the respective dimensions for mass transfer coefficients (cf. Fig. 7.7) and resistances

References

Balsamo M, Montagnaro F (2015) Fractal-like Vermeulen kinetic equation for the description of diffusion-controlled adsorption dynamics. J Phys Chem C 119:8781–8785. https://doi.org/10.1021/acs.jpcc.5b01783

Bird RB, Stewart WE, Lightfoot EN (2007) Transport phenomena. Wiley, New York

Fick A (1855) Ueber diffusion. Ann Phys 170:59–86. https://doi.org/10.1002/andp.18551700105

Green DW, Perry RH (2008) Perry's chemical engineers' handbook. McGraw-Hill, New York

Inglezakis VJ, Balsamo M, Montagnaro F (2020) Liquid–solid mass transfer in adsorption systems–an overlooked resistance? Ind Eng Chem Res 59:22007–220016. https://doi.org/10.1021/acs.iecr.0c05032
McCabe WL, Smith JC, Harriott P (2001) Unit operations of chemical engineering. McGraw-Hill, Singapore
Montagnaro F (2024a) Sustainable energy production using solid materials. Springer, Cham
Montagnaro F (2024b) Sherwood (Sh) number in chemical engineering applications–a brief review. Energies 17:4342. https://doi.org/10.3390/en17174342
Schmidt E (1926) Die Wärmeübertragung durch Rippen. ZVDI 70:885–889
Sherwood TK, Pigford RL, Wilke CR (1975) Mass transfer. McGraw-Hill, New York
Whitman WG (1923) The two-film theory of gas absorption. Chem Met Eng 29:146–148

Absorption Columns 8

Abstract

The unit operation that applies the concepts seen on mass transport phenomena is represented here by gas-liquid absorption columns. Referring to processes in the absence of chemical reaction and for sufficiently diluted solutions, packed absorption columns are presented and, with the method of the operating lines, the approach for determining the flow rates is illustrated. The text then delves into the aspects of the correct design of the cross-section of the absorption column, which must be such as to avoid flooding phenomena. The reference to mass transfer coefficients, liquid-gas equilibrium relations, and two-film theory, allows us to identify, through mass balances, the design equations for the height of the column, highlighting the role of the height of transfer unit (for which some literature relations are provided) and the number of transfer units (for which the analytical solutions are obtained).

Nomenclature

a	Packing specific surface area [1/length]
D	Column internal diameter [length]
$\mathcal{D}_G$	Diffusion coefficient in gas [length^2/time]
$\mathcal{D}_L$	Diffusion coefficient in liquid [length^2/time]
G	Molar flow rate of gas [moles/time]
g	Gravitational acceleration [length/time^2]
H_{ox}	Height of transfer unit (global liquid side) [length]
H_{oy}	Height of transfer unit (global gas side) [length]
H_x	Height of transfer unit (local liquid side) [length]
H_y	Height of transfer unit (local gas side) [length]
$H_y^{NH_3}$	Height of transfer unit (local gas side) for ammonia absorption [length]
K_x	Liquid phase global mass transfer coefficient [moles/(length^2·time)]

F. Montagnaro et al., *Elements of Unit Operations and Chemical Reactors*,
https://doi.org/10.1007/978-3-032-18046-9_8

K_y	Gas phase global mass transfer coefficient [moles/(length^2·time)]
k_m	Convective mass transfer coefficient [length/time]
k_x	Liquid phase mass transfer coefficient [moles/(length^2·time)]
k_y	Gas phase mass transfer coefficient [moles/(length^2·time)]
L	Molar flow rate of liquid [moles/time]
L^{min}	Minimum molar flow rate of liquid [moles/time]
m	Henry's equilibrium constant [−]
m	Molar mass [mass/moles]
m_L	Molar mass of the liquid phase [mass/moles]
N_{ox}	Number of transfer units (global liquid side) [−]
N_{oy}	Number of transfer units (global gas side) [−]
N_x	Number of transfer units (local liquid side) [−]
N_y	Number of transfer units (local gas side) [−]
$\dot{N}_A$	Molar flux of A species [moles/(length^2·time)]
r	Absorption rate [moles/(volume·time)]
Re	Reynolds number [−]
S	Absorption tower cross-section [length^2]
S_{min}	Minimum critical value for absorption tower cross-section [length^2]
Sc	Schmidt number [−]
Sc_G	Schmidt number for gas [−]
Sc_L	Schmidt number for liquid [−]
Sh	Sherwood number [−]
W_G	Mass flow rate of gas [mass/time]
W_L	Mass flow rate of liquid [mass/time]
x_A	Molar fraction of A in liquid phase [−]
x_A^i	Molar fraction of A in liquid phase at interface [−]
x_A^{in}	Molar fraction of A in inlet liquid [−]
x_A^{out}	Molar fraction of A in outlet liquid [−]
$x_A^{out,max}$	Maximum molar fraction of A in outlet liquid [−]
x_A^*	Virtual molar fraction of A in liquid phase [−]
y_A	Molar fraction of A in gas phase [−]
y_A^i	Molar fraction of A in gas phase at interface [−]
y_A^{in}	Molar fraction of A in inlet gas [−]
y_A^{out}	Molar fraction of A in outlet gas [−]
y_A^*	Virtual molar fraction of A in gas phase [−]
Z	Absorption column height [length]
z	Absorption column longitudinal coordinate [length]
ε	Degree of void for packing [−]
μ_G	Dynamic viscosity of gas [mass/(length·time)]
μ_L	Dynamic viscosity of liquid [mass/(length·time)]
μ_{water}	Dynamic viscosity of water [mass/(length·time)]
ρ_G	Density of gas [mass/volume]
ρ_L	Density of liquid [mass/volume]

8.1 Gas-Liquid Absorption

Gas-liquid absorption (the transfer of a compound, named A, from a gaseous phase to a liquid phase) is a relevant unit operation involving the mass transfer between phases, and which finds applications, e.g., in the environmental field, for the removal of polluting gaseous compounds from effluents of industrial processes. We will treat this operation under some simplifying assumptions, while literature can be referred to (e.g., Astarita 1967; Levenspiel 1999; McCabe et al. 2001) for extensions:

- absorption is carried out in vertically arranged columns (towers), where gas and liquid streams exchange matter in correspondence of the *packing* (an inert solid interface surface, which encourages transfer phenomena). Our reference is therefore to *packed absorption columns*. Other typologies, such as plate columns (as those we have seen in distillation), are not examined here;
- let us consider *stationary* systems, and the *only transfer of matter* is that of the compound A from the bulk of the gaseous phase to the bulk of the liquid phase according to the two-film theory (Fig. 7.6), for which all the relations presented in the scheme of Fig. 7.7 are valid. We therefore neglect the absorption of the solvent gas (in which A is solute) into the liquid phase, and we neglect the desorption of the liquid solvent (in which A is destined to be absorbed) towards the gaseous phase;
- absorption occurs in the absence of chemical reactions (so-called *physical absorption*);
- we will assume *isothermal and isobaric absorption*. Typically, absorption of a gas into a liquid is preferred at low temperatures (e.g., room temperature). In relation to pressure, conceptually, gas-liquid transfer would be favoured at high pressure, but absorption at atmospheric pressure is common indeed, for reasons of economy and safety;
- in particular with reference to environmental applications (see the following example), we are faced with gaseous streams that are already quantitatively very poor in A, but still with a concentration above the legal emission standards. Therefore, while from an environmental point of view it is relevant to reduce this concentration below the legal limits, from a macroscopic point of view this does not significantly alter the molar flow rate of the two currents (gas and liquid) that exchange matter in the absorption column. Thus, the *hypothesis of diluted solutions* will be valid here (literature adopts this approach for fluids with $<10\%$ of the A compound; the lower this concentration, the more accurate this hypothesis is), which has as its main implication (many others then follow) the *constancy of the molar flow rates* that circulate in the column. *The molar flow rates for liquid and gas will be taken constant and equal to their inlet values.*

Example: Validity of the Hypothesis of Diluted Solutions
Consider a gaseous stream coming from an industrial process, with a molar flow rate of 100 mol/h and containing SO_2 (the A species to be absorbed) in a molar fraction of 0.002. This content, corresponding to 0.2% or 0.2 mol/h of SO_2, is objectively low but, given the polluting nature of the compound, it is higher than the emission standards. Before sending the gaseous stream into the atmosphere, therefore, it must be purified. Let us imagine that we want to reduce the molar fraction of SO_2 to 0.0001 (i.e., 0.01%). The operation is certainly relevant, since we want to reduce SO_2 with an efficiency of (0.002–0.0001)/0.002 = 95%.

However, let us evaluate the situation from the point of view of the gaseous stream as a whole. At the beginning of the process, it consists of 0.2 mol/h of SO_2 (and 99.8 mol/h of an inert). At the end, 99.8 mol/h of the inert will remain. In addition, the molar flow rate of uncaptured SO_2 will be "x" such that $x/(99.8 + x) = 0.0001$, i.e. 0.009981 mol/h. Therefore, in the extreme condition (i.e., at the end of the process), the gaseous flow rate will equal 99.809981 mol/h and, during its passage through the absorption column, the gaseous stream will actually see its flow rate gradually decrease from 100 mol/h to 99.809981 mol/h, with variations that can therefore certainly be considered negligible on the scale of the absorption column design, although on the scale of SO_2 removal a significant operation has been carried out.

Concluding, the hypothesis of diluted solutions holds (this must be verified also for the liquid that absorbs A but, apart from special cases, it is normally true) and, along the absorption tower, we will consider the molar flow rates constant (and equal to the inlet values).

8.2 Absorption Packed Towers

Figure 8.1 shows a diagram of an absorption column, with the two fluids flowing in *countercurrent*. The absorption column is fed with a gas containing A to be purified, and with a liquid poor in A (or, a pure liquid not containing A). At the exit, we will have a gaseous stream purified in A (that is, with a concentration below the legal emission standards or below the value we have set), and a liquid richer in A, which subsequently, for example, can be disposed of or treated for the recovery of A (a compound that, in this case, can be disposed of or reused in other processes).

When dealing with absorption columns, we will often notice *analogies with heat exchangers*. For example, here the countercurrent feeding follows the same principles as those seen for heat exchangers. In this case, in fact, at one end of the column we will have both the gas (inlet) and the liquid (outlet) richer in A, and at the other end we will have both the gas (outlet) and the liquid (inlet) poorer in A. This means that the driving force of the process, along the column, does not undergo

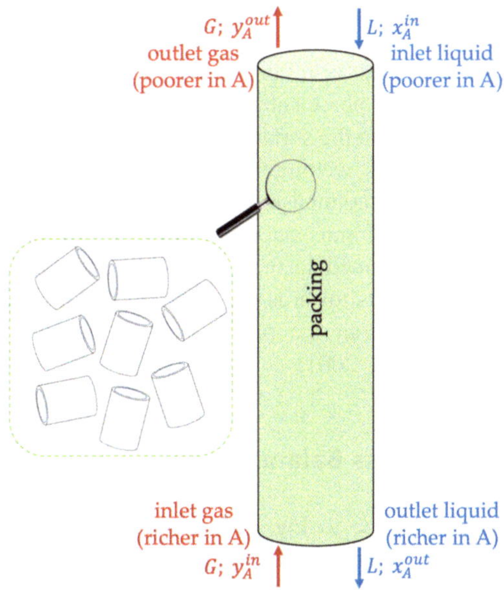

Fig. 8.1 Scheme of a packed absorption column, with relevant flow rates and molar fractions. As example of packing, it has been reported the case of "Raschig" type rings

major variations, unlike what would happen for an equicurrent feeding (which in this case, in fact, we will not consider). The reasons that suggest, once the countercurrent flow has been chosen, to send the gas from the bottom to the top, and the liquid from the top to the bottom, are obvious. Continuing with the analogy with heat exchangers, also in this case it is possible to identify a *process fluid* (the gas) and a *service fluid* (the liquid). Unlike heat exchangers, here there is no need to define two geometrically separated ducts, and we will not refer to a logarithmic mean driving force (although some literature makes use of it, cf. for example McCabe et al. 2001, we are going to propose the punctual and analytical integration of the design equation, with reference to actual driving forces).

The key element of absorption towers is its *packing*. These are pieces of solid material (with average dimensions canonically between units and hundreds of mm), generally randomly poured into the absorber, which offer a high contact area between the two phases, thus encouraging the transfer of mass from gas to liquid. This is therefore the *interface of separation* between the two phases, already discussed in the two-film theory (Fig. 7.6). The packing can be thought of as metallic, ceramic, plastic material, with shapes such as rings, saddles, or much more complex (for an in-depth overview, see e.g. McCabe et al. 2001; Green and Perry 2008), the choice of which also depends on the compatibility of the materials with the chemical nature of the fluids involved. For the sake of simplicity, the example of "Raschig" type rings has been illustrated in Fig. 8.1. Two fundamental characteristics of a packing material, which are given in tables according to commercial packings (an example is given in McCabe et al. 2001), are the degree of void ε (typically, values >0.55 are encountered, up to 0.9 and even more) and a, the specific surface area offered to the contact between liquid and gas. This parameter is defined as the ratio between the total surface area offered by the packing material and the total volume of the packed column, and therefore it has dimensions [1/length]. Typical values for a range between 90–620 m^{-1}.

The liquid phase enters the column at its top, and it is uniformly distributed (e.g., by nozzles) on the surface of the packed filling. It then flows downwards following tortuous paths, wetting the packing material and exchanging matter with the gas that flows up the column in countercurrent. Liquid redistribution systems are possible along the column, since the liquid could tend to follow preferential paths thus showing channelling (e.g. near the wall, where it finds more space; redistribution from the walls towards the centre, e.g. with diverters/baffles, allows for improved contact between the phases, and therefore it increases the process efficiency, McCabe et al. 2001).

8.3 Mass Balance

With reference to Fig. 8.1, the gaseous stream circulates in the column with a constant molar flow rate that we will indicate, for simplicity, with the symbol G. The molar fraction of A in the inlet stream is y_A^{in}, while that in the outlet stream is y_A^{out} ($<y_A^{in}$). The liquid stream circulates in the column with a constant molar flow rate that we will indicate, for simplicity, with the symbol L. The molar fraction of A in the inlet stream is x_A^{in} (this value can also be zero, in the case of an absorbent liquid without A at its inlet), while that in the outlet stream is x_A^{out} ($>x_A^{in}$).

Let us write the mass balance around the entire column, and with reference to A (the molar flow rate transferred from the gaseous phase is equivalent to that absorbed in the liquid phase):

$$\boxed{G\left(y_A^{in} - y_A^{out}\right) = L\left(x_A^{out} - x_A^{in}\right)} \tag{8.1}$$

8.4 Operating Lines

The approach with the method of the operating lines is similar to that presented for heat exchangers in Fig. 5.6 (molar fractions take on the role of temperatures, molar flow rates are used instead of mass flow rates, the role of process and service fluid is here obvious). In Eq. 8.1 we note that, for the process fluid (the gas), the molar flow rate (G), the inlet A molar fraction (y_A^{in}) and the outlet A molar fraction (y_A^{out}) must be known (the latter is likely to be determined by law limits, i.e. the emission standards). While, for the service fluid (the liquid), we assume that only the inlet A molar fraction is known (x_A^{in}, e.g. equal to zero or to a low value for an absorbent, respectively, "pure" or recycled after separation from a previous absorption stage). Therefore, we have two unknown quantities for the service fluid (the molar flow rate L and the outlet A molar fraction x_A^{out}). These two unknown quantities are linked to

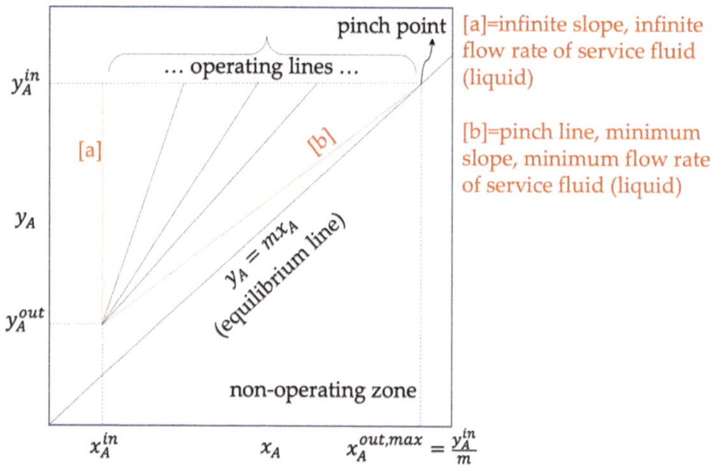

Fig. 8.2 Operating chart for gas-liquid absorption column

each other. Determining the value of the molar flow rate of the liquid (L) will allow us, via Eq. 8.1, to calculate its outlet A molar fraction (x_A^{out}):

$$x_A^{out} = x_A^{in} + \frac{G\left(y_A^{in} - y_A^{out}\right)}{L} \tag{8.2}$$

In more detail, if the molar flow rate of the service fluid (L) is increased, its outlet A molar fraction (x_A^{out}) will decrease (the inlet/outlet difference in A molar fraction for the liquid will be less marked, if a greater flow rate absorbs the same amount of A), and vice versa (Eq. 8.2).

To address the problem of choosing the molar flow rate value of the service fluid (L), consider the plane of Fig. 8.2 which reports the A molar fraction in gas (y_A) against the A molar fraction in liquid (x_A). Draw the *equilibrium line*, which is $y_A = mx_A$ according to the Henry's law valid for diluted solutions. Below the equilibrium line (for a fixed A molar fraction in gas, the A molar fraction in liquid would be higher than its maximum amount dictated by the equilibrium conditions), there is the non-working zone. First of all, it can be sensed that *the greater the distance between the operating conditions and the equilibrium conditions, the smaller the volume (or, for a fixed absorber cross-section, the column height) necessary for the operation.*

Now rewrite Eq. 8.1 as:

$$\frac{y_A^{in} - y_A^{out}}{x_A^{out} - x_A^{in}} = \frac{L}{G} \tag{8.3}$$

which represents the *operating line of the absorber*, passing through the points (x_A^{in} ;y_A^{out}), with known coordinates, and (x_A^{out};y_A^{in}), where we only know the value of the ordinate y_A^{in}. This operating line has slope *L*/*G*, and it is the locus of points having coordinates x_A (the A molar fraction in liquid) and y_A (the A molar fraction in gas), where x_A and y_A are the A molar fraction values of the two fluids, respectively, that exchange mass in a given (generic) section of the absorber. The operating line will therefore pass through the known point of coordinates (x_A^{in};y_A^{out}), which represents the section of inlet of the liquid and outlet of the gas (both A molar fractions are in fact known). Its slope will instead depend on the value chosen for L (all the other values being known), in the sense that the greater this flow rate, the greater the slope.

For infinite molar flow rate of service fluid (L), a vertical line will be obtained (Fig. 8.2, line [a]). This means that the operation will occur without variation in A molar fraction for the liquid (obviously $x_A^{out} = x_A^{in}$ is the only possible condition, the product between an infinite flow rate and a zero difference in A molar fraction will give a finite amount of A transferred from gas to liquid; in other words, if an infinite flow rate of the service fluid absorbs the amount of A released by the process fluid, the service fluid will not see its A molar fraction change, cf. Eq. 8.2) and the distance between operating and equilibrium conditions will be the maximum possible. Therefore, *with maximum driving force, this would allow a minimum column height*, but this cannot be considered an operating condition for obvious reasons.

As the molar flow rate of the service fluid (L) decreases (Fig. 8.2), the slope will decrease and, clearly, x_A^{out} will increase (if the amount of A transferred from the process fluid is absorbed by a progressively lower flow rate of the service fluid, the latter will progressively see its A molar fraction increase), with savings on the flow rate of the service fluid but an increase in the column height (since the distance between operating and equilibrium conditions decreases).

At the other conceptual extreme, by imposing the *pinch condition* $x_A^{out,max} = y_A^{in}/m$ (operating point belonging to the equilibrium line), an operating line with minimum slope (Fig. 8.2, line [b]) will be obtained (i.e., the minimum flow rate of service fluid, L, to be used). However, this is not an operating condition either, as an infinite column height must be foreseen.

Thus, the actual operating line will be comprised between [a] and [b] in Fig. 8.2. It is necessary to mediate between the effects, first of all by finding *the minimum value for the flow rate of the service fluid*, which is, from Eq. 8.1 for $x_A^{out,max} = y_A^{in}/m$:

$$L^{min} = \frac{G\left(y_A^{in} - y_A^{out}\right)}{x_A^{out,max} - x_A^{in}} \tag{8.4}$$

The value of L^{min} must then be appropriately increased, to obtain a value that results in not too high absorber height, but without having to work with a value that is too high for the flow rate of the service fluid (that, in addition, would result in too high costs for, e.g., further separation of A from the liquid). The determination of this flow rate allows, by entering the mass balance equation (Eq. 8.2), to calculate the operating value for x_A^{out}, and therefore to determine the operating line on the chart in Fig. 8.2.

Example: Optimal Operation of an Absorption Column

We consider the absorption of CO_2, from a gas stream ($G = 20$ kmol/h) containing 3% CO_2 ($y_A^{in} = 0.03$), with the aim of arriving at $y_A^{out} = 0.0003$ (i.e., decreasing the CO_2 molar fraction in the gas phase by two orders of magnitude). The absorbent is pure liquid water ($x_A^{in} = 0$). The operation is carried out at 25 °C and 1 atm, for which thermodynamics (Green and Perry 2008) dictates $m = 1.6$. The maximum concentration of CO_2 in water is $x_A^{out,max} = y_A^{in}/m = 0.01875$ (i.e., the liquid can not be richer in CO_2 than about 1.9%, to confirm the hypothesis of dilute solutions also on liquid side). The liquid minimum (pinch) molar flow rate is, Eq. 8.4, $L^{min} = 31.68$ kmol/h. Under an economic analysis that suggests operating with $L = 1.5L^{min} = 47.52$ kmol/h, Eq. 8.2 gives $x_A^{out} = 0.0125$.

8.5 Design of the Absorber Cross-Section

The determination of the cross-section S of a packed column must consider the allowable pressure drop upon absorption. The issue can be discussed with reference to Fig. 8.3, where we qualitatively report (on a log-log chart) the pressure drop (usually normalised for the packing height) as a function of the gas flux, with parametric curves in the liquid flux.

We observe the following:

- for dry system, i.e. in the absence of liquid and with only gas circulating in the packed column, the pressure drop increases with the power of the gas flux, where

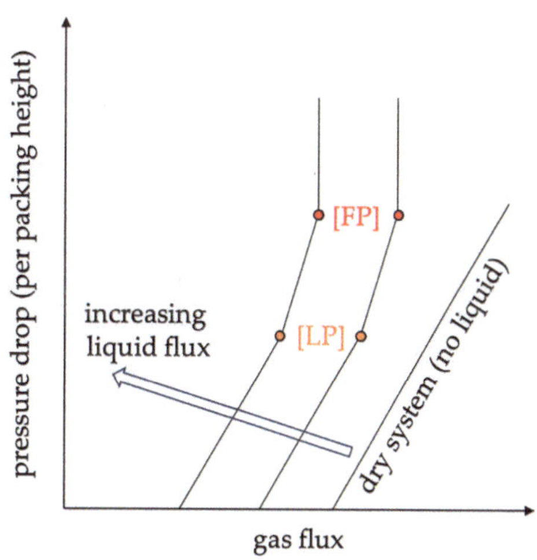

Fig. 8.3 Qualitative trend of pressure drop (per absorber packing height) vs. gas flux, on a log-log chart with curves parametric in liquid flux; [LP] is the loading point and [FP] is the flooding point

the exponent is equal to approximately 1.8–2 (therefore, we observe a straight line on the log-log plot, following the discussion in Fig. 3.14 for the Moody's chart);

- as (more and more) liquid is added from the column top, at a fixed gas flux, the pressure drop increases (therefore, the curves shift to the left), since the presence of the liquid fills the voids in the packing and reduces the space available for the passage of the gas;
- considering a given flux of liquid, as the gas flux increases, the pressure drop at a certain point shows a variation in slope (made explicit in Fig. 8.3, although it is not always so clearly visible, cf. McCabe et al. 2001). This condition, which corresponds to an increase in the holdup of liquid (which is unable to flow in an orderly fashion, due to the ever-increasing gas flux), is called the *loading point* ([LP] in Fig. 8.3). This can be understood as a condition in which the packing material begins to retain the liquid, due to the high countercurrent gas flux. If the gas flux is further increased, a condition is reached in which the entire column fills with liquid: this is the *flooding point* ([FP] in Fig. 8.3), whereby the pressure drop shows a substantially vertical trend which makes the operation no longer possible, with serious obvious dangers.

There are numerous relationships, also depending on the type of packing material, to estimate the flooding conditions (McCabe et al. 2001; Piché et al. 2001). The criterion that is generally adopted is to design a column with a *larger section than the minimum one* (S_{min}), which is the *critical section* that would induce flooding for given liquid and gas flow rates. In this way, we act so as to decrease liquid and gas velocities and *move away from the flooding conditions*, thus decreasing the pressure drop. Having a section only slightly larger than the critical flooding section (S_{min}) would present advantages in economic terms, but it would be too risky given the empirical, and not always extremely precise, nature of the relationships that identify the flooding conditions. On the other hand, it is necessary to avoid designing a column with a section that is too large. Practice suggests working with a section S equal to *double the critical flooding section* (S_{min}):

$$S = 2S_{min} \tag{8.5}$$

Example: Evaluation of the Critical Flooding Section

As mentioned, there are many literature correlations for estimating flooding conditions. An overview can be found, e.g., in Piché et al. 2001. In this example, we propose to use the generalised relationship by Lobo et al. (1945), indicated in literature (Piché et al. 2001; Sawistowski 1957) and here reworked as:

(continued)

$$S_{min} = \frac{W_G}{\sqrt{\frac{g\rho_G\rho_L\varepsilon^3}{a}\left(\frac{\mu_{water}}{\mu_L}\right)^{0.2}\exp\left[-4\left(\frac{W_L}{W_G}\right)^{0.25}\left(\frac{\rho_G}{\rho_L}\right)^{0.125}\right]}} \quad (8.6)$$

where we find: the mass flow rates of liquid (W_L) and gas (W_G), their densities (ρ_L and ρ_G, respectively), the viscosity of the liquid (μ_L) and that of water (μ_{water})—noting that for a liquid absorbent with the properties of water, it is $\mu_{water}/\mu_L = 1$—the gravitational acceleration g (9.81 m/s^2), and the two characteristics of the packing material (i.e. the degree of void ε and the specific surface area a).

Equation 8.6 (or one or the many others reported in literature, cf. McCabe et al. 2001; Piché et al. 2001) allows to estimate the critical flooding section S_{min}. By doubling this value according to Eq. 8.5, it can be obtained the value for S so to work in a safe condition by moving away from the flooding conditions. Naturally, if it were a verification problem for an already existing column, it must be ascertained that the given value for S is sufficiently larger than S_{min} calculated via Eq. 8.6.

Finally, we note in Eq. 8.6 that packing materials with a greater degree of void allows to work with a smaller cross-section (the greater the degree of void, the less likely the onset of flooding), while packings with a greater specific surface area act in the opposite direction, that is, they make us approach the flooding conditions by opposing the flow, just as the viscosity of the liquid absorbent acts similarly.

Example: Design of the Absorber Cross-Section and Diameter

Let us recall a previous example, in which, for the absorption of CO_2 from a gaseous stream ($G = 20$ kmol/h) at 25 °C and 1 atm, we calculated a flow rate of liquid absorbent (water) equal to $L = 47.52$ mol/h. To standardise the measure units, we express the mass in [kg], the time in [s], and the length in [m]. We assume that the gas has the properties of nitrogen (N_2, molar mass approximated at $m = 28$ g/mol), so $G = 20$ kmol/h becomes $W_G = (G/3600)$ $28 = 0.156$ kg/s, while $\rho_G = 1.14$ kg/m^3 from the ideal gas law Eq. 3.11. For liquid absorbent with properties of H_2O ($m = 18$ g/mol), $L = 47.52$ kmol/h becomes $W_L = (L/3600)18 = 0.238$ kg/s, while we can assume $\rho_L =$ 1000 kg/m^3.

Let the absorption be conducted with a saddle packing (of the "Berl" type, with a nominal size equal to 0.50 inches), for which the literature tables report $\varepsilon = 0.63$ and $a = 465$ 1/m. Application of Eq. 8.6 allows to calculate the

(continued)

critical flooding section S_{min} (where, in our case $\mu_{water}/\mu_L = 1$). We get $S_{min} = 0.165\ \text{m}^2$, from which we will design, Eq. 8.5:

$$S = 2S_{min} = 0.330\ \text{m}^2 = \frac{\pi D^2}{4} \tag{8.7}$$

where D is the column internal diameter:

$$D = \sqrt{\frac{4S}{\pi}} = 0.648\ \text{m} \tag{8.8}$$

8.6 Absorption Rate, Mass Transfer Coefficients and Interfacies Conditions

Here we define the *absorption rate*, generally indicated by r [moles of A transferred by absorption/(volume of the absorption column·time)]. We first recall the 4 expressions (equivalent to each other) for the molar flux of A transferred by absorption (as per the synoptic diagram in Fig. 7.7 in the hypotheses of validity of the two-film theory):

$$\begin{cases} \dot{N}_A = k_y\left(y_A - y_A^i\right) & \text{gas phase, local driving force and mass transfer coefficient} \\ \dot{N}_A = k_x\left(x_A^i - x_A\right) & \text{liquid phase, local driving force and mass transfer coefficient} \\ \dot{N}_A = K_y\left(y_A - y_A^*\right) & \text{gas phase, global driving force and mass transfer coefficient} \\ \dot{N}_A = K_x\left(x_A^* - x_A\right) & \text{liquid phase, global driving force and mass transfer coefficient} \end{cases} \tag{8.9}$$

The absorption rate is seen as the molar flux of A transferred by absorption, multiplied by the specific area of the packing material (a). In fact, the molar flux has dimensions [moles of A transferred/(transfer surface·time)], and a has dimensions [transfer surface/volume of the absorber]. So we will write, with 4 equivalent formulations:

$$\boxed{r = \dot{N}_A a = \begin{cases} k_y a\left(y_A - y_A^i\right) \\ k_x a\left(x_A^i - x_A\right) \\ K_y a\left(y_A - y_A^*\right) \\ K_x a\left(x_A^* - x_A\right) \end{cases}} \tag{8.10}$$

We can now recall the relationships between local and global transfer coefficients, taking up Fig. 7.7 and rewriting as a function of the specific surface area of the packing material, a:

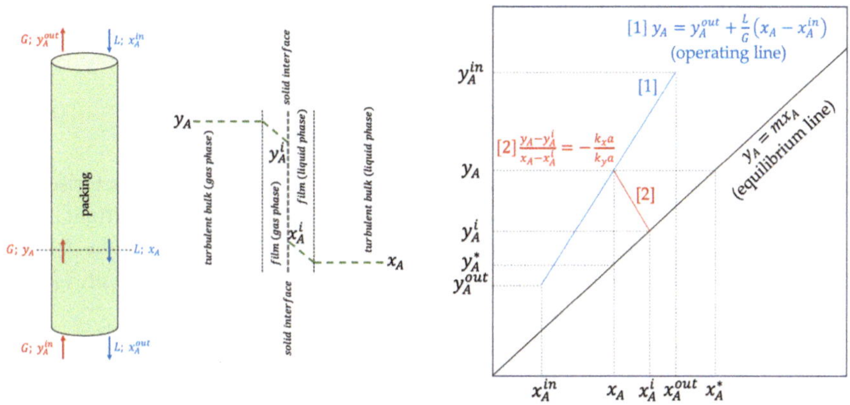

Fig. 8.4 All the molar fractions of interest: on the left, a diagram of an absorption column with the indication of a generic section, where a liquid (with bulk composition x_A) and a gas (with bulk composition y_A) meet; in the centre, a reminder of the two-film theory which also highlights the interface compositions (in equilibrium with each other); on the right, on the operating chart, having arbitrarily fixed the operating line [1] and the representative point of the generic section (x_A;y_A), the virtual values x_A^* and y_A^* are highlighted, as well as the interfacies values x_A^i and y_A^i obtained by using the auxiliary line [2]

$$\boxed{\begin{cases} \dfrac{1}{K_y a} = \dfrac{1}{k_y a} + \dfrac{m}{k_x a} \\ \dfrac{1}{K_x a} = \dfrac{1}{m k_y a} + \dfrac{1}{k_x a} \end{cases}} \tag{8.11}$$

We note that these mass transfer coefficients (here expressed as $k_y a$, $k_x a$, $K_y a$, $K_x a$) are, Fig. 7.7, ultimately a function of the transfer coefficient k_m which, in turn, is present in the number of *Sh* (Eq. 7.10), expressible as a function of *Sc* and *Re* as for example in Eq. 7.11. In addition to the geometric characteristics of the system (given) and the properties of the fluids involved (given), in a relation of the type *Sh*(*Re*,*Sc*), the velocities, or flow rates, of the fluids will appear (specifically, in the *Re* number). In the hypotheses of dilute solutions, the flow rates along the column do not vary (*Re* is constant, as well as *Sc*). Therefore, we will assume the number of *Sh*, then k_m, then (Fig. 7.7) *the 4 mass transfer coefficients to be constant along the absorption tower*. As we will see, often the relationships that give an estimate for the transfer coefficients refer more properly to the product between the transfer coefficient and *a*, which is why we do not simplify the value of the specific area of the packing material in Eq. 8.11.

Let us take the graph of Fig. 8.2 as in Fig. 8.4, where we have chosen a possible operating line. Remember that this line has the expression as in Eq. 8.3, but it can also be rewritten based on a mass balance, for example, between a generic section of the column (where a gas with a bulk mole fraction of A equal to y_A meets a liquid with a bulk mole fraction of A equal to x_A) and the top section of the column (where the gas exits, and the liquid enters):

$$G\left(y_A - y_A^{out}\right) = L\left(x_A - x_A^{in}\right) \Rightarrow \boxed{y_A = y_A^{out} + \frac{L}{G}\left(x_A - x_A^{in}\right)} \tag{8.12}$$

Therefore, in Fig. 8.4 we identify the generic section of the column characterised by bulk mole fractions y_A and x_A. Furthermore, the values of $y_A^* = mx_A$ (the virtual molar fraction in the gas phase that would stay in equilibrium with x_A) and $x_A^* = y_A/m$ (the virtual molar fraction in the liquid phase that would stay in equilibrium with y_A) can be obtained. Moreover, combining the first two relations in Eq. 8.10:

$$k_y a\left(y_A - y_A^i\right) = k_x a\left(x_A^i - x_A\right) \Rightarrow \boxed{\frac{y_A - y_A^i}{x_A - x_A^i} = -\frac{k_x a}{k_y a}} \tag{8.13}$$

This equation represents a straight line passing through the characteristic point of our generic section (x_A;y_A) and the point (x_A^i;y_A^i), that is, the one that identifies the conditions in which, in that generic section, the liquid and the gas meet on an interface offered by the packing material. This straight line, which clearly has a negative slope, can be started from the point (x_A;y_A) and must be intersected with the equilibrium line, which we can also write as:

$$y_A^i = mx_A^i \tag{8.14}$$

since we know that one of the founding hypotheses of the two-film theory is that the interfacies values are in equilibrium with each other. This allows us to identify x_A^i and y_A^i, and we can note how the 4 driving forces defined in Eq. 8.10 are now all determined and all positive ($y_A > y_A^i$, $x_A^i > x_A$, $y_A > y_A^*$, $x_A^* > x_A$). Of course, the calculation of the interfacies values can also be obtained analytically, by joining Eqs. 8.13 and 8.14:

$$\begin{cases} y_A^i = \dfrac{y_A + \dfrac{k_x a}{k_y a} x_A}{1 + \dfrac{k_x a}{m k_y a}} \\ x_A^i = \dfrac{y_A + \dfrac{k_x a}{k_y a} x_A}{m + \dfrac{k_x a}{k_y a}} \end{cases} \tag{8.15}$$

At this point, we are ready to develop the design equation for an absorption packed tower.

8.7 Design Equations

Consider the infinitesimal control volume, Sdz, around which to write the mass balance on the to-be-transferred species, A. In Fig. 8.5, by z we mean the main direction of motion, arbitrarily assumed to be positive from the bottom ($z = 0$, gas inlet and liquid outlet section) upwards ($z = Z$, gas outlet and liquid inlet section; Z is the *column height*, namely the variable to be obtained by means of the design equation, once known S by means of the approach previously presented).

The balance equations are as follows:

$$\begin{cases} Gy_A(z) - Gy_A(z+dz) = rSdz \\ Lx_A(z) - Lx_A(z+dz) = rSdz \end{cases} \tag{8.16}$$

Let us consider the balance, [moles of A/time], on the gas side (first of Eq. 8.16): the molar flow rate of A entering at z, minus that leaving at $z + dz$ (this difference is >0), is transferred by absorption, so it is equal to the product of the absorption rate and the differential control volume. Let us now consider a similar balance on the liquid side (second of Eq. 8.16): the molar flow rate of A leaving at z, minus that entering at $z + dz$ (this difference is >0), comes by absorption from the gas phase, so it is again equal to the product of the absorption rate and the differential control volume. These equations are then rewritten in differential terms, Eq. 8.17, and subsequently reformulated as in Eq. 8.18:

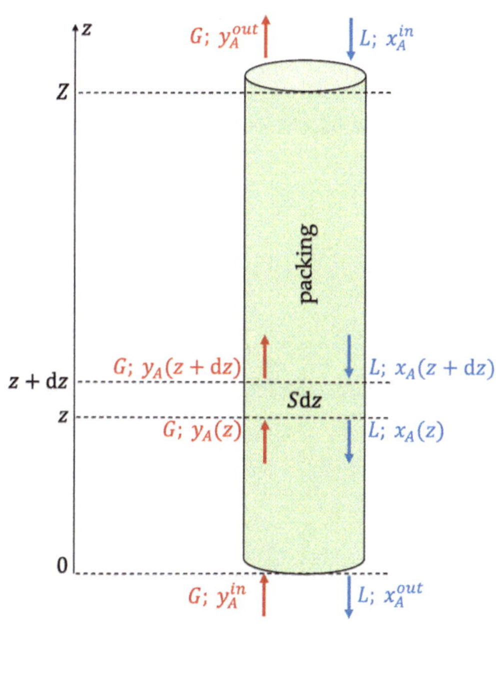

Fig. 8.5 Scheme of a packed absorption column, with relevant flow rates and molar fractions, and indication of the infinitesimal control volume (Sdz). The z-axis illustrates the main direction of motion, arbitrarily assumed to be positive from the bottom upwards

$$\begin{cases} -G\mathrm{d}y_A(z) = rS\mathrm{d}z \\ -L\mathrm{d}x_A(z) = rS\mathrm{d}z \end{cases} \tag{8.17}$$

$$\begin{cases} -\dfrac{G}{S}\dfrac{\mathrm{d}y_A}{\mathrm{d}z} = r = \begin{cases} k_y a\left(y_A - y_A^i\right) \\ K_y a\left(y_A - y_A^*\right) \end{cases} \\ -\dfrac{L}{S}\dfrac{\mathrm{d}x_A}{\mathrm{d}z} = r = \begin{cases} k_x a\left(x_A^i - x_A\right) \\ K_x a\left(x_A^* - x_A\right) \end{cases} \end{cases} \tag{8.18}$$

That is, we have isolated the absorption rate r which, in turn, has been written with the aid of Eq. 8.10. For the gas-side balance equation, the two expressions for r have been reported referring to transfer coefficients, and driving forces, gas side (local and global, respectively). For the liquid-side balance equation, the two expressions for r have been reported referring to transfer coefficients, and driving forces, liquid side (local and global, respectively). We have therefore arrived at 4 equivalent formulations of our mass balance. We then rewrite as:

$$\begin{cases} \mathrm{d}z = \begin{cases} -\dfrac{\frac{G}{S}}{k_y a}\dfrac{\mathrm{d}y_A}{y_A - y_A^i} \\ -\dfrac{\frac{G}{S}}{K_y a}\dfrac{\mathrm{d}y_A}{y_A - y_A^*} \end{cases} \\ \mathrm{d}z = \begin{cases} -\dfrac{\frac{L}{S}}{k_x a}\dfrac{\mathrm{d}x_A}{x_A^i - x_A} \\ -\dfrac{\frac{L}{S}}{K_x a}\dfrac{\mathrm{d}x_A}{x_A^* - x_A} \end{cases} \end{cases} \tag{8.19}$$

that is, separating the variables, and integrating, as:

$$\begin{cases} \displaystyle\int_0^Z \mathrm{d}z = -\frac{\frac{G}{S}}{k_y a}\int_{y_A^{in}}^{y_A^{out}} \frac{1}{y_A - y_A^i}\mathrm{d}y_A \\ \displaystyle\int_0^Z \mathrm{d}z = -\frac{\frac{G}{S}}{K_y a}\int_{y_A^{in}}^{y_A^{out}} \frac{1}{y_A - y_A^*}\mathrm{d}y_A \\ \displaystyle\int_0^Z \mathrm{d}z = -\frac{\frac{L}{S}}{k_x a}\int_{x_A^{out}}^{x_A^{in}} \frac{1}{x_A^i - x_A}\mathrm{d}x_A \\ \displaystyle\int_0^Z \mathrm{d}z = -\frac{\frac{L}{S}}{K_x a}\int_{x_A^{out}}^{x_A^{in}} \frac{1}{x_A^* - x_A}\mathrm{d}x_A \end{cases} \tag{8.20}$$

We have left out of the integral what can be considered constant for dilute solutions, that is, the flow rates and the mass transfer coefficients. We can rework as in:

$$\begin{cases} Z = \dfrac{\frac{G}{S}}{k_y a} \displaystyle\int_{y_A^{out}}^{y_A^{in}} \frac{1}{y_A - y_A^i} dy_A \\ Z = \dfrac{\frac{G}{S}}{K_y a} \displaystyle\int_{y_A^{out}}^{y_A^{in}} \frac{1}{y_A - y_A^*} dy_A \\ Z = \dfrac{\frac{L}{S}}{k_x a} \displaystyle\int_{x_A^{in}}^{x_A^{out}} \frac{1}{x_A^i - x_A} dx_A \\ Z = \dfrac{\frac{L}{S}}{K_x a} \displaystyle\int_{x_A^{in}}^{x_A^{out}} \frac{1}{x_A^* - x_A} dx_A \end{cases} \tag{8.21}$$

where we have elided the minus sign, inverting the integration extremes, a formally more coherent operation since $y_A^{out} < y_A^{in}$ and $x_A^{in} < x_A^{out}$. *These are the 4 equivalent expressions for the design equation of an absorption packed column.*

These design equations express the height of the absorption column as the product of two factors. The first is the one outside the integral, which we will call *height of transfer unit* and indicate with the letter H [length], having subscript y or x if we refer (respectively) to gas side or liquid side with local coefficients, while we will use the subscript oy or ox ("o" = overall) if we refer (respectively) to gas side or liquid side with global coefficients:

$$\begin{cases} H_y = \dfrac{\frac{G}{S}}{k_y a} \\ H_{oy} = \dfrac{\frac{G}{S}}{K_y a} \\ H_x = \dfrac{\frac{L}{S}}{k_x a} \\ H_{ox} = \dfrac{\frac{L}{S}}{K_x a} \end{cases} \tag{8.22}$$

The second factor is dimensionless and is represented by the integral, which we will call *number of transfer units* and indicate with the letter N, having subscript y or x if we refer (respectively) to gas side or liquid side with local driving forces, while we will use the subscript oy or ox if we refer (respectively) to gas side or liquid side with global driving forces:

$$\begin{cases} N_y = \int_{y_A^{out}}^{y_A^{in}} \frac{1}{y_A - y_A^i} \mathrm{d}y_A \\ N_{oy} = \int_{y_A^{out}}^{y_A^{in}} \frac{1}{y_A - y_A^*} \mathrm{d}y_A \\ N_x = \int_{x_A^{in}}^{x_A^{out}} \frac{1}{x_A^i - x_A} \mathrm{d}x_A \\ N_{ox} = \int_{x_A^{in}}^{x_A^{out}} \frac{1}{x_A^* - x_A} \mathrm{d}x_A \end{cases} \tag{8.23}$$

In other words, the following design equations are valid:

$$\boxed{\begin{cases} Z = H_y N_y \\ Z = H_{oy} N_{oy} \\ Z = H_x N_x \\ Z = H_{ox} N_{ox} \end{cases}} \tag{8.24}$$

that is, the height of the absorption column (Z) is seen *as the product between the height of a transfer unit and the number of such units*. This number (N) includes all the (infinite in number) driving forces (whatever they may be) that are realised, section by section, along the column, and the integral takes this aspect into account. In fact, returning to Fig. 8.4, we observe that for each of the infinite sections in which the column develops, there exists a pair of values y_A and x_A, from which all the other molar fractions of interest can be identified (y_A^i and x_A^i, y_A^* and x_A^*). Each of the driving forces under the integral in Eq. 8.23 can be determined, and this must be done for each section of the column. The presence of the integral resolves the issue with formal correctness.

Furthermore, the height of the column (Z) is, as expected, inversely proportional to the transfer coefficients, to the packing specific area, and to the driving forces. The height Z, instead, is directly proportional to the gas and liquid fluxes circulating in the column (G/S and L/S), and to the extent of the separation, i.e. the "in/out" difference in terms of molar fractions (the greater this difference, the greater the distance between the integration extremes, the greater the integral value).

8.8 Height of Transfer Unit

The 4 expressions for the height of transfer unit, Eq. 8.22, can be related to each other by recalling Eq. 8.11 for the mass transfer coefficients:

$$\begin{cases} H_{oy} = \dfrac{\frac{G}{S}}{K_y a} = \dfrac{\frac{G}{S}}{k_y a} + \dfrac{m\frac{G}{S}}{k_x a} = \dfrac{\frac{G}{S}}{k_y a} + \dfrac{m\frac{G}{L}\frac{L}{S}}{k_x a} = H_y + m\dfrac{G}{L}H_x \\ H_{ox} = \dfrac{\frac{L}{S}}{K_x a} = \dfrac{\frac{L}{S}}{mk_y a} + \dfrac{\frac{L}{S}}{k_x a} = \dfrac{\frac{L}{G}\frac{G}{S}}{mk_y a} + \dfrac{\frac{L}{S}}{k_x a} = \dfrac{1}{m}\dfrac{L}{G}H_y + H_x \end{cases} \tag{8.25}$$

Each of the two overall heights of transfer unit can be expressed as the sum of the local height of transfer unit referred to the phase under consideration, plus the local height of transfer unit referred to the other phase (corrected with the mG/L group, or its reciprocal).

Knowing any value for the height of transfer unit (Eq. 8.22) requires knowing the appropriate mass transfer coefficients (which should be evaluated according to the already recalled scheme in Fig. 7.7). It is rare to find exact relationships in the literature, which instead proposes (McCabe et al. 2001) several semi-empirical expressions, often in graphical form, capable of providing the transfer coefficients and/or directly the values for the height of transfer unit. These expressions will be a function of the operating conditions, the gas and liquid fluxes, and their properties.

For example, for absorption of ammonia (NH_3) from air in water at room conditions (ceramic packing made of "Raschig" type rings with nominal size of 1.5 inches), for $W_L/S = 0.7$, 2.0 and 6.1 kg/(m^2·s), approximate values, the following data have been respectively reported for the height of transfer unit (local gas side):

- for $W_G/S = 0.3$ kg/(m^2·s), $H_y = 0.55$ m (at $W_L/S = 0.7$ kg/(m^2·s)), $H_y = 0.27$ m (at $W_L/S = 2.0$ kg/(m^2·s)), $H_y = 0.18$ m (at $W_L/S = 6.1$ kg/(m^2·s));
- for $W_G/S = 0.5$ kg/(m^2·s), $H_y = 0.70$ m (at $W_L/S = 0.7$ kg/(m^2·s)), $H_y = 0.40$ m (at $W_L/S = 2.0$ kg/(m^2·s)), $H_y = 0.21$ m (at $W_L/S = 6.1$ kg/(m^2·s));
- for $W_G/S = 0.8$ kg/(m^2·s), $H_y = 0.91$ m (at $W_L/S = 0.7$ kg/(m^2·s)), $H_y = 0.46$ m (at $W_L/S = 2.0$ kg/(m^2·s)), $H_y = 0.25$ m (at $W_L/S = 6.1$ kg/(m^2·s)).

Although these data are referred to a particular system, they can be of reference for the evaluation of, e.g., H_y. Naming $H_y^{NH_3}$ the values for the (local gas side) height of transfer unit just reported, the following expression has been proposed to estimate the absorption of other gases in water:

$$H_y = H_y^{NH_3}\sqrt{\frac{Sc_G}{0.66}} \tag{8.26}$$

where Sc_G is the Schmidt number of the gas (0.66 is its value for the reference air/water/ammonia system):

$$Sc_G = \frac{\mu_G}{\rho_G \mathcal{D}_G} \tag{8.27}$$

with μ_G, ρ_G and $\mathcal{D}_G$, respectively, the gas viscosity, gas density, and gas diffusion coefficient of the species being absorbed. For reference, with a gas having properties of air at 20 °C, we learn from Table 3.1 that $\mu_G = 0.018 \times 10^{-3}$ Pa·s and $\rho_G = 1.2$ kg/m^3. For absorption in water of a species having the properties of O_2 (to give an example different from ammonia), we can assume $\mathcal{D}_G = 2 \times 10^{-5}$ m^2/s. Thus, $Sc_G = 0.75$ in this case.

There is another expression, valid on the local liquid side, that we can report:

$$H_x = \gamma \left(\frac{\frac{W_L}{S}}{\mu_L} \right)^n \sqrt{Sc_L} \tag{8.28}$$

This must be used with the following measure units: W_L/S, the liquid mass flux, is in [kg/(m^2·h)] and the liquid viscosity μ_L (typically, 1 cP = 10^{-3} Pa·s for water) is in [kg/(m·h)], that is, for water, $\mu_L = 3.6$ kg/(m·h). The Schmidt number of the liquid phase contains the liquid viscosity, the liquid density ρ_L, and the liquid diffusion coefficient of the species being adsorbed $\mathcal{D}_L$:

$$Sc_L = \frac{\mu_L}{\rho_L \mathcal{D}_L} \tag{8.29}$$

For reference, with a liquid having properties of water at 20 °C, we assume $\rho_L = 1000$ kg/m^3. For absorption in water of a species having the properties of O_2, we can take $\mathcal{D}_L = 2.4 \times 10^{-9}$ m^2/s. Thus, Sc_L is about 415 in this case. In Eq. 8.28, the empirical constants γ and n depend on the packing: a good estimate is $\gamma = 0.001$–0.003 and $n = 0.2$–0.3 for typical saddle- and ring-type packings. The value obtained in Eq. 8.28 for H_x is in [m], and the expression has been reported to be valid for W_L/S from 2000 to 75,000 kg/(m^2·h).

Application of Eq. 8.25 then allows us to obtain the overall values for the height of transfer units, once known the local ones. In any case, it is recalled that it is not necessary, for the evaluation of Z, to calculate all 4 values for H: in Eq. 8.24, a formulation is chosen (the four are equivalent each other), and it will be sufficient to calculate only the value of H, and the corresponding value of N, relative to the chosen formulation.

8.9 Number of Transfer Units

Once one of the 4 heights of transfer unit is known, the design equations (Eq. 8.24) require the coupling of the respective value of the number of transfer units, for the calculation of the height Z of the absorption column. What is now proposed, is the analytical solution for the 4 integrals in Eq. 8.23.

8.9.1 Local Gas Side

To solve the integral:

$$N_y = \int_{y_A^{out}}^{y_A^{in}} \frac{1}{y_A - y_A^i} \mathrm{d}y_A \tag{8.30}$$

we need a relation $y_A^i(y_A)$, which we take from the first of Eq. 8.15:

$$y_A^i = \frac{y_A + \frac{k_x a}{k_y a} x_A}{1 + \frac{k_x a}{m k_y a}} \tag{8.31}$$

This is not yet the relation we are looking for, since x_A is present, which we put in function of y_A by reworking the equation of the operating line, Eq. 8.12:

$$x_A = \frac{G}{L} y_A + x_A^{in} - \frac{G}{L} y_A^{out} \tag{8.32}$$

Substituting Eq. 8.32 in Eq. 8.31, we obtain:

$$\begin{aligned} y_A^i &= \frac{y_A + \frac{k_x a}{k_y a}\left(\frac{G}{L} y_A + x_A^{in} - \frac{G}{L} y_A^{out}\right)}{1 + \frac{k_x a}{m k_y a}} = \frac{1 + \frac{k_x a}{k_y a}\frac{G}{L}}{1 + \frac{k_x a}{m k_y a}} y_A + \frac{\frac{k_x a}{k_y a}\left(x_A^{in} - \frac{G}{L} y_A^{out}\right)}{1 + \frac{k_x a}{m k_y a}} \\ &= \frac{\frac{k_y a}{k_x a} + \frac{G}{L}}{\frac{k_y a}{k_x a} + \frac{1}{m}} y_A + \frac{x_A^{in} - \frac{G}{L} y_A^{out}}{\frac{k_y a}{k_x a} + \frac{1}{m}} \end{aligned} \tag{8.33}$$

which we rewrite as:

$$\begin{cases} y_A^i = \alpha y_A + \beta \\ \alpha = \dfrac{\dfrac{k_y a}{k_x a} + \dfrac{G}{L}}{\dfrac{k_y a}{k_x a} + \dfrac{1}{m}} \\ \beta = \dfrac{x_A^{in} - \dfrac{G}{L} y_A^{out}}{\dfrac{k_y a}{k_x a} + \dfrac{1}{m}} \end{cases} \tag{8.34}$$

Substituting Eq. 8.34 in Eq. 8.30, we obtain the solution for N_y:

$$\begin{aligned} N_y &= \int_{y_A^{out}}^{y_A^{in}} \frac{1}{y_A - y_A^i} \mathrm{d}y_A = \int_{y_A^{out}}^{y_A^{in}} \frac{1}{(1-\alpha) y_A - \beta} \mathrm{d}y_A \\ &= \frac{1}{1-\alpha} \ln\left[(1-\alpha) y_A - \beta\right]_{y_A^{out}}^{y_A^{in}} = \frac{1}{1-\alpha} \ln\left[\frac{(1-\alpha) y_A^{in} - \beta}{(1-\alpha) y_A^{out} - \beta}\right] \end{aligned} \tag{8.35}$$

8.9.2 Local Liquid Side

To solve the integral:

$$N_x = \int_{x_A^{in}}^{x_A^{out}} \frac{1}{x_A^i - x_A} \mathrm{d}x_A \tag{8.36}$$

we need a relation $x_A^i(x_A)$, which we take from the second of Eq. 8.15:

$$x_A^i = \frac{y_A + \frac{k_x a}{k_y a} x_A}{m + \frac{k_x a}{k_y a}} \tag{8.37}$$

This is not yet the relation we are looking for, since y_A is present, which we put in function of x_A by reworking the equation of the operating line, Eq. 8.12:

$$y_A = \frac{L}{G} x_A + y_A^{out} - \frac{L}{G} x_A^{in} \tag{8.38}$$

Substituting Eq. 8.38 in Eq. 8.37, we obtain:

$$x_A^i = \frac{\frac{L}{G} x_A + y_A^{out} - \frac{L}{G} x_A^{in} + \frac{k_x a}{k_y a} x_A}{m + \frac{k_x a}{k_y a}} = \frac{\frac{k_x a}{k_y a} + \frac{L}{G}}{m + \frac{k_x a}{k_y a}} x_A + \frac{y_A^{out} - \frac{L}{G} x_A^{in}}{m + \frac{k_x a}{k_y a}} = \frac{\frac{k_x a}{k_y a} + \frac{L}{G}}{\frac{k_x a}{k_y a} + m} x_A$$
$$+ \frac{y_A^{out} - \frac{L}{G} x_A^{in}}{\frac{k_x a}{k_y a} + m} \tag{8.39}$$

which we rewrite as:

$$\begin{cases} x_A^i = \alpha x_A + \beta \\ \alpha = \dfrac{\dfrac{k_x a}{k_y a} + \dfrac{L}{G}}{\dfrac{k_x a}{k_y a} + m} \\ \beta = \dfrac{y_A^{out} - \dfrac{L}{G} x_A^{in}}{\dfrac{k_x a}{k_y a} + m} \end{cases} \tag{8.40}$$

Substituting Eq. 8.40 in Eq. 8.36, we obtain the solution for N_x:

$$N_x = \int_{x_A^{in}}^{x_A^{out}} \frac{1}{x_A^i - x_A} \mathrm{d}x_A = \int_{x_A^{in}}^{x_A^{out}} \frac{1}{(\alpha - 1) x_A + \beta} \mathrm{d}x_A$$
$$= \frac{1}{\alpha - 1} \ln \left[(\alpha - 1) x_A + \beta\right]_{x_A^{in}}^{x_A^{out}} = \frac{1}{\alpha - 1} \ln \left[\frac{(\alpha - 1) x_A^{out} + \beta}{(\alpha - 1) x_A^{in} + \beta}\right] \tag{8.41}$$

The passionate reader will not have missed a certain degree of symmetry between the constants α and β in the two cases.

8.9.3 Global Gas Side

To solve the integral:

$$N_{oy} = \int_{y_A^{out}}^{y_A^{in}} \frac{1}{y_A - y_A^*} \mathrm{d}y_A \tag{8.42}$$

we need a relation $y_A^*(y_A)$, via Eq. 8.32:

$$y_A^* = mx_A = m\left(\frac{G}{L}y_A + x_A^{in} - \frac{G}{L}y_A^{out}\right) \tag{8.43}$$

That is:

$$y_A^* = m\frac{G}{L}y_A + m\left(x_A^{in} - \frac{G}{L}y_A^{out}\right) \tag{8.44}$$

which we rewrite as:

$$\begin{cases} y_A^* = \alpha y_A + \beta \\ \alpha = m\dfrac{G}{L} \\ \beta = m\left(x_A^{in} - \dfrac{G}{L}y_A^{out}\right) \end{cases} \tag{8.45}$$

Substituting Eq. 8.45 in Eq. 8.42, we obtain the solution for N_{oy}:

$$\begin{aligned} N_{oy} &= \int_{y_A^{out}}^{y_A^{in}} \frac{1}{y_A - y_A^*} \mathrm{d}y_A = \int_{y_A^{out}}^{y_A^{in}} \frac{1}{(1-\alpha)y_A - \beta} \mathrm{d}y_A \\ &= \frac{1}{1-\alpha} \ln\left[(1-\alpha)y_A - \beta\right]_{y_A^{out}}^{y_A^{in}} = \frac{1}{1-\alpha} \ln\left[\frac{(1-\alpha)y_A^{in} - \beta}{(1-\alpha)y_A^{out} - \beta}\right] \end{aligned} \tag{8.46}$$

8.9.4 Global Liquid Side

To solve the integral:

$$N_{ox} = \int_{x_A^{in}}^{x_A^{out}} \frac{1}{x_A^* - x_A} \mathrm{d}x_A \tag{8.47}$$

we need a relation $x_A^*(x_A)$, via Eq. 8.38:

$$x_A^* = \frac{y_A}{m} = \frac{1}{m}\left(\frac{L}{G}x_A + y_A^{out} - \frac{L}{G}x_A^{in}\right) \tag{8.48}$$

That is:

$$x_A^* = \frac{1}{m}\frac{L}{G}x_A + \frac{1}{m}\left(y_A^{out} - \frac{L}{G}x_A^{in}\right) \tag{8.49}$$

which we rewrite as (again, with symmetry with respect to the overall gas side):

$$\begin{cases} x_A^* = \alpha x_A + \beta \\ \alpha = \dfrac{1}{m}\dfrac{L}{G} \\ \beta = \dfrac{1}{m}\left(y_A^{out} - \dfrac{L}{G}x_A^{in}\right) \end{cases} \tag{8.50}$$

Substituting Eq. 8.50 in Eq. 8.47, we obtain the solution for N_{ox}:

$$\begin{aligned} N_{ox} &= \int_{x_A^{in}}^{x_A^{out}} \frac{1}{x_A^* - x_A}\mathrm{d}x_A = \int_{x_A^{in}}^{x_A^{out}} \frac{1}{(\alpha - 1)x_A + \beta}\mathrm{d}x_A \\ &= \frac{1}{\alpha - 1}\ln\left[(\alpha - 1)x_A + \beta\right]_{x_A^{in}}^{x_A^{out}} = \frac{1}{\alpha - 1}\ln\left[\frac{(\alpha - 1)x_A^{out} + \beta}{(\alpha - 1)x_A^{in} + \beta}\right] \end{aligned} \tag{8.51}$$

8.10 Overall View

To conclude, Fig. 8.6 synoptically illustrates the four equivalent design equations for the calculus of the column height (Z), under the hypotheses of validity of the two-film theory for dilute solutions. This figure complements the information given in Fig. 7.7.

Example: Design of an Absorption Packed Tower

A nitrogen-rich stream has flow rate (G) of 160 kmol/h. This value can be translated into $G = 44.4$ mol/s, or $W_G = 1.243$ kg/s (assuming the gas with the properties of N_2, i.e. $m = 28$ g/mol). This gaseous stream contains CO_2 ($y_A^{in} = 0.03$), and must be treated in a packed absorption tower (ceramic Raschig rings, nominal size of 1.5 inches; manufacturer data: $\varepsilon = 0.75$ and $a = 126$ 1/m) using water (CO_2-free, i.e. $x_A^{in} = 0$) as liquid absorbent. The process is carried out at 25 °C and 1 atm, and we assume the equilibrium parameter $m = 1.6$. We want to achieve 98% CO_2 capture:

(continued)

Gas side	Liquid side
$k_m = \left[\frac{\text{length}}{\text{time}}\right]$ in $N_A = k_m\lvert\Delta C_A\rvert = \left[\frac{\text{moles}}{\text{length}^2\cdot\text{time}}\right]$	
Local mass transfer coefficients	
$k_y = k_m C_{TOT} = k_m \frac{\mathcal{P}}{\Re T} = \left[\frac{\text{moles}}{\text{length}^2\cdot\text{time}}\right]$ in $N_A = k_y(y_A - y_A^i)$	$k_x = k_m C_{TOT} = \left[\frac{\text{moles}}{\text{length}^2\cdot\text{time}}\right]$ in $N_A = k_x(x_A^i - x_A)$
Global mass transfer coefficients	
$K_y = \left[\frac{\text{moles}}{\text{length}^2\cdot\text{time}}\right]$ in $N_A = K_y(y_A - y_A^*)$ where $y_A^* = mx_A$	$K_x = \left[\frac{\text{moles}}{\text{length}^2\cdot\text{time}}\right]$ in $N_A = K_x(x_A^* - x_A)$ where $x_A^* = y_A/m$
Relations between global and local mass transfer coefficients	
$\frac{1}{K_y} = \frac{1}{k_y} + \frac{m}{k_x}$	$\frac{1}{K_x} = \frac{1}{mk_y} + \frac{1}{k_x}$
Design equations	
$Z = H_y N_y$ $\begin{cases} H_y = \frac{\frac{G}{S}}{k_y a} = [\text{length}] \\ N_y = \int_{y_A^{out}}^{y_A^{in}} \frac{1}{y_A - y_A^i} dy_A = [-] \end{cases}$	$Z = H_x N_x$ $\begin{cases} H_x = \frac{\frac{L}{S}}{k_x a} = [\text{length}] \\ N_x = \int_{x_A^{in}}^{x_A^{out}} \frac{1}{x_A^i - x_A} dx_A = [-] \end{cases}$
$Z = H_{oy} N_{oy}$ $\begin{cases} H_{oy} = \frac{\frac{G}{S}}{K_y a} = [\text{length}] \\ N_{oy} = \int_{y_A^{out}}^{y_A^{in}} \frac{1}{y_A - y_A^*} = [-] \end{cases}$	$Z = H_{ox} N_{ox}$ $\begin{cases} H_{ox} = \frac{\frac{L}{S}}{K_x a} = [\text{length}] \\ N_{ox} = \int_{x_A^{in}}^{x_A^{out}} \frac{1}{x_A^* - x_A} dx_A = [-] \end{cases}$
Expressions for the number of transfer units	
$N_y = \frac{1}{1-\alpha} \ln\left[\frac{(1-\alpha)y_A^{in} - \beta}{(1-\alpha)y_A^{out} - \beta}\right]$ $\begin{cases} \alpha = \frac{\frac{k_y a}{k_x a} + \frac{G}{L}}{\frac{k_y a}{k_x a} + \frac{1}{m}} \\ \beta = \frac{x_A^{in} - \frac{G}{L} y_A^{out}}{\frac{k_y a}{k_x a} + \frac{1}{m}} \end{cases}$	$N_x = \frac{1}{\alpha-1} \ln\left[\frac{(\alpha-1)x_A^{out} + \beta}{(\alpha-1)x_A^{in} + \beta}\right]$ $\begin{cases} \alpha = \frac{\frac{k_x a}{k_y a} + \frac{L}{G}}{\frac{k_x a}{k_y a} + m} \\ \beta = \frac{y_A^{out} - \frac{L}{G} x_A^{in}}{\frac{k_x a}{k_y a} + m} \end{cases}$
$N_{oy} = \frac{1}{1-\alpha} \ln\left[\frac{(1-\alpha)y_A^{in} - \beta}{(1-\alpha)y_A^{out} - \beta}\right]$ $\begin{cases} \alpha = m\frac{G}{L} \\ \beta = m\left(x_A^{in} - \frac{G}{L} y_A^{out}\right) \end{cases}$	$N_{ox} = \frac{1}{\alpha-1} \ln\left[\frac{(\alpha-1)x_A^{out} + \beta}{(\alpha-1)x_A^{in} + \beta}\right]$ $\begin{cases} \alpha = \frac{1}{m}\frac{L}{G} \\ \beta = \frac{1}{m}\left(y_A^{out} - \frac{L}{G} x_A^{in}\right) \end{cases}$

Fig. 8.6 Synoptic illustration for the design equations (four equivalent ones) applied to a gas-liquid absorption packed column, two-film theory, dilute solutions, with indication of local and global mass transfer coefficients

$$\frac{Gy_A^{in} - Gy_A^{out}}{Gy_A^{in}} = 0.98 \tag{8.52}$$

From Eq. 8.52, we obtain $y_A^{out} = (1–0.98)y_A^{in} = 0.0006$.

Liquid flow rate

Using Eq. 8.4, it is possible determine the minimum value for the flow rate of the service fluid:

$$L^{min} = \frac{G\left(y_A^{in} - y_A^{out}\right)}{x_A^{out,\,max} - x_A^{in}} = 250.88\ \text{kmol/h} \tag{8.53}$$

where $x_A^{out,\,max} = y_A^{in}/m = 0.01875$. Deciding to operate with a liquid flow rate twice the pinch value, it is $L = 2L^{min} = 501.76$ kmol/h (corresponding to 139.38 mol/s, or $W_L = 2.509$ kg/s for $m = 18$ g/mol). The actual value for the CO_2 molar fraction in the liquid phase, at the absorber exit, is, from Eq. 8.2, $x_A^{out} = 0.009375$ (please observe that, both gas- and liquid-side, the two streams can be considered under the hypotheses of dilute solution).

Design of the absorber cross-section

(continued)

Let us evaluate the critical flooding section, Eq. 8.6:

$$S_{min} = \frac{W_G}{\sqrt{\frac{g\rho_G\rho_L\varepsilon^3}{a}\left(\frac{\mu_{water}}{\mu_L}\right)^{0.2}\exp\left[-4\left(\frac{W_L}{W_G}\right)^{0.25}\left(\frac{\rho_G}{\rho_L}\right)^{0.125}\right]}} = 0.56\ \text{m}^2 \quad (8.54)$$

assuming, as densities, $\rho_L = 1000$ kg/m^3 and $\rho_G = 1.14$ kg/m^3 (see a previous example). Deciding to operate with an actual absorber tower cross-section $S = 2.8S_{min}$, it is $S = 1.568$ m^2.

<u>*Evaluation of the height of transfer unit*</u>

For the evaluation of the height of the absorption column to be designed, we choose one of the four equivalent expressions in Eq. 8.24:

$$Z = H_{oy}N_{oy} \quad (8.55)$$

The expression for the height of transfer unit, gas side and overall, is, Eq. 8.25:

$$H_{oy} = H_y + m\frac{G}{L}H_x \quad (8.56)$$

That is, we need to calculate, separately, the local height of transfer unit both gas side and liquid side. Start with Eq. 8.26:

$$H_y = H_y^{NH_3}\sqrt{\frac{Sc_G}{0.66}} \quad (8.57)$$

where, being W_G/S = 0.79 kg/(m^2·s) and W_L/S = 1.60 kg/(m^2·s), we can assume $H_y^{NH_3}$ = 0.6 m (calculated from data presented in the text where Eq. 8.26 has been introduced; please note that the packing is the same). Now, from Eq. 8.27:

$$Sc_G = \frac{\mu_G}{\rho_G \mathcal{D}_G} \quad (8.58)$$

where the viscosity of N_2 is μ_G = 1.77·10^{-5} kg/(m·s) as calculated from Eq. 3.10:

$$\mu_G = \frac{C_1T^{C_2}}{1 + \frac{C_3}{T} + \frac{C_4}{T^2}} \quad (8.59)$$

with the following set of constants (Green and Perry 2008):

(continued)

$$\begin{cases} C_1 = 6.5592 \cdot 10^{-7} \\ C_2 = 0.6081 \\ C_3 = 54.714 \\ C_4 = 0 \end{cases} \tag{8.60}$$

The value of the CO_2 gas diffusion coefficient is assumed (Green and Perry) as $\mathcal{D}_G = 1.6{\cdot}10^{-5}$ m²/s. The Schmidt number, gas side, is 0.97 (from Eq. 8.58) and, finally, Eq. 8.57, $H_y = 0.73$ m.

Let us continue with the liquid side, Eq. 8.28:

$$H_x = \gamma \left(\frac{\frac{W_L}{S}}{\mu_L} \right)^n \sqrt{Sc_L} \tag{8.61}$$

For our packing, literature reports $\gamma = 0.0026$ and $n = 0.22$. Use Eq. 8.61 with $W_L/S = 5760$ kg/(m²·h). The viscosity of water is calculated according to Eq. 3.10, with the constants given in Table 3.1:

$$\mu_L = \exp\left[C_1 + \frac{C_2}{T} + C_3 \ln(T) + C_4 T^{C_5}\right] \tag{8.62}$$

It is $\mu_L = 3.3$ kg/(m·h), the measure units needed in Eq. 8.61. Moreover, from Eq. 8.29:

$$Sc_L = \frac{\mu_L}{\rho_L \mathcal{D}_L} \tag{8.63}$$

where the value of the CO_2 liquid diffusion coefficient is assumed (Green and Perry 2008) as $\mathcal{D}_L = 2.0{\cdot}10^{-9}$ m²/s. Thus, it is $Sc_L = 458$ and, Eq. 8.61, $H_x = 0.29$ m.

We are ready to come back to Eq. 8.56: $H_{oy} = 0.88$ m.

Number of transfer units

From Eq. 8.46, it is:

$$N_{oy} = \frac{1}{1-\alpha} \ln\left[\frac{(1-\alpha)y_A^{in} - \beta}{(1-\alpha)y_A^{out} - \beta}\right] = 6.56 \tag{8.64}$$

since, Eq. 8.45:

$$\alpha = m\frac{G}{L} = 0.51 \tag{8.65}$$

$$\beta = m\left(x_A^{in} - \frac{G}{L} y_A^{out}\right) = -0.00031 \tag{8.66}$$

Design of the absorber height

Finally, from Eq. 8.55, the column height is:

(continued)

$$Z = H_{oy} N_{oy} = 5.77 \text{ m} \tag{8.67}$$

that is, an absorption column with a volume of about 9 m^3.

Example: Verification of an Absorption Packed Tower

Consider a $Z = 2.5$ m, $S = 0.5$ m^2 packed absorption tower (1 inch-stainless steel Pall rings, $a = 217$ 1/m), proposed to treat a gaseous (air-based) stream with mass flow rate $W_G = 300$ kg/h, containing SO_2 ($y_A^{in} = 0.05$). Sulphur dioxide must be absorbed so to obtain a gaseous stream with $y_A^{out} = 0.005$. Assuming the properties for air ($m = 29$ g/mol), it is $G = 2.87$ mol/s.

The process is carried out using liquid water, where $x_A^{in} = 0.0001$ (as water entering the tower comes from a previous absorption cycle followed by purification, that leaves traces of SO_2 in the liquid). Absorption is carried out at 25 °C and 1 atm. For the system at hand, the equilibrium constant is taken at $m = 40$. We need to verify the feasibility of the process.

Liquid flow rate

From Eq. 8.4, it is:

$$L^{min} = \frac{G\left(y_A^{in} - y_A^{out}\right)}{x_A^{out,max} - x_A^{in}} = 112.30 \text{ mol/s} \tag{8.68}$$

where $x_A^{out,max} = y_A^{in}/m = 0.00125$. We start with the hypothesis of operating with a liquid flow rate twice the pinch value, $L = 2L^{min} = 224.60$ mol/s (corresponding to $W_L = 14{,}554.08$ kg/h). The actual value for the CO_2 molar fraction in the liquid phase, at the absorber exit, is, from Eq. 8.2, $x_A^{out} = 0.000675$.

Evaluation of the height of transfer unit

Let us choose one of the four equivalent expressions in Eq. 8.24:

$$Z = H_x N_x \tag{8.69}$$

The expression for the height of transfer unit, local liquid side, is, Eq. 8.22:

$$H_x = \frac{\frac{L}{S}}{k_x a} \tag{8.70}$$

For the evaluation of $k_x a$, we propose the following expression, obtained appropriately reworking a relationship published in Rejl et al. (2009):

(continued)

$$k_x a = 0.05 \frac{\rho_L^{0.372}}{m_L} a^{0.747} \left(\frac{W_L}{S}\right)^{0.628} \tag{8.71}$$

with ρ_L in [kg/m^3], approximated at 1000 kg/m^3 as usual, a in [1/m], W_L in [kg/h], S in [m^2], and m_L the molar mass of the liquid phase (assuming the properties of water, it is 18 g/mol as usual). The expressions returns $k_x a$ in [mol/(m^3·s)]. For our system, it is $k_x a$ = 1283.6 mol/(m^3·s) and, Eq. 8.70, $H_x = 0.35$ m.

Number of transfer units

From Eq. 8.41, it is:

$$N_x = \frac{1}{\alpha - 1} \ln\left[\frac{(\alpha - 1)x_A^{out} + \beta}{(\alpha - 1)x_A^{in} + \beta}\right] = 6.11 \tag{8.72}$$

where $k_y a$ = 36.1 mol/(s·m^3) is assigned by experimental studies on the column, and, Eq. 8.40:

$$\alpha = \frac{\frac{k_x a}{k_y a} + \frac{L}{G}}{\frac{k_x a}{k_y a} + m} = 1.51 \tag{8.73}$$

$$\beta = \frac{y_A^{out} - \frac{L}{G} x_A^{in}}{\frac{k_x a}{k_y a} + m} = -0.0000374 \tag{8.74}$$

Absorber height

Finally, from Eq. 8.69, the column height is:

$$Z = H_x N_x = 2.14 \text{ m} \tag{8.75}$$

This means that our column, with $Z = 2.5$ m, is more than sufficient to carry out the required operation. A trial-and-error procedure can help, for example, in saving the liquid flow rate while keeping the specifics of the process. It is verified that taking L at about $L = 1.75L^{min}$ = 196.52 mol/s, allows to operate with a column of the given height (2.5 m), while saving ca. 12.5% of liquid.

8.11 Background Information

- Gas-liquid absorption packed columns are presented and solved under the main hypotheses of dilute solutions and validity of the two-film theory.
- The approach presented in Fig. 8.2 and Eq. 8.4 allows to individuate the minimum and, then, the operating value for the flow rate of the liquid absorbent.

- The determination of the absorber cross-section must be carried out so as to avoid the onset of flooding conditions. Eq. 8.6 gives a possible example for the evaluation of the minimum cross-section, that will be then appropriately (e.g., by 2 times) increased.
- All the molar fractions of interest are illustrated in Fig. 8.4.
- Four equivalent expressions for the absorber design equation are presented in Eq. 8.24, as the product between height of transfer unit and number of transfer units. Once chosen if we want to refer to local gas side, local liquid side, overall gas side, or overall liquid side, Eq. 8.24 allows for the determination of the column height.
- Expressions for the evaluation of the height of transfer unit are given, e.g., in Eqs. 8.26, 8.28, 8.25, while the analytical solutions for the number of transfer units are presented in Eqs. 8.35, 8.41, 8.46, 8.51.
- The reader could find useful the synoptic illustration in Fig. 8.6.

References

Astarita G (1967) Mass transfer with chemical reaction. Elsevier, Amsterdam

Green DW, Perry RH (2008) Perry's chemical engineers' handbook. McGraw-Hill, New York

Levenspiel O (1999) Chemical reaction engineering. Wiley, New York

Lobo WE, Friend L, Hashmall F, Zenz F (1945) Limiting capacity of dumped tower packings. Trans Am Inst Chem Eng 41:693–710

McCabe WL, Smith JC, Harriott P (2001) Unit operations of chemical engineering. McGraw-Hill, Singapore

Piché S, Larachi F, Grandjean BPA (2001) Flooding capacity in packed towers: database, correlations, and analysis. Ind Eng Chem Res 40:476–487. https://doi.org/10.1021/ie000486s

Rejl JF, Linek V, Moucha T, Valenz L (2009) Methods standardization in the measurement of mass-transfer characteristics in packed absorption columns. Chem Eng Res Des 87:695–704. https://doi.org/10.1016/j.cherd.2008.09.009

Sawistowski H (1957) Flooding velocities in packed columns operating at reduced pressures. Chem Eng Sci 6:138–140. https://doi.org/10.1016/0009-2509(57)85007-6

Humidification Processes

9

Abstract

We end the part of the book devoted to unit operations with this chapter, which discusses humidification and dehumidification processes, focusing on the mass and energy transfer between a gas and a pure liquid in which it is practically insoluble. The air/water system is examined in detail, with definition of key hygrometric quantities such as absolute, saturation, percentage, and relative humidity. Psychrometric charts and analytical relationships for enthalpy are introduced as essential tools for system analysis. The design and operation of cooling towers are analysed, including the determination of minimum gas flux, packing height, section and interfacial conditions using the two-film theory.

Nomenclature

A_b	Base area of the cooling tower [length^2]
a	Packing specific surface area [1/length]
C_D	Duty coefficient [−]
C_P	Performance coefficient [−]
$\mathbb{C}_g$	Specific heat of dry gas [energy/(mass·temperature)]
$\mathbb{C}_l$	Specific heat of liquid [energy/(mass·temperature)]
$\mathbb{C}_p$	Average specific heat of gas/vapour mixture [energy/(mass·temperature)]
$\mathbb{C}_s$	Humid heat [energy/(mass·temperature)]
$\mathbb{C}_v$	Specific heat of vapour [energy/(mass·temperature)]
D_b	Base diameter of the cooling tower [length]
H	Specific enthalpy [energy/mass]
H_g	Specific enthalpy of gas/vapour mixture [energy/mass]
H_l	Specific enthalpy of liquid [energy/mass]
H_{g1}	Specific enthalpy of gas/vapour mixture feed [energy/mass]

F. Montagnaro et al., *Elements of Unit Operations and Chemical Reactors*,
https://doi.org/10.1007/978-3-032-18046-9_9

H_{g2}	Specific enthalpy of gas/vapour mixture exiting the cooling tower [energy/mass]
H_{l1}	Specific enthalpy of liquid exiting the cooling tower [energy/mass]
H_{l2}	Specific enthalpy of liquid feed [energy/mass]
H_s	Saturation specific enthalpy [energy/mass]
$\mathcal{H}$	Absolute humidity [−]
$\mathcal{H}_p$	Percentage humidity [−]
$\mathcal{H}_r$	Relative humidity [−]
$\mathcal{H}_s$	Saturation humidity [−]
h_g	Heat transfer coefficient in the gaseous film [energy/(length^2·time·temperature)]
h_l	Heat transfer coefficient in the liquid film [energy/(length^2·time·temperature)]
k_m	Mass transfer coefficient in the gaseous film [length/time]
Le	Lewis number [−]
m_g	Molar mass of dry gas [mass/moles]
m_v	Molar mass of vapour [mass/moles]
P_v^0	Partial pressure of vapour in saturated gas [mass/(length·time^2)]
p_v	Partial pressure of vapour in gas [mass/(length·time^2)]
T	Temperature [temperature]
T_d	Dew point temperature [temperature]
T_{db}	Dry bulb temperature [temperature]
T_g	Temperature of gas/vapour mixture [temperature]
T_{g1}	Temperature of gas/vapour mixture feed [temperature]
T_{g2}	Temperature of gas/vapour mixture exiting the cooling tower [temperature]
T_l	Temperature of liquid [temperature]
T_{l1}	Temperature of liquid exiting the cooling tower [temperature]
T_{l2}	Temperature of liquid feed [temperature]
T_{ref}	Reference temperature [temperature]
T_s	Saturation temperature [temperature]
T_{wb}	Wet bulb temperature [temperature]
W_L	Liquid mass flow rate [mass/time]
$\dot{W}_G$	Gas/vapour mixture mass flux [mass/(length^2·time)]
$\dot{W}_L$	Water mass flux [mass/(length^2·time)]
$\dot{W}_G^{min}$	Minimum gas flux [mass/(length^2·time)]
Z	Packing height of the cooling tower [length]
z	Longitudinal coordinate of the cooling tower [length]
z_t	Total height of the cooling tower [length]
$\Delta H'$	Overall change of air enthalpy [energy/mass]
ΔT	Overall change of water temperature [temperature]
$\Delta T'$	Difference between the temperature of air leaving the packing and dry bulb temperature of inlet air [temperature]
$\lvert\lambda_{ref}\rvert$	Latent heat of vaporisation of liquid at T_{ref} [energy/mass]
π	Total pressure [mass/(length·time^2)]
ρ	Average density of gas/vapour mixture [mass/volume]

9.1 Generalities

Humidification and *dehumidification* operations generally refer to the mass and energy transfer that occur when a gas is brought into contact with a pure liquid in which it is practically insoluble (Coulson and Richardson 2017; Green and Perry 2008; McCabe et al. 2001; Treybal 1981). The most common application involves the air/water system. Specifically, in humidification, the vapour content in the gas stream is increased, whereas, in dehumidification, the vapour is partially condensed. Humidification and dehumidification can be achieved, for example, by contacting the gas with a liquid that partially evaporates or with a cold liquid (or another cold surface) that causes the partial condensation of the vapour, respectively. These processes are used for air conditioning, removal of H_2O from gases (e.g., from Cl_2 to prevent corrosion of plants/pipelines), and recovery of solvent vapour (e.g., acetone) from gas streams.

An important application of gas/liquid contact processes is the cooling of water used as a refrigerant in power plants and chemical/petrochemical facilities. This operation is carried out in *cooling towers*. In these units, hot water from the industrial process is sprayed at the top of the tower, flows downward over packing materials, and is cooled by a countercurrent air stream. The cooling process is based on the partial evaporation of water. The cooled water is collected at the bottom of the tower and recirculated back into the process loop, while the warm, humid air is discharged into the atmosphere from the top of the tower.

Cooling towers can be classified based on how the air draught (circulation) is achieved:

- *natural draught* systems (typically with a hyperbolic geometry) exploit the density difference between the warm internal air and the external air to generate the air flow. They are often used in nuclear and geothermal power plants. They have the advantage of low operating costs (absence of fans) but require high investment costs, and their operation is influenced by weather conditions;
- *mechanical draught* towers use fans to move the air. They can be based on either forced or induced draught, where the fan is located at the base (pushing air upward) or at the top (sucking air upward), respectively. These systems allow for greater control over the air velocity and have smaller size compared to natural draught towers, but with associated higher energy consumption.

Cooling towers are packed with filling materials in order to increase the air/water contact surface (as discussed for packed absorption towers). Wooden slats were used originally, while structured packings (made of plastic or metal) are nowadays largely adopted since they provide larger mass and heat exchange surfaces and lower pressure drops.

9.2 Hygrometric Quantities and Psychrometric Charts

The key parameters of interest for humidification operations are now presented. We start with the *absolute humidity* $\mathcal{H}$, which is the mass of vapour (water vapour or other solvent) carried by the unit mass of dry gas (air or other carrier gas):

$$\boxed{\mathcal{H} = \frac{p_v m_v}{(\pi - p_v) m_g}} \tag{9.1}$$

where π and p_v are the total pressure and partial pressure of vapour in the gas, respectively, while m_v and m_g represent the molar mass of vapour and dry gas, respectively.

The *saturated gas* is the one in which the vapour is in equilibrium with the liquid at the system temperature. The associated *saturation humidity* $\mathcal{H}_s$ is:

$$\boxed{\mathcal{H}_s = \frac{P_v^0 m_v}{\left(\pi - P_v^0\right) m_g}} \tag{9.2}$$

with P_v^0 representing the partial pressure of vapour in the saturated gas (equal to the vapour pressure of the liquid at the gas temperature).

The *percentage humidity* $\mathcal{H}_p$ and the *relative humidity* $\mathcal{H}_r$, at a given gas temperature, read respectively:

$$\boxed{\mathcal{H}_p = 100 \frac{\mathcal{H}}{\mathcal{H}_s} = 100 \frac{p_v\left(\pi - P_v^0\right)}{P_v^0(\pi - p_v)}} \tag{9.3}$$

$$\boxed{\mathcal{H}_r = 100 \frac{p_v}{P_v^0}} \tag{9.4}$$

The *humid heat* $\mathbb{C}_s$ is the heat required to increase by 1 K the temperature of the unit mass of humid gas (i.e. dry gas plus the vapour it contains):

$$\boxed{\mathbb{C}_s = \mathbb{C}_g + \mathcal{H}\mathbb{C}_v} \tag{9.5}$$

where $\mathbb{C}_g$ and $\mathbb{C}_v$ are the specific heats of the dry gas and vapour, respectively. Obviously, the humid heat corresponds to the specific heat of dry gas ($\mathbb{C}_s = \mathbb{C}_g$) if the vapour content in the gas is zero ($\mathcal{H} = 0$).

The specific enthalpies of the gas/vapour mixture H_g and pure liquid H_l are given by, respectively:

$$H_g = \mathbb{C}_s\left(T_g - T_{ref}\right) + \mathcal{H}\,|\,\lambda_{ref}\,| \tag{9.6}$$

$$H_l = \mathbb{C}_l\left(T_l - T_{ref}\right) \tag{9.7}$$

where T_{ref}, T_g and T_l are the reference temperature (273 K for the air/water system), the one of humid gas and liquid, respectively; $\mathcal{C}_l$ is the specific heat of the liquid while $|\lambda_{ref}|$ is the latent heat of vaporisation of liquid, taken at at T_{ref}.

Finally, the *dew point temperature* T_d is the temperature at which a mixture of gas and vapour becomes saturated on cooling; the *dry bulb temperature* T_{db} is the temperature of a vapour/gas mixture measured by a bulb thermometer; and the *wet bulb temperature* T_{wb} is the steady-state temperature reached by a small amount of liquid immersed in a gaseous stream under adiabatic conditions.

Example: Hygrometric Calculation for the Air/Water System

An air stream at $\pi = 1$ bar total pressure and 310 K contains water vapour at partial pressure $p_v = 0.031$ bar. Calculate $\mathcal{H}$, $\mathcal{H}_s$, $\mathcal{H}_p$, $\mathcal{H}_r$ and $\mathcal{C}_s$.

The molar mass of water and air can be taken, as usual, at $m_v = 18$ kg/kmol and $m_g = 29$ kg/kmol, respectively. Thus, we can compute the absolute humidity (Eq. 9.1): $\mathcal{H} = 0.020$ kg H_2O/kg air. The vapour pressure of water at 310 K can be calculated by the Antoine's formula (cf., e.g., Green and Perry 2008), giving $P_v^0 = 0.062$ bar. We can now evaluate the saturation humidity (Eq. 9.2): $\mathcal{H}_s = 0.041$ kg H_2O/kg air. The percentage humidity (Eq. 9.3) is $\mathcal{H}_p = 48.78\%$ and the relative humidity (Eq. 9.4) is $\mathcal{H}_r = 50\%$. Assuming $\mathcal{C}_g = 1.00$ kJ/(kg·K) and $\mathcal{C}_v = 1.88$ kJ/(kg·K) for air and water vapour, respectively, we obtain, Eq. 9.5, $\mathcal{C}_s = 1.04$ kJ/(kg·K).

Psychrometric charts (or *humidity charts*) are diagrams reporting hygrometric and thermodynamic properties (e.g., humidity, humid heat and enthalpy) of mixtures of a permanent gas and a condensable vapour at a fixed pressure. The most common system is the air/water one, for which different charts are available (cf., e.g., Green and Perry 2008). As an illustrative example, Fig. 9.1 reports the absolute humidity vs. temperature trend of air/water mixtures at different relative humidity levels.

Enthalpy data of the saturated gas, H_s, are fundamental for the design of cooling towers (as it will be shown later), and they can be retrieved from humidity charts. In the common case of the air/water system, the following useful H_s vs. T_s (saturation temperature) relationship has been here derived from data fitting in the range 278–333 K:

$$\begin{cases} H_s = C_1 T_s^3 - C_2 T_s^2 + C_3 T_s - C_4 \\ C_1 = 0.0026845; C_2 = 2.2958658; C_3 = 657.0727254; C_4 = 62,891.2830545 \end{cases} \tag{9.8}$$

with H_s and T_s expressed in [kJ/kg dry air] and [K], respectively.

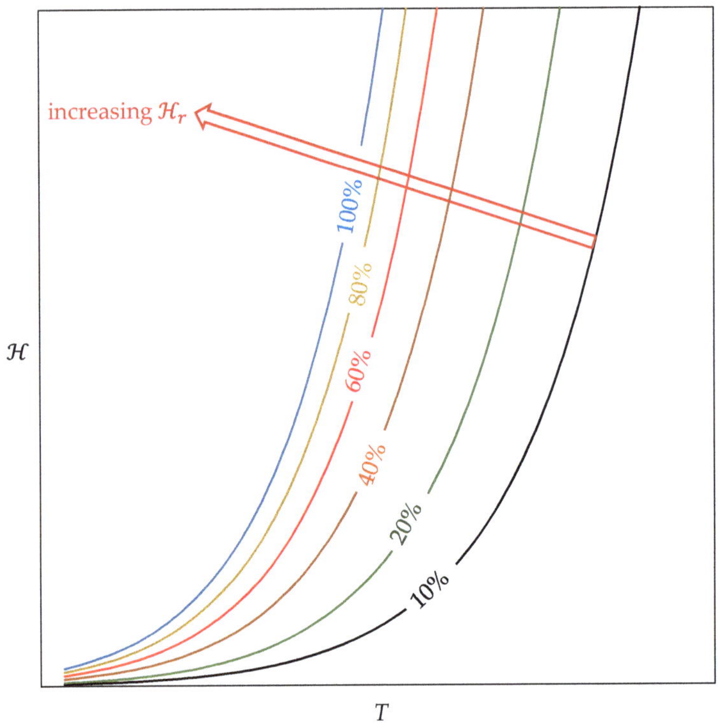

Fig. 9.1 Humidity vs. temperature chart for the air/water system

Example: Enthalpy Calculation for the Air/Water System
Determine the enthalpy of an air stream at 1 bar total pressure and 323 K containing water vapour at partial pressure $p_v = 0.121$ bar. The vapour pressure of water at the gas temperature calculated from Antoine's equation is: $P_v^0 = 0.121$ bar. It follows that the gas is saturated ($p_v = P_v^0$) and Eq. 9.8 can be applied to obtain $H_s = 280.8$ kJ/kg.

9.3 Design of Cooling Towers

9.3.1 Section of Natural Draught Tower

Natural draught towers of hyperbolic type, adopted for water cooling with air, are designed so that the average gas velocity above the packing material is in the range 1–2 m/s, and the ratio between the total height of the equipment z_t (not limited to the packing material) and the base diameter D_b is typically $z_t/D_b = 1.5$ (Coulson and

Richardson 2017). The base area of the hyperbolic tower A_b is related to z_t through the following relationship proposed in literature (Chilton 1952), where A_b is expressed in [m^2] and z_t in [m]:

$$A_b = 0.051 \frac{C_P^{1.5} C_D}{z_t^{0.5}} \tag{9.9}$$

where C_P is a performance coefficient (typically 5.2 for wooden slats), while C_D is a duty coefficient that can be evaluated via the following equation:

$$C_D = \frac{W_L}{C_1 \frac{\Delta H'}{\Delta T} \left(\Delta T' + C_2 \Delta H'\right)^{0.5}} \tag{9.10}$$

where W_L is the water mass flow rate in [kg/s], $\Delta H'$ is the overall change of air enthalpy in [kJ/kg], ΔT is the overall change of water temperature in [K], and $\Delta T'$ is the difference between the temperature of air leaving the packing and the dry bulb temperature of the air entering the tower, in [K]; the constants value $C_1 = 0.00369$ and $C_2 = 0.0752$.

Example: Water Flow Rate Calculation for Natural Draught Cooling Tower

A hyperbolic natural draught tower with maximum height $z_t = 80$ m has to be adopted to cool a water stream from 308 to 293 K using air (i.e., $\Delta T = 15$ K). Determine the water flow rate that can be processed in the equipment, in the case that $\Delta H' = 40$ kJ/kg and $\Delta T' = 5$ K.

The design criterion $z_t/D_b = 1.5$ allows to determine the base diameter $D_b = 53.3$ m, corresponding to a circular base area $A_b = 2230.1$ m^2. Given $C_P = 5.2$ for wooden slats, the duty coefficient is (from Eq. 9.9) $C_D = 32{,}983.3$. Finally, we get the liquid mass flow rate via Eq. 9.10; it is $W_L = 918.4$ kg/s.

9.3.2 Operating and Equilibrium Conditions

Figure 9.2 depicts a schematic representation of a continuous countercurrent cooling tower with relevant quantities:

- the packing height of the tower is Z, with 1 and 2 denoting the bottom and top section, respectively;
- water to be cooled enters the top of the column (hot water) with mass flux $\dot{W}_L$, temperature T_{l2} and specific enthalpy H_{l2}. The cold water exits the bottom of the column with the same mass flux, temperature T_{l1} (with $T_{l1} < T_{l2}$) and specific enthalpy H_{l1};

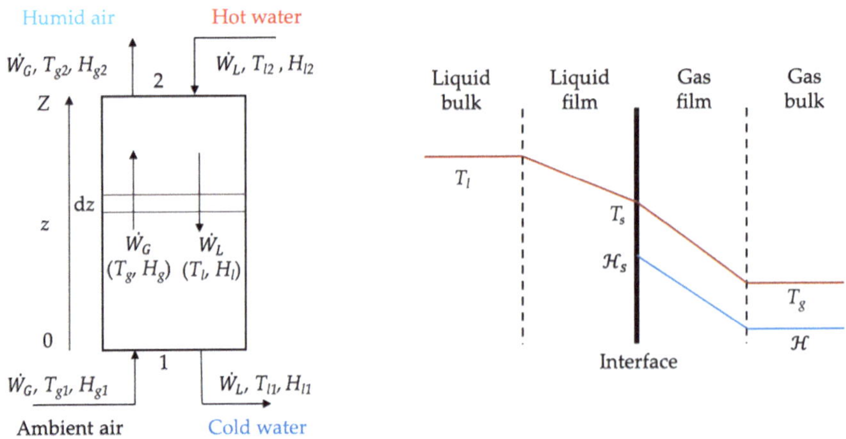

Fig. 9.2 Cooling tower. On the left: general operating scheme and relevant streams. On the right: temperature and humidity profiles according to the two-film theory (case of $T_l > T_g$)

- ambient air enters the bottom of the tower with mass flux $\dot{W}_G$, temperature T_{g1} and specific enthalpy H_{g1}. The air exits the top of the column (humid air) with the same mass flux, temperature T_{g2} (with $T_{g2} > T_{g1}$) and specific enthalpy H_{g2}.

It is therefore assumed that the mass flux for both gas and liquid is constant in the column, as typically only about 1–5% of the total fed water evaporates (Coulson and Richardson 2017), using an approach similar to that highlighted for absorption towers.
The cooling process occurs through combined mass and energy transfer phenomena, and a schematic representation of humidity and temperature profiles according to the two-film theory is reported in Fig. 9.2. It is highlighted that the case of liquid temperature greater than the gas one is reported only as an example (indeed, $T_l < T_g$ can also occur as a function of the tower longitudinal coordinate z). It is assumed that *at the air/water interface an equilibrium condition exists, and the air is saturated*; this implies that the temperature at interface is the saturation one (T_s) and (T_s;$\mathcal{H}_s$) lies on the saturation curve of the humidity chart as in Fig. 9.1 (Coulson and Richardson 2017; McCabe et al. 2001).

Operating conditions in the cooling tower are determined by applying energy balances usually written in terms of enthalpy. In particular, the heat flux removed from the water to be cooled is transferred to the air, which is heated. Equations 9.11 and 9.12 represent, respectively, the enthalpy balance for the entire apparatus and that between a generic section and the bottom of the tower (please remember Eq. 9.7):

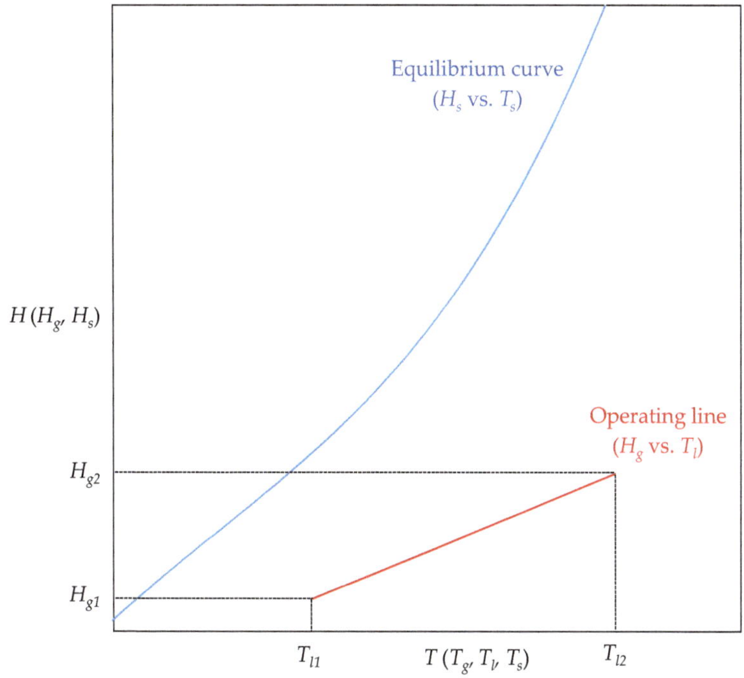

Fig. 9.3 Operating and equilibrium conditions in a cooling tower

$$\dot{W}_L(H_{l2} - H_{l1}) = \dot{W}_G(H_{g2} - H_{g1}) \Rightarrow \dot{W}_L C_l(T_{l2} - T_{l1})$$
$$= \dot{W}_G(H_{g2} - H_{g1}) \Rightarrow H_{g2} = H_{g1} + \frac{\dot{W}_L}{\dot{W}_G} C_l(T_{l2} - T_{l1}) \tag{9.11}$$

$$\boxed{H_g = H_{g1} + \frac{\dot{W}_L}{\dot{W}_G} C_l(T_l - T_{l1})} \tag{9.12}$$

Equation 9.12 (or, equivalently, Eq. 9.11) represents the operating conditions in the cooling tower and it is a straight line on a H_g vs. T_l plane (whose slope is proportional to $\dot{W}_L/\dot{W}_G$), as shown in Fig. 9.3. The enthalpy of saturated air H_s vs. T_s (from psychrometric chart or Eq. 9.8), representing the equilibrium curve, is also reported.

9.3.3 Minimum Gas Flux

An increase in the gas mass flux $\dot{W}_G$ (at fixed inlet gas temperature and humidity, and inlet and outlet liquid temperatures) results in a decrease in the slope of the operating line. The *minimum gas flux condition* (pinch, in analogy with unit

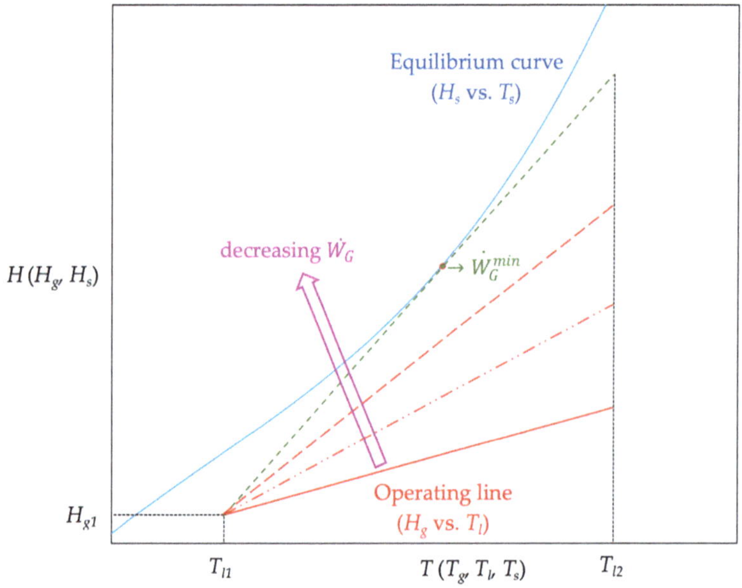

Fig. 9.4 Minimum gas flux in a cooling tower

operations described in previous chapters) for the cooling tower, $\dot{W}_G^{min}$, is that for which the operating and equilibrium conditions coincide in a section of the cooling tower: the operation would be achievable only with a tower of theoretically infinite height. This critical condition is determined by drawing, from a known point (T_{l1}; H_{g1}), the operating line tangent to the equilibrium curve (cf. Figure 9.4). Therefore, the operation is feasible with a finite packing height for $\dot{W}_G > \dot{W}_G^{min}$ (i.e., as also seen for other unit operations, with flow rate of the service fluid properly increased vs. its minimum value, according to feasibility and economic considerations: the higher this flux, the smaller the tower but at higher operating costs—an optimal condition will apply; please remember the concept of "distance" between operating and equilibrium curves already discussed in other chapters).

9.3.4 Packing Height: Design Equation

To determine the packing height Z of a cooling tower, mass and energy transport equations for the liquid/gas system are written in a differential element dz of the tower, in accordance with the two-film theory (cf. Fig. 9.2). The heat flux from the liquid bulk to the gas/liquid interface, determining a change in the liquid enthalpy, is:

$$h_l a \mathrm{d}z(T_l - T_s) = \dot{W}_L \mathbb{C}_l \mathrm{d}T_l \Rightarrow \frac{\mathrm{d}T_l}{T_l - T_s} = \frac{h_l a \mathrm{d}z}{\dot{W}_L \mathbb{C}_l} \tag{9.13}$$

where h_l is the heat transfer coefficient in the liquid film, and a represents the packing specific surface area, as usual. The heat flux from the gas/liquid interface to the bulk gas is:

$$h_g a \mathrm{d}z\left(T_s - T_g\right) = \dot{W}_G \mathbb{C}_s \mathrm{d}T_g \Rightarrow \frac{\mathrm{d}T_g}{T_s - T_g} = \frac{h_g a \mathrm{d}z}{\dot{W}_G \mathbb{C}_s} \tag{9.14}$$

where h_g is the heat transfer coefficient in the gaseous film. The mass flux of water vapour from the gas/liquid interface to the bulk gas is:

$$k_m a \rho \mathrm{d}z(\mathcal{H}_s - \mathcal{H}) = \dot{W}_G \mathrm{d}\mathcal{H} \Rightarrow \frac{\mathrm{d}\mathcal{H}}{\mathcal{H}_s - \mathcal{H}} = \frac{k_m a \rho \mathrm{d}z}{\dot{W}_G} \tag{9.15}$$

where k_m is the mass transfer coefficient in the gaseous film, while ρ is the average density of the gas/vapour mixture. It is highlighted that correlations for heat and mass transfer coefficients (appearing in Eqs. 9.13–9.15) as a function of $\dot{W}_L$ and $\dot{W}_G$ are available for different kinds of packing materials (cf., e.g., Coulson and Richardson 2017; Goshayshi and Missenden 2000).

Now we introduce the Lewis number Le [named after Warren Kendall Lewis (1882–1975), U.S. scientist, Professor at Massachusetts Institute of Technology; see Walker et al. 1923] that, for such applications, is expressed as (Coulson and Richardson 2017):

$$\boxed{Le = \frac{h_g}{k_m \rho \mathbb{C}_p}} \tag{9.16}$$

where $\mathbb{C}_p$ is the average specific heat of the gas/water vapour mixture. *For the air/water system, $Le \cong 1$*, and, for low values of $\mathcal{H}$, it is $\mathbb{C}_p \cong \mathbb{C}_s$; hence, Eq. 9.16 can be rewritten as:

$$h_g \cong k_m \rho \mathbb{C}_s \tag{9.17}$$

Using Eq. 9.17 in combination with Eqs. 9.13–9.15 and relationships between gas enthalpy and humidity (the scholar may consult Coulson and Richardson 2017 for a detailed demonstration), literature reports:

$$\frac{\mathrm{d}H_g}{H_s - H_g} = \frac{k_m a \rho \mathrm{d}z}{\dot{W}_G} \tag{9.18}$$

The integration of Eq. 9.18 between the bottom and top of the column provides the total height of the packing or, in other words, the *design equation*:

$$Z = \frac{\dot{W}_G}{k_m a \rho} \int_{H_{g1}}^{H_{g2}} \frac{dH_g}{H_s - H_g} \tag{9.19}$$

9.3.5 Interface Conditions and Gas Temperature Profile

From mass and heat transport equations, the following relationship among liquid temperature, air enthalpy and interface (saturation) conditions can be obtained (Coulson and Richardson 2017):

$$\frac{H_g - H_s}{T_l - T_s} = -\frac{h_l a}{k_m a \rho} \tag{9.20}$$

Equation 9.20 represents a line with slope $-h_l a/(k_m a \rho)$ in the enthalpy vs. temperature diagram (cf. Fig. 9.5). This line with negative slope joins the point $(T_l;H_g)$ with the point $(T_s;H_s)$. We operate as follows: first, we draw the actual operating line, connecting H_g and T_l for the top and bottom section of the tower; for a given section of the tower, i.e. for a given point on the operating line $(T_l;$ $H_g)$, we draw the line in Eq. 9.20; the intersection of this line with the equilibrium curve gives, for that section, the values for H_s and Ts (this has been done, in Fig. 9.5 for section "1" of the tower). Of course, this procedure must be iterated for each

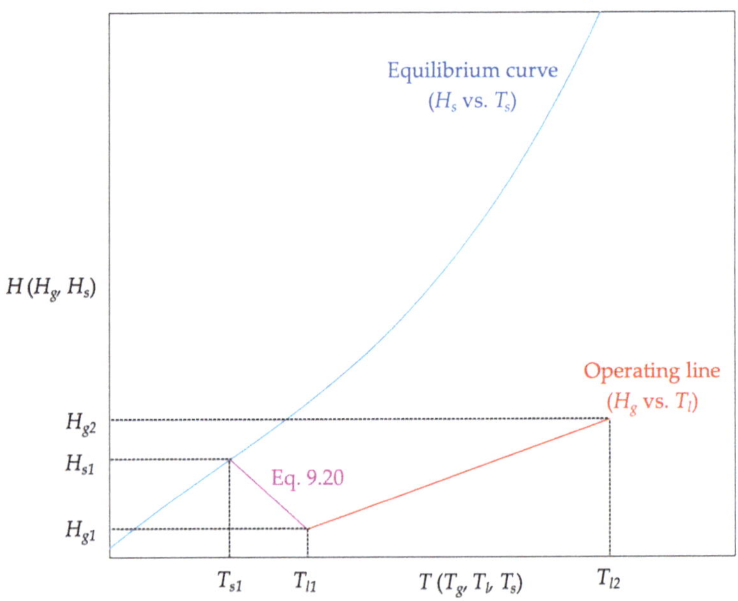

Fig. 9.5 Conditions at the gas/liquid interface in a cooling tower

section of the tower. The knowledge, for each section and, therefore, for each H_g-value, of the corresponding H_s-value, will finally allow to use Eq. 9.19 for the evaluation of the total height. The reader has already noticed some similarity with the approach followed in absorption towers, Fig. 8.4.

In conclusion, it is highlighted that the determination of the gas temperature profile within the tower is important to verify the absence of fog formation conditions, that is, the formation of water droplets where the air becomes saturated. This, of course, would nullify the driving force for heat exchange. Therefore, it is necessary to verify that the gas temperature profile lies below the equilibrium curve at every section of the tower. This aspect, that goes beyond the scope of the present chapter, is treated in detail in specialised textbooks (e.g., Coulson and Richardson 2017).

Example: Design of a Cooling Tower

A packed cooling tower has to be operated to cool a water stream with flux $\dot{W}_L = 3.50$ kg/(m^2·s) from $T_{l2} = 323$ K to $T_{l1} = 303$ K, with ambient air fed at $T_{g1} = 293$ K and relative humidity $\mathcal{H}_r = 10\%$. Under the hypothesis of $\dot{W}_G = 2.04\dot{W}_G^{min}$ (optimal value from economic analyses), calculate the total packing height Z. It can be assumed that, cf. Eq. 9.19, $\dot{W}_G/(k_m a\rho) = 2$ m.

Firstly, we can represent the equilibrium curve on the enthalpy vs. temperature diagram (cf. Fig. 9.6) by using Eq. 9.8 or psychrometric diagrams. The absolute humidity of air at 293 K and $\mathcal{H}_r = 10\%$ can be determined from Eqs. 9.4 and 9.1, and by using the Antoine's equation; it is $\mathcal{H} = 1.4 \cdot 10^{-3}$ kg H_2O/kg air. Through Eq. 9.5, we verify that at this (low) humidity content the humid heat is practically equivalent to the specific heat of dry air; hence, we assume $\mathcal{C}_s = \mathcal{C}_g = 1$ kJ/(kg·K). The latent heat of vaporisation of water at the reference temperature $T_{ref} = 273$ K is $|\lambda_{ref}| = 2502$ kJ/kg (cf., e.g., steam tables in Coulson and Richardson 2023). Now, we can apply Eq. 9.6 to calculate the inlet air specific enthalpy; we get $H_{g1} = 23.5$ kJ/kg. This enables us to locate, on the H vs. T diagram of Fig. 9.6, the point A ($T_{l1} = 303$ K; $H_{g1} = 23.5$ kJ/kg) representing the operating conditions at the tower bottom.

Now, the minimum gas flux $\dot{W}_G^{min}$ is determined by drawing, from point A, the operating line tangent to the equilibrium curve (condition represented by point C in Fig. 9.6; of course, the condition of tangency leading to point C can also be found analytically). We read on the plot $H_{g2}(\dot{W}_G^{min}) = 278.0$ kJ/kg at $T_{l2} = 323$ K (point D in Fig. 9.6). So, we derive $\dot{W}_G^{min} = 1.15$ kg/(m^2·s) by applying Eq. 9.11 (with $\mathcal{C}_l = 4.18$ kJ/(kg·K)), rewritten as:

(continued)

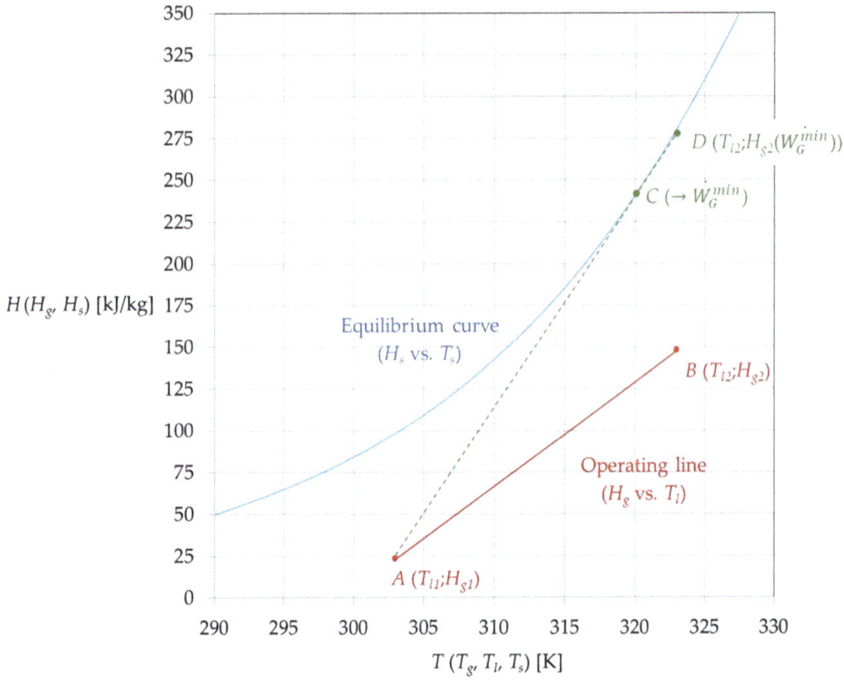

Fig. 9.6 Design of a cooling tower

Table 9.1 Data for the approximated solution of the design equation

H_g [kJ/kg]	T [K] Eq. 9.12	H_s [kJ/kg] Eq. 9.8	H_s–H_g [kJ/kg]
31.8	304.3	105.3	73.5
40.1	305.7	113.4	73.3
48.4	307.0	121.5	73.1
56.7	308.3	130.2	73.5
65.0	309.7	140.2	75.2
73.3	311.0	150.3	77.0
81.6	312.3	161.0	79.4
89.9	313.7	173.4	83.5
98.2	315.0	185.7	87.5
106.5	316.3	198.9	92.4
114.8	317.7	214.0	99.2
123.1	319.0	228.9	105.8
131.4	320.3	244.8	113.4
139.7	321.7	262.9	123.2
148.0	323.0	280.8	132.8

$$\dot{W}_G = \dot{W}_L \frac{C_l(T_{l2} - T_{l1})}{H_{g2} - H_{g1}} \tag{9.21}$$

That is, $\dot{W}_G = 2.35$ kg/(m^2·s). We have the slope of the operating line (Eq. 9.11, or 9.12, or 9.21: they are equivalent), and we get $H_{g2} = 148.0$ kJ/kg. So, the point B ($T_{l2} = 323$ K;$H_{g2} = 148.0$ kJ/kg) can be located on the plot and represents the operating conditions at the tower top. The operating line is drawn.

Now, the design equation. The approach we propose is to approximate the integral in Eq. 9.19 to a summation with constant interval ΔH_g (for example, take $\Delta H_g = 8.3$ kJ/kg):

$$Z \cong \frac{\dot{W}_G}{k_m a \rho} \Delta H_g \sum_{H_{g1}}^{H_{g2}} \frac{1}{H_s - H_g} \tag{9.22}$$

We can now build a table (Table 9.1) where, for each H_g interval from H_{g1} to H_{g2} (for example, choose to calculate at the higher bound of the interval, so the first value will be $H_{g1} + \Delta H_g$ and so on, up to H_{g2}), we list the value of temperature on the operating line, calculated from Eq. 9.12. The knowledge of temperature, which is the abscissa of our plot in Fig. 9.6, allows to calculate, from Eq. 9.8, H_s. The last column of the table is the difference H_s–H_g. By application of the design equation, it is:

$$Z \cong \frac{\dot{W}_G}{k_m a \rho} \Delta H_g \left(\frac{1}{73.5} + \ldots + \frac{1}{132.8} \right) = 2.8 \text{ m} \tag{9.23}$$

Graphically, although it is a less precise procedure, we could read the values reported in Table 9.1 directly on the map in Fig. 9.6.

9.4 Background Information

- Key hygrometric quantities (absolute humidity, saturation humidity, percentage humidity, relative humidity, humid heat, and specific enthalpies) are defined (Eqs. 9.1–9.7), also with the aid of the psychrometric diagram (Fig. 9.1).
- The section of a natural draught tower is determined through the Chilton's relationships (Eqs. 9.9 and 9.10).
- Cooling towers (Fig. 9.2) operating conditions are determined using enthalpy balances (Eqs. 9.11 and 9.12).
- The minimum gas flux (Fig. 9.4) corresponds to a pinch condition where the operating line is tangent to the equilibrium curve, requiring an infinite packing height.

- The packing height design equation (Eq. 9.19) is determined by integrating the differential mass and energy transfer equations.
- Interfacial conditions are described by Eq. 9.20 and represented graphically in Fig. 9.5.

References

Chilton H (1952) Performance of natural-draught cooling towers. Proc IEE Part II 99:440–452. https://doi.org/10.1049/pi-2.1952.0105

Coulson JM, Richardson JF (2017) Coulson and Richardson's chemical engineering: heat and mass transfer, vol 1B. Butterworth-Heinemann, Oxford

Coulson JM, Richardson JF (2023) Coulson and Richardson's chemical engineering: separation processes, vol 2B. Butterworth-Heinemann, Oxford

Goshayshi HR, Missenden JF (2000) The investigation of cooling tower packing in various arrangements. Appl Therm Eng 20:69–80. https://doi.org/10.1016/S1359-4311(99)00011-3

Green DW, Perry RH (2008) Perry's chemical engineers' handbook. McGraw-Hill, New York

McCabe WL, Smith JC, Harriott P (2001) Unit operations of chemical engineering. McGraw-Hill, Singapore

Treybal RE (1981) Mass-transfer operations. McGraw-Hill, Singapore

Walker WH, Lewis WK, McAdams WH (1923) Principles of chemical engineering. McGraw-Hill, York

Ideal Chemical Reactors 10

Abstract

With this chapter, a new part of the book begins, dedicated to chemical reactors. They should be seen as strictly interconnected to the unit operations (with related transport phenomena) we have seen so far, since it is plausible that a reactor includes unit operations (e.g., various separations, heat exchanges) both upstream (for the preparation of the reacting feed) and downstream (for obtaining the desired products in their final form). This chapter introduces the general concepts related to reaction rate and degree of conversion, and then it presents the three ideal chemical reactors (discontinuous with perfect mixing, continuous with perfect axial segregation, continuous with perfect mixing), each with its fluid dynamic aspects and design equation. Reactors are then accompanied by considerations related to productivity in the desired species.

Nomenclature

C_{TOT}	Total concentration [moles/volume]
C_i	Concentration of i species [moles/volume]
C_{i0}	Concentration of i species at reaction beginning [moles/volume]
C_i^e	Concentration of i species at reactor exit [moles/volume]
$\mathcal{D}$	Diffusion coefficient [length^2/time]
Da	Damköhler number [−]
E	Activation energy [energy/moles]
F_i	Molar flow rate of i species [moles/time]
F_{i0}	Molar flow rate of i species at reaction beginning [moles/time]
F_i^e	Molar flow rate of i species at reactor exit [moles/time]
k	Kinetic constant [(1/time)·(moles/volume)$^{1-n}$]
k_0	Arrhenius' pre-exponential [(1/time)·(moles/volume)$^{1-n}$]
L	Reactor length [length]

F. Montagnaro et al., *Elements of Unit Operations and Chemical Reactors*,
https://doi.org/10.1007/978-3-032-18046-9_10

n	Total reaction order [−]
n_{TOT}	Total moles [moles]
n_i	Moles of i species [moles]
n_{i0}	Moles of i species at reaction beginning [moles]
$\mathcal{P}$	Pressure [mass/(length·time^2)]
P_P	Productivity of the desired species [moles/time]
Pe_a	Péclet number (axial) [−]
Pe_r	Péclet number (radial) [−]
Q	Volumetric flow rate [volume/time]
Q^e	Volumetric flow rate at reactor exit [volume/time]
Q_0	Volumetric flow rate at reaction beginning [volume/time]
R	Reactor radius [length]
$\mathfrak{R}$	Gas constant [energy/(moles·temperature)]
r_A	Rate of reaction [moles/(volume·time)]
S	Reactor cross-section area [length^2]
T	Temperature [temperature]
t	Time [time]
t^*	Optimal reaction time for maximum productivity [time]
t_R	Characteristic reaction time [time]
$t_{d,a}$	Characteristic time for axial diffusive phenomena [time]
$t_{d,r}$	Characteristic time for radial diffusive phenomena [time]
t_m	Dead time [time]
V	Volume [volume]
V_0	Volume at reaction beginning [volume]
v	Fluid velocity [length/time]
v_0	Fluid velocity at reactor inlet [length/time]
x_A	Degree of conversion [−]
x_A^e	Exit degree of conversion [−]
z	Reactor axial coordinate [length]
α	Stoichiometric coefficient for the limiting species [−]
β	Stoichiometric coefficient for the desired species [−]
γ	Normalised stoichiometric coefficient [−]
ε	Expansion coefficient [−]
τ	Space-time for continuous reactor [time]

10.1 General Definitions and Concepts

The discussion begins by exposing some general aspects. For further considerations and approaches on chemical reactors, scholars can consult other textbooks, such as those by Hill and Root 2014; Jakobsen 2014; Levenspiel 1999; Li et al. 2017; Roberts 2009; Westerterp et al. 1988. Reference will be made to chemical reactions (extensions will be discussed later in this book):

- *single;*
- *isothermal and isobaric;*
- *homogeneous (liquid or ideal gas phase).*

We will generally write a reaction as:

$$A + \ldots \Rightarrow P + \ldots \tag{10.1}$$

where "A" is the *limiting reactant* (often in presence of other reactants), and "P" is the *desired product* (often in presence of other products). When needed, stoichiometric coefficients will be declared. And, when it is not relevant to declare the presence of other reactants/products, we will simplify Eq. 10.1 (for example, by writing A⇒P). The reaction in Eq. 10.1 has been written as an irreversible process (equilibrium thermodynamic constant much greater than 1), without affecting here the generality of the concepts.

10.1.1 Reaction Rate

The reaction in Eq. 10.1 is described by its *rate*, r_A, i.e. the moles of A that react per unit time and per unit volume of the reacting phase, which in a discontinuous system can be expressed as:

$$r_A = -\frac{dC_A}{dt} \tag{10.2}$$

i.e. the variation in time t of the concentration of A species, C_A (the minus sign is to make the reaction rate positive, since the derivative has an intrinsically negative sign as C_A decreases over time). The most common analytical expression to describe the reaction rate is the so-called *power law*, translating the concept that the higher the concentration of A, the higher the reaction rate:

$$r_A = kC_A^n \tag{10.3}$$

where n is the *reaction order* and k the *kinetic constant* (actually constant under isothermal conditions). The dimensions of k depend on the reaction order; generally, k is expressed in [(1/time)·(moles/volume)$^{1-n}$]. In fact, these dimensions guarantee the correctness of Eq. 10.3. Examples are given in Table 10.1 and Fig. 10.1, where we also consider, although less frequent, the special case of $n = 0$ (constant reaction rate until A completely disappears).

Table 10.1 Examples of power law for the rate of reaction, Eq. 10.3

n	r_A	Dimensions for k
0	k ($\forall C_A \geq 0$)	[(1/time)·(moles/volume)]
0.5	$kC_A^{0.5}$	[(1/time)·(moles/volume)$^{0.5}$]
1	kC_A	[1/time]
2	kC_A^2	[(1/time)·(volume/moles)]

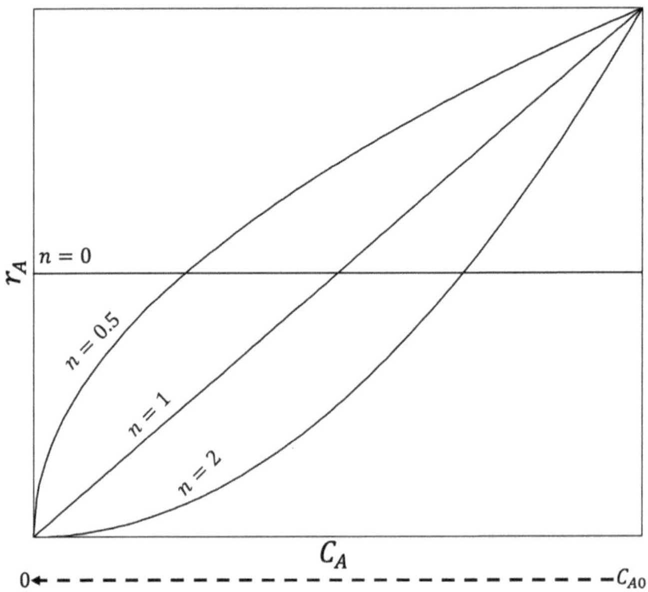

Fig. 10.1 Examples (general qualitative trends) of power law for the rate of reaction, Eq. 10.3. The arrow indicates the direction of occurrence of the chemical reaction, from $C_A = C_{A0}$ (initial conditions) to $C_A = 0$

For *elementary reactions*, the reaction orders coincide with the stoichiometric coefficients. It is therefore possible to derive, from the stoichiometry, the kinetic law. As examples:

$$\left\{\begin{array}{c} A \Rightarrow P \\ 2A \Rightarrow P \\ A+2B \Rightarrow P \end{array}\right. \Rightarrow \left\{\begin{array}{c} r_A = kC_A \\ r_A = kC_A^2 \\ r_A = kC_A C_B^2 \end{array}\right. \tag{10.4}$$

In the last case of Eq. 10.4, we will define partial reaction orders (1 for A, 2 for B), while $n = 3$ is the total reaction order (with respect to which the dimensions of k are calculated).

For *non-elementary reactions* (i.e. those whose stoichiometry hides a series of intermediate stages), the kinetic law must be known through experiments or literature. For example:

$$\left\{\begin{array}{c} A \Rightarrow P \\ A+B \Rightarrow P \\ A \Rightarrow P \end{array}\right. \Rightarrow \left\{\begin{array}{c} r_A = kC_A^2 \\ r_A = kC_A C_B^2 \\ r_A = k \end{array}\right. \tag{10.5}$$

There may also be cases of reaction rates not expressed by power law, such as:

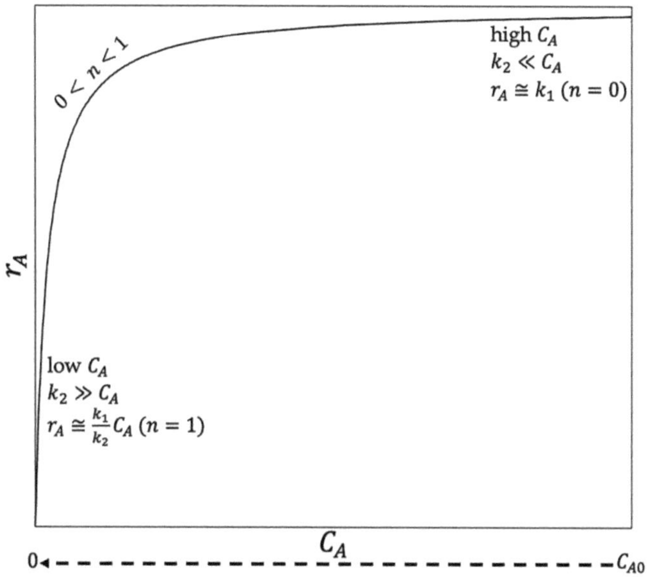

Fig. 10.2 Example of Michaelis–Menten kinetic law, Eq. 10.6, with shifting order

- enzymatic catalysis/growth of microorganisms (A⇒P), described by the Michaelis–Menten/Monod law (Michaelis and Menten 1913, further translated in Johnson and Goody 2011; Monod 1942):

$$r_A = \frac{k_1 C_A}{k_2 + C_A} \Rightarrow \begin{cases} k_2 \ll C_A \Rightarrow r_A \cong k_1\ (n = 0) \\ k_2 \gg C_A \Rightarrow r_A \cong \dfrac{k_1}{k_2} C_A\ (n = 1) \end{cases} \tag{10.6}$$

  a case, as also illustrated in Fig. 10.2, of a "shifting order" reaction (cf. Westerterp et al. 1988). Here, k_1 has the same dimensions of r_A, while k_2 is homodimensional to a concentration, so as to respect the dimensional correctness of Eq. 10.6;
- reactions of microbiological interests (A⇒P) for which it is:

$$r_A = \frac{k_1 C_A C_P}{k_2 + C_A} \tag{10.7}$$

  where, again, k_2 is homodimensional to a concentration, while k_1 is in [1/time]. Expressions for the rate of reaction, such as in Eq. 10.7, directly depending also on the concentration of the product (C_P), have an autocatalytic behaviour and will be investigated further in a next chapter;
- chain reactions in gas phase (A + B ⇒2P), where:

$$r_A = \frac{k_1 C_A C_B^{0.5}}{k_2 + \frac{C_P}{C_B}} \tag{10.8}$$

with k_1 expressed in [(1/time)·(volume/moles)$^{0.5}$] and k_2 non-dimensional.

10.1.2 Degree of Conversion and Effect of Volume Changes

The *degree of conversion* of A (sometimes abbreviated in "conversion"), x_A, is defined for discontinuous processes as:

$$x_A = \frac{n_{A0} - n_A}{n_{A0}} = \frac{C_{A0}V_0 - C_A V}{C_{A0}V_0} \tag{10.9}$$

with n_A and V the number of moles of A and the volume of reactant phase ("0" indicates the initial conditions for $t = 0$), respectively. x_A varies between 0 (at the beginning of the reaction, when $n_A = n_{A0}$) and 1 (a virtual value achieved when A completely disappears, i.e. $n_A = 0$). For continuous processes, the number of moles of A is replaced by the concept of molar flow rate of A, F_A [moles of A/time], and the reacting volume is replaced by the total volumetric flow rate Q circulating in the reactor [volume/time]. It is $F_A = QC_A$ and:

$$x_A = \frac{F_{A0} - F_A}{F_{A0}} = \frac{C_{A0}Q_0 - C_A Q}{C_{A0}Q_0} \tag{10.10}$$

For discontinuous processes (nothing changes for continuous processes, provided that the appropriate substitutions of V with Q, and of n_A with F_A, are made), if the reaction occurs at constant V, the equations can be simplified into:

$$\boxed{\begin{cases} x_A = \dfrac{C_{A0} - C_A}{C_{A0}} \\ C_A = C_{A0}(1 - x_A) \end{cases}} \tag{10.11}$$

which offers a first set of equations relating x_A and C_A. It is worth reminding that Eq. 10.11 holds for *isochoric processes* (constant volume), such as we assume reactions in the liquid phase, or reactions in the gas phase for which there is no overall variation in the number of moles—remembering that if temperature and pressure are constant by hypothesis, V must be constant according to the ideal gas law. It has been seen that it is possible to express $r_A(C_A)$, e.g. in Eq. 10.3. Even in cases where the concentration of other species appears (examples in Eqs. 10.4, 10.5, 10.7 and 10.8), by mass balances (with A being the limiting reactant) we can always obtain a $r_A(C_A)$ function. Using Eq. 10.11 in the form of $C_A(x_A)$, we get $r_A(x_A)$ as detailed by reworking, e.g., Eq. 10.3:

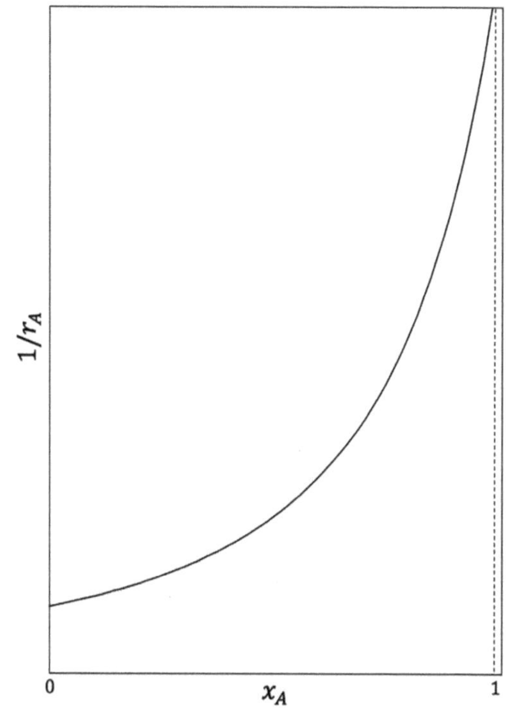

Fig. 10.3 Qualitative aspect for the Levenspiel kinetic plot—canonical kinetic law

$$r_A = kC_A^n = kC_{A0}^n(1 - x_A)^n \tag{10.12}$$

Figure 10.3 presents the kinetic plot (Levenspiel 1999) as the reciprocal of r_A (for reasons that will become clear later) vs. x_A. In canonical cases such the one in Eq. 10.12, the reaction rate is maximum for $x_A = 0$ (i.e., when C_A is maximum), then it decreases until it vanishes (its reciprocal tends to infinity) for complete conversion ($C_A = 0$). It should be remembered that this trend does not apply to zero-order kinetics, for which $1/r_A$ is constant ($= 1/k$ if, in Eq. 10.3, $n = 0$). In this last case, in reality, although we assume (for simplicity) $r_A = k$ until $x_A = 1$ is reached, when x_A approaches 1 the reaction rate begins to depend on C_A, so as to maintain the formal concept that r_A is zeroed when C_A is zeroed (i.e., when $x_A = 1$), and to take into account the physical phenomenon that, for conversion degrees close to unity, the reaction rate rapidly slows down given the lack of concentration of the limiting species.

The discussion is now extended to *non-isochoric systems*: those, for example, in the gas phase with an overall variation in the number of moles following the chemical reaction. Keeping the reference to discontinuous processes, now V varies with the degree of conversion and the following linear relationship is assumed:

$$V = V_0(1 + \varepsilon x_A) \tag{10.13}$$

which returns $V = V_0$ for $x_A = 0$ while, for $x_A = 1$, it allows us to isolate—by defining it—the *expansion coefficient* ε:

$$\varepsilon = \frac{V|_{x_A=1} - V_0}{V_0} \tag{10.14}$$

that can therefore be seen as the maximum possible variation in volume (i.e. when $x_A = 1$) with respect to the initial volume. Later on, examples will show how to calculate ε. Equations 10.13 and 10.14 are still valid for continuous processes, by the usual substitution of V with Q:

$$Q = Q_0(1 + \varepsilon x_A) \tag{10.15}$$

$$\varepsilon = \frac{Q|_{x_A=1} - Q_0}{Q_0} \tag{10.16}$$

A linear relationship (Eq. 10.13) has been postulated between the volume of a reacting gaseous system (where an overall variation in the number of moles occurs) and the degree of conversion of the limiting species (A). This is now demonstrated. First, we recall that, for a gas at constant pressure $\mathcal{P}$ and temperature T (extensions on ε for variable temperature/pressure are given in, e.g., Levenspiel 1999), the overall volume is linearly proportional to the total number of moles n_{TOT}:

$$V = \frac{n_{TOT}\Re T}{\mathcal{P}} \tag{10.17}$$

with $\Re$ the ideal gas constant. Take for example a chemical reaction in the gas phase of the type aA + bB⇒pP + cC. Given a certain degree of conversion, the actual moles of A can be expressed from Eq. 10.9 as:

$$n_A = n_{A0}(1 - x_A) \tag{10.18}$$

Moreover, by simple stoichiometric calculations, it is possible to obtain the actual moles of the other species:

$$\begin{cases} n_B = n_{B0} - \dfrac{b}{a}(n_{A0} - n_A) = n_{B0} - \dfrac{b}{a}n_{A0}x_A \\ n_P = n_{P0} + \dfrac{p}{a}(n_{A0} - n_A) = n_{P0} + \dfrac{p}{a}n_{A0}x_A \\ n_C = n_{C0} + \dfrac{c}{a}(n_{A0} - n_A) = n_{C0} + \dfrac{c}{a}n_{A0}x_A \end{cases} \tag{10.19}$$

Now, since:

$$n_{TOT} = n_A + n_B + n_P + n_C \tag{10.20}$$

and since each addend is linear with x_A as per Eqs. 10.18 and 10.19, n_{TOT} (and therefore V, as stated in Eq. 10.17) will also be linear with x_A, *q.e.d.*

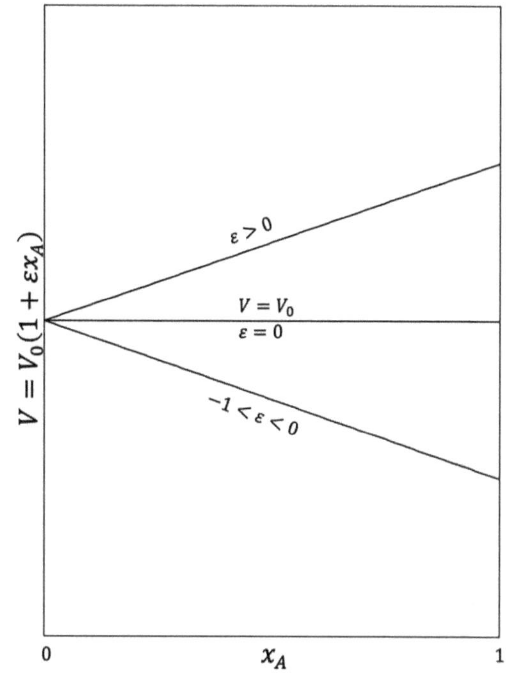

Fig. 10.4 General trend of reaction volume vs. degree of conversion for isochoric ($\varepsilon = 0$) and non-isochoric systems

The expansion coefficient ε assumes positive values for reactions that actually occur with expansion (e.g.: reactions in the gas phase with an overall increase in the number of moles), and values strictly between -1 and 0 for reactions that occur with contraction, meaning "negative expansion" (e.g.: reactions in the gas phase with an overall decrease in the number of moles, but in any case V, per $x_A = 1$, cannot be null, therefore ε cannot be equal to -1 nor be less than it, see Eq. 10.14). A graphical interpretation of the role of ε is given in Fig. 10.4. In the case of $\varepsilon > 0$, the volume V, equal to V_0 for $x_A = 0$, increases linearly up to $V_0(1+\varepsilon)$ when $x_A = 1$, Eq. 10.13. For a generic value of x_A, the effect of ε will be damped, and it will be $V = V_0(1+\varepsilon x_A)$. For example, if $\varepsilon = 0.5$, V will be $1.5V_0$ for $x_A = 1$, but only $1.15V_0$ for $x_A = 0.3$. The same applies for $-1 < \varepsilon < 0$, with V decreasing linearly up to $V_0(1+\varepsilon)$ for $x_A = 1$. For example, if $\varepsilon = -0.2$, V will be $0.8V_0$ for $x_A = 1$ and $0.94V_0$ for $x_A = 0.3$.

Substituting Eq. 10.13 into Eq. 10.9 (or, Eq. 10.15 into Eq. 10.10):

$$x_A = \frac{C_{A0}V_0 - C_A V}{C_{A0}V_0}$$

$$= \frac{C_{A0}V_0 - C_A V_0(1+\varepsilon x_A)}{C_{A0}V_0} \Rightarrow \boxed{\begin{cases} x_A = \dfrac{C_{A0} - C_A}{C_{A0} + \varepsilon C_A} \\ C_A = \dfrac{C_{A0}(1 - x_A)}{1 + \varepsilon x_A} \end{cases}} \tag{10.21}$$

we obtain the general relations between x_A and C_A for $\varepsilon \neq 0$ (that lead back to those in Eq. 10.11 for systems at constant volume, i.e. $\varepsilon = 0$). The expression of $r_A(x_A)$, e.g. by power law, reads now as:

$$r_A = kC_A^n = k\frac{C_{A0}^n(1 - x_A)^n}{(1 + \varepsilon x_A)^n} \tag{10.22}$$

and the qualitative trend on the $1/r_A$ vs. x_A graph is the same as in Fig. 10.3.

Example: Evaluation of the Expansion Coefficient for Gas Phase Reactions

The numerical calculation for ε is done by applying the ideal gas law, given (1) the stoichiometry of the reaction and (2) the composition of the feed (with possible presence of inert species). The calculation of ε, on the other hand, does not depend on the type of reactor or on the kinetic law. Equation 10.14 is reworked as (considering Eq. 10.17):

$$\varepsilon = \frac{V|_{x_A=1} - V_0}{V_0} = \frac{\frac{n_{TOT}|_{x_A=1}\mathfrak{R}T}{\mathcal{P}} - \frac{n_{TOT}|_{x_A=0}\mathfrak{R}T}{\mathcal{P}}}{\frac{n_{TOT}|_{x_A=0}\mathfrak{R}T}{\mathcal{P}}}$$

$$= \frac{n_{TOT}|_{x_A=1} - n_{TOT}|_{x_A=0}}{n_{TOT}|_{x_A=0}} \tag{10.23}$$

Thus, ε can be evaluated by simple analysis of the gas total moles at the beginning of the process ($x_A = 0$) and if the conversion were 1.

Consider the following gas phase reaction: 2A $\Rightarrow$ P + 3C. We can anticipate that $\varepsilon > 0$, as the reaction proceeds with an overall increase in the moles number (i.e., an increase in volume). For a feed composed by 70% A and 30% I (inert species, often present for safety reasons), by choosing as a basis $n_{TOT}|_{x_A=0} = 100$ (this is a free choice, as in Eq. 10.23 we look at ratios between moles number), we can populate Table 10.2 (if the conversion were complete, the 70 moles of A would disappear, generating 35 moles of P and 105 moles of C, with the moles of inert remaining constant) and calculate $\varepsilon = (170–100)/100 = 0.7$.

(continued)

Table 10.2 Evaluation of the expansion coefficient for the gas reaction: 2A$\Rightarrow$P + 3C (feed: 70% A + 30% I)

Species	$n_i\|_{x_A=0}$	$n_i\|_{x_A=1}$
A	70	0
P	0	35
C	0	105
I	30	30
TOT	$n_{TOT}\|_{x_A=0}=100$	$n_{TOT}\|_{x_A=1}=170$

Table 10.3 Evaluation of the expansion coefficient for the gas reaction: A + 2B⟹2P + 2C (feed: 20% A + 70% B + 10% I)

Species	$n_i\|_{x_A=0}$	$n_i\|_{x_A=1}$
A	20	0
B	70	30
P	0	40
C	0	40
I	10	10
TOT	$n_{TOT}\|_{x_A=0}=100$	$n_{TOT}\|_{x_A=1}=120$

Table 10.4 Evaluation of the expansion coefficient for the gas reaction: A + 2B ⟹ P + C (feed: 20% A + 70% B + 10% I)

Species	$n_i\|_{x_A=0}$	$n_i\|_{x_A=1}$
A	20	0
B	70	30
P	0	20
C	0	20
I	10	10
TOT	$n_{TOT}\|_{x_A=0}=100$	$n_{TOT}\|_{x_A=1}=80$

Another example of gas phase reaction with $\varepsilon > 0$ is: A + 2B⟹2P + 2C. For a feed composed by 20% A (limiting), 70% B (in excess) and 10% I, we can fill Table 10.3 and calculate $\varepsilon = (120–100)/100 = 0.2$. It is evident that, changing the composition of the feed (e.g., by adding or subtracting inert, and/or in presence of products in the feed), the value of ε will change accordingly.

The last example, A + 2B⟹ P + C, is for a gas reaction with decrease in the overall number of moles. Then, we expect a decrease in volume with $-1 < \varepsilon < 0$. For a feed composed again by 20% A (limiting), 70% B (in excess) and 10% I, Table 10.4 allows to calculate $\varepsilon = (80–100)/100 = -0.2$.

10.1.3 Damköhler Number

Given a reaction, its characteristic time t_R can be defined as the ratio of a characteristic value for C_A (the greater C_A, the longer t_R) to one of r_A (the greater r_A, the shorter t_R). In terms of orders of magnitude (O), to obtain a time-scale along which the reaction proceeds, we can take the initial values and express t_R as:

$$t_R = O\left(\frac{C_{A0}}{r_A(x_A = 0)}\right) \tag{10.24}$$

For a discontinuous process considered as example, the ratio between the reaction time t and t_R represents the Damköhler number [Da; named after Gerhard Damköhler (1908–1944), German chemist, Professor at University of Göttingen; see Damköhler 1940; Inger 2001]:

$$Da = \frac{t}{t_R} \tag{10.25}$$

Reactor engineering is developed around unitary values for Da (well-sized reactor). Values of $Da \gg 1$ and $Da \ll 1$ would mean, respectively, an oversized reactor (effective reaction time much longer than the characteristic time) or an undersized reactor (vice versa).

Example: Expressions for the Da Number for Basic Kinetic Laws
If we consider the power law kinetics in Eq. 10.3, Eq. 10.24 can be reworked as:

$$t_R = O\left(\frac{C_{A0}}{r_A(x_A = 0)}\right) = O\left(\frac{C_{A0}}{kC_{A0}^n}\right) = O\left(\frac{1}{kC_{A0}^{n-1}}\right) = \begin{cases} n = 0 \Rightarrow O\left(\frac{C_{A0}}{k}\right) \\ n = 1 \Rightarrow O\left(\frac{1}{k}\right) \\ n = 2 \Rightarrow O\left(\frac{1}{kC_{A0}}\right) \end{cases} \tag{10.26}$$

Therefore, by assuming the values given by the orders of magnitude, we can express Da as:

$$Da = \frac{t}{t_R} = ktC_{A0}^{n-1} = \begin{cases} n = 0 \Rightarrow \frac{kt}{C_{A0}} \\ n = 1 \Rightarrow kt \\ n = 2 \Rightarrow kC_{A0}t \end{cases} \tag{10.27}$$

10.2 Batch (Discontinuous) Reactor with Perfect Mixing (STR = Stirred Tank Reactor)

The first ideal reactor that we define is the *stirred tank reactor* (STR): it indicates a discontinuous (batch) system, i.e. without inlet and outlet flow, perfectly mixed (Fig. 10.5). The intensive properties of the reacting system (those not depending on the size of the system such as, in particular here, the chemical composition, i.e. x_A) are, for a fixed reaction time t, uniform in space by hypothesis of perfect mixing. To observe, inside the reactor, different values for x_A, it will be necessary to observe the reactor at a different time t. While the reactors can be kept isothermal, for example, by means of thermostatic systems, STR can maintain constant pressure even in the case of gaseous mixtures with variable density ($\varepsilon \neq 0$), by equipping it with a mobile

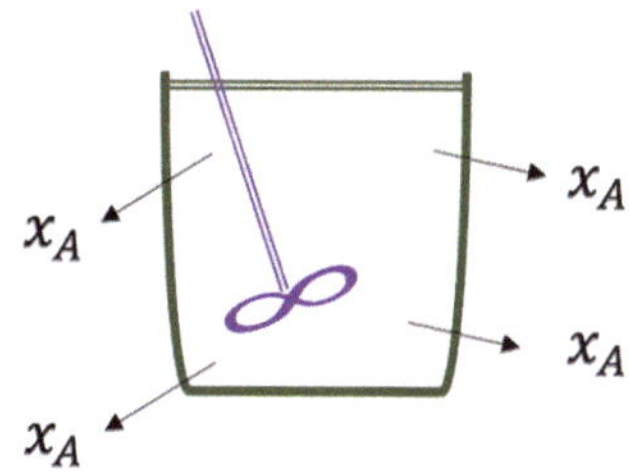

Fig. 10.5 Schematic illustration for an STR, indicating that, at fixed time, the perfect mixing induces uniformity in composition (and, then, in conversion degree) for the reacting solution

wall, which allows the volume of the reactor to vary (V must increase if there is expansion, and vice versa) while leaving $\mathcal{P}$ constant.

We now write a mass balance to obtain the STR design equation. Generally speaking, we know that the following points must be addressed when preparing a balance:

- consideration, as a guide, of the balance law ("*what enters plus what is generated is equal to what exits plus what is accumulated*"):

$$\boxed{\mathrm{IN} + \mathrm{GEN} = \mathrm{OUT} + \mathrm{ACC}} \tag{10.28}$$

- definition of the "subject" of the balance (in our case, the limiting species A);
- determination of the dimensions against which the balance is written (in our case, [moles/time]);
- individuation of the control volume (we will here use the entire reactor as control volume, as STR is perfectly mixed, and there are no spatial gradients for intensive properties, so there is no need to consider infinitesimal volumes).

Since we are dealing with a batch reactor, in Eq. 10.28 it is IN = 0 and OUT = 0, while the remainder can be rewritten as ACC = GEN, or (since it is a balance on a reacting species) as in Eq. 10.29, i.e. "decumulation = disappearance", both terms >0. The decumulation of A is represented by the variation in time of the moles of A (with the minus sign), while the disappearance of A (consumption by chemical reaction) is the product between the reaction rate (the reacting moles per unit time and unit volume) and the reaction volume, given a generic chemical reaction described by a reaction rate function of the conversion degree, $r_A(x_A)$. This is represented in Eq. 10.30, further manipulated through Eqs. 10.9 and 10.13, see Eq. 10.31:

$$-\mathrm{ACC} = -\mathrm{GEN} \tag{10.29}$$

$$-\frac{\mathrm{d}n_A}{\mathrm{d}t} = r_A(x_A)V \tag{10.30}$$

$$-\frac{d[n_{A0}(1-x_A)]}{dt} = n_{A0}\frac{dx_A}{dt} = r_A(x_A)V_0[1+\varepsilon x_A] \tag{10.31}$$

Equation 10.31 can be integrated by the Fourier method (separation of variables), between $t = 0$ ($x_A = 0$) and a generic value of t (with the corresponding x_A):

$$\int_0^t dt = \frac{n_{A0}}{V_0}\int_0^{x_A}\frac{1}{r_A(x_A)[1+\varepsilon x_A]}dx_A \tag{10.32}$$

We obtain the *design equation for STR*:

$$\boxed{t = C_{A0}\int_0^{x_A}\frac{1}{r_A(x_A)[1+\varepsilon x_A]}dx_A} \tag{10.33}$$

Relevant considerations:

- first, we note the importance of the approach presented above, aimed at expressing r_A as a function of x_A, which allows the function $r_A(x_A)$ to be properly present under integral (which is defined, and must be solved, as a function of the degree of conversion);
- the design equation is a *biunivocal relationship* between t and x_A: it returns, for example, the time necessary to reach a given conversion;
- t is an increasing function of x_A (upper bound of integration; the higher the required conversion, the longer the time needed), while higher values of r_A allow us to obtain the same x_A for lower t (in the expression of a time, as in classical physics, the rate of the process is at the denominator);
- the presence of the integral arises from the need to consider infinite instants of t between the start and the end of the process, and therefore infinite values for x_A and, eventually, for $r_A(x_A)$. The fluid elements will experience all these values between the start and the end of the process, and the integral is present to take this into account.

Example: Solution of the STR Design Equation for Particular Cases

The STR design equation (Eq. 10.33) can be solved for power law kinetics (Eq. 10.22):

$$\begin{aligned} t &= C_{A0}\int_0^{x_A}\frac{1}{r_A(x_A)[1+\varepsilon x_A]}dx_A \\ &= C_{A0}\int_0^{x_A}\frac{1}{k\frac{C_{A0}^n(1-x_A)^n}{(1+\varepsilon x_A)^n}(1+\varepsilon x_A)}dx_A \\ &= \frac{1}{kC_{A0}^{n-1}}\int_0^{x_A}\frac{(1+\varepsilon x_A)^{n-1}}{(1-x_A)^n}dx_A \end{aligned} \tag{10.34}$$

For $\varepsilon = 0$, it simplifies into:

(continued)

$$t = C_{A0}\int_0^{x_A}\frac{1}{r_A(x_A)}\mathrm{d}x_A = \frac{1}{kC_{A0}^{n-1}}\int_0^{x_A}\frac{1}{(1-x_A)^n}\mathrm{d}x_A \qquad (10.35)$$

Note the dependence on C_{AO}. In fact, if t is inversely proportional to C_{A0}^{n-1}, this means that t is proportional to C_{AO} (the more A to convert, the longer time needed) but inversely proportional to C_{A0}^{n} (faster reaction when C_{AO} is greater): if $n = 1$, the two effects cancel each other out and t is independent of C_{AO}. If $n > 1$, the second effect will prevail and therefore it will be convenient (as far as economically possible) to increase C_{AO}. If $n < 1$, the first effect will prevail and therefore it will be convenient (if no problems arise in separating the products from the solvent at the end of the process, therefore as far as reasonably possible) to decrease C_{AO} when the objective is to minimise the time needed to reach a given degree of conversion.

As also reported in Table 10.5, we can solve Eq. 10.35 for $n = 1$ (Eq. 10.36) and $n = 2$ (Eq. 10.37):

$$\begin{cases} t=\dfrac{1}{k}\displaystyle\int_0^{x_A}\frac{1}{1-x_A}\mathrm{d}x_A=-\frac{1}{k}[\ln(1-x_A)]_0^{x_A}=-\frac{1}{k}\ln(1-x_A)\Leftrightarrow kt=Da=-\ln(1-x_A) \\ x_A=1-\exp(-kt)=1-\exp(-Da) \end{cases} \qquad (10.36)$$

$$\begin{cases} t=\dfrac{1}{kC_{A0}}\displaystyle\int_0^{x_A}\frac{1}{(1-x_A)^2}\mathrm{d}x_A=\frac{1}{kC_{A0}}\left[\frac{1}{1-x_A}\right]_0^{x_A}=\frac{1}{kC_{A0}}\frac{x_A}{1-x_A}\Leftrightarrow kC_{A0}t=Da=\frac{x_A}{1-x_A} \\ x_A=\dfrac{kC_{A0}t}{1+kC_{A0}t}=\dfrac{Da}{1+Da} \end{cases} \qquad (10.37)$$

In solving, we have recalled the expressions for Da (Eq. 10.27). Moreover, the mutual relationships between t and x_A are observed: when $x_A = 0$, $t = 0$; at increasing x_A, t increases; for $x_A = 1$, t would tend to infinite as the reaction rate becomes progressively slower when the degree of conversion approaches unity (cf. Figure 10.3). The solution for constant kinetics ($n = 0$) is an exception:

$$t=\frac{C_{A0}}{k}\int_0^{x_A}\mathrm{d}x_A=\frac{C_{A0}}{k}x_A\Leftrightarrow\frac{kt}{C_{A0}}=Da=x_A \qquad (10.38)$$

There is a maximum (but not infinite) value of t ($= C_{AO}/k$) for which $x_A = 1$ (the reaction rate, in fact, does not tend to zero but is constant). In these cases it makes no sense to investigate the process for times longer than C_{AO}/k, since at that point A will have disappeared.

(continued)

Table 10.5 Synoptical view for the solutions of the STR design equation (Eq. 10.33), for canonical power law kinetics. The case at constant volume has been elaborated for $n = 0$, 1, and 2. Solutions have been also expressed in terms of the Damköhler number

any ε: $r_A(x_A) = kC_A^n = k\frac{C_{A0}^n(1-x_A)^n}{(1+\varepsilon x_A)^n}$		
$t = \frac{1}{kC_{A0}^{n-1}}\int_0^{x_A}\frac{(1+\varepsilon x_A)^{n-1}}{(1-x_A)^n}\mathrm{d}x_A$		
$\varepsilon = 0$		
$t = \frac{1}{kC_{A0}^{n-1}}\int_0^{x_A}\frac{1}{(1-x_A)^n}\mathrm{d}x_A$		
$n = 0$	$n = 1$	$n = 2$
$t = \frac{C_{A0}}{k}x_A$	$t = -\frac{1}{k}\ln(1-x_A)$	$t = \frac{1}{kC_{A0}}\frac{x_A}{1-x_A}$
$x_A = \frac{kt}{C_{A0}}$	$x_A = 1 - \exp(-kt)$	$x_A = \frac{kC_{A0}t}{1+kC_{A0}t}$
$Da = \frac{kt}{C_{A0}}$	$Da = kt$	$Da = kC_{A0}t$
$Da = x_A$	$Da = -\ln(1-x_A) \Leftrightarrow x_A = 1 - \exp(-Da)$	$Da = \frac{x_A}{1-x_A} \Leftrightarrow x_A = \frac{Da}{1+Da}$

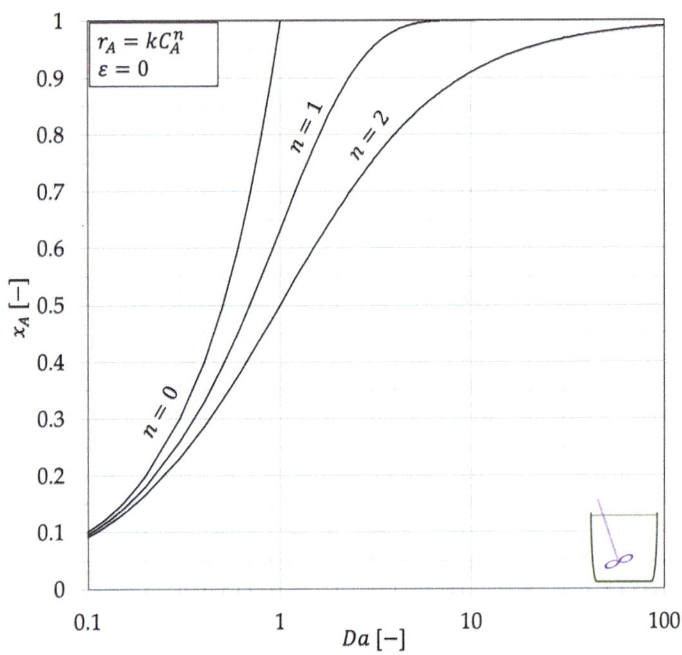

Fig. 10.6 Relations between degree of conversion and Damköhler number as solutions of the STR design equation—case of power law kinetics and system at constant volume. Semi-log plot

Finally, solutions have been plotted in the non-dimensional form of x_A vs. Da (then, independent of k and C_{A0}), general graph in Fig. 10.6. We can observe the increasing trend of x_A with Da (x_A is equal to 1 for $Da = 1$ in the case of constant kinetics; in the other cases, it tends to 1 as Da increases).

Example: Ozone Decomposition in STR

We consider the gas phase decomposition reaction of ozone:

$$O_3 \Rightarrow O_2 + O \tag{10.39}$$

schematised as A ⇒ P + C. Be the reaction carried out at $T = 400$ K, in an STR. Assume the following first-order kinetic law (with kinetic constant given by a classical Arrhenius law):

$$\begin{cases} r_A = kC_A \\ k = 7.6 \cdot 10^{12} \exp\left(-\frac{12{,}329}{T}\right) \ 1/\mathrm{s} \end{cases} \tag{10.40}$$

with $k = 0.31$ 1/s at $T = 400$ K. The general solution in Eq. 10.34 reduces, for $n = 1$, to that in Eq. 10.36:

$$t = -\frac{1}{k}\ln(1 - x_A) \tag{10.41}$$

It is now possible to calculate the time needed to reach a certain (desired) degree of conversion. For example, for $x_A = 50\%$, 70% and 90%, it is, respectively, $t = 2.24$ s, 3.88 s and 7.43 s.

Example: Solution of the Design Equation with the Partial Fraction Decomposition Method

Take now a general gas phase reaction:

$$A + 2B \Rightarrow P + C \tag{10.42}$$

The knowledge of the feed composition ($C_{A0} = 0.01$ M; $C_{B0} = 0.03$ M, excess reactant; $C_{I0} = 0.01$ M) allows to calculate $\varepsilon = (4\text{–}5)/5 = -0.2$ as per Table 10.6, that considers the system contraction upon chemical reaction. We have chosen $n_{TOT}|_{x_A=0} = 5$ as a basis (for simplicity, given the problem data).

Assume the following kinetic law ($n = 2$, first order respect to A and first order respect to B), with an ideal value for k:

$$\begin{cases} r_A = kC_AC_B \\ k = 1 \ \dfrac{\mathrm{L}}{\mathrm{mol} \cdot \mathrm{min}} \end{cases} \tag{10.43}$$

The solution of the design equation must now be retrieved. First, we need to express $r_A(x_A)$; from the fundamental Eq. 10.21:

(continued)

Table 10.6 Evaluation of the expansion coefficient for the gas reaction: A + 2B ⇒ P + C (feed: $C_{AO} = 0.01$ M; $C_{BO} = 0.03$ M; $C_{IO} = 0.01$ M)

Species	$n_i\|_{x_A=0}$	$n_i\|_{x_A=1}$
A	1	0
B	3	1
P	0	1
C	0	1
I	1	1
TOT	$n_{TOT}\|_{x_A=0} = 5$	$n_{TOT}\|_{x_A=1} = 4$

$$C_A = \frac{C_{A0}(1 - x_A)}{1 + \varepsilon x_A} \tag{10.44}$$

To express $C_B(x_A)$, consider the following mass balance (cf. Eqs. 10.9 and 10.13):

$$\begin{aligned} n_{B0} - n_B &= 2(n_{A0} - n_A) = 2n_{A0}x_A \Rightarrow C_{B0}V_0 - C_B V \\ &= 2C_{A0}V_0 x_A \Rightarrow C_{B0}V_0 - C_B V_0(1 + \varepsilon x_A) = 2C_{A0}V_0 x_A \end{aligned} \tag{10.45}$$

That is, after manipulation:

$$C_B = \frac{C_{B0} - 2C_{A0}x_A}{1 + \varepsilon x_A} \tag{10.46}$$

Inspection of Eqs. 10.44 and 10.46 reveals the following: the concentration of a species can be expressed as the ratio between a numerator (which corresponds to a "mass balance on concentration", valid for systems at constant volume and usually quickly obtainable) and the denominator $(1 + \varepsilon x_A)$. This trick will be used also in other cases. Now, substituting into Eq. 10.43, it is:

$$r_A(x_A) = k\frac{C_{A0}(1 - x_A)(C_{B0} - 2C_{A0}x_A)}{(1 + \varepsilon x_A)^2} \tag{10.47}$$

and the design equation, Eq. 10.33, becomes:

$$\begin{aligned} t &= C_{A0}\int_0^{x_A} \frac{1}{r_A(x_A)[1 + \varepsilon x_A]} \mathrm{d}x_A \\ &= \frac{1}{k}\int_0^{x_A} \frac{1 + \varepsilon x_A}{(1 - x_A)(C_{B0} - 2C_{A0}x_A)} \mathrm{d}x_A \end{aligned} \tag{10.48}$$

This is solved by the method of partial fraction decomposition (PFD, a concept developed, among the others, by Bernoulli, Leibniz, in the early eighteenth century), by posing:

(continued)

$$\frac{1+\varepsilon x_A}{(1-x_A)(C_{B0}-2C_{A0}x_A)}=\frac{C_1}{1-x_A}+\frac{C_2}{C_{B0}-2C_{A0}x_A} \quad (10.49)$$

Thus:

$$t=\frac{1}{k}\left(\int_0^{x_A}\frac{C_1}{1-x_A}\mathrm{d}x_A+\int_0^{x_A}\frac{C_2}{C_{B0}-2C_{A0}x_A}\mathrm{d}x_A\right)=\frac{1}{k}$$
$$\times\left\{-C_1[\ln(1-x_A)]_0^{x_A}-\frac{C_2}{2C_{A0}}[\ln(C_{B0}-2C_{A0}x_A)]_0^{x_A}\right\}=\frac{1}{k}$$
$$\times\left[-C_1\ln(1-x_A)-\frac{C_2}{2C_{A0}}\ln(C_{B0}-2C_{A0}x_A)+\frac{C_2}{2C_{A0}}\ln(C_{B0})\right]$$
$$=\frac{1}{k}\left[-C_1\ln(1-x_A)+\frac{C_2}{2C_{A0}}\ln\left(\frac{C_{B0}}{C_{B0}-2C_{A0}x_A}\right)\right] \quad (10.50)$$

This is the solution of the design equation, once Eq. 10.49 is further manipulated to obtain the constants C_1 and C_2:

$$\frac{1+\varepsilon x_A}{(1-x_A)(C_{B0}-2C_{A0}x_A)}=\frac{C_1}{1-x_A}+\frac{C_2}{C_{B0}-2C_{A0}x_A}$$
$$=\frac{C_{B0}C_1-2C_{A0}x_AC_1+C_2-C_2x_A}{(1-x_A)(C_{B0}-2C_{A0}x_A)}$$
$$=\frac{(C_{B0}C_1+C_2)+(-2C_{A0}C_1-C_2)x_A}{(1-x_A)(C_{B0}-2C_{A0}x_A)} \quad (10.51)$$

Equation 10.51 is valid if:

$$\left\{\begin{array}{l} C_{B0}C_1+C_2=1 \\ -2C_{A0}C_1-C_2=\varepsilon \end{array}\right. \Rightarrow \left\{\begin{array}{l} C_1=\dfrac{1+\varepsilon}{C_{B0}-2C_{A0}}=80 \\ C_2=-2C_{A0}\left(\dfrac{1+\varepsilon}{C_{B0}-2C_{A0}}\right)-\varepsilon=-1.4 \end{array}\right. \quad (10.52)$$

By substituting data in Eq. 10.50, we can calculate the time needed to reach a certain (desired) degree of conversion. For example, for $x_A = 40\%$, 60% and 80%, it is, respectively, $t = 19.16$ min, 37.55 min and 75.40 min.

10.2.1 Productivity for STR

Given a reaction $\alpha\mathrm{A} \Rightarrow \beta\mathrm{P}$, with $\gamma = \beta/\alpha$, the *productivity* of the desired product P is defined as the ratio between the moles of P produced and the time needed to produce them (with the aid of Eq. 10.9):

$$P_P(t) = \frac{n_P(t) - n_{P0}}{t + t_m} = \frac{\gamma[n_{A0} - n_A(t)]}{t + t_m} = \frac{\gamma n_{A0} x_A(t)}{t + t_m} \tag{10.53}$$

where t_m is the "dead time" to be considered for the unloading, cleaning and reloading operations between one cycle and another. The expression for the productivity takes into account the possible presence of P in the reactor feed (which is subtracted from the moles of P present in STR at time t, so to consider only P actually produced by the chemical reaction in the STR cycle, equal to the reacting moles of A corrected by the normalised stoichiometric coefficient γ). Moreover, it is recalled that it is possible to express $x_A(t)$ by solving the STR design equation, so to consider P_P as a function of t only (vice versa, P_P could be also expressed as a function of x_A only).

For canonical cases, as depicted in Fig. 10.7, $P_P(t = 0) = 0$ and $P_P(t \to \infty) = 0$. If the reaction time equals 0, the reactor will have been prepared after the previous cycle, therefore time (t_m) will have been spent, but the conversion will be zero. That is, Eq. 10.53, there is no productivity. If one prolongs the reaction for virtually infinite time t, even if x_A will tend to 1, the products will never be extracted from the reactor. The denominator in Eq. 10.53 is infinite, and the productivity is, again, zero. As P_P is intrinsically positive, it will have a maximum value: the calculation of the zero derivative condition of $P_P(t)$ will provide the value of t (and, therefore, of x_A) for which P_P is maximum (this time is called t^* in Fig. 10.7). Alternatively, t^* can be obtained by trial-and-error procedures. The time that makes P_P maximum could

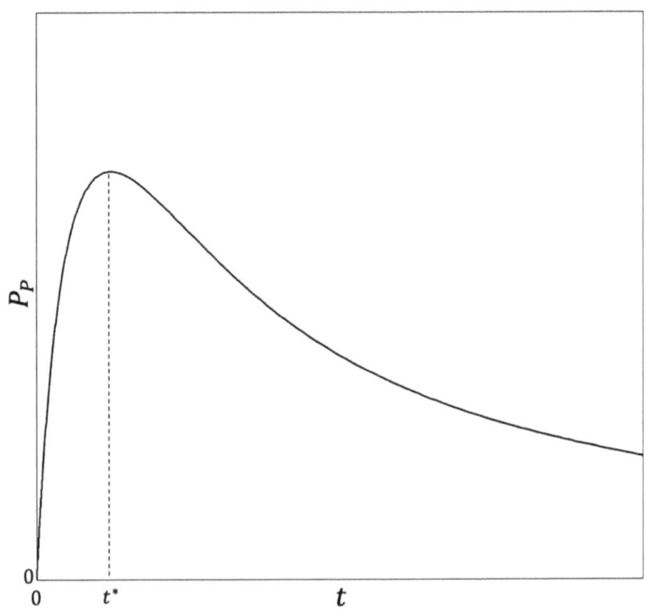

Fig. 10.7 Qualitative, canonical, trend of productivity vs. time in STR

correspond to not high values for x_A (and, therefore, it will be necessary to spend on the downstream separation). Vice versa, high purity requirements (high x_A, long t) can determine lower values for P_P. Only economic considerations will help to determine the best operating conditions in a general sense, and, for the correct design of the batch reactor, it has to be known (case by case) the correct "objective function" (degree of conversion, productivity, etc.).

Example: Productivity for Ozone Decomposition in STR

Take again the previous example on ozone decomposition (schematised as A ⟹ P + C), for which we obtained Eq. 10.41. We add here the knowledge of the feed composition (80% A + 10% P + 10% inert), which allows to calculate ε = (180–100)/100 = 0.8 as per Table 10.7, that considers the system expansion upon chemical reaction.

Be $\mathcal{P}$ = 1 atm. This allows to get, by the ideal gas law, the total gas concentration:

$$C_{TOT} = \frac{n_{TOT}}{V} = \frac{\mathcal{P}}{\Re T} \quad (10.54)$$

that is, 0.03 mol/L per $\Re$ = 0.08206 atm·L/(mol·K), from which it is $C_{A0} = 0.8C_{TOT} = 0.024$ mol/L, $C_{P0} = C_{I0} = 0.1C_{TOT} = 0.003$ mol/L.

Consider an STR with initial volume V_O = 1 L and t_m = 20 s. The productivity in P (O_2 species) is (being $n_{AO} = C_{AO}V_O = 0.024$ mol):

$$P_P(t) = \frac{n_{A0}x_A(t)}{t + t_m} = \frac{C_{A0}V_0[1 - \exp(-kt)]}{t + t_m} \quad (10.55)$$

The study of the $P_P(t)$ function, Fig. 10.8 (the solution can be also found by trial-and-error procedures), gives the conditions for the maximum productivity at t^* = 7.2 s (corresponding to x_A = 89.3%), valuing $7.9 \cdot 10^{-4}$ mol/s.

Once fixed the operating conditions, it is possible to get $Da = kt = 2.2 = O$

(1) from Eq. 10.27, i.e. well-sized reactor. Finally, we can speciate the gas mixture at the end of the process, by means of the considerations developed in another example above (see, e.g., Eq. 10.46):

(continued)

Table 10.7 Evaluation of the expansion coefficient for the gas reaction: A⟹P + C (feed: 80% A + 10% P + 10% I)

Species	$n_i\vert_{x_A=0}$	$n_i\vert_{x_A=1}$
A	80	0
P	10	90
C	0	80
I	10	10
TOT	$n_{TOT}\vert_{x_A=0} = 100$	$n_{TOT}\vert_{x_A=1} = 180$

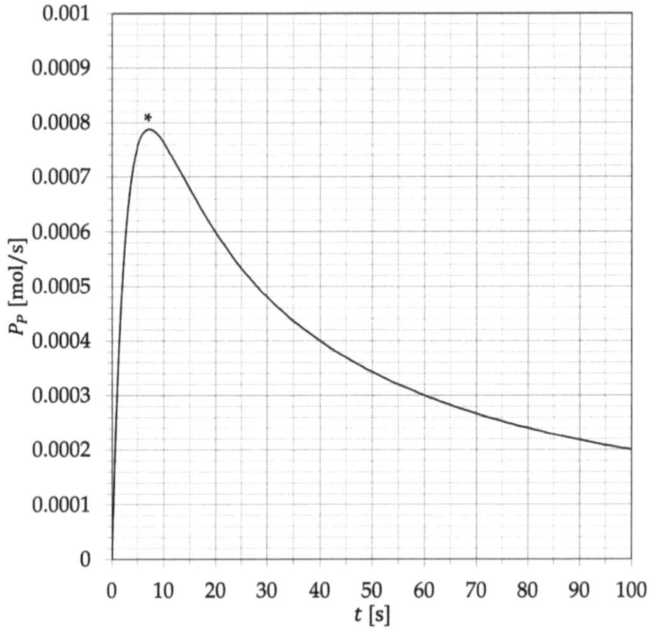

Fig. 10.8 Productivity vs. time for the reaction under investigation (ozone decomposition). The asterisk marks the condition of maximum productivity

$$\begin{cases} C_A = \dfrac{C_{A0}(1 - x_A)}{1 + \varepsilon x_A} = 0.00150 \text{ mol/L} \\ C_P = \dfrac{C_{P0} + C_{A0}x_A}{1 + \varepsilon x_A} = 0.01425 \text{ mol/L} \\ C_C = \dfrac{C_{C0} + C_{A0}x_A}{1 + \varepsilon x_A} = 0.01250 \text{ mol/L} \\ C_I = \dfrac{C_{I0}}{1 + \varepsilon x_A} = 0.00175 \text{ mol/L} \end{cases} \tag{10.56}$$

The reader will observe that the sum of these concentrations equals $C_{TOT} = 0.03$ mol/L (that is, the reaction occurs with variation in density, but the total concentration must remain constant at constant temperature and pressure per the ideal gas law). Moreover, although I is an inert species, its concentration changes accordingly to the variation in volume of the reacting solution. In fact, mass balance on I gives:

$$n_{I0} = n_I \Rightarrow C_{I0}V_0 = C_I V \Rightarrow C_{I0}V_0 = C_I V_0(1 + \varepsilon x_A) \Rightarrow q.e.d. \tag{10.57}$$

Example: Productivity for a Gas Phase Reaction in STR

In a previous example, we have obtained the solution Eq. 10.50 for the STR design equation, in the case of the gas phase A + 2B ⟹ P + C reaction with given kinetic law. Consider an STR with initial volume $V_0 = 1$ L and $t_m = 5$ min (this data adds to those already declared in the previous example), and individuate the conditions for maximum productivity. In this case, it is easier to express P_P as a function of x_A (with $n_{A0} = C_{A0}V_0 = 0.01$ mol):

$$P_P(x_A) = \frac{n_{A0}x_A}{t(x_A) + t_m}$$
$$= \frac{C_{A0}V_0x_A}{\frac{1}{k}\left[-C_1\ln(1-x_A) + \frac{C_2}{2C_{A0}}\ln\left(\frac{C_{B0}}{C_{B0} - 2C_{A0}x_A}\right)\right] + t_m} \tag{10.58}$$

The study of the $P_P(x_A)$ function, Fig. 10.9 (again, the solution can be also found by trial-and-error procedures), gives the conditions for the maximum productivity at $x_A = 33\%$ (corresponding to $t^* = 14.6$ min), valuing $1.7 \cdot 10^{-4}$ mol/min.

(continued)

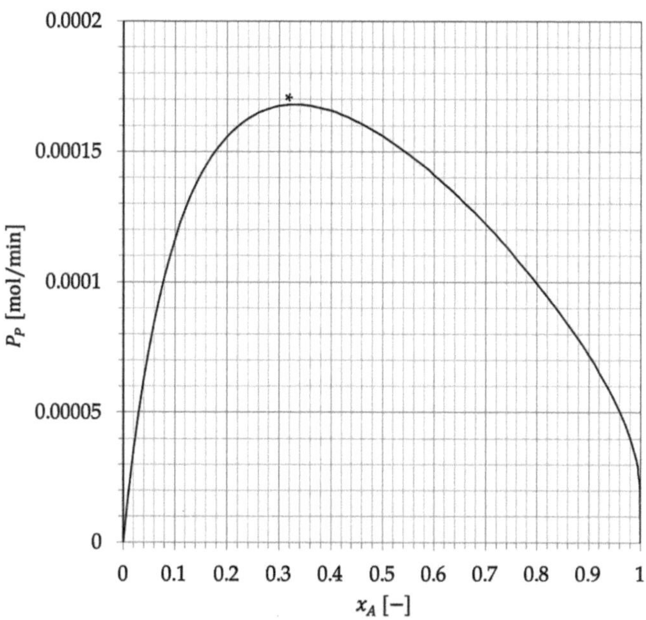

Fig. 10.9 Productivity vs. degree of conversion for the reaction under investigation (A + 2B⟹P + C). The asterisk marks the condition of maximum productivity

For the evaluation of Da in this case, we observe that:

$$t_R = O\left(\frac{C_{A0}}{r_A(x_A = 0)}\right) = O\left(\frac{C_{A0}}{kC_{A0}C_{B0}}\right) = O\left(\frac{1}{kC_{B0}}\right) \tag{10.59}$$

$$Da = \frac{t}{t_R} = ktC_{B0} \tag{10.60}$$

It is $Da = 0.4$, that we can still assume around 1.

10.3 Continuous Reactors

10.3.1 PFR = Plug Flow Reactor (Perfect Axial Segregation)

Consider now the tubular system in Fig. 10.10. The *plug flow reactor* (PFR) is defined as continuous (there are inlet and outlet flows) and is assumed to be stationary (ACC = 0 in Eq. 10.28), i.e. with behaviour not dependent on time. Plug flow means:

- first of all, to postulate fluid dynamic conditions of *perfect axial segregation*. Fluid elements that are located at different values of the z-axis (the main direction of motion) do not interact with each other (there is no backmixing, diffusive phenomena, vortices, etc. along the axis, where instead convective motions prevail). Each fluid element will retain its intensive properties (in particular x_A, for reacting systems), without mixing with fluid elements that are located "before" or "after" along z;
- that there is *perfect radial mixing*. Fluid elements that are located at the same value of z, but at different radial heights, are indistinguishable in their intensive properties (e.g., they have the same value of x_A). Furthermore, they are

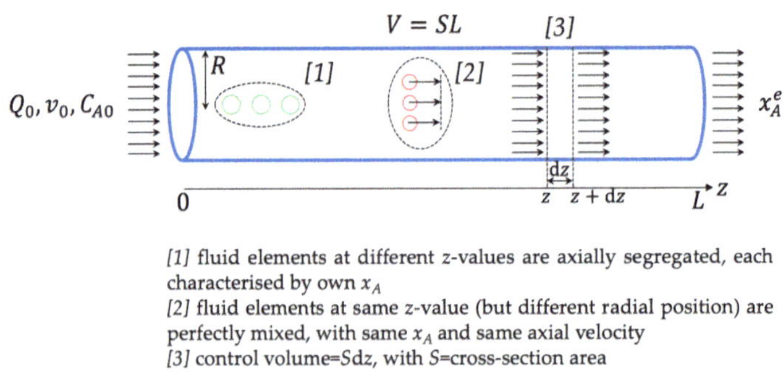

Fig. 10.10 Schematic illustration for a PFR, with main fluid dynamic features

characterised by the same axial component of the velocity vector (*flat velocity profile*), in a way that is completely opposite to what we have seen, in Chap. 3, for laminar motion (parabolic velocity profile, see Fig. 3.7). It is like considering the plug flow as an ideal extreme of the turbulent motion seen in Fig. 3.13 (plug flow, ideally, corresponds to Reynolds number tending to infinity, as we have anticipated in Chap. 3).

The plug flow hypotheses imply that each fluid element entering the system has the *same residence time* inside it, and it is immaterial whether the fluid element is in a position more or less close to the axis or to the wall (flat velocity profile). It is like proposing the image of a perfectly ordered "parade", in which the participants in a row are perfectly aligned (radial mixing), and furthermore they do not interact with those preceding or following them (axial segregation). The residence time is given by the ratio between the length to be traveled, L, and the average velocity, v. In turn, v can be expressed as Q/S, i.e. as the ratio between the average volumetric flow rate and the cross-section area. For constant density systems ($Q = Q_O$, $v = v_O$), it is possible to express the residence time as the *space-time* τ of the reactor:

$$\tau = \frac{L}{v_0} = \frac{LS}{Q_0} = \frac{V}{Q_0} \tag{10.61}$$

τ (dimensionally, a time) is, therefore, the ratio between the reactor volume $V = LS$ and the inlet volumetric flow rate Q_O. This last definition of τ is also considered a reference for variable density systems, even though it no longer represents, strictly speaking, the residence time. The term "space-time" recalls the concept that this quantity gives an indication of the residence time of the fluid elements in the reactor (seen from a "*Lagrangian perspective*": if we impersonated a fluid element and started a stopwatch on our wrist at the entrance to the reactor, halting it at the exit, the time marked by the stopwatch would be the residence time), but an external observer ("*Eulerian perspective*") perceives this as the space travelled by the fluid elements between reactor entrance and exit.

In order to ensure that only convective motions are effectively present along the axis (no diffusive backmixing), first we can define the characteristic time for diffusive phenomena along the axis ($t_{d,a}$), recalling a previous vision (see Chap. 7) of the diffusion coefficient $\mathcal{D}$ as the molecular velocity of the random walker:

$$t_{d,a} = O\left(\frac{L^2}{\mathcal{D}}\right) \tag{10.62}$$

In other words, while a characteristic time for convective motion will be the ratio between a length and a velocity, for diffusive motion it will be the ratio between the square of a length and a diffusion coefficient expressed in [length^2/time]. Then, we define the axial Péclet number [named after Jean Claude Eugène Péclet (1793–1857), French physicist, Professor at Central School of Paris; see Péclet 1827] as:

$$Pe_a = \frac{t_{d,a}}{\tau} = \frac{\frac{L^2}{\mathcal{D}}}{\frac{L}{v_0}} = \frac{Lv_0}{\mathcal{D}} \tag{10.63}$$

We require that Pe_a be $\gg 1$. In this way, the elements of fluid will not have time to experience diffusive phenomena along z, since in the meantime they will have already left the system (the characteristic time of axial diffusive phenomena is, in this case, much longer than the residence time of the fluid elements).

Similarly, the characteristic time for diffusion phenomena along the radius ($t_{d,r}$) can be expressed as:

$$t_{d,r} = O\left(\frac{R^2}{\mathcal{D}}\right) \tag{10.64}$$

where R is the radius of the tubular reactor. The radial Péclet number:

$$Pe_r = \frac{t_{d,r}}{\tau} = \frac{\frac{R^2}{\mathcal{D}}}{\frac{L}{v_0}} = \frac{R^2 v_0}{L\mathcal{D}} \tag{10.65}$$

is required to be $\ll 1$. In this way, the fluid elements will mix along the radial coordinate long before exiting, and radial mixing is one of the required characteristics of the plug flow (the characteristic time of radial diffusion phenomena is, in this case, much shorter than the residence time of the fluid elements).

The conditions of very large axial Péclet number and very small radial Péclet number can be translated as in:

$$\frac{Pe_a}{Pe_r} = \frac{t_{d,a}}{t_{d,r}} = \left(\frac{L}{R}\right)^2 \gg 1 \tag{10.66}$$

This, therefore, justifies the implicit assumption (Fig. 10.10) of tubular geometry (with length much higher than radius) for a PFR.

For the PFR design equation, as done in the case of STR, we start with a mass balance. Equation 10.28 is rewritten as (continuous and stationary system):

$$\mathrm{IN} + \mathrm{GEN} = \mathrm{OUT} \Rightarrow \mathrm{IN} - \mathrm{OUT} = -\mathrm{GEN} \tag{10.67}$$

The balance, whose subject is the limiting species A and whose terms are expressed in [moles/time], reads: the difference between entering and exiting A equals to A disappeared (consumed) per chemical reaction. Of course, we have to individuate the control volume which the terms in Eq. 10.67 refer to. Here, we consider Sdz as the control volume (since the system is axially segregated and radially mixed, therefore it is necessary to consider the local variations along z but not along the radius, Fig. 10.10). Therefore, IN and OUT respectively refer to z and $z + dz$. The balance is:

$$\begin{cases} \mathrm{IN} = F_A(z) = Q(z)C_A(z) \\ \mathrm{OUT} = F_A(z+\mathrm{d}z) = Q(z+\mathrm{d}z)C_A(z+\mathrm{d}z) \\ -\mathrm{GEN} = r_A(x_A)S\mathrm{d}z \\ F_A(z) - F_A(z+\mathrm{d}z) = Q(z)C_A(z) - Q(z+\mathrm{d}z)C_A(z+\mathrm{d}z) = r_A(x_A)S\mathrm{d}z \end{cases} \tag{10.68}$$

$$-\frac{Q(z+\mathrm{d}z)C_A(z+\mathrm{d}z) - Q(z)C_A(z)}{\mathrm{d}z} = -\frac{\mathrm{d}(QC_A)}{\mathrm{d}z} = r_A(x_A)S \tag{10.69}$$

By Eq. 10.10:

$$-\frac{\mathrm{d}(QC_A)}{\mathrm{d}z} = -\frac{\mathrm{d}[Q_0C_{A0}(1-x_A)]}{\mathrm{d}z} = Q_0C_{A0}\frac{\mathrm{d}x_A}{\mathrm{d}z} = r_A(x_A)S \tag{10.70}$$

$$\frac{S}{Q_0}\mathrm{d}z = C_{A0}\frac{\mathrm{d}x_A}{r_A(x_A)} \tag{10.71}$$

Equation 10.71 can be integrated by separation of variables, between $z = 0$ ($x_A = 0$) and $z = L$ (the reactor exit), which corresponds to the exit degree of conversion, x_A^e:

$$\frac{S}{Q_0}\int_0^L \mathrm{d}z = \frac{SL}{Q_0} = \frac{V}{Q_0} = C_{A0}\int_0^{x_A^e} \frac{1}{r_A(x_A)}\mathrm{d}x_A \tag{10.72}$$

Namely, we obtain the *design equation for PFR*:

$$\boxed{\tau = \frac{V}{Q_0} = C_{A0}\int_0^{x_A^e} \frac{1}{r_A(x_A)}\mathrm{d}x_A} \tag{10.73}$$

where the space-time can be seen as the volume needed, given the value for Q_0, or the value of Q_0 to be supplied, given V, to obtain a desired conversion.

Relevant considerations are similar to those developed when commenting the STR design equation:

- the design equation is a *biunivocal relationship* between τ and x_A^e, i.e., between a desired conversion and the residence time needed by the fluid elements to reach that conversion;
- τ is an increasing function of x_A^e, while higher values of r_A allow us to obtain the same x_A^e for lower τ;
- the presence of the integral arises from the need to consider infinite axial positions between reactor inlet and outlet, and therefore infinite values for x_A and, eventually, for $r_A(x_A)$. The fluid elements will experience all these values between inlet and outlet, and the integral is present to take this into account.

Using the canonical kinetic plot in Fig. 10.11 (but this would be true for any kinetics), we note that the area under the curve between 0 and x_A^e indicates a measure for τ: it is zero for zero conversion, then increases until it tends to infinity for x_A^e tending to 1. The space-time is seen as the residence time of the elements of fluid:

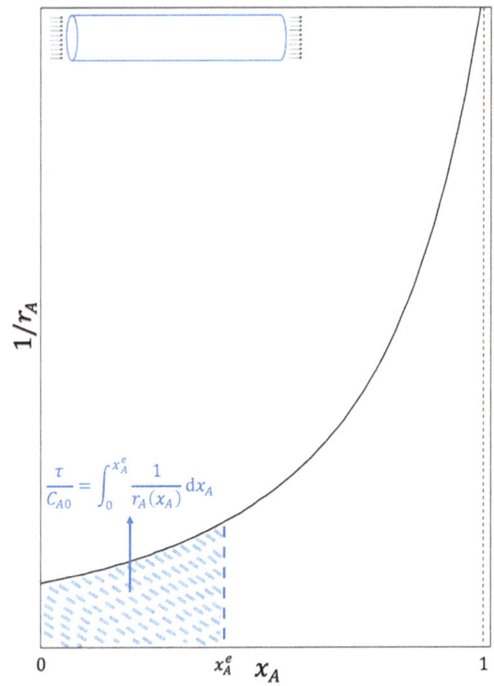

Fig. 10.11 Graphical interpretation of the PFR design equation: the area under the curve represents a measure of the space-time required for a given degree of conversion (see Eq. 10.73)

reasoning in a Lagrangian perspective, this is a true "time" for which the system experiences the chemical reaction. This time increases (thereby allowing greater conversion) if the reactor is larger or if the volumetric flow rate (and, therefore, the convective velocity) is lower, and in turn a longer residence time allows the elements of fluid to convert more.

Except for the volume correction in the movable-wall STR, and therefore reasoning for $\varepsilon = 0$, the two design equations (Eqs. 10.33 and 10.73) coincide. Even if STR is discontinuous (PFR is instead continuous and stationary) and mixed (PFR is instead axially segregated), *the two reactors can be seen as analogous.* In STR, for a fixed time, the intensive properties are uniform in space. To read a different value for x_A, it is necessary to move in time. Conversely: for PFR, if the fluid elements passing through a fixed spatial position (z) are observed as time varies, the value of x_A will not vary. To read a different value of composition, it will be necessary to move in space. It is therefore like saying that "*what happens in space for PFR, happens in time for STR*" or, it is like interpreting a PFR as the sum of an infinite number of (infinitesimal) STRs in series—each operating at a specific time t which in reality represents the Lagrangian time of the fluid elements—Fig. 10.12.

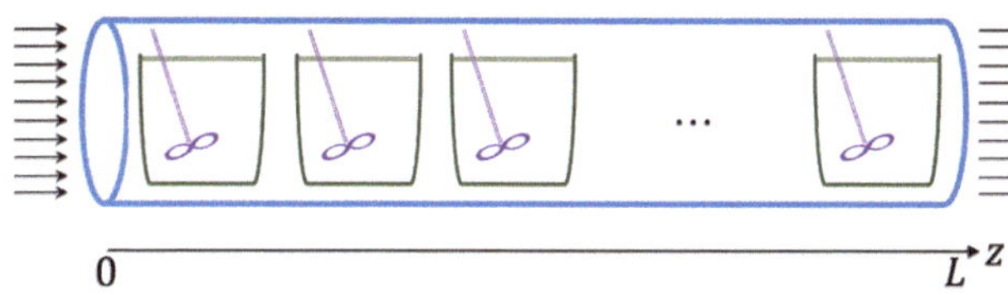

Fig. 10.12 Point of view: PFR as series of an infinite number of (infinitesimal) STRs, each operating at time t corresponding to the residence time (and, therefore, to the achieved degree of conversion) of the fluid elements in that position

Example: Solution of the PFR Design Equation for Particular Cases

The PFR design equation can be solved for power law kinetics (Eq. 10.22):

$$\tau = C_{A0}\int_0^{x_A^e} \frac{1}{r_A(x_A)}\mathrm{d}x_A = C_{A0}\int_0^{x_A^e} \frac{1}{k\frac{C_{A0}^n(1-x_A)^n}{(1+\varepsilon x_A)^n}}\mathrm{d}x_A$$
$$= \frac{1}{kC_{A0}^{n-1}}\int_0^{x_A^e} \frac{(1+\varepsilon x_A)^n}{(1-x_A)^n}\mathrm{d}x_A \quad (10.74)$$

For $\varepsilon = 0$, it simplifies into:

$$\tau = \frac{1}{kC_{A0}^{n-1}}\int_0^{x_A^e} \frac{1}{(1-x_A)^n}\mathrm{d}x_A \quad (10.75)$$

which is, as already observed, the same as Eq. 10.35. Solutions for $n = 0$, $n = 1$ and $n = 2$ have been already given in Eqs. 10.38, 10.36 and 10.37, respectively. Also the concept of Da is still valid (and Fig. 10.6 still holds), provided that t is replaced by the residence time τ in Eq. 10.27. Results are synoptically illustrated in Table 10.8.

Example: Sizing a PFR

Take the reaction in liquid phase ($\varepsilon = 0$):

$$\mathrm{A} + \mathrm{B} \Rightarrow \mathrm{P} \quad (10.76)$$

A stream with $Q_0 = 350$ L/min containing A ($C_{A0} = 0.2$ M) and B ($C_{B0} = 0.4$ M, excess reactant) is sent to a PFR where $x_A^e = 0.99$ is required. We must individuate the volume of the reactor, knowing that:

(continued)

Table 10.8 Synoptical view for the solutions of the PFR design equation (Eq. 10.73), for canonical power law kinetics. The case at constant volume has been elaborated for $n = 0$, 1, and 2. Solutions have been also expressed in terms of the Damköhler number

any $\boldsymbol{\varepsilon}$: $r_A(x_A) = kC_A^n = k\frac{C_{A0}^n(1-x_A)^n}{(1+\varepsilon x_A)^n}$		
$\tau = \frac{1}{kC_{A0}^{n-1}}\int_0^{x_A^e}\frac{(1+\varepsilon x_A)^n}{(1-x_A)^n}\mathrm{d}x_A$		
$\boldsymbol{\varepsilon = 0}$		
$\tau = \frac{1}{kC_{A0}^{n-1}}\int_0^{x_A^e}\frac{1}{(1-x_A)^n}\mathrm{d}x_A$		
$\boldsymbol{n = 0}$	$\boldsymbol{n = 1}$	$\boldsymbol{n = 2}$
$\tau = \frac{C_{A0}}{k}x_A^e$	$\tau = -\frac{1}{k}\ln(1-x_A^e)$	$\tau = \frac{1}{kC_{A0}}\frac{x_A^e}{1-x_A^e}$
$x_A^e = \frac{k\tau}{C_{A0}}$	$x_A^e = 1 - \exp(-k\tau)$	$x_A^e = \frac{kC_{A0}\tau}{1+kC_{A0}\tau}$
$Da = \frac{k\tau}{C_{A0}}$	$Da = k\tau$	$Da = kC_{A0}\tau$
$Da = x_A^e$	$Da = -\ln(1-x_A^e) \Leftrightarrow x_A^e = 1 - \exp(-Da)$	$Da = \frac{x_A^e}{1-x_A^e} \Leftrightarrow x_A^e = \frac{Da}{1+Da}$

$$\begin{cases} r_A = kC_AC_B \\ k = 150\ \dfrac{\mathrm{L}}{\mathrm{mol \cdot min}} \end{cases} \tag{10.77}$$

The sizing of the reactor requires the knowledge of τ. Being, for systems at constant density (i.e., we can balance on concentrations observing that the concentration of B must equal that of the feed minus that reacted of A):

$$\begin{cases} C_A = C_{A0}(1-x_A) \\ C_B = C_{B0} - C_{A0}x_A \end{cases} \tag{10.78}$$

it is:

$$r_A(x_A) = kC_{A0}(1-x_A)(C_{B0} - C_{A0}x_A) \tag{10.79}$$

and the design equation becomes:

$$\tau = \frac{V}{Q_0} = C_{A0}\int_0^{x_A^e}\frac{1}{r_A(x_A)}\mathrm{d}x_A = \frac{1}{k}\int_0^{x_A^e}\frac{1}{(1-x_A)(C_{B0} - C_{A0}x_A)}\mathrm{d}x_A \tag{10.80}$$

Solve by PFD:

$$\frac{1}{(1-x_A)(C_{B0} - C_{A0}x_A)} = \frac{C_1}{1-x_A} + \frac{C_2}{C_{B0} - C_{A0}x_A} \tag{10.81}$$

Thus:

(continued)

$$\tau=\frac{1}{k}\left(\int_0^{x_A^e}\frac{C_1}{1-x_A}\mathrm{d}x_A+\int_0^{x_A^e}\frac{C_2}{C_{B0}-C_{A0}x_A}\mathrm{d}x_A\right)=\frac{1}{k}$$
$$\times\left\{-C_1[\ln(1-x_A)]_0^{x_A^e}-\frac{C_2}{C_{A0}}[\ln(C_{B0}-C_{A0}x_A)]_0^{x_A^e}\right\}=\frac{1}{k}$$
$$\times\left[-C_1\ln\left(1-x_A^e\right)-\frac{C_2}{C_{A0}}\ln\left(C_{B0}-C_{A0}x_A^e\right)+\frac{C_2}{C_{A0}}\ln(C_{B0})\right]$$
$$=\frac{1}{k}\left[-C_1\ln\left(1-x_A^e\right)+\frac{C_2}{C_{A0}}\ln\left(\frac{C_{B0}}{C_{B0}-C_{A0}x_A^e}\right)\right] \quad (10.82)$$

This is the solution of the design equation, once Eq. 10.81 is further manipulated to obtain the constants C_1 and C_2:

$$\frac{1}{(1-x_A)(C_{B0}-C_{A0}x_A)}=\frac{C_1}{1-x_A}+\frac{C_2}{C_{B0}-C_{A0}x_A}$$
$$=\frac{C_{B0}C_1-C_{A0}x_AC_1+C_2-x_AC_2}{(1-x_A)(C_{B0}-C_{A0}x_A)}$$
$$=\frac{(C_{B0}C_1+C_2)+(-C_{A0}C_1-C_2)x_A}{(1-x_A)(C_{B0}-C_{A0}x_A)} \quad (10.83)$$

Equation 10.83 is valid if:

$$\begin{cases} C_{B0}C_1+C_2=1 \\ -C_{A0}C_1-C_2=0 \end{cases} \Rightarrow \begin{cases} C_1=\dfrac{1}{C_{B0}-C_{A0}}=5 \\ C_2=\dfrac{-C_{A0}}{C_{B0}-C_{A0}}=-1 \end{cases} \quad (10.84)$$

By substituting data in Eq. 10.82, it is τ = 0.13 min. To ensure that the elements of fluid live 0.13 min with an inlet flow rate of 350 L/min, the reactor volume must be $V = Q_O\tau$ = 45.5 L. Of course, there will be infinite combinations of reactor length (L) and radius (R) giving that value for V. But, from Eq. 10.66 we know that:

$$\left(\frac{L}{R}\right)^2 \gg 1 \quad (10.85)$$

Assuming for example $(L/R)^2$ = 100, it is L = 10R. Then, for cylindrical geometry:

$$V=\pi R^2L=10\pi R^3 \quad (10.86)$$

(continued)

namely, for $V = 45.5\ \text{dm}^3$, it will be, approximately, $R = 0.113$ m and $L = 1.13$ m. To conclude, evaluate Da. Being the kinetics the same as in a previous example, Eqs. 10.59 and 10.60 are still valid (provided that t is replaced by τ) and $Da = k\tau C_{B0} = 7.8$, a value having order in between 1 and 10, justified by the need of achieving a near-unitary conversion (that requires a slightly oversized reactor).

10.3.2 CSTR = Continuous Stirred Tank Reactor (Perfect Mixing)

The *continuous stirred tank reactor* (CSTR) is, as PFR, continuous and stationary, but, as STR, it can be seen as a *perfectly mixed* tank reactor. All the elements of fluid have the same intensive properties that (unlike in STR) are now constantly equal to the exit values, Fig. 10.13. These assumptions imply that the elements of fluid, as soon as they step in the reactor, are directly brought at their exit conversion value for their whole residence time (also in this case, we define $\tau = V/Q_0$). Therefore, in the CSTR, the elements of fluid experience only one value of x_A (i.e., x_A^e) and only one value of r_A (the one corresponding to x_A^e).

Actually, in systems like these, the fluid elements at the inlet spend a certain residence time before undergoing the effects of perfect mixing and adapting to the properties of the fluid at the outlet. This transition is all the more negligible the better the mixing effect is, and, for an ideal reactor, it will not be taken into consideration. By its nature, CSTR is therefore an opposite reactor to PFR. *It represents the opposite paradigmatic extreme.* These two paradigms (*segregated reactor* and *mixed reactor*) will also serve as a reference for the study of non-ideal reactors, which can more or less approach one or the other behavioural extreme, as we will see further on in this book.

The mass balance on A [moles/time] is, again:

$$\text{IN} + \text{GEN} = \text{OUT} \Rightarrow \text{IN} - \text{OUT} = -\text{GEN} \tag{10.87}$$

Unlike the PFR, however, here the entire reactor of volume V can be chosen as the control volume, as there are no intensive property gradients within the system. The material balance is then written as (cf. Eq. 10.10):

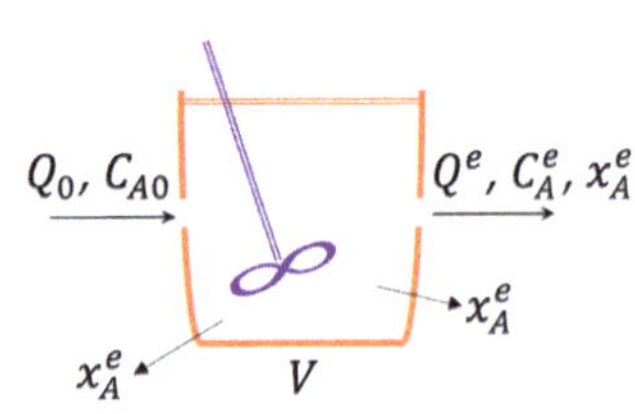

Fig. 10.13 Schematic illustration for a CSTR, indicating that the perfect mixing induces uniformity in composition (constantly equal to the exit conditions)

$$\begin{cases} \text{IN} = F_{A0} = Q_0 C_{A0} \\ \text{OUT} = F_A^e = Q^e C_A^e \\ -\text{GEN} = r_A\left(x_A^e\right) V \\ Q_0 C_{A0} - Q^e C_A^e = Q_0 C_{A0} x_A^e = r_A\left(x_A^e\right) V \end{cases} \tag{10.88}$$

where it has been pointed out that the reaction rate value—the only one experienced by the fluid elements—is calculated at exit conversion. The *design equation for CSTR*, algebraic and not integral as there is only one value for the reaction rate, reads:

$$\boxed{\tau = \frac{V}{Q_0} = \frac{C_{A0} x_A^e}{r_A\left(x_A^e\right)}} \tag{10.89}$$

representing again a *biunivocal relationship* between τ and x_A^e. The graphical interpretation is given in Fig. 10.14 (canonical kinetics, but the concept is valid for any kinetics): the area of the rectangle, having as its base the desired degree of conversion and as its height the reciprocal of r_A calculated at that conversion, measures τ (note: the area of the rectangle is zero for zero conversion, and then increases to infinity for unitary conversion).

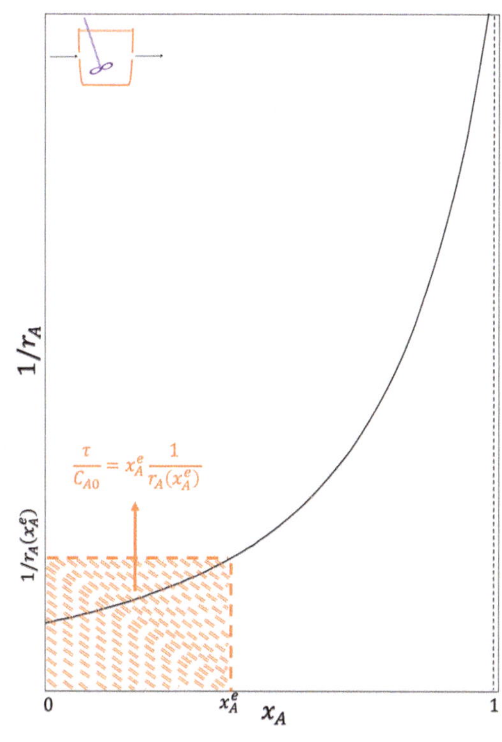

Fig. 10.14 Graphical interpretation of the CSTR design equation: the area of the rectangle represents a measure of the space-time required for a given degree of conversion (see Eq. 10.89)

Example: Solution of the CSTR Design Equation for Particular Cases

The CSTR design equation can be solved for power law kinetics (Eq. 10.22):

$$\tau = \frac{C_{A0} x_A^e}{r_A\left(x_A^e\right)} = \frac{1}{k C_{A0}^{n-1}} \frac{x_A^e\left(1+\varepsilon x_A^e\right)^n}{\left(1-x_A^e\right)^n} \tag{10.90}$$

For $\varepsilon = 0$, it simplifies into:

$$\tau = \frac{1}{k C_{A0}^{n-1}} \frac{x_A^e}{\left(1-x_A^e\right)^n} \tag{10.91}$$

As also reported in Table 10.9, we can solve Eq. 10.91 for $n = 1$ (Eq. 10.92) and $n = 2$ (Eq. 10.93):

$$\begin{cases} \tau = \dfrac{1}{k} \dfrac{x_A^e}{1-x_A^e} \Leftrightarrow k\tau = Da = \dfrac{x_A^e}{1-x_A^e} \\ x_A^e = \dfrac{k\tau}{1+k\tau} = \dfrac{Da}{1+Da} \end{cases} \tag{10.92}$$

$$\begin{cases} \tau = \dfrac{1}{k C_{A0}} \dfrac{x_A^e}{\left(1-x_A^e\right)^2} \Leftrightarrow k C_{A0}\tau = Da = \dfrac{x_A^e}{\left(1-x_A^e\right)^2} \\ x_A^e = \dfrac{(2kC_{A0}\tau+1)-\sqrt{4kC_{A0}\tau+1}}{2kC_{A0}\tau} = \dfrac{(2Da+1)-\sqrt{4Da+1}}{2Da} \end{cases} \tag{10.93}$$

As usual, τ increases from 0 to infinity as the degree of conversion between 0 and 1 increases. The case of kinetics of order 0 is an exception:

$$\tau = \frac{C_{A0}}{k} x_A^e \Leftrightarrow \frac{k\tau}{C_{A0}} = \mathrm{Da} = x_A^e \tag{10.94}$$

and it is identical to the case of the PFR. Reaction rates independent of the fluid dynamics of the system (since they are independent of C_A as in the case of $n = 0$) will, in fact, determine reactor equivalence between PFR and CSTR.

Finally, solutions have been plotted in the non-dimensional form of x_A vs. Da (then, independent of k and C_{AO}), which gives the general graph in Fig. 10.15. We can observe the increasing trend of x_A with Da (x_A is equal to 1 for $Da = 1$ in the case of constant kinetics; in the other cases, it tends to 1 as Da increases).

Table 10.9 Synoptical view for the solutions of the CSTR design equation (Eq. 10.89), for canonical power law kinetics. The case at constant volume has been elaborated for $n = 0$, 1, and 2. Solutions have been also expressed in terms of the Damköhler number

any ε: $r_A(x_A) = kC_A^n = k\frac{C_{A0}^n(1-x_A)^n}{(1+\varepsilon x_A)^n}$		
$\tau = \frac{1}{kC_{A0}^{n-1}}\frac{x_A^e(1+\varepsilon x_A^e)^n}{(1-x_A^e)^n}$		
$\varepsilon = 0$		
$\tau = \frac{1}{kC_{A0}^{n-1}}\frac{x_A^e}{(1-x_A^e)^n}$		
$n = 0$	$n = 1$	$n = 2$
$\tau = \frac{C_{A0}}{k}x_A^e$	$\tau = \frac{1}{k}\frac{x_A^e}{1-x_A^e}$	$\tau = \frac{1}{kC_{A0}}\frac{x_A^e}{(1-x_A^e)^2}$
$x_A^e = \frac{k\tau}{C_{A0}}$	$x_A^e = \frac{k\tau}{1+k\tau}$	$x_A^e = \frac{(2kC_{A0}\tau+1)-\sqrt{4kC_{A0}\tau+1}}{2kC_{A0}\tau}$
$Da = \frac{k\tau}{C_{A0}}$	$Da = k\tau$	$Da = kC_{A0}\tau$
$Da = x_A^e$	$Da = \frac{x_A^e}{1-x_A^e} \Leftrightarrow x_A^e = \frac{Da}{1+Da}$	$Da = \frac{x_A^e}{(1-x_A^e)^2} \Leftrightarrow x_A^e = \frac{(2Da+1)-\sqrt{4Da+1}}{2Da}$

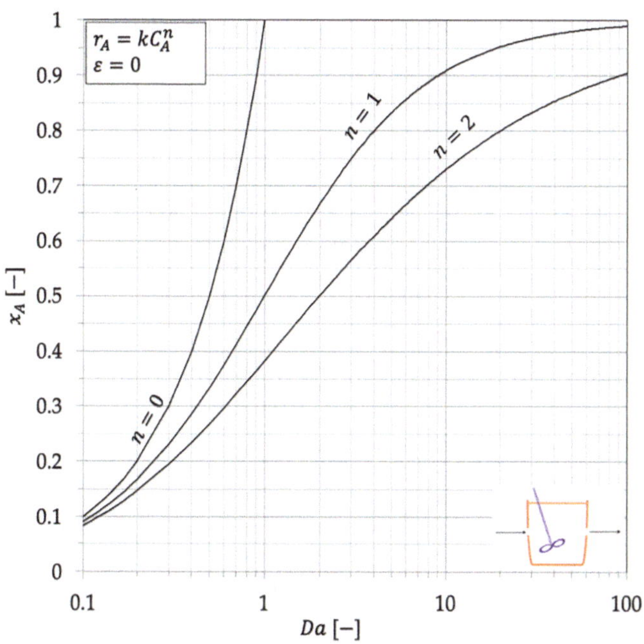

Fig. 10.15 Relations between degree of conversion and Damköhler number as solutions of the CSTR design equation—case of power law kinetics and system at constant volume. Semi-log plot

Example: Determination of the Inlet Flow Rate for a CSTR, Case of Naphthalene Partial Oxidation

We consider the gas phase naphthalene partial oxidation for the production of hydrogen:

$$C_{10}H_8 + 5O_2 \Rightarrow 4H_2 + 10CO \qquad (10.95)$$

schematised as A + 5B ⇒ 4P + 10C. The feed is composed by 3% A + 20.37% B + 76.63% inert (nitrogen), thus it is confirmed that naphthalene is the limiting reactant. We can now calculate $\varepsilon = (124\text{–}100)/100 = 0.24$ as per Table 10.10, that considers the system expansion upon chemical reaction.

Be the reaction carried out at $\mathcal{P} = 1$ atm and $T = 1000$ K, in an existing CSTR having $V = 1\ \text{m}^3$, up to $x_A^e = 2/3$. The knowledge of operating temperature and pressure allows to calculate, by the ideal gas law, the total gas concentration:

$$C_{TOT} = \frac{\mathcal{P}}{\Re T} \qquad (10.96)$$

that is, 0.01219 mol/L, from which it is $C_{A0} = 0.03C_{TOT} = 3.657 \times 10^{-4}$ mol/L, $C_{B0} = 0.2037C_{TOT} = 2.483 \times 10^{-3}$ mol/L, $C_{I0} = 0.7663C_{TOT} = 9.341 \times 10^{-3}$ mol/L. Assume the following kinetic law:

$$\begin{cases} r_A = kC_AC_B \\ k = k_0 \exp\left(-\dfrac{E}{\Re T}\right) \end{cases} \qquad (10.97)$$

For k_0 (which brings the dimensions to k) equal to 10^{12} L/(mol·s), $E = 166{,}280$ J/mol and using here $\Re = 8.31$ J/(mol·K), it is $k = 2041.4$ L/(mol·s).

Being:

$$C_A = \frac{C_{A0}(1 - x_A)}{1 + \varepsilon x_A} \qquad (10.98)$$

$$C_B = \frac{C_{B0} - 5C_{A0}x_A}{1 + \varepsilon x_A} \qquad (10.99)$$

it is:

$$r_A(x_A) = k\frac{C_{A0}(1 - x_A)(C_{B0} - 5C_{A0}x_A)}{(1 + \varepsilon x_A)^2} \qquad (10.100)$$

and the design equation, Eq. 10.89, becomes:

(continued)

Table 10.10 Evaluation of the expansion coefficient for the gas reaction: A + 5B⟹4P + 10C (feed: 3% A + 20.37% B + 76.63% I). Being the process continuous, Eq. 10.23 has been reworked in terms of molar flow rates

Species	$F_i\vert_{x_A=0}$	$F_i\vert_{x_A=1}$
A	3	0
B	20.37	5.37
P	0	12
C	0	30
I	76.63	76.63
TOT	$F_{TOT}\vert_{x_A=0}=100$	$F_{TOT}\vert_{x_A=1}=124$

$$\tau = \frac{C_{A0}x_A^e}{k\frac{C_{A0}\left(1-x_A^e\right)\left(C_{B0}-5C_{A0}x_A^e\right)}{\left(1+\varepsilon x_A^e\right)^2}} \tag{10.101}$$

By substituting data, it is $\tau = 1.0$ s, namely $Q_0 = V/\tau = 1$ m^3/s.

Finally, we can speciate the gas mixture at the end of the process:

$$\begin{cases} C_A^e = \dfrac{C_{A0}\left(1-x_A^e\right)}{1+\varepsilon x_A^e} = 0.1051 \text{ mmol/L} \\ C_B^e = \dfrac{C_{B0}-5C_{A0}x_A^e}{1+\varepsilon x_A^e} = 1.0897 \text{ mmol/L} \\ C_P^e = \dfrac{C_{P0}+4C_{A0}x_A^e}{1+\varepsilon x_A^e} = 0.8407 \text{ mmol/L} \\ C_C^e = \dfrac{C_{C0}+10C_{A0}x_A^e}{1+\varepsilon x_A^e} = 2.1017 \text{ mmol/L} \\ C_I^e = \dfrac{C_{I0}}{1+\varepsilon x_A^e} = 8.0526 \text{ mmol/L} \end{cases} \tag{10.102}$$

with $C_{TOT} = 0.0122$ mol/L (constant).

10.4 Productivity for Continuous Reactors

Let us consider any continuous reactor. The *productivity* now represents the molar flow rate of the desired product P at the exit, formed by chemical reaction (therefore, net of the molar flow rate of P possibly entering), cf. also Eq. 10.10:

$$\boxed{P_P = Q^e C_P^e - Q_0 C_{P0} = \gamma\left(Q_0 C_{A0} - Q^e C_A^e\right) = \gamma Q_0 C_{A0} x_A^e} \tag{10.103}$$

Productivity is a value that can only be calculated (i.e., there is no degree of freedom on its optimisation) if both V and Q_0 are known: in fact, since at this point τ is known

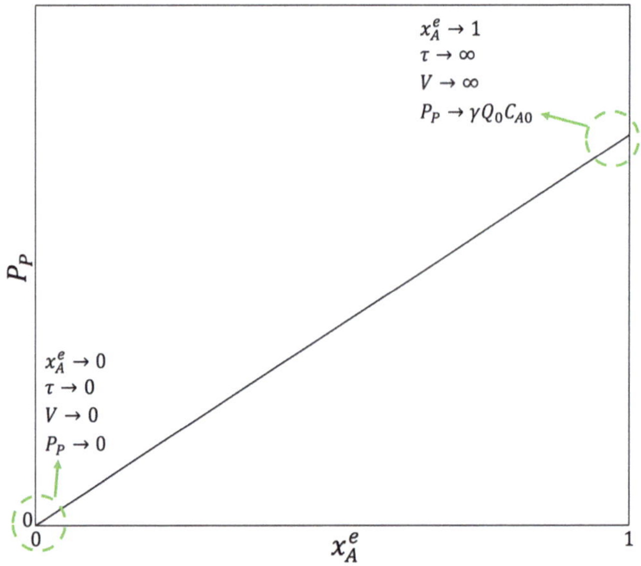

Fig. 10.16 Trend of productivity vs. degree of conversion for canonical kinetics and continuous reactor with known feed rate and unknown volume

(and therefore, from the design equations, x_A^e uniquely), Eq. 10.103 does not present degrees of freedom.

If, instead, Q_O is known (but not V), the productivity is a function, linearly increasing, of x_A^e, Fig. 10.16. $P_P = 0$ if $x_A^e = 0$ ($\tau = 0$ therefore $V = 0$, obviously the productivity is zero if the reactor does not exist), while P_P assumes its maximum value ($\gamma Q_O C_{AO}$) for $x_A^e = 1$. But, for canonical cases, this implies a residence time $\tau \longrightarrow \infty$, i.e. a reactor of infinite volume (known Q_O). It is therefore evident that the optimal solution will pass through an economic mediation: values of x_A^e will be chosen as sufficiently high to guarantee high productivity without forcing V to be designed towards excessively large values.

For systems with known volume, recalling the definition of τ, Eq. 10.103 can be rewritten as:

$$P_P = \gamma \frac{V}{\tau\left(x_A^e\right)} C_{A0} x_A^e \tag{10.104}$$

Since τ is a unique function of x_A^e, P_P is again a function of a single variable, namely x_A^e. Varying the output conversion degree will mean having to change the residence time of the fluid elements inside the reactor and, therefore, determining the value for Q_O (known V) for such τ to realise. We now deal with the problem of individuating the best operating conditions for maximising the productivity in this case (reactor with known volume). Let us start by observing that, differently from before, for $x_A^e = 1$, P_P is not maximum (actually, it is zero). In fact, since the elements of fluid need

τ tending to infinity to reach unitary conversion (in canonical situations), this can happen (in a reactor of known V) only by canceling Q_0. Therefore, there can be no productivity (see also the general Eq. 10.103, for $Q_0 = 0$).

It is now necessary to distinguish the case of CSTR from that of PFR. In fact, in Eq. 10.104 the correct expression for the space-time must be inserted. For CSTR (cf. Eq. 10.89), the productivity is the product of V (known) and the unique reaction rate (calculated in the exit conditions):

$$P_P = \gamma \frac{V}{\frac{C_{A0} x_A^e}{r_A\left(x_A^e\right)}} C_{A0} x_A^e = \gamma V r_A\left(x_A^e\right) \quad \text{for CSTR at known } V \tag{10.105}$$

Experiencing the fluid elements this unique value of rate, and representing the rate as the number of moles reacted per unit of time and volume, the product of this unique value of rate and the volume represents the productivity in P (once given the stoichiometric correction, as usual). Therefore, it can be said that, for CSTR with known volume, *productivity is maximum if the value of the unique reaction rate experienced by the fluid elements is maximum.*

For PFR, Eq. 10.104 is rewritten as (cf. Eq. 10.73):

$$\begin{aligned} P_P &= \gamma \frac{V}{C_{A0} \int_0^{x_A^e} \frac{1}{r_A(x_A)} \mathrm{d}x_A} C_{A0} x_A^e \\ &= \gamma V \frac{x_A^e}{\int_0^{x_A^e} \frac{1}{r_A(x_A)} \mathrm{d}x_A} \quad \text{for PFR at known } V \end{aligned} \tag{10.106}$$

P_P is maximum if the ratio between x_A^e and the integral is maximum. Maximising this ratio is equivalent to minimising its reciprocal, which represents the average value (between 0 and x_A^e) of the function $1/r_A$:

$$\frac{\int_0^{x_A^e} \frac{1}{r_A(x_A)} \mathrm{d}x_A}{x_A^e - 0} = \overline{\frac{1}{r_A(x_A)}}\Bigg|_0^{x_A^e} \tag{10.107}$$

This is essentially equivalent to saying that, while to maximise P_P in a CSTR with known V it is necessary to maximise the only value of the reaction rate, for PFR with known V *it is necessary to maximise the average value of* r_A among the infinities that the fluid elements experience between entrance and exit—since it is not the exit rate value that counts individually, but the average of all the reaction rate values experienced by the fluid elements in the reactor with perfect axial segregation.

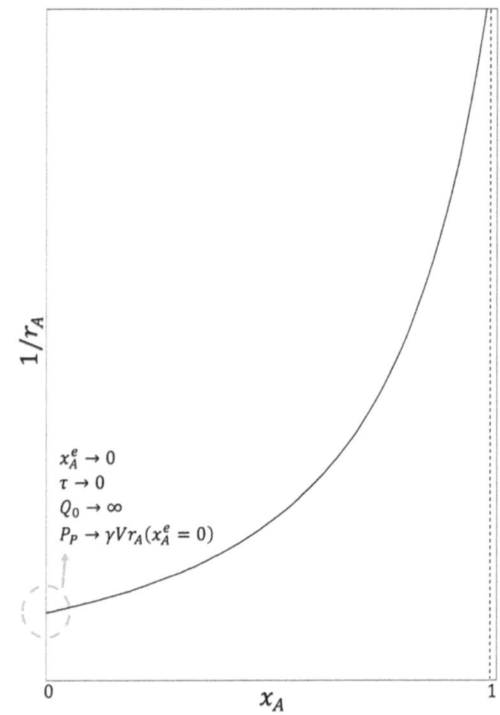

Fig. 10.17 Differential operating conditions for canonical kinetics and continuous reactor with known volume and unknown feed rate

10.4.1 The Differential Reactor

Consider now a canonical kinetic law, represented as in Fig. 10.17: we know that the reaction rate is maximum for zero conversion, and then it decreases. Whether it is a CSTR with a known V (we need to maximise the only value for r_A) or a PFR with a known V (we need to maximise the average value for r_A), it is clear that this occurs when the concentration of A is maximum, i.e. for $x_A = 0$. This condition is virtually of interest if, for a reactor with a known volume, the objective is the maximisation of productivity and not, for example, the purity of the output product (which instead could be prevalent if the product is difficult to separate from the reagent). Reactors that operate with zero conversion are ideally called *differential reactors*: in fact, a zero ("differential") residence time is required, which in a reactor with a known V is obtained by feeding the system a virtually infinite flow rate Q_0.

Returning to Eq. 10.103 for P_P, it is expressed as the product of Q_0 (now infinite) and x_A^e (now infinitesimal): the product of infinity and infinitesimal will here give a finite value (*vide infra*), which represents the maximum productivity value for canonical cases. That is, it is true that a "differential" conversion is obtained, but within an infinite volumetric flow rate, and this allows for the virtual realisation of a non-zero productivity and, in reality, maximum in these cases.

For differential CSTR, taking up Eq. 10.105, the maximum productivity is:

$$P_P = \gamma V r_A\left(x_A^e = 0\right) \tag{10.108}$$

For differential PFR, from Eq. 10.106 it is:

$$P_P = \gamma V \lim_{x_A^e \to 0}\left[\frac{x_A^e}{\displaystyle\int_0^{x_A^e} \frac{1}{r_A(x_A)}\,\mathrm{d}x_A}\right] = \gamma V \lim_{x_A^e \to 0}\left[\frac{x_A^e}{\displaystyle\frac{1}{r_A(x_A)}\int_0^{x_A^e}\mathrm{d}x_A}\right]$$

$$= \gamma V \lim_{x_A^e \to 0}\left[\frac{x_A^e}{\frac{1}{r_A(x_A)}x_A^e}\right] = \gamma V r_A\left(x_A^e = 0\right) \tag{10.109}$$

since, if the integral in the denominator extends between 0 and 0, the function $1/r_A$ is constant and can be placed outside the integral. Therefore, the maximum productivity assumes the same expression as for differential CSTR. It is not correct, in fact, to refer to a "differential CSTR or PFR". *The differential reactor is uniquely determined*: if the elements of fluid are destined to live inside the reactor for a time →0, the different fluid dynamics of PFR and CSTR will not be able to influence the performance of the system, since the elements of fluid will not have time to make experience of segregated/mixed conditions.

Of course, purely differential operating conditions would not be possible (infinite feed rate, recovering the maximum productivity inside an infinitely diluted output stream with respect to the product, then infinite separation costs). We therefore operate with x_A^e low enough (e.g., 5–10%) to allow obtaining a high productivity with low (but non-zero) τ values, i.e. feeding a high (but not infinite) Q_0. Only an economic balance between advantages linked to productivity, and costs linked to the feed and to the recovery of the product at the exit, can guide us towards the optimal operating conditions.

Example: Productivity for Power Law Kinetics and Continuous Reactors at Known Volume

Refer to a canonical A⟹P reaction described by:

$$r_A = kC_A^n = kC_{A0}^n(1 - x_A)^n \tag{10.110}$$

Take C_{A0} = 1 mol/L and a continuous reactor (either PFR or CSTR) with V = 2 L. For n = 1 (assume k = 2 1/s), the general expression for the productivity is, for CSTR (Eq. 10.105):

$$P_P = V r_A\left(x_A^e\right) = VkC_{A0}\left(1 - x_A^e\right) \tag{10.111}$$

while, for PFR (Eq. 10.106):

(continued)

$$P_P = V\frac{x_A^e}{\int_0^{x_A^e}\frac{1}{r_A(x_A)}\mathrm{d}x_A} = V\frac{x_A^e}{\frac{1}{kC_{A0}}\int_0^{x_A^e}\frac{1}{1-x_A}\mathrm{d}x_A} = \frac{VkC_{A0}x_A^e}{-\ln\left(1-x_A^e\right)} \quad (10.112)$$

For both reactors, productivity is maximum for $x_A^e = 0$. Equation 10.111 becomes (cf. Eq. 10.108):

$$P_P = VkC_{A0} \quad (10.113)$$

and Eq. 10.112 reads (cf. Eq. 10.109):

$$P_P = VkC_{A0}\lim_{x_A^e\to 0}\left[\frac{x_A^e}{-\ln\left(1-x_A^e\right)}\right] = VkC_{A0}\lim_{x_A^e\to 0}\left(1-x_A^e\right) = VkC_{A0} \quad (10.114)$$

by application of the de l'Hôpital's rule. The two values, as expected, coincide. Then, the maximum productivity is 4 mol/s. Any other value for x_A would result in lower productivity, as it can be easily verified using Eqs. 10.111 and 10.112, up to zero productivity per unit conversion.

Instead, for $n = 0$, the non-canonical case ends up into a different result. Assume $k = 1$ mol/(L s). The general expression for the productivity is, for CSTR:

$$P_P = Vr_A\left(x_A^e\right) = Vk \quad (10.115)$$

and, for PFR:

$$P_P = V\frac{x_A^e}{\int_0^{x_A^e}\frac{1}{r_A(x_A)}\mathrm{d}x_A} = V\frac{x_A^e}{\frac{1}{k}\int_0^{x_A^e}\mathrm{d}x_A} = Vk \quad (10.116)$$

That is, the productivity is constant (= 2 mol/s) whatever the conversion, and there is no difference between the two reactors as, for kinetics independent of concentration, segregation and mixing have no role over the whole conversion range. This suggests working (with any reactor) at maximum x_A^e (= 1) which, for the same P_P (= Vk), allows designing the maximum value for τ (= C_{A0}/k = 1 s, see Tables 10.8 and 10.9) and therefore the minimum value for Q_0 (= V/τ = 2 L/s). This will permit saving on both feed and downstream separation costs.

10.5 Background Information

- The reaction rate can be expressed as a function of the degree of conversion of the limiting reactant (e.g., Eq. 10.22), for the practical application of the reactors design equations.
- The three ideal reactors have each its own design equation: Eq. 10.33 for STR, Eq. 10.73 for PFR, Eq. 10.89 for CSTR, where each design equation represents a relationship between time (or space-time for PFR and CSTR continuous reactors) and degree of conversion.
- Fluid dynamics of PFR (perfect axial segregation) implies a strong similarity with STR (discontinuous reactor), while CSTR (perfect mixing) is a conceptually different reactor.
- The productivity for a continuous reactor is a parameter which shows different optimisation patterns according to what is known and what is to design.

References

Damköhler G (1940) Der Einfluss der Turbulenz auf die Flammengeschwindigkeit in Gasgemischen. Z Elektrochem Angew Phys Chem 46:601–626. https://doi.org/10.1002/bbpc.19400461102

Hill CG, Root TW (2014) Introduction to chemical engineering kinetics & reactor design. Wiley, Hoboken

Inger GR (2001) Scaling nonequilibrium-reacting flows: the legacy of Gerhard Damköhler. J Spacecr Rocket 38:185–190. https://doi.org/10.2514/2.3692

Jakobsen HA (2014) Chemical reactor modeling. Springer, Cham

Johnson KA, Goody RS (2011) The original Michaelis constant: translation of the 1913 Michaelis–Menten paper. Biochemistry 50:8264–8269. https://doi.org/10.1021/bi201284u

Levenspiel O (1999) Chemical reaction engineering. Wiley, New York

Li S, Xin F, Li L (2017) Reaction engineering. Butterworth-Heinemann, Oxford

Michaelis L, Menten ML (1913) Die Kinetik der Invertinwirkung. Biochem Z 49:333–369

Monod J (1942) Recherches sur la croissance des cultures bactériennes. Hermann & Cie, Paris

Péclet JCE (1827) Traité de l'éclairage. Librairie Scientifique et Industrielle. Malher et Cie, Paris

Roberts GW (2009) Chemical reactions and chemical reactors. Wiley, Hoboken

Westerterp KR, van Swaaij WPM, Beenackers AACM (1988) Chemical reactor design and operation. Wiley, Hoboken

Comparison Between Chemical Reactors 11

Abstract

This chapter addresses the issue of which continuous reactor to choose, starting from the case of canonical kinetics, and then extending the analysis to autocatalytic kinetics. This last scenario is also analysed in relation to the conditions of reactor equivalence, and how the choice of reactor varies as a function of the desired conversion degree.

Nomenclature

C_i	Concentration of i species [moles/volume]
C_{i0}	Concentration of i species at reaction beginning [moles/volume]
Da	Damköhler number [−]
k	Kinetic constant [(1/time)·(moles/volume)$^{1-n}$]
k_d	Kinetic constant for direct reaction [(1/time)·(moles/volume)$^{1-n}$]
k_i	Kinetic constant for inverse reaction [(1/time)·(moles/volume)$^{1-n}$]
n	Total reaction order [−]
P_P	Productivity of the desired species [moles/time]
Q_0	Volumetric flow rate at reaction beginning [volume/time]
r_A	Rate of reaction [moles/(volume·time)]
V	Volume [volume]
x_A	Degree of conversion [−]
x_A^{EQ}	Equilibrium degree of conversion [−]
x_A^e	Exit degree of conversion [−]
x_A^*	Degree of conversion for maximum reaction rate [−]
x_A'	Conversion degree such that $r_A(x_A') = r_A(0)$ [−]
τ	Space-time for continuous reactor [time]
τ_{CSTR}	Space-time for CSTR [time]
τ_{PFR}	Space-time for PFR [time]

F. Montagnaro et al., *Elements of Unit Operations and Chemical Reactors*,
https://doi.org/10.1007/978-3-032-18046-9_11

11.1 PFR Vs. CSTR

The discontinuous reactor is generally used for the analysis of a process that must be then brought to a large scale. In fact, the batch STR is a reactor simpler to use, and more flexible, than PFR/CSTR, with the possibility of changing the reaction environment between one cycle and another. However, although with some exceptions, this is not the reactor normally used for large-scale production, due to the intrinsic presence of dead times between one cycle and another, which determine a reduction in productivity (cf. Eq. 10.53). For further extensions on this topic, the reader can refer to, e.g., Westerterp et al. 1988.

For large-scale processes, continuous reactors (PFR/CSTR) are therefore chosen, even if they do not have the high flexibility characteristics as for STR (e.g., the "shut-down" and subsequent "start-up" operations of a continuous reactor, sometimes necessary for maintenance or to change operating parameters, are much longer). The subject of the following discussion will be the comparison PFR vs. CSTR.

To begin with, we recall the design equations for PFR and CSTR, respectively:

$$\tau_{PFR} = \frac{V}{Q_0} = C_{A0} \int_0^{x_A^e} \frac{1}{r_A(x_A)} \mathrm{d}x_A \tag{11.1}$$

$$\tau_{CSTR} = \frac{V}{Q_0} = \frac{C_{A0} x_A^e}{r_A\left(x_A^e\right)} \tag{11.2}$$

The graphical counterpart—for canonical kinetics—is illustrated in Fig. 11.1: given a desired value for the degree of conversion, the space-time is measured by the area under the curve (PFR), or by the area of the rectangle (CSTR).

It is therefore evident, given a fixed value for x_A^e, and referring to canonical kinetics, *how the PFR requires less space-time than a CSTR, and it is therefore more convenient.* The elements of fluid in a PFR need less time (in Lagrangian terms) to reach a given conversion, compared to what happens in a CSTR. This is because the fluid elements in a PFR experience all the infinite values of r_A, from those corresponding to high C_A (low x_A), which are the highest, to the worst, i.e. those corresponding to low C_A (high x_A). In the CSTR, the elements of fluid are instead destined to always react in conditions equivalent to x_A^e: this corresponds to the worst possible value for r_A, and the fluid elements in a CSTR will therefore react to a value of r_A lower than the average value of r_A experienced by the fluid elements in a PFR. This is based on the different fluid dynamics of PFR (perfect axial segregation) and CSTR (perfect mixing). Reasons other than those exposed (e.g., availability of reactors, need to keep the reacting mixture stirred) could lead to CSTR being preferred even in canonical cases such as those we are discussing here.

The ratio between τ_{CSTR} and τ_{PFR} is therefore, in this scenario, greater than 1. The gap increases (and thus the PFR becomes increasingly more convenient compared to the CSTR):

- *as the exit degree of conversion increases* (the only possible value for r_A, in the case of CSTR, will always be worse for increasingly higher x_A, while for PFR all values for r_A between 0 and x_A^e will be experienced anyway);

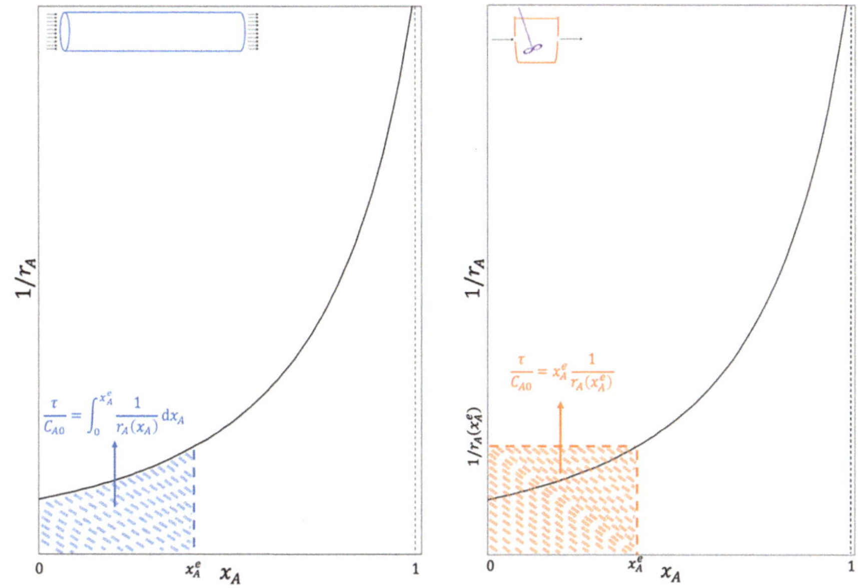

Fig. 11.1 Graphical interpretation of the design equations. Left: PFR (the area under the curve represents a measure of the space-time required for a given degree of conversion). Right: CSTR (the area of the rectangle represents a measure of the space-time required for a given degree of conversion)

- *as the reaction order increases* (if r_A depends more strongly on C_A, for low values of C_A, i.e. for exit conditions, r_A will be increasingly lower, worsening the performance of CSTR even more);
- *as the expansion coefficient increases* (for greater expansion, A is diluted more, and this generally worsens the values for r_A and in particular the lowest value of r_A, i.e. that at x_A^e, the only possible one for CSTR, which will be progressively lower as systems with increasing expansion are considered).

Example: Comparison Between Continuous Reactors for Power Law Kinetics

A first way is to construct graphs (Levenspiel 1999) where, on the ordinate, τ_{CSTR} / τ_{PFR} is reported on a logarithmic scale, to better appreciate the small deviations from the value $\tau_{CSTR} / \tau_{PFR} = 1$ (we remember that, for canonical cases, this ratio is always greater than 1). On the abscissa, $1 - x_A^e$ is reported on a logarithmic scale as well, so as to better appreciate the differences for high values of x_A^e (if x_A^e had been placed on a logarithmic scale, the differences for low values of x_A^e would have been appreciated better, and it would not have been of as much interest). The values on the abscissa axis can vary, e.g., between 0.01 ($x_A^e = 0.99$, as of course it is not possible to place values equal to 0 on a logarithmic scale) and 1 ($x_A^e = 0$).

(continued)

Take as reference a canonical kinetic law with no volume effect:

$$r_A = kC_A^n = kC_{A0}^n(1 - x_A)^n \tag{11.3}$$

By solving Eqs. 11.1 and 11.2, the ratio between space-times is written as in Eq. 11.4 for $n = 1$ and $n = 2$ (cf. Tables 10.8 and 10.9):

$$\frac{\tau_{CSTR}}{\tau_{PFR}} = \begin{cases} \dfrac{\dfrac{1}{k}\dfrac{x_A^e}{(1-x_A^e)}}{-\dfrac{1}{k}\ln(1-x_A^e)} = -\dfrac{x_A^e}{(1-x_A^e)\ln(1-x_A^e)} & n=1 \\ \dfrac{\dfrac{1}{kC_{A0}}\dfrac{x_A^e}{(1-x_A^e)^2}}{\dfrac{1}{kC_{A0}}\dfrac{x_A^e}{(1-x_A^e)}} = \dfrac{1}{1-x_A^e} & n=2 \end{cases} \tag{11.4}$$

Graph this ratio as in Fig. 11.2, which is suggested to be read "from right to left". For $1 - x_A^e = 1$ ($x_A^e = 0$), we fall back into the differential exercise: therefore, there can be no difference between PFR and CSTR, and in fact the

(continued)

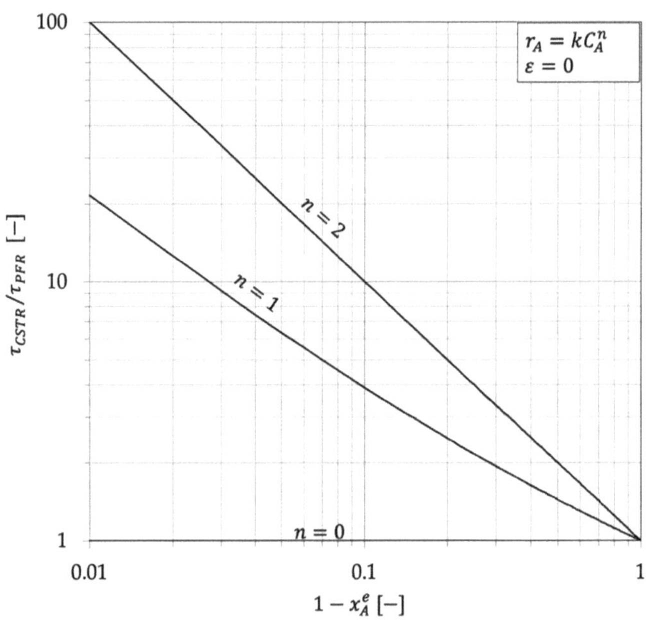

Fig. 11.2 Ratio between CSTR and PFR space-time vs. degree of conversion—case of power law kinetics and system at constant volume. Log-log plot

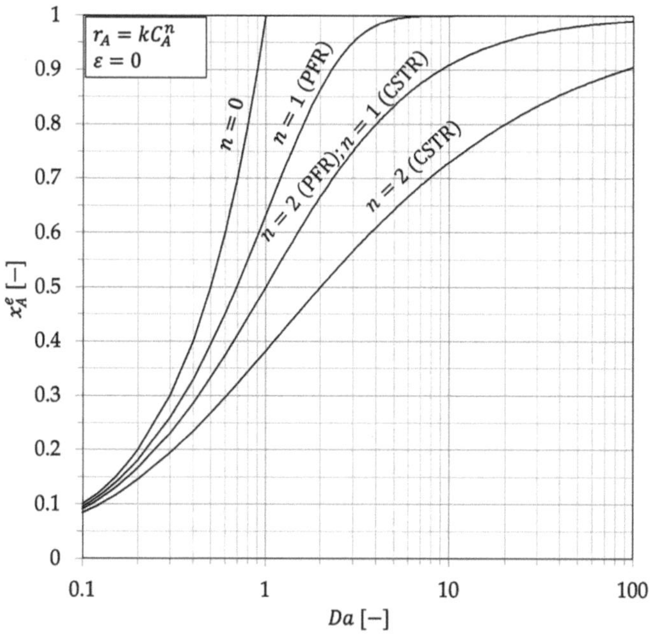

Fig. 11.3 Comparison between PFR and CSTR in terms of degree of conversion vs. Damköhler number—case of power law kinetics and system at constant volume. Semi-log plot

ratio between space-times is 1 whatever n is. As $1- x_A^e$ decreases (i.e. as x_A^e increases), the gap between the two reactors increases (in favour of PFR), and, for a fixed x_A^e, this gap increases as n increases. Moreover, we note how the ratio on the ordinate does not depend on the values of k and C_{A0}.

The careful scholar notes that, for the solution in Eq. 11.4 with $n = 1$, when $1- x_A^e = 1$, i.e. for $x_A^e = 0$, the ratio on the ordinate is expressed by an indeterminate form (0/0). Applying de l'Hôpital's rule, it is verified that the ratio is 1:

$$\frac{\tau_{CSTR}}{\tau_{PFR}} = \lim_{x_A^e \to 0} \left[-\frac{\frac{x_A^e}{1-x_A^e}}{\ln\left(1-x_A^e\right)} \right] = \lim_{x_A^e \to 0} \left[-\frac{\frac{1}{\left(1-x_A^e\right)^2}}{-\frac{1}{1-x_A^e}} \right] = \lim_{x_A^e \to 0} \left[\frac{1}{1-x_A^e} \right] \to 1 \quad (11.5)$$

Furthermore, for the solution in Eq. 11.4 with $n = 2$, the ratio on the ordinate is expressed on a logarithmic basis as $\ln(\tau_{CSTR} / \tau_{PFR}) = -\ln(1- x_A^e)$, i.e. a straight line with slope − 1 on a log-log plot. When $1- x_A^e = 1$, $\tau_{CSTR} / \tau_{PFR} = 1$. When $1- x_A^e = 0.01$, $\tau_{CSTR} / \tau_{PFR} = 100$. Obviously, for $n = 0$, the ratio between τ is identically 1, since the kinetics do not depend on fluid dynamic conditions, cf. again Tables 10.8 and 10.9.

The issue can be equivalently viewed in the light of the trends of x_A^e versus Da, an aspect presented in Fig. 11.3 by reworking Figs. 10.6 and 10.15. Apart

(continued)

from the particular case of $n = 0$ (for which the two reactors are equivalent), it is noted that, for a fixed n, and a fixed Da (i.e., for a fixed space-time, cf. Tables 10.8 and 10.9), the degree of conversion obtainable in a PFR is always greater. For example, for $n = 1$ and $Da = 1$, $x_A^e = 63.2\%$ (PFR) and $x_A^e = 50\%$ (CSTR); for $n = 2$ and $Da = 10$, $x_A^e = 90.9\%$ (PFR) and $x_A^e = 73.0\%$ (CSTR).

Example: Choice of Reactor for a Canonical Equilibrium Reaction

We consider the following equilibrium reaction between acetic acid and ethanol, to give ethyl acetate:

$$CH_3COOH + C_2H_5OH \Leftrightarrow CH_3COOC_2H_5 + H_2O \tag{11.6}$$

schematised as A + B ⇔ P + C. We assume the following kinetics:

$$\begin{cases} r_A = k_d C_A C_B - k_i C_P \\ k_d = 0.0027 \dfrac{\mathrm{L}}{\mathrm{mol \cdot min}} \quad k_i = 0.0046\ 1/\mathrm{min} \end{cases} \tag{11.7}$$

that is, we encounter two kinetic constants, one for the direct reaction (k_d) and the other for the inverse reaction (k_i). We desire $x_A^e = 50\%$, and we have to choose between a PFR and a CSTR, both available with volume $V = 1\ \mathrm{m}^3$. The feed is composed by A and B in stoichiometric ratio, $C_{AO} = C_{BO} = 8.3$ mol/L. This allows to formally simplify the reaction rate as:

$$r_A = k_d C_A^2 - k_i C_P \tag{11.8}$$

First, put $r_A(x_A)$:

$$r_A(x_A) = k_d C_{A0}^2 (1 - x_A)^2 - k_i C_{A0} x_A \tag{11.9}$$

For an equilibrium reaction, the degree of conversion is limited by a maximum value that is not 1, but is dictated by equilibrium constraints (x_A^{EQ}). To check this (and, therefore, to verify that $x_A^e = 50\%$ is a correct request), we re-write Eq. 11.9 at equilibrium, where the reaction rate must be zero:

$$0 = k_d C_{A0}^2 \left(1 - x_A^{EQ}\right)^2 - k_i C_{A0} x_A^{EQ} \tag{11.10}$$

It is solved as a regular quadratic equation:

(continued)

$$k_d C_{A0} \left(x_A^{EQ}\right)^2 - (2k_d C_{A0} + k_i) x_A^{EQ} + k_d C_{A0} = 0 \tag{11.11}$$

with acceptable solution $x_A^{EQ} = 0.638 > x_A^e$. So, we can proceed.

Back to Eq. 11.9, we see that it has a canonical trend, with r_A decreasing as x_A increases. The situation is that schematised in Fig. 11.1 (with the exception that the asymptote is at x_A^{EQ} and not at $x_A = 1$), and we certainly prefer the segregated reactor (PFR). The design equation is:

$$\begin{aligned}
\tau_{PFR} &= \frac{V}{Q_0} = C_{A0} \int_0^{x_A^e} \frac{1}{r_A(x_A)} \mathrm{d}x_A \\
&= \int_0^{x_A^e} \frac{1}{k_d C_{A0}(1 - x_A)^2 - k_i x_A} \mathrm{d}x_A \\
&= \int_0^{x_A^e} \frac{1}{k_d C_{A0}(x_A)^2 - (2k_d C_{A0} + k_i) x_A + k_d C_{A0}} \mathrm{d}x_A \\
&= \frac{1}{k_d C_{A0}} \int_0^{x_A^e} \frac{1}{(x_A)^2 - \left(\frac{2k_d C_{A0} + k_i}{k_d C_{A0}}\right) x_A + 1} \mathrm{d}x_A \\
&= \frac{1}{k_d C_{A0}} \int_0^{x_A^e} \frac{1}{(s_1 - x_A)(s_2 - x_A)} \mathrm{d}x_A
\end{aligned} \tag{11.12}$$

where $s_1 = 0.638$ and $s_2 = 1.567$ are the solutions for the polynomial in the denominator under integral, expressed in a canonical form (i.e., with the coefficient of $(x_A)^2$ equal to 1). Solve by PFD:

$$\frac{1}{(s_1 - x_A)(s_2 - x_A)} = \frac{C_1}{s_1 - x_A} + \frac{C_2}{s_2 - x_A} \tag{11.13}$$

Thus:

$$\begin{aligned}
\tau_{PFR} &= \frac{1}{k_d C_{A0}} \left(\int_0^{x_A^e} \frac{C_1}{s_1 - x_A} \mathrm{d}x_A + \int_0^{x_A^e} \frac{C_2}{s_2 - x_A} \mathrm{d}x_A \right) = \frac{1}{k_d C_{A0}} \\
&\times \left\{ -C_1 [\ln(s_1 - x_A)]_0^{x_A^e} - C_2 [\ln(s_2 - x_A)]_0^{x_A^e} \right\} = \frac{1}{k_d C_{A0}} \\
&\times \left[-C_1 \ln\left(\frac{s_1 - x_A^e}{s_1}\right) - C_2 \ln\left(\frac{s_2 - x_A^e}{s_2}\right) \right]
\end{aligned} \tag{11.14}$$

This is the solution of the design equation, once Eq. 11.13 is further manipulated to obtain the constants C_1 and C_2:

(continued)

$$\frac{1}{(s_1-x_A)(s_2-x_A)}=\frac{C_1}{s_1-x_A}+\frac{C_2}{s_2-x_A}$$
$$=\frac{s_2C_1-x_AC_1+s_1C_2-x_AC_2}{(s_1-x_A)(s_2-x_A)}$$
$$=\frac{(s_2C_1+s_1C_2)+(-C_1-C_2)x_A}{(s_1-x_A)(s_2-x_A)} \tag{11.15}$$

Equation 11.15 is valid if:

$$\begin{cases} s_2C_1+s_1C_2=1 \\ -C_1-C_2=0 \end{cases} \Rightarrow \begin{cases} C_1=\dfrac{1}{s_2-s_1}=1.076 \\ C_2=-C_1=-\dfrac{1}{s_2-s_1}=-1.076 \end{cases} \tag{11.16}$$

Therefore, Eq. 11.14 reads:

$$\tau_{PFR}=\frac{C_1}{k_dC_{A0}}\ln\left[\frac{s_1\left(s_2-x_A^e\right)}{s_2\left(s_1-x_A^e\right)}\right] \tag{11.17}$$

and it is $\tau_{PFR}=55.1$ min. We already know, from theory, that $\tau_{PFR}<\tau_{CSTR}$. To quantify this gap:

$$\tau_{CSTR}=\frac{V}{Q_0}=\frac{C_{A0}x_A^e}{r_A\left(x_A^e\right)}=\frac{x_A^e}{k_dC_{A0}\left(1-x_A^e\right)^2-k_ix_A^e}=151.4\ \text{min} \tag{11.18}$$

i.e., $\tau_{CSTR}\,/\,\tau_{PFR}=2.75$. Once fixed the PFR as our choice, the knowledge of V allows to calculate Q_0 = 18.1 L/min. Since this value is higher than that corresponding to a CSTR with equal volume (being $\tau_{PFR}<\tau_{CSTR}$, the PFR allows the same conversion in shorter residence time, so we can feed a greater flow rate), the productivity in ethyl acetate will be higher: $P_P=Q_0C_{A0}\,x_A^e=$ 75.1 mol/min.

Example: Reaction Rate Reducing to Constant Kinetics

Take a general reaction: A ⇒ P, described by a Michaelis–Menten/Monod law (Eq. 10.6):

$$\begin{cases} r_A=\dfrac{k_1C_A}{k_2+C_A} \\ k_1=5\ \dfrac{\text{mol}}{\text{L}\cdot\text{s}} \\ k_2=0.1\ \dfrac{\text{mol}}{\text{L}} \end{cases} \tag{11.19}$$

(continued)

We aim at $x_A^e = 0.8$, and we have to choose between a PFR and a CSTR, both available at lab-scale with volume $V = 1$ L. The feed is composed by A at $C_{AO} = 12.5$ mol/L. First, we observe that the lowest value for C_A (corresponding to x_A^e) will be $C_{AO}(1 - x_A^e) = 2.5$ mol/L. We can therefore assume that, in the whole range of reactor operation, $k_2 \ll C_A$. We can then simplify $r_A = k_1$. For zero-order kinetics, there is no difference between a CSTR and a PFR (see Figs. 11.2 and 11.3). It is (cf. Tables 10.8 and 10.9):

$$\tau = \frac{V}{Q_0} = \frac{C_{A0}}{k_1} x_A^e = 2 \text{ s} \tag{11.20}$$

Then, $Q_O = 0.5$ L/s and $P_P = Q_O C_{AO} x_A^e = 5$ mol/s.

11.2 Autocatalytic Reactions

11.2.1 Kinetics

Let us consider a reaction of the type:

$$\text{A} + \text{P} \Rightarrow 2\text{P} \tag{11.21}$$

Here the species A needs, in order to react, the product P, to form more product P. It is therefore a reaction that "catalyses itself" (*autocatalytic*). From a stoichiometric point of view, the reaction can also be reduced to: A $\Rightarrow$ P, to indicate that, net, 1 mole of A reacts to form 1 mole of P. The canonical kinetic expression with which an autocatalytic reaction can be represented is:

$$r_A = kC_A C_P \tag{11.22}$$

Being:

$$\begin{cases} C_A = C_{A0}(1 - x_A) \\ C_P = C_{P0} + C_{A0} x_A \end{cases} \tag{11.23}$$

It is:

$$r_A(x_A) = kC_{A0}(1 - x_A)(C_{P0} + C_{A0} x_A) \tag{11.24}$$

In particular, we have explicitly considered the case in which $C_{PO} \neq 0$, i.e. having a feed current that already contains some initiator (C_{PO}) for the chemical reaction.

By plotting the reciprocal of r_A as a function of x_A, a non-canonical kinetic graph is obtained (Fig. 11.4). It is observed that:

Fig. 11.4 Qualitative aspect for the kinetic plot—autocatalytic kinetic law in the form of Eq. 11.24

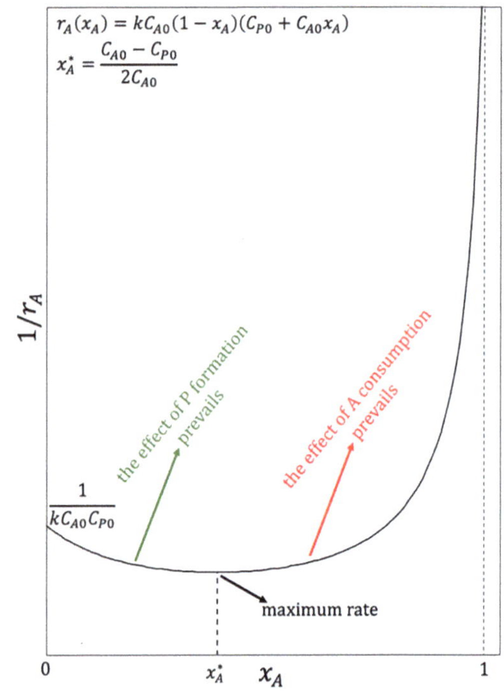

- as already intuited from Eq. 11.24, the trend is non-monotonic and parabolic. When the reaction progresses (x_A increases), C_A decreases (negative effect on r_A) but, at the same time, C_P increases (positive effect on r_A);
- r_A is equal to $kC_{A0}C_{P0}$ in $x_A = 0$, then the reaction accelerates because, for low x_A, the positive effect of P that is formed prevails. The disappearance of A as x_A increases, in this case, is suffered less (since A is still present in abundance for low x_A) compared to the advantage of having, as x_A increases, more P that catalyses the reaction. It should be remembered that, typically, C_{P0} is present in modest quantities compared to C_{A0}—species P being then produced by autocatalysis;
- as x_A further increases, r_A *passes through a maximum* (x_A^*). Once x_A^* is exceeded, for even greater x_A, the system suffers more from the disappearance of A (now A is scarce) than from the advantage of having more and more P (now already abundant). There is no doubt, in fact (Eq. 11.24), that the fate of r_A is to cancel itself out for $x_A = 1$ ($C_A = 0$ even in the presence of the maximum concentration for P).

For this kinetics, the value for x_A^* can be calculated by:

$$\frac{\mathrm{d}r_A}{\mathrm{d}x_A} = \frac{\mathrm{d}[kC_{A0}(1-x_A)(C_{P0}+C_{A0}x_A)]}{\mathrm{d}x_A}$$
$$=0 \Rightarrow \frac{\mathrm{d}[(1-x_A)(C_{P0}+C_{A0}x_A)]}{\mathrm{d}x_A}$$
$$= \frac{\mathrm{d}\left(C_{P0}+C_{A0}x_A - C_{P0}x_A - C_{A0}x_A^2\right)}{\mathrm{d}x_A} = 0 \tag{11.25}$$

$$C_{A0} - C_{P0} - 2C_{A0}x_A^* = 0 \Rightarrow x_A^* = \frac{C_{A0}-C_{P0}}{2C_{A0}} \tag{11.26}$$

In the limiting case of $C_{PO} = 0$ (feed without P), it would be $x_A^* = 0.5$. Otherwise, $x_A^* < 0.5$ and the more P is present in the feed, the smaller x_A^* is (the more C_{PO} is, the less extensive is the zone in which the course of the reaction will favour r_A, since the positive effect of P that is formed will be "saturated" earlier). We also observe that the condition of maximum rate (Eq. 11.26) can be equivalently expressed as (cf. Eq. 10.11 and remembering that $C_P – C_{PO} = C_{AO} – C_A$):

$$\frac{C_{A0}-C_A}{C_{A0}} = \frac{C_{A0}-C_{P0}}{2C_{A0}} \Rightarrow C_{A0} - C_A = \frac{C_{A0}-C_{P0}}{2} \Rightarrow C_{A0} - 2C_A =$$
$$-C_{P0} = C_{A0} - C_A - C_P \Rightarrow C_A = C_P \tag{11.27}$$

i.e., the rate is maximum when the concentrations of the two species equal each other. By using Eq. 11.22, the maximum rate is:

$$r_A\left(x_A^*\right) = kC_A^2 = kC_{A0}^2\left(1-x_A^*\right)^2 \tag{11.28}$$

Example: Finding the Maximum Reaction Rate for Autocatalytic Kinetics

Our reference scenario is the autocatalytic reaction: A + P $\Rightarrow$ 2P, described by $r_A = kC_AC_P$. Consider the following data: $C_{AO} = 1$ mol/L; $C_{PO} = 0.2$ mol/L; $k = 0.5$ L/(mol·min). Applying Eqs. 11.26 and 11.28, the maximum rate is located at $x_A^* = 0.4$ and values 0.18 mol/(L·min).

11.2.2 Reactor Choice

First, we begin with a particular case, i.e. when the feed is product-free ($C_{PO} = 0$). Here, r_A in Eq. 11.24 reduces to:

$$r_A(x_A) = kC_{A0}^2 x_A(1-x_A) \tag{11.29}$$

as represented in Fig. 11.5—a parabolic symmetric trend, i.e. $r_A = 0$ also for $x_A = 0$ (since there is no P in the feed, the reaction "does not start"), besides being equal to zero for $x_A = 1$. In this case, $x_A^* = 0.5$ (cf. Eq. 11.26).

It is evident that a product-free feed stream cannot be treated in a segregated reactor: the A species, in the absence of P, could not react, and the PFR would be reduced to a mere conduit where the feed flows without reacting, due to the hypotheses of axial segregation. Also note how, for each chosen value of x_A^e, the area under the curve between 0 and x_A^e (which represents a measure for τ_{PFR}) would always be infinite (since the function asymptotes also for $x_A = 0$), indicating the impossibility of operating with a PFR. *In these cases, it will be necessary to work with a CSTR* where the reactant A, entering the reactor without P, can react with P already present in the reactor, given the hypotheses of perfect mixing that place the elements of fluid to work directly at the exit conditions. All this is conceptually possible if, in the start-up phase of the CSTR, a little P is added to allow the reacting system to reach the required level (i.e.: to allow the reaction to "start"), and then a stream without P is fed where, however, A can react encountering the product P previously formed, and so on. Graphically, in Fig. 11.5 we can see how, for each value of x_A^e, the area of the rectangle (which represents a measure for τ_{CSTR}) has a finite value (except, obviously, for $x_A^e = 1$), indicating the possibility of working with a CSTR.

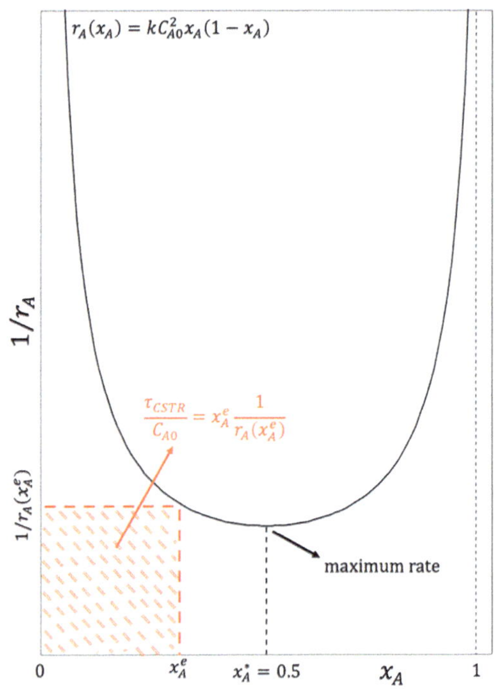

Fig. 11.5 Qualitative aspect for the kinetic plot—autocatalytic kinetic law in the form of Eq. 11.29, with product-free feed; it is conceptually possible to operate with a CSTR, and not with a PFR

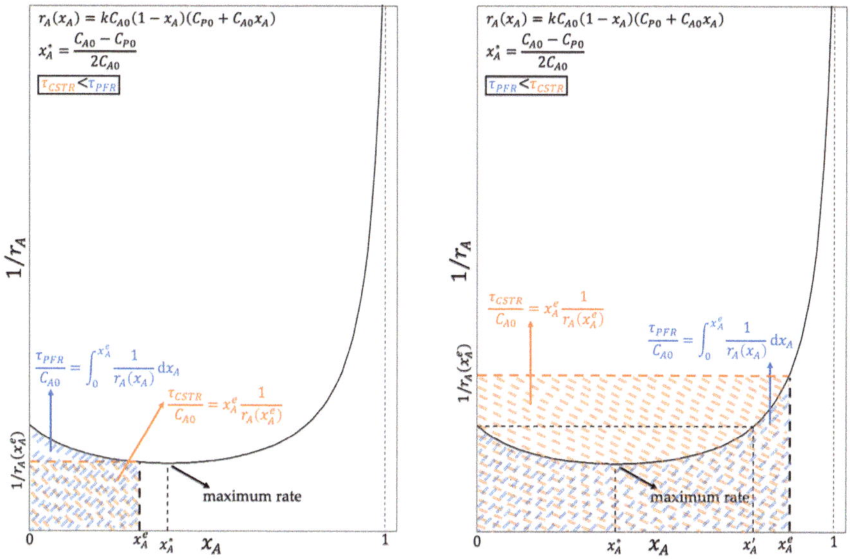

Fig. 11.6 Qualitative aspect for the kinetic plot—autocatalytic kinetic law. Left: $x_A^e \leq x_A^*$; CSTR is better. Right: $x_A^e \geq x_A'$; PFR is better. The subtended area measures τ_{PFR}; the rectangle area measures τ_{CSTR}

Returning to the case where there is P in the feed, consider a value of $x_A^e \leq x_A^*$ (Fig. 11.6, on the left). The elements of fluid in the CSTR will react at the output value for r_A, which is the best of all between 0 and x_A^e, since we are in the area where r_A is favoured by the reaction progress. Using a PFR, instead, the elements of fluid will experience the infinite values of r_A between 0 and x_A^e, and therefore on average they will react at a lower r_A. Graphically, we observe that $\tau_{CSTR} < \tau_{PFR}$: the area of the rectangle is inscribed in the area subtended between 0 and x_A^e. *We will therefore choose a CSTR.*

Let x_A' be the conversion degree such that r_A assumes the same value as at $x_A = 0$ (Fig. 11.6, on the right). Let a sufficiently high conversion value be considered, $x_A^e \geq x_A'$. The elements of fluid in the CSTR will react at the output value for r_A, which is now the worst of all between 0 and x_A^e, since we are in the area where r_A is disadvantaged by the course of the reaction. Using a PFR, instead, the elements of fluid will experience the infinite values of r_A between 0 and x_A^e, and therefore on average they will react at a higher r_A. Graphically, we observe that $\tau_{PFR} < \tau_{CSTR}$: the area of the rectangle inscribes the area subtended between 0 and x_A^e. *We will therefore choose a PFR.*

The situation is more complex if $x_A^* < x_A^e < x_A'$ (Fig. 11.7). In these cases, it is not possible to say a priori which of the two areas is inferior, that is, which of the two space-times is lower. This happens because, among the infinite values of r_A experienced by the elements of fluid in a PFR (between 0 and x_A^e), there are values both

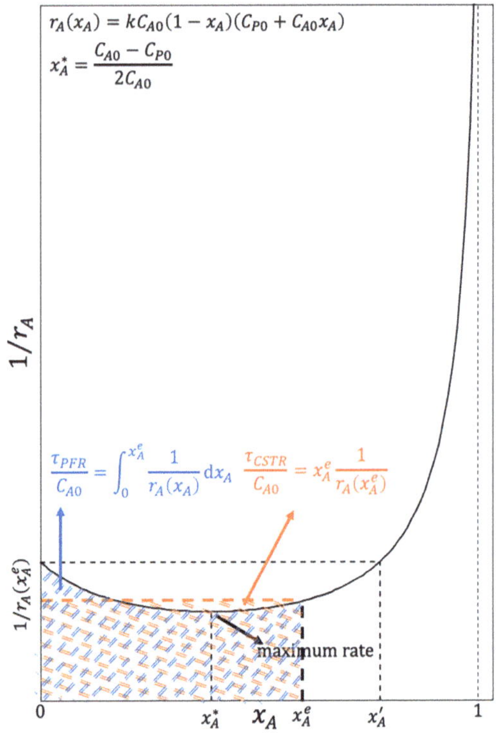

Fig. 11.7 Qualitative aspect for the kinetic plot—autocatalytic kinetic law, for $x_A^* < x_A^e < x_A'$. The subtended area measures τ_{PFR}; the rectangle area measures τ_{CSTR}

higher and lower than the single value of r_A that the elements of fluid in a CSTR are destined for. In fact, comparing the areas, one notices areas that are disadvantageous either for the PFR (where r_A is lower than the single valid value for the CSTR), or for the CSTR (where the opposite occurs). Only a comparison between the two τ-values can resolve the issue.

It is clear that, since in general at low x_A^e the CSTR is convenient, and at high x_A^e the PFR, there will be a value of x_A^e strictly included between x_A^* and x_A' for which the two τ-values will be equal (*condition of reactor equivalence* for which $\tau_{CSTR} = \tau_{PFR}$, Balsamo and Montagnaro 2018). *For conversion values beyond this condition, the PFR will be convenient. Vice versa, the CSTR*. The condition can be investigated by equaling the two space-times:

$$\frac{C_{A0}x_A^e}{r_A\left(x_A^e\right)} = C_{A0}\int_0^{x_A^e} \frac{1}{r_A(x_A)} dx_A \Rightarrow \frac{1}{r_A\left(x_A^e\right)} = \frac{\int_0^{x_A^e} \frac{1}{r_A(x_A)} dx_A}{x_A^e - 0} = \overline{\frac{1}{r_A(x_A)}}\Bigg|_0^{x_A^e} \tag{11.30}$$

The condition of reactor equivalence is substantially achieved when the only reaction rate for CSTR equals the average value of the infinite reaction rates for PFR. And, this can happen only for x_A^e strictly included between x_A^* and x_A', since for

$x_A^e \leq x_A^*$ the only value of r_A for CSTR would always be higher, and for $x_A^e \geq x_A'$ always lower, than the average value of r_A achieved in PFR.

Furthermore, condition in Eq. 11.30 has also another meaning, when we consider a PFR with known volume (any kinetics): we know that (cf. Eq. 10.107) the maximum productivity is attained when the ratio in Eq. 10.107 is maximum, i.e. (recalling the quotient rule and that we can confuse x_A^e with x_A, they have the same meaning):

$$\frac{\mathrm{d}\left[\frac{\int_0^{x_A^e}\frac{1}{r_A(x_A)}\mathrm{d}x_A}{x_A^e}\right]}{\mathrm{d}x_A} = \frac{\frac{1}{r_A(x_A)}x_A^e - \int_0^{x_A^e}\frac{1}{r_A(x_A)}\mathrm{d}x_A}{\left(x_A^e\right)^2} = 0 \tag{11.31}$$

This imposes:

$$\frac{1}{r_A(x_A)}x_A^e = \int_0^{x_A^e}\frac{1}{r_A(x_A)}\mathrm{d}x_A \tag{11.32}$$

which is the same as Eq. 11.30 although in a different context. We will come back on this result in an example at the end of this chapter.

Example: Condition of Reactor Equivalence

We keep the canonical reaction: A + P ⇒ 2P, described by $r_A = kC_AC_P$. In Eq. 11.30, we have an integral whose general solution is given here. Start with (cf. Eq. 11.24):

$$\int_0^{x_A^e}\frac{1}{r_A(x_A)}\mathrm{d}x_A = \frac{1}{kC_{A0}}\int_0^{x_A^e}\frac{1}{(1-x_A)(C_{P0}+C_{A0}x_A)}\mathrm{d}x_A \tag{11.33}$$

Solve by PFD, by posing:

$$\frac{1}{(1-x_A)(C_{P0}+C_{A0}x_A)} = \frac{C_1}{1-x_A} + \frac{C_2}{C_{P0}+C_{A0}x_A} \tag{11.34}$$

Thus:

$$\int_0^{x_A^e}\frac{1}{r_A(x_A)}\mathrm{d}x_A = \frac{1}{kC_{A0}} \times \left(\int_0^{x_A^e}\frac{C_1}{1-x_A}\mathrm{d}x_A + \int_0^{x_A^e}\frac{C_2}{C_{P0}+C_{A0}x_A}\mathrm{d}x_A\right) = \frac{1}{kC_{A0}} \times \left\{-C_1[\ln(1-x_A)]_0^{x_A^e} + \frac{C_2}{C_{A0}}[\ln(C_{P0}+C_{A0}x_A)]_0^{x_A^e}\right\} \tag{11.35}$$

Manipulate Eq. 11.34 to obtain the constants C_1 and C_2:

(continued)

$$\frac{1}{(1-x_A)(C_{P0}+C_{A0}x_A)} = \frac{C_1}{1-x_A} + \frac{C_2}{C_{P0}+C_{A0}x_A}$$
$$= \frac{C_{P0}C_1 + C_{A0}x_AC_1 + C_2 - C_2x_A}{(1-x_A)(C_{P0}+C_{A0}x_A)}$$
$$= \frac{(C_{P0}C_1 + C_2) + (C_{A0}C_1 - C_2)x_A}{(1-x_A)(C_{P0}+C_{A0}x_A)} \tag{11.36}$$

Valid if:

$$\begin{cases} C_{P0}C_1 + C_2 = 1 \\ C_{A0}C_1 - C_2 = 0 \end{cases} \Rightarrow \begin{cases} C_1 = \dfrac{1}{C_{A0}+C_{P0}} \\ C_2 = \dfrac{C_{A0}}{C_{A0}+C_{P0}} \end{cases} \tag{11.37}$$

Back to Eq. 11.35:

$$\int_0^{x_A^e} \frac{1}{r_A(x_A)} \mathrm{d}x_A = \frac{1}{kC_{A0}(C_{A0}+C_{P0})}$$
$$\times \left\{ -[\ln(1-x_A)]_0^{x_A^e} + [\ln(C_{P0}+C_{A0}x_A)]_0^{x_A^e} \right\}$$
$$= \frac{1}{kC_{A0}(C_{A0}+C_{P0})} \left[\ln\left(\frac{C_{P0}+C_{A0}x_A}{1-x_A} \right) \right]_0^{x_A^e} \tag{11.38}$$

We finally obtain the following solution, which will be useful also further:

$$\int_0^{x_A^e} \frac{1}{r_A(x_A)} \mathrm{d}x_A = \frac{1}{kC_{A0}(C_{A0}+C_{P0})} \ln\left[\frac{C_{P0}+C_{A0}x_A^e}{C_{P0}(1-x_A^e)} \right] \tag{11.39}$$

We can now apply Eq. 11.30:

$$\frac{x_A^e}{(1-x_A^e)(C_{P0}+C_{A0}x_A^e)} = \frac{1}{C_{A0}+C_{P0}} \ln\left[\frac{C_{P0}+C_{A0}x_A^e}{C_{P0}(1-x_A^e)} \right] \tag{11.40}$$

Using the same data as in an above example: $C_{A0} = 1$ mol/L; $C_{P0} = 0.2$ mol/L; $k = 0.5$ L/(mol·min), Eq. 11.40 is solved (by trial-and-error) at $x_A^e = 0.617$. That is, below this conversion a CSTR would be preferred ($\tau_{CSTR} < \tau_{PFR}$), and vice versa.

Example: Condition of Reactor Equivalence for Non-canonical Autocatalytic Kinetics

Consider now an autocatalytic process (simply, A $\Rightarrow$ P) described by a different kinetics, i.e. the one presented in Eq. 10.7:

$$r_A = \frac{k_1 C_A C_P}{k_2 + C_A} = \frac{k_1 C_{A0}(1 - x_A)(C_{P0} + C_{A0}x_A)}{k_2 + C_{A0}(1 - x_A)} \tag{11.41}$$

This could be the case of a microbial fermentation, where the substrate A is converted into the microbial product P in presence of the latter in the role of auto-catalyst (a more extended discussion is offered in Balsamo and Montagnaro 2018). To determine the condition of reactor equivalence in Eq. 11.30, first we solve the integral:

$$\int_0^{x_A^e} \frac{1}{r_A(x_A)} \mathrm{d}x_A = \frac{1}{k_1 C_{A0}} \int_0^{x_A^e} \frac{k_2 + C_{A0}(1 - x_A)}{(1 - x_A)(C_{P0} + C_{A0}x_A)} \mathrm{d}x_A \tag{11.42}$$

Apply the PFD method:

$$\frac{k_2 + C_{A0}(1 - x_A)}{(1 - x_A)(C_{P0} + C_{A0}x_A)} = \frac{C_1}{1 - x_A} + \frac{C_2}{C_{P0} + C_{A0}x_A} \tag{11.43}$$

Thus:

$$\begin{aligned}\int_0^{x_A^e} \frac{1}{r_A(x_A)} \mathrm{d}x_A &= \frac{1}{k_1 C_{A0}} \\ &\times \left(\int_0^{x_A^e} \frac{C_1}{1 - x_A} \mathrm{d}x_A + \int_0^{x_A^e} \frac{C_2}{C_{P0} + C_{A0}x_A} \mathrm{d}x_A \right) = \frac{1}{k_1 C_{A0}} \\ &\times \left\{ - C_1 [\ln(1 - x_A)]_0^{x_A^e} + \frac{C_2}{C_{A0}} [\ln(C_{P0} + C_{A0}x_A)]_0^{x_A^e} \right\}\end{aligned} \tag{11.44}$$

Manipulate Eq. 11.43 to obtain the constants C_1 and C_2:

$$\begin{aligned}\frac{k_2 + C_{A0} - C_{A0}x_A}{(1 - x_A)(C_{P0} + C_{A0}x_A)} &= \frac{C_1}{1 - x_A} + \frac{C_2}{C_{P0} + C_{A0}x_A} \\ &= \frac{C_{P0}C_1 + C_{A0}x_A C_1 + C_2 - C_2 x_A}{(1 - x_A)(C_{P0} + C_{A0}x_A)} \\ &= \frac{(C_{P0}C_1 + C_2) + (C_{A0}C_1 - C_2)x_A}{(1 - x_A)(C_{P0} + C_{A0}x_A)}\end{aligned} \tag{11.45}$$

Valid if:

(continued)

$$\begin{cases} C_{P0}C_1 + C_2 = k_2 + C_{A0} \\ C_{A0}C_1 - C_2 = -C_{A0} \end{cases} \Rightarrow \begin{cases} C_1 = \dfrac{k_2}{C_{A0} + C_{P0}} \\ C_2 = \dfrac{C_{A0}(k_2 + C_{A0} + C_{P0})}{C_{A0} + C_{P0}} \end{cases} \tag{11.46}$$

Back to Eq. 11.44:

$$\begin{aligned} \int_0^{x_A^e} \frac{1}{r_A(x_A)} \mathrm{d}x_A &= \frac{1}{k_1 C_{A0}} \left\{ -\frac{k_2}{C_{A0} + C_{P0}} [\ln(1 - x_A)]_0^{x_A^e} \right. \\ &\quad \left. + \frac{k_2 + C_{A0} + C_{P0}}{C_{A0} + C_{P0}} [\ln(C_{P0} + C_{A0}x_A)]_0^{x_A^e} \right\} \\ &= \frac{1}{k_1 C_{A0}} \left\{ -\frac{k_2}{C_{A0} + C_{P0}} [\ln(1 - x_A)]_0^{x_A^e} \right. \\ &\quad + \frac{k_2}{C_{A0} + C_{P0}} [\ln(C_{P0} + C_{A0}x_A)]_0^{x_A^e} \\ &\quad \left. + [\ln(C_{P0} + C_{A0}x_A)]_0^{x_A^e} \right\} \\ &= \frac{1}{k_1 C_{A0}} \left\{ \frac{k_2}{C_{A0} + C_{P0}} \left[\ln\left(\frac{C_{P0} + C_{A0}x_A}{1 - x_A} \right) \right]_0^{x_A^e} \right. \\ &\quad \left. + [\ln(C_{P0} + C_{A0}x_A)]_0^{x_A^e} \right\} \\ &= \frac{1}{k_1 C_{A0}} \left\{ \frac{k_2}{C_{A0} + C_{P0}} \ln\left[\frac{C_{P0} + C_{A0}x_A^e}{C_{P0}(1 - x_A^e)} \right] + \ln\left(\frac{C_{P0} + C_{A0}x_A^e}{C_{P0}} \right) \right\} \end{aligned} \tag{11.47}$$

We can now apply Eq. 11.30:

$$\frac{x_A^e \left[k_2 + C_{A0}\left(1 - x_A^e\right) \right]}{\left(1 - x_A^e\right)\left(C_{P0} + C_{A0}x_A^e\right)} = \frac{k_2}{C_{A0} + C_{P0}} \ln\left[\frac{C_{P0} + C_{A0}x_A^e}{C_{P0}\left(1 - x_A^e\right)} \right] + \ln\left(\frac{C_{P0} + C_{A0}x_A^e}{C_{P0}} \right) \tag{11.48}$$

Consider the following reference data: $C_{AO} = 1$ mol/L; $C_{PO} = 0.2$ mol/L; $k_1 = 0.4$ 1/h; $k_2 = 5.5$ mol/L. Solving Eq. 11.48 by trial-and-error, $x_A^e = 0.655$. Again, below this conversion a CSTR would be preferred, and vice versa.

Example: Comparison of Known-Volume Reactors for Autocatalytic Process

Back to our base-case scenario: A + P $\Rightarrow$ 2P; $r_A = kC_AC_P$; $C_{A0} = 1$ mol/L; $C_{P0} = 0.2$ mol/L; $k = 0.5$ L/(mol·min). We know, from an above example, that $x_A^* = 0.4$.

Start with a CSTR with known volume ($V = 2$ L). The design equation is:

$$\tau_{CSTR} = \frac{V}{Q_0} = \frac{C_{A0}x_A^e}{r_A\left(x_A^e\right)} = \frac{x_A^e}{k\left(1 - x_A^e\right)\left(C_{P0} + C_{A0}x_A^e\right)} \tag{11.49}$$

while the formula for the productivity (cf. Eq. 10.105) is, when we have a CSTR with known volume:

$$P_P = Vr_A\left(x_A^e\right) = VkC_{A0}\left(1 - x_A^e\right)\left(C_{P0} + C_{A0}x_A^e\right) \tag{11.50}$$

As, for CSTR with known volume, productivity is maximum if the value of the unique reaction rate is maximum, we will suggest working at $x_A^e = x_A^* = 0.4$. The plot for the productivity is given in Fig. 11.8: with the assigned data, the maximum productivity values 0.36 mol/min, letting the CSTR work at $\tau_{CSTR} = 2.22$ min (from Eq. 11.49), that is, feeding $Q_0 = V/\,\tau_{CSTR} = 0.90$ L/min.

Now consider a PFR with the same volume. The design equation is (cf. Eq. 11.39):

$$\begin{aligned}\tau_{PFR} &= \frac{V}{Q_0} = C_{A0}\int_0^{x_A^e}\frac{1}{r_A(x_A)}\mathrm{d}x_A \\ &= \frac{1}{k(C_{A0} + C_{P0})}\ln\left[\frac{C_{P0} + C_{A0}x_A^e}{C_{P0}\left(1 - x_A^e\right)}\right]\end{aligned} \tag{11.51}$$

while the formula for the productivity (cf. Eq. 10.106) is, when we have a PFR with known volume:

$$P_P = V\frac{x_A^e}{\int_0^{x_A^e}\frac{1}{r_A(x_A)}\mathrm{d}x_A} = V\frac{x_A^e}{\frac{1}{kC_{A0}(C_{A0}+C_{P0})}\ln\left[\frac{C_{P0}+C_{A0}x_A^e}{C_{P0}\left(1-x_A^e\right)}\right]} \tag{11.52}$$

For PFR with known volume, productivity is maximum when we maximise the average value of r_A among the infinities that the fluid elements experience between entrance and exit. We will then choose to work at a value of x_A^e greater than x_A^*, capable of maximising the average value for r_A. Looking at Fig. 11.8, this is attained at $x_A^e = 0.617$, which also corresponds to the conditions of reactor equivalence (cf. an above example and Eqs. 11.30 and 11.32). With the

(continued)

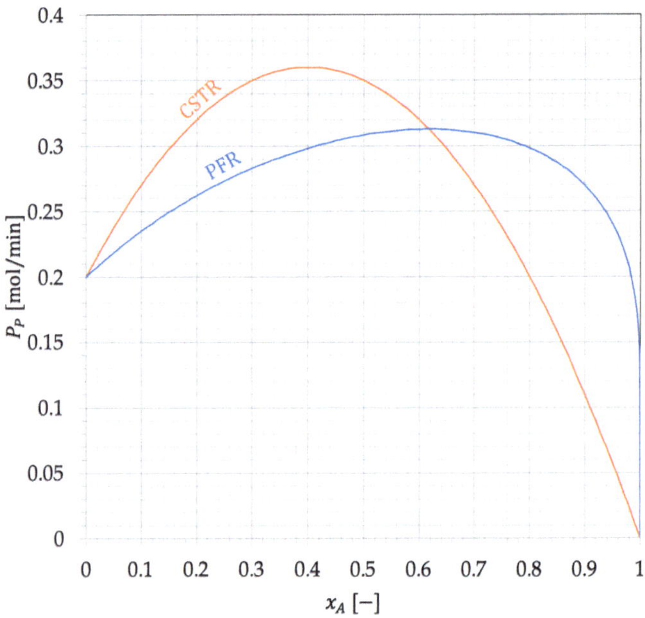

Fig. 11.8 Productivity vs. degree of conversion for the autocatalytic reaction under investigation, for CSTR and PFR with known volume

assigned data, the maximum productivity values 0.31 mol/min, letting the PFR work at $\tau_{PFR} = 3.94$ min (from Eq. 11.51), that is, feeding $Q_0 = V/\tau_{PFR} = 0.51$ L/min.

Of course, conditions that maximise the unique reaction rate for CSTR ($x_A^e = x_A^*$) are better than conditions that maximise the average value of the reaction rates for PFR. No matter how optimised the PFR may be (in terms of maximum productivity with known reactor volume), no condition will be as attractive as having the fluid elements always work at the maximum reaction rate for an autocatalytic process—which is possible in a CSTR. Therefore, if the two reactors have the same volume, and if the objective is to maximise productivity, the best solution to choose will be the CSTR working at x_A^*.

11.3 Background Information

- For canonical kinetics, the PFR is preferred since the average reaction rate is greater than the single reaction rate experienced by the fluid elements in a CSTR.
- Autocatalytic reactions show a maximum reaction rate which is positioned for degrees of conversion other than zero.

- In the autocatalytic case, for sufficiently low conversion degrees, a CSTR is preferred, and vice versa a PFR; for intermediate conversion degrees, it will be necessary to verify the reactor equivalence condition (Eq. 11.30) from which it will be possible to identify which reactor to choose between PFR and CSTR.

References

Balsamo M, Montagnaro F (2018) On the performance of continuous stirred tank reactor and plug flow reactor for chemical reactions characterised by non-elementary kinetics. React Kinet Mech Catal 125:449–469. https://doi.org/10.1007/s11144-018-1475-0

Levenspiel O (1999) Chemical reaction engineering. John Wiley & Sons, New York

Westerterp KR, van Swaaij WPM, Beenackers AACM (1988) Chemical reactor design and operation. John Wiley & Sons, Hoboken

Combination of Chemical Reactors, and Recycle Reactor

12

Abstract

This chapter extends the concepts on ideal reactors, first analysing some configurations of interest (parallel and series reactors), highlighting the design equations and the optimisation procedures of the proposed reactor schemes. Furthermore, the role of mixing is explored in depth, underlining when it is convenient to anticipate or postpone mixing in a series composed of reactors with different nature. The chapter ends with the detailed analysis of the PFR with recycle, which will be shown to be a reactor that encompasses both the PFR and the CSTR case: the design equation is obtained, and the case study of autocatalytic reactions is presented, for which it is possible to identify a condition for the optimisation of the recycle ratio.

Nomenclature

C_{TOT}	Total concentration [moles/volume]
C_i	Concentration of i species [moles/volume]
C_{i0}	Concentration of i species at reaction beginning [moles/volume]
C_{iR}	Concentration of i species in recycled stream [moles/volume]
C_{ij}	Concentration of i species exiting a reactor "j" [moles/volume]
C_{ik}	Concentration of i species in position "k" [moles/volume]
C_i^e	Concentration of i species at reactor exit [moles/volume]
F_{i0}	Molar flow rate of i species at reaction beginning [moles/time]
F_{iR}	Molar flow rate of i species in recycled stream [moles/time]
F_{ik}	Molar flow rate of i species in position "k" [moles/time]
F'_{A0}	Virtual A molar flow rate at reaction beginning [moles/time]
F_i^e	Molar flow rate of i species at reactor exit [moles/time]
k	Kinetic constant [(1/time)·(moles/volume)$^{1-n}$]
L	Reactor length [length]

F. Montagnaro et al., *Elements of Unit Operations and Chemical Reactors*,
https://doi.org/10.1007/978-3-032-18046-9_12

N	Number of reactors [–]
n	Total reaction order [–]
$\mathcal{P}$	Pressure [mass/(length·time^2)]
P_P	Productivity of the desired species [moles/time]
Q	Volumetric flow rate [volume/time]
Q^e	Volumetric flow rate at reactor exit [volume/time]
Q_0	Volumetric flow rate at reaction beginning [volume/time]
Q_R	Recycled volumetric flow rate [volume/time]
Q_j	Volumetric flow rate entering a reactor "*j*" (if in parallel) or exiting a reactor "*j*" (if in series) [volume/time]
Q_k	Volumetric flow rate in position "*k*" [volume/time]
Q'_0	Virtual volumetric flow rate at reaction beginning [volume/time]
R	Recycle ratio [–]
$\mathfrak{R}$	Gas constant [energy/(moles·temperature)]
r_A	Rate of reaction [moles/(volume·time)]
S	Reactor cross-section area [length^2]
T	Temperature [temperature]
V	Volume [volume]
V_{CSTR}	Volume of CSTR [volume]
V_{PFR}	Volume of PFR [volume]
V_j	Volume of a reactor "*j*" [volume]
x_A	Degree of conversion [–]
x_{AR}	Degree of conversion in recycled stream [–]
x_{Aj}	Degree of conversion for a reactor "*j*" [–]
x_{Ak}	Degree of conversion in position "*k*" [–]
x_A^e	Exit degree of conversion [–]
x_A^*	Degree of conversion for maximum reaction rate [–]
z	Reactor axial coordinate [length]
ε	Expansion coefficient [–]
τ	Space-time for continuous reactor [time]
τ'	Virtual space-time for continuous reactor [time]
τ_{CSTR}	Space-time for CSTR [time]
τ_{PFR}	Space-time for PFR [time]
τ_{TOT}	Sum of space-times [time]
τ_j	Space-time for a reactor "*j*" [time]

12.1 General Premise

We learned how to choose the best continuous reactor depending, for example, on the specific kinetic law that governs the reaction, the desired conversion degree, the function to be optimised (e.g. productivity, space-time). However, the ideal choice of a single reactor is not always practicable, or the best possible. Consider, for example, the case in which some reactors are available while we, ideally, would like others (in terms of type and/or volume), or the situation in which the best choice is actually constituted by a *combination of ideal reactors*. Or, again, consider the case in which we have a single reactor in operation, but at a certain point there are needs to increase the potentiality of the plant: rather than decommissioning the existing reactor and designing a new one, it may be convenient to add a new reactor to the existing one. More complex cases in which a combination of ideal reactors is of interest, are those, e.g., of a system made up of several reactors (even catalytic), each operating at a certain temperature (f.i., due to the activity requirements of the catalyst), with the interposition of heat exchangers, lateral inlet streams, recycles. In this textbook, the combination of ideal reactors is presented with reference to some configurations of interest (extensions are possible starting from the concepts provided here).

12.2 Reactors in Parallel

We begin with the analysis of a system consisting of *reactors in parallel*: for simplicity, we refer to two reactors (termed R1 and R2, respectively); however, it is important to point out that the results we will arrive at have a more general validity. The system is presented in Fig. 12.1, and it is considered, generally, at variable density:

- the inlet stream has a volumetric flow rate Q_O and a concentration of A equal to C_{AO}, and it is divided into two sub-streams, each with concentration C_{AO}, having flow rates Q_I and $Q_O - Q_I$ respectively;
- the flow rate Q_I is fed to the reactor R1 of volume V_I, while the flow rate $Q_O - Q_I$ to the reactor R2 of volume V_2; the space-time for the two reactors is, respectively, $\tau_1 = V_I/Q_I$ and $\tau_2 = V_2/(Q_O - Q_I)$;
- at the exit of R1, the stream is converted to x_{AI}, characterised by volumetric flow rate (see Eq. 10.15) $Q_I(1 + \varepsilon x_{AI})$ and concentration of A equal to (see Eq. 10.21) $C_{AI} = C_{AO}(1 - x_{AI})/(1 + \varepsilon x_{AI})$; similarly, at the exit of R2, it is x_{A2}, $(Q_O - Q_I)$

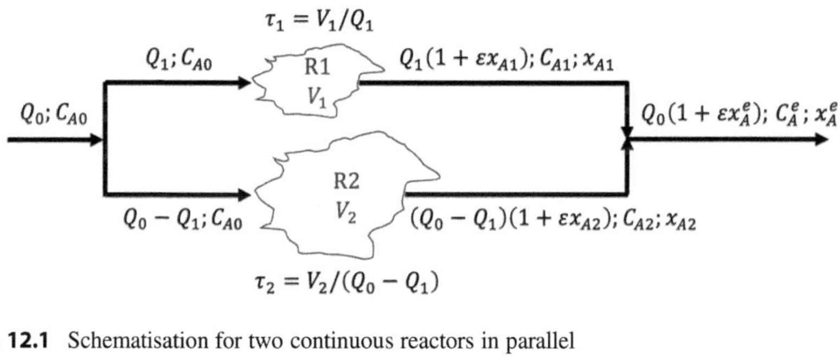

Fig. 12.1 Schematisation for two continuous reactors in parallel

$(1+\varepsilon x_{A2})$ and $C_{A2} = C_{A0}(1 - x_{A2})/(1 + \varepsilon x_{A2})$ for conversion degree, volumetric flow rate and concentration of A, respectively;

- the two converted streams mix into a single exit stream, characterised by x_A^e, $Q_0(1 + \varepsilon x_A^e)$ and $C_A^e = C_{A0}(1 - x_A^e)/(1 + \varepsilon x_A^e)$.

To obtain an expression for x_A^e, a material balance must be written on A species, around the outlet mixing node, equating the incoming molar flow rates with that leaving the node:

$$Q_1(1+\varepsilon x_{A1})C_{A1} + (Q_0 - Q_1)(1+\varepsilon x_{A2})C_{A2} = Q_0\left(1+\varepsilon x_A^e\right)C_A^e \tag{12.1}$$

$$Q_1 C_{A0}(1 - x_{A1}) + (Q_0 - Q_1)C_{A0}(1 - x_{A2}) = Q_0 C_{A0}\left(1 - x_A^e\right) \tag{12.2}$$

$$Q_1 - Q_1 x_{A1} + Q_0 - Q_1 - Q_0 x_{A2} + Q_1 x_{A2} = Q_0 - Q_0 x_A^e \tag{12.3}$$

$$Q_0 x_A^e = Q_1 x_{A1} + (Q_0 - Q_1)x_{A2} \tag{12.4}$$

$$x_A^e = \frac{Q_1 x_{A1} + (Q_0 - Q_1)x_{A2}}{Q_0} \tag{12.5}$$

It is noted that Eq. 12.1 is converted into Eq. 12.2 by substituting C_{A1}, C_{A2} and C_A^e with their above-referred expressions. The important result is that *the exit conversion degree is the weighted average (via the volumetric flow rates) of the conversion degrees of the streams leaving each reactor*. This result can be clearly generalised to a parallel of a generic number N of reactors.

A system of reactors in parallel (e.g., N CSTR or N PFR) is optimised by making *all reactors operate with the same value of space-time*, as highlighted in the literature (Levenspiel 1999):

$$\boxed{\tau_1 = \tau_2 = \ldots = \tau_N} \tag{12.6}$$

This will be equivalent to obtain, at the output of each of the N reactors of the same nature (all CSTR, or all PFR) placed in parallel, *the same degree of conversion* (given the bijective correspondence between space-time and degree of conversion):

$$x_{A1} = x_{A2} = \ldots = x_{AN} \tag{12.7}$$

Therefore:

$$x_A^e = x_{A1} = x_{A2} = \ldots = x_{AN} \tag{12.8}$$

If the parallel of N reactors works in an optimised way, the reactor system can be equivalently treated as a single reactor, with a volume equal to the sum of the volumes of the N reactors, which processes the entire volumetric flow rate Q_0 up to conversion degree x_A^e. These aspects are now clarified by some case studies.

Example: Optimisation of 2 PFR in Parallel, Linear Kinetic Law, Constant Density

We now apply the scheme of Fig. 12.1 to two PFR in parallel. The reaction, for simplicity, is A⟹P, described by $r_A = kC_A$, and $\varepsilon = 0$. For the first (R1) PFR, the solution of the design equation is (cf. Table 10.8):

$$x_{A1} = 1 - \exp(-k\tau_1) = 1 - \exp\left(-k\frac{V_1}{Q_1}\right) \tag{12.9}$$

Similarly, for the second (R2) PFR:

$$x_{A2} = 1 - \exp(-k\tau_2) = 1 - \exp\left(-k\frac{V_2}{Q_0 - Q_1}\right) \tag{12.10}$$

Thus, it is (cf. Eq. 12.5):

$$x_A^e = \frac{Q_1\left[1 - \exp\left(-k\frac{V_1}{Q_1}\right)\right] + (Q_0 - Q_1)\left[1 - \exp\left(-k\frac{V_2}{Q_0 - Q_1}\right)\right]}{Q_0} \tag{12.11}$$

Consider the case in which our degree of freedom is $Q_1 \in [0, Q_0]$, i.e. the choice on the way in which we decide to divide the inlet stream Q_0 between the two reactors (known the volumes of the two reactors). If $Q_1 = 0$, only the second PFR works, processing the entire flow rate (in fact, Eq. 12.5, $x_A^e = x_{A2}$). If $Q_1 = Q_0$, only the first PFR works, processing the entire flow rate ($x_A^e = x_{A1}$). It is clear that, by exploiting both reactors, better performances will be obtained than in the case of using only the first or only the second reactor. Therefore, there is an optimal value of Q_1 such that x_A^e is maximum. It is necessary to set the derivative of x_A^e with respect to Q_1 equal to 0. Maths follow:

$$x_A^e = \frac{Q_1 - Q_1\exp\left(-k\frac{V_1}{Q_1}\right) + Q_0 - Q_0\exp\left(-k\frac{V_2}{Q_0 - Q_1}\right) - Q_1 + Q_1\exp\left(-k\frac{V_2}{Q_0 - Q_1}\right)}{Q_0} \tag{12.12}$$

$$\frac{dx_A^e}{dQ_1} = \frac{-\exp\left(-k\frac{V_1}{Q_1}\right) - \frac{kV_1}{Q_1}\exp\left(-k\frac{V_1}{Q_1}\right) + \frac{Q_0 kV_2}{(Q_0 - Q_1)^2}\exp\left(-k\frac{V_2}{Q_0 - Q_1}\right) + \exp\left(-k\frac{V_2}{Q_0 - Q_1}\right) - \frac{Q_1 kV_2}{(Q_0 - Q_1)^2}\exp\left(-k\frac{V_2}{Q_0 - Q_1}\right)}{Q_0} = 0 \tag{12.13}$$

$$\begin{aligned} &- \exp(-k\tau_1) - k\tau_1\exp(-k\tau_1) + \frac{Q_0}{Q_0 - Q_1}k\tau_2\exp(-k\tau_2) \\ &+ \exp(-k\tau_2) - \frac{Q_1}{Q_0 - Q_1}k\tau_2\exp(-k\tau_2) = 0 \end{aligned} \tag{12.14}$$

(continued)

$$(1+k\tau_1)\exp(-k\tau_1) = \left(1+\frac{Q_0}{Q_0-Q_1}k\tau_2-\frac{Q_1}{Q_0-Q_1}k\tau_2\right)\exp(-k\tau_2) = (1+k\tau_2)\exp(-k\tau_2) \tag{12.15}$$

Condition in Eq. 12.15 is verified if:

$$\tau_1 = \tau_2 \tag{12.16}$$

which demonstrates the assertion of Eq. 12.6 of general validity. That is, Q_0 must be subdivided into two streams, with flow rates Q_1 and $Q_0 - Q_1$ respectively, such that $\tau_1 = \tau_2$, i.e., the two flow rates must be subdivided proportionally to the reactors volume. Formally, Eq. 12.16 becomes:

$$\frac{V_1}{Q_1} = \frac{V_2}{Q_0-Q_1} \Rightarrow Q_1 = \frac{V_1}{V_1+V_2}Q_0 \tag{12.17}$$

For example: if $Q_0 = 6$ L/h, $V_1 = 1$ L and $V_2 = 2$ L, R1 has a volume equal to 1/3 of the sum $V_1 + V_2$. Therefore, it will be necessary to send a flow rate $Q_1 = Q_0/3 = 2$ L/h to R1, and a flow rate $Q_0 - Q_1 = 4$ L/h to R2. Thus, $\tau_1 = V_1/Q_1 = 0.5$ h $= \tau_2 = V_2/(Q_0 - Q_1)$. Of course, this will verify Eq. 12.8 (for $N = 2$), and the value of x_A^e will be the highest possible.

As already observed, this system can be equivalently treated as a single reactor (Fig. 12.2), with a volume equal to the sum of the volumes of the 2 reactors, which processes the entire volumetric flow rate Q_0 up to conversion degree x_A^e. It is, for the equivalent system, $\tau = (V_1 + V_2)/Q_0$ (equal to $\tau_1 = \tau_2 = 0.5$ h in our case), and $x_A^e = x_{A1} = x_{A2} = 1 - \exp(-k\tau)$.

It is emphasised that the parallel could also be operated with a different partition of flow rates, but only in the conditions of uniformisation of space-times (and conversion degrees) will the highest possible value for x_A^e be obtained. Therefore, only in the optimisation conditions it is possible to use the convenient equivalence of Fig. 12.2 (in all other cases, it is necessary to solve the two PFR separately). Operationally, assuming that the PFR parallel works in optimal conditions, we use the equivalent model of Fig. 12.2, and then we indicate how to divide the flow rate Q_0 respect to the optimisation condition in Eqs. 12.16 and 12.17.

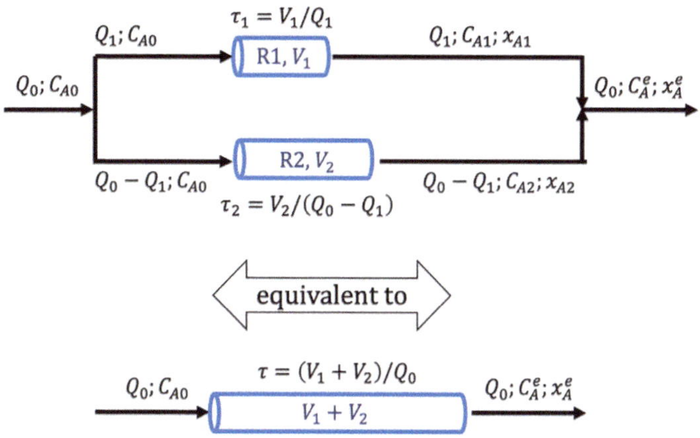

Fig. 12.2 Equivalence of 2 PFR in parallel (optimised conditions) with one virtual PFR, for the case under analysis

Example: Optimisation of 2 PFR in Parallel for Methyl Group Oxidation
We consider the gas phase reaction (with $\varepsilon = 0$):

$$CH_3 + O_2 \Rightarrow CH_3O + O \tag{12.18}$$

schematised as A + B⟹P + C. The feed composition is: 10% A + 20% B + inert. Be the reaction carried out at $\mathcal{P} = 1$ atm and $T = 650$ K, that is:

$$C_{TOT} = \frac{\mathcal{P}}{\Re T} = 0.0188 \text{ mol/L} \tag{12.19}$$

$$C_{A0} = 0.00188 \text{ mol/L} \tag{12.20}$$

$$C_{B0} = 0.00376 \text{ mol/L} \tag{12.21}$$

We assume the kinetic law:

$$\begin{cases} r_A = kC_AC_B = kC_{A0}(1 - x_A)(C_{B0} - C_{A0}x_A) \\ k = 4.77 \dfrac{\text{L}}{\text{mol} \cdot \text{s}} \end{cases} \tag{12.22}$$

Our system is composed by two PFR in parallel ($V_1 = 50$ L and $V_2 = 150$ L). The feed rate is $Q_0 = 1$ L/s. For an optimised system (cf. Figure 12.2), the equivalent PFR will have $\tau = (V_1 + V_2)/Q_0 = 200$ s. Solve the design equation (this has been already done in Eqs. 10.82 and 10.84):

(continued)

$$\tau = \frac{1}{k}\left[-C_1 \ln\left(1 - x_A^e\right) + \frac{C_2}{C_{A0}} \ln\left(\frac{C_{B0}}{C_{B0} - C_{A0}x_A^e}\right)\right] \tag{12.23}$$

$$\begin{cases} C_1 = \dfrac{1}{C_{B0} - C_{A0}} \\ C_2 = \dfrac{-C_{A0}}{C_{B0} - C_{A0}} \Rightarrow \dfrac{C_2}{C_{A0}} = -C_1 \end{cases} \tag{12.24}$$

$$\tau = \frac{1}{k(C_{B0} - C_{A0})} \ln\left[\frac{C_{B0} - C_{A0}x_A^e}{C_{B0}\left(1 - x_A^e\right)}\right] \tag{12.25}$$

Solve per x_A^e:

$$x_A^e = \frac{\exp[k\tau(C_{B0} - C_{A0})] - 1}{\exp[k\tau(C_{B0} - C_{A0})] - \frac{C_{A0}}{C_{B0}}} = 0.909 \tag{12.26}$$

This is equivalent to obtain the maximum productivity for $P_P = Q_0 C_{A0} x_A^e = 0.0017$ mol/s. Operationally, this is achieved by posing:

$$Q_1 = \frac{V_1}{V_1 + V_2} Q_0 = 0.25 \frac{\text{L}}{\text{s}} \Rightarrow \tau_1 = \frac{V_1}{Q_1} = 200 \text{ s} \Rightarrow x_{A1} = 0.909 \tag{12.27}$$

$$Q_0 - Q_1 = 0.75 \frac{\text{L}}{\text{s}} \Rightarrow \tau_2 = \frac{V_2}{Q_0 - Q_1} = 200 \text{ s} \Rightarrow x_{A2} = 0.909 \tag{12.28}$$

Example: Optimisation of 2 CSTR in Parallel, Linear Kinetic Law, Constant Density

The optimisation condition (Eq. 12.6), previously demonstrated for two PFR, can be now discussed for the case of two CSTR in parallel. The reaction (A⇒P, $r_A = kC_A$, $\varepsilon = 0$) is the same, and similar is the approach. Only the mathematical details differ. The case is schematised in Fig. 12.3.

For the first (R1) and the second (R2) CSTR, the solution of the design equation is, respectively (cf. Table 10.9):

$$x_{A1} = \frac{k\tau_1}{1 + k\tau_1} = \frac{k\frac{V_1}{Q_1}}{1 + k\frac{V_1}{Q_1}} \tag{12.29}$$

(continued)

$$x_{A2} = \frac{k\tau_2}{1 + k\tau_2} = \frac{k\frac{V_2}{Q_0 - Q_1}}{1 + k\frac{V_2}{Q_0 - Q_1}} \tag{12.30}$$

Thus, Eq. 12.5:

$$x_A^e = \frac{Q_1 \frac{k\frac{V_1}{Q_1}}{1+k\frac{V_1}{Q_1}} + (Q_0 - Q_1)\frac{k\frac{V_2}{Q_0-Q_1}}{1+k\frac{V_2}{Q_0-Q_1}}}{Q_0} = \frac{\frac{kV_1}{1+k\frac{V_1}{Q_1}} + \frac{kV_2}{1+k\frac{V_2}{Q_0-Q_1}}}{Q_0} \tag{12.31}$$

Our degree of freedom is, again, $Q_1 \in [0, Q_0]$, with the same considerations developed for the case of two PFR. Therefore, there is an optimal value of Q_1 such that x_A^e is maximum. Set the derivative of x_A^e with respect to Q_1 equal to 0:

$$\frac{dx_A^e}{dQ_1} = \frac{kV_1 \frac{\frac{kV_1}{Q_1^2}}{\left(1+k\frac{V_1}{Q_1}\right)^2} - kV_2 \frac{\frac{kV_2}{(Q_0-Q_1)^2}}{\left(1+k\frac{V_2}{Q_0-Q_1}\right)^2}}{Q_0} = k^2\left[\frac{\frac{\frac{V_1^2}{Q_1^2}}{\left(1+k\frac{V_1}{Q_1}\right)^2} - \frac{\frac{V_2^2}{(Q_0-Q_1)^2}}{\left(1+k\frac{V_2}{Q_0-Q_1}\right)^2}}{Q_0}\right] = 0 \tag{12.32}$$

$$\frac{\frac{V_1^2}{Q_1^2}}{\left(1 + k\frac{V_1}{Q_1}\right)^2} = \frac{\frac{V_2^2}{(Q_0 - Q_1)^2}}{\left(1 + k\frac{V_2}{Q_0 - Q_1}\right)^2} \Leftrightarrow \frac{\tau_1^2}{(1 + k\tau_1)^2} = \frac{\tau_2^2}{(1 + k\tau_2)^2} \tag{12.33}$$

Verified if:

$$\tau_1 = \tau_2 \tag{12.34}$$

with the same considerations developed above on the subdivision of the Q_0 stream between the two reactors. As already observed, this system can be equivalently treated as a single reactor (Fig. 12.3), with volume equal to the sum of the volumes of the 2 reactors, which processes the entire volumetric flow rate Q_0 up to conversion degree x_A^e. It is, for the equivalent system, $\tau = (V_1 + V_2)/Q_0$, and $x_A^e = x_{A1} = x_{A2} = k\tau/(1 + k\tau)$.

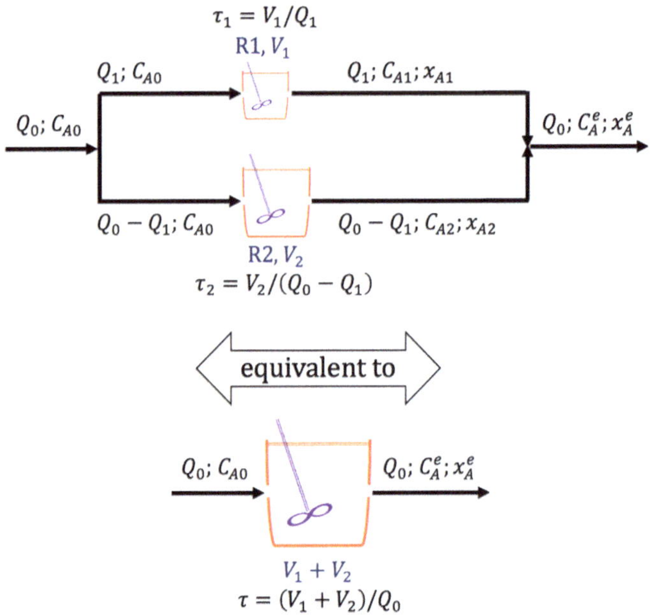

Fig. 12.3 Equivalence of 2 CSTR in parallel (optimised conditions) with one virtual CSTR, for the case under analysis

12.3 Reactors in Series

For the sake of simplicity, let us consider $N = 2$ continuous reactors (R1 and R2) arranged *in series*. For the moment, it is immaterial whether they are PFR or CSTR (Fig. 12.4). The conversion degree for the stream exiting R1 is, as usual, $x_{A1} = (Q_0C_{A0} - Q_1C_{A1})/Q_0C_{A0}$. Doubts may arise for the definition of the conversion degree x_{A2} characterising the stream exiting R2. One choice could be to reset x_A between R1 and R2, but it is more convenient to continue to refer to the stream (Q_0; C_{A0}) *entering the entire system*. Therefore, x_A is a variable that increases from reactor to reactor and, in the example in question, it will be $x_{A2} = (Q_0C_{A0} - Q_2C_{A2})/Q_0C_{A0}$. Being $Q_2C_{A2} < Q_1C_{A1}$, it must be $x_{A2} > x_{A1}$. With obvious extensions, the discussion is valid for a series of any number N of reactors:

$$x_{AN} = \frac{Q_0C_{A0} - Q_NC_{AN}}{Q_0C_{A0}} \tag{12.35}$$

If, regardless of the nature of R1, the second one were a CSTR (Fig. 12.5), the mass balance (using the usual approach, cf. Eq. 10.88) would be written as:

$$Q_1C_{A1} - Q_2C_{A2} = r_A(x_{A2})V_2 \tag{12.36}$$

Being:

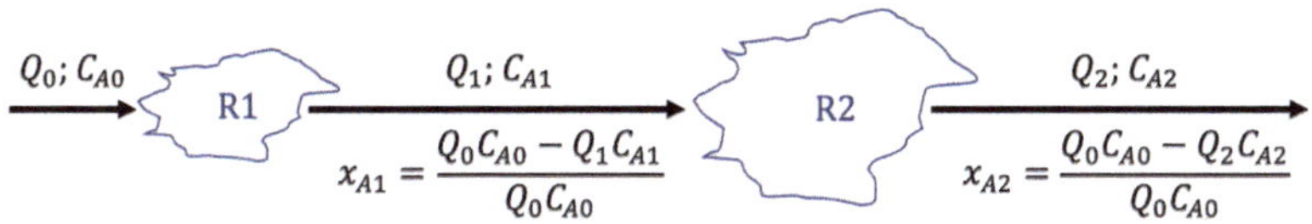

Fig. 12.4 Schematisation for two continuous reactors in series

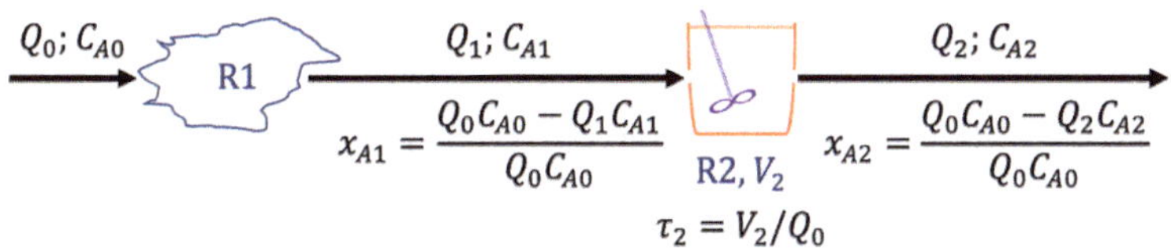

Fig. 12.5 Reactors in series: CSTR following another continuous reactor

$$Q_1C_{A1} = Q_0C_{A0}(1 - x_{A1}) \tag{12.37}$$

$$Q_2C_{A2} = Q_0C_{A0}(1 - x_{A2}) \tag{12.38}$$

it is:

$$Q_0C_{A0}(x_{A2} - x_{A1}) = r_A(x_{A2})V_2 \tag{12.39}$$

That is, the design equation:

$$\tau_2 = \frac{V_2}{Q_0} = \frac{C_{A0}(x_{A2} - x_{A1})}{r_A(x_{A2})} \tag{12.40}$$

Generalising the approach for a generic CSTR "*j*" in a series:

$$\boxed{\tau_j = \frac{V_j}{Q_0} = \frac{C_{A0}\left(x_{Aj} - x_{Aj-1}\right)}{r_A\left(x_{Aj}\right)}} \tag{12.41}$$

This means that the design equation has a classical expression, but paying attention to:

- the space-time is defined as the ratio between the volume of the CSTR and the volumetric flow rate entering the *entire system* (which does not necessarily coincide with the volumetric flow rate entering the CSTR "*j*");
- the concentration that appears in the design equation, C_{A0}, is that referred to the stream entering the *entire system* (and, in fact, the basis to calculate the conversion degree is the stream entering the entire system);
- at the numerator (right-hand side, Eq. 12.41), we have to consider the difference $x_{A2} - x_{A1}$ (and not only x_{A2}) or, generalising, the difference $x_{Aj} - x_{Aj-1}$, which is

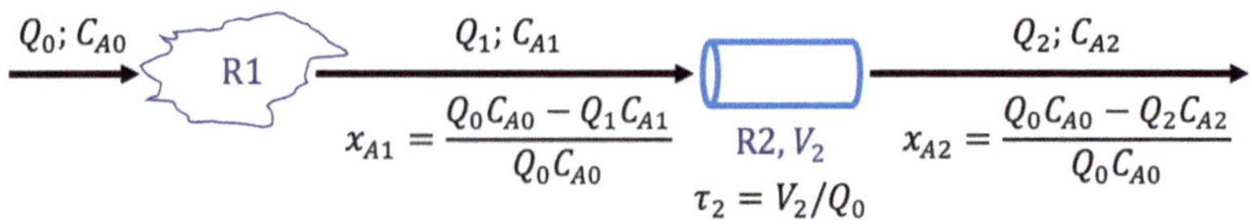

Fig. 12.6 Reactors in series: PFR following another continuous reactor

actually the increase in conversion degree attributable to the use of the CSTR in question.

If instead, regardless of the nature of R1, the second one were a PFR (Fig. 12.6), the mass balance (using the usual approach, cf. Eq. 10.70) would be written as:

$$-\frac{\mathrm{d}(QC_A)}{\mathrm{d}z} = r_A(x_A)S \tag{12.42}$$

Being, according to our choice:

$$QC_A = Q_0C_{A0}(1 - x_A) \tag{12.43}$$

it is:

$$-\frac{\mathrm{d}[Q_0C_{A0}(1 - x_A)]}{\mathrm{d}z} = Q_0C_{A0}\frac{\mathrm{d}x_A}{\mathrm{d}z} = r_A(x_A)S \tag{12.44}$$

$$\frac{S}{Q_0}\int_0^L \mathrm{d}z = \frac{SL}{Q_0} = \frac{V_2}{Q_0} = C_{A0}\int_{x_{A1}}^{x_{A2}} \frac{1}{r_A(x_A)}\mathrm{d}x_A \tag{12.45}$$

That is, the design equation:

$$\tau_2 = \frac{V_2}{Q_0} = C_{A0}\int_{x_{A1}}^{x_{A2}} \frac{1}{r_A(x_A)}\mathrm{d}x_A \tag{12.46}$$

Generalising the approach for a generic PFR "*j*" in a series:

$$\boxed{\tau_j = \frac{V_j}{Q_0} = C_{A0}\int_{x_{Aj-1}}^{x_{Aj}} \frac{1}{r_A(x_A)}\mathrm{d}x_A} \tag{12.47}$$

Similar considerations respect to what stated above (for a CSTR in a series) can be here developed. In particular, it is observed that the integral must be extended between x_{A1} (and not 0) and x_{A2} (or, between x_{Aj-1} and x_{Aj}), i.e. the actual range of x_A-values experienced by the fluid elements in the PFR.

We now consider the case of *a series of (N) PFR*. Observe the expressions for the space-times:

$$\begin{cases} \tau_1 = \dfrac{V_1}{Q_0} = C_{A0}\displaystyle\int_0^{x_{A1}} \frac{1}{r_A(x_A)} \mathrm{d}x_A \\ \quad \cdots \\ \tau_j = \dfrac{V_j}{Q_0} = C_{A0}\displaystyle\int_{x_{Aj-1}}^{x_{Aj}} \frac{1}{r_A(x_A)} \mathrm{d}x_A \\ \quad \cdots \\ \tau_N = \dfrac{V_N}{Q_0} = C_{A0}\displaystyle\int_{x_{AN-1}}^{x_{AN}} \frac{1}{r_A(x_A)} \mathrm{d}x_A \end{cases} \tag{12.48}$$

If we take a single (virtual) PFR with a volume equal to the sum of the volumes ($V = V_1 + \ldots + V_j + \ldots + V_N$), which converts the entire flow rate Q_0 up to conversion x_{AN}, there would be no difference. From a physical point of view, given the assumptions of perfect axial segregation for PFR, adding a second PFR to a first one is equivalent to simply considering a (virtual) PFR with a volume equal to the sum of the volumes. And so on, up to the N-th reactor. From a mathematical point of view, the space-time of this single (virtual) PFR can be obtained by summing the design equations for the N PFR placed in series:

$$\begin{aligned} \tau &= \frac{V_1 + \ldots + V_j + \ldots + V_N}{Q_0} = C_{A0}\int_0^{x_{AN}} \frac{1}{r_A(x_A)} \mathrm{d}x_A \\ &= C_{A0}\left[\int_0^{x_{A1}} \frac{1}{r_A(x_A)} \mathrm{d}x_A + \ldots + \int_{x_{Aj-1}}^{x_{Aj}} \frac{1}{r_A(x_A)} \mathrm{d}x_A + \ldots + \int_{x_{AN-1}}^{x_{AN}} \frac{1}{r_A(x_A)} \mathrm{d}x_A\right] \\ &= \sum_{j=1}^{N} \tau_j \end{aligned} \tag{12.49}$$

Therefore: *a combination of any number of PFR in series works as a single PFR with a volume equal to the sum of the volumes*, and no optimisation is needed. This is schematised in Fig. 12.7 for the case of $N = 2$.

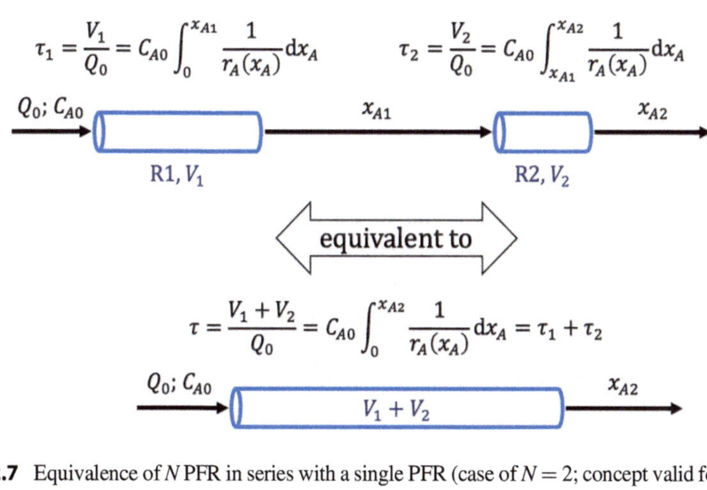

Fig. 12.7 Equivalence of N PFR in series with a single PFR (case of $N = 2$; concept valid for any N)

Fig. 12.8 Kinetic plot (canonical reaction rate) for, as example, $N = 3$ CSTR in series. The graphical interpretation is based on Eq. 12.41

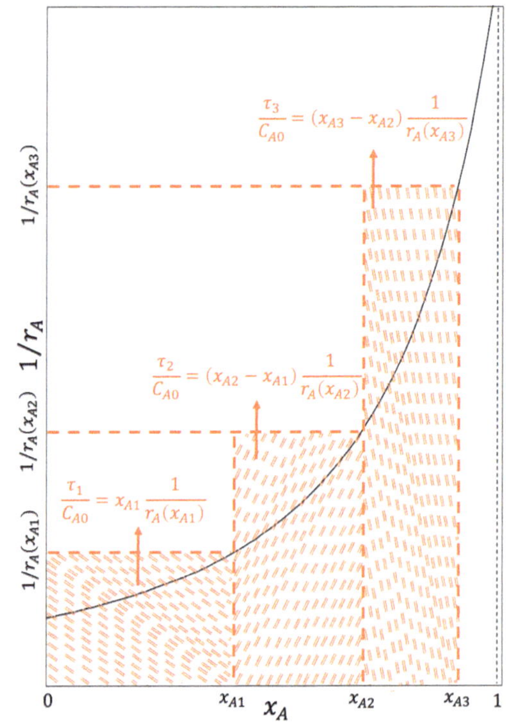

More complex is the case of *CSTR in series*. From a physical point of view, having two (or more) CSTR in series means, at least partially, "segregating" the flow, that is, tending, as much as possible, to the configuration of a PFR (which corresponds to the preferred one for canonical kinetics). Then, if we want to tend, with CSTR, to a "PFR-type" behaviour, we should (conceptually) place infinite CSTR in series. This, in the limit, would "segregate" the flow: in a kinetic plot $1/r_A$ vs. x_A, the sum of the infinitesimal areas of the infinite rectangles corresponding to the infinite CSTR would be equivalent to the area under the curve, characteristic of a PFR. From a practical point of view, we could have, for example, $N = 3$–4 CSTR in series (Fig. 12.8). For higher number of in-series CSTR, the expense of having more CSTR often no longer justifies the advantage in terms of space-time, an advantage that becomes less and less relevant when we add more and more CSTR (while, instead, the advantage is maximum when we add a second CSTR to the first one: this is the case discussed and optimised in an example below).

Compared to the case of (i) PFR in parallel, (ii) CSTR in parallel, (iii) PFR in series, the case of CSTR in series is the only one in which the system cannot be made equivalent to a single reactor (which is instead possible for PFR or CSTR in optimised parallel, and always possible for PFR in series). We can therefore conclude by observing:

- if a PFR-type behaviour is desired, and two (or more) PFR are available, they can be arranged in series, or in optimised parallel;
- if a PFR-type behaviour is desired, and two (or more) CSTR are available, they can be arranged in series (with an optimisation procedure that, for the case of $N = 2$, is illustrated in the example below; more complex cases are treated in, e.g., Levenspiel 1999; Westerterp et al. 1988);
- if a CSTR-type behaviour is desired, and two (or more) CSTR are available, they can be arranged in optimised parallel.

Example: Optimisation of 2 CSTR in Series, Canonical Kinetic Law

Let us now consider the case of $N = 2$ CSTR in series, Fig. 12.9. Having the expressions for τ_1 and τ_2, Eq. 12.41:

$$\tau_1 = \frac{V_1}{Q_0} = \frac{C_{A0}x_{A1}}{r_A(x_{A1})} \tag{12.50}$$

$$\tau_2 = \frac{V_2}{Q_0} = \frac{C_{A0}(x_{A2} - x_{A1})}{r_A(x_{A2})} \tag{12.51}$$

we define τ_{TOT} as:

$$\tau_{TOT} = \tau_1 + \tau_2 = \frac{C_{A0}x_{A1}}{r_A(x_{A1})} + \frac{C_{A0}(x_{A2} - x_{A1})}{r_A(x_{A2})} = \frac{C_{A0}x_{A2}}{r_A(x_{A2})} - C_{A0}x_{A1}\left[\frac{1}{r_A(x_{A2})} - \frac{1}{r_A(x_{A1})}\right] < \frac{C_{A0}x_{A2}}{r_A(x_{A2})} \tag{12.52}$$

If we set a value for, e.g., x_{A2}, it is observed that τ_{TOT} (which is a quantity to be minimised) is an exclusive function of x_{A1}, i.e. of the intermediate conversion degree, which represents the degree of freedom for the designer in this case. Analysing Eq. 12.52, there is a term, $C_{A0}x_{A2}/r_A(x_{A2})$, which can be seen as the space-time that would be required if, working up to x_{A2}, there were a single CSTR. Then there is a term in square brackets which, being for canonical kinetics (the one considered here, Fig. 12.10) $r_A(x_{A2}) < r_A(x_{A1})$, represents a

(continued)

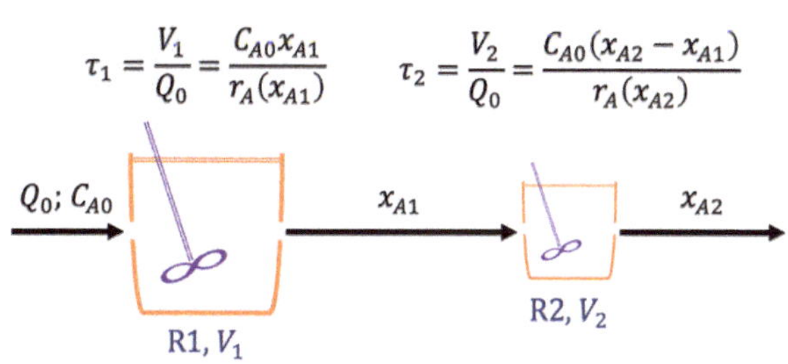

Fig. 12.9 Scheme for 2 CSTR in series

difference > 0. Therefore, τ_{TOT} is lower (better) than the value of the space-time that a single CSTR would have to bring the mixture to the same conversion (x_{A2}).

The issue is also evident from a graphical point of view: given a canonical kinetic plot, having two CSTR means making the stream react partly at x_{A1} and partly at x_{A2}, which is less penalising than making the stream react entirely at x_{A2} (case of a single CSTR). By putting two CSTR in series vs. one single CSTR, it is saved, in terms of space-time, the area represented in purple in Fig. 12.10. In fact, recalling the canonical graphical interpretation of the design equation for CSTR, the space-time of the first CSTR (Eq. 12.50) is measured by the area of the rectangle having x_{A1} as its base and the reciprocal of $r_A(x_{A1})$ as its height; the space-time of the second CSTR (Eq. 12.51) is measured by the area of the rectangle having $(x_{A2} - x_{A1})$ as its base and the reciprocal of $r_A(x_{A2})$ as its height. The sum of these two areas is less than the area of the rectangle with base x_{A2} and height the reciprocal of $r_A(x_{A2})$, case of the single CSTR. As already highlighted, having two CSTR in series means, at least partially, "segregating" the flow, and this is desired since, in this case (canonical kinetics), the solution that we would theoretically prefer is a PFR

(continued)

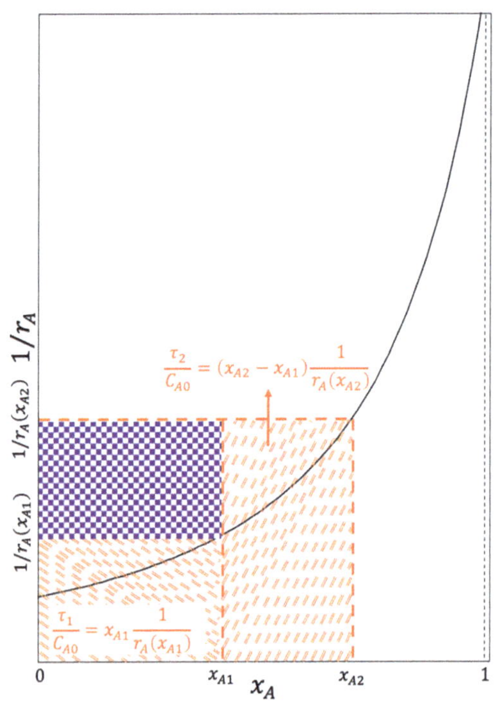

Fig. 12.10 Kinetic plot (canonical reaction rate) for 2 CSTR in series

(but we have only CSTR available). Then, we have decided to put the two CSTR in series.

It has already been said that, Eq. 12.52, τ_{TOT} is an exclusive function of x_{A1}. We can therefore plot $\tau_{TOT}(x_{A1})$ as in Fig. 12.11, considering that $x_{A1} \in [0, x_{A2}]$:

- if $x_{A1} = 0$, this is equivalent to making only the second CSTR work in Fig. 12.9, and in fact, Eq. 12.52, $\tau_{TOT} = C_{A0}x_{A2}/r_A(x_{A2})$;
- if $x_{A1} = x_{A2}$, this is equivalent to making only the first CSTR work. The situation is identical to the previous one: in fact, Eq. 12.52 (the term in square brackets cancels out), $\tau_{TOT} = C_{A0}x_{A2}/r_A(x_{A2})$;
- for each intermediate value of x_{A1}, there will actually be two CSTR in series, and τ_{TOT} can only be smaller (better) than the case of a single CSTR that converts the mixture up to x_{A2}. Therefore, by setting the derivative of τ_{TOT} with respect to x_{A1} equal to 0, the optimal intermediate conversion degree can be found, i.e. we will obtain x_{A1} (indicated by an asterisk in Fig. 12.11) such that the sum of the two space-times is, given x_{A2}, the minimum possible. Once this is done, it will be possible to determine the values of τ_1 and τ_2 (which, for example if the value of Q_0 is known, will allow the calculation of V_1 and V_2 to be designed).

(continued)

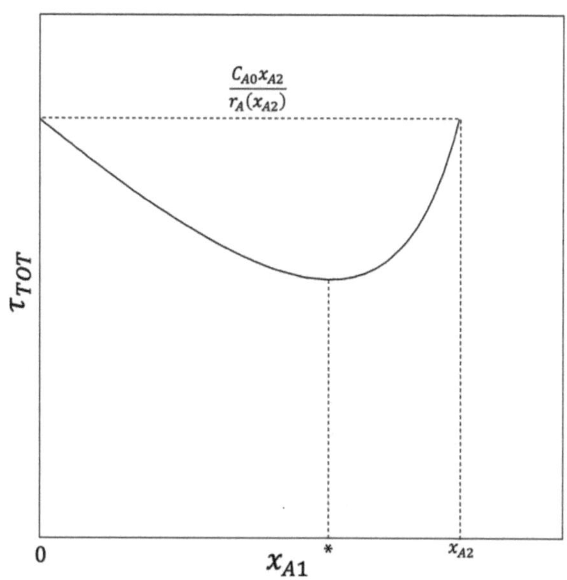

Fig. 12.11 Case of 2 CSTR in series, canonical kinetics: trend of the sum of the two space-times as a function of the intermediate conversion degree, once set the value for x_{A2}. The asterisk marks the optimal operating conditions

This optimal condition is defined by:

$$\frac{d\tau_{TOT}}{dx_{A1}} = \frac{d\tau_1}{dx_{A1}} + \frac{d\tau_2}{dx_{A1}} = 0 \Leftrightarrow |d\tau_1| = |d\tau_2| \Leftrightarrow \tau_1 = \tau_2 \tag{12.53}$$

i.e., by letting the two CSTR in series work with the same space-time. Evidently, at the design stage, this can only translate into designing 2 CSTR of equal volume, given the volumetric flow rate to be processed.

Apply to the case: A⟹P, $r_A = kC_A$, $\varepsilon = 0$. The optimal condition is found when (cf. Eqs. 12.50 and 12.51):

$$\tau_1 = \tau_2 \Leftrightarrow \frac{x_{A1}}{(1 - x_{A1})} = \frac{x_{A2} - x_{A1}}{(1 - x_{A2})} \Leftrightarrow (x_{A1})^2 - 2x_{A1} + x_{A2} = 0 \tag{12.54}$$

whose solution (physically acceptable) is:

$$x_{A1} = 1 - \sqrt{1 - x_{A2}} \tag{12.55}$$

If, for example, we pose $x_{A2} = 0.8$, the optimal value for x_{A1} is 0.553. That is, giving $k = 0.5$ 1/min, $\tau_1 = \tau_2 = 2.47$ min, and $\tau_{TOT} = 4.94$ min is the minimum possible. To make a comparison, with just one CSTR ($x_{A1} = 0$), it would have been $\tau_{TOT} = 8$ min. Using 2 CSTR in optimised series allows an advantage of about 38% in terms of space-time. While, a series of infinite CSTR would tend to a PFR, where (Table 10.8) $\tau = -(1/k)\ln(1 - x_A^e)$ that results equal to 3.22 min for $x_A^e = 0.8$ (−35% circa, vs. the optimised series of 2 CSTR).

12.4 The Role of Mixing

Now consider the case in which we have reactors of different nature (e.g., CSTR and PFR) to be arranged in series (since, for example, we want to tend to the behaviour of a PFR and, not having a single PFR of the desired volume, but having multiple reactors available, we have seen that it is preferred to arrange the reactors in series). *What is the role of mixing?* Or, to put the question differently: should we position CSTR or PFR first? To answer this question, let us analyse not, as usual, $r_A(x_A)$, but $r_A(C_A)$. In fact we know that, given A as the limiting reactant, the $r_A(C_A)$ function can always be defined.

Plot the reaction rate on a plane r_A vs. C_A, Fig. 12.12. If the function is concave upwards (that is, convex), i.e. if the second derivative of $r_A(C_A)$ is >0, consider, on the C_A-axis, two equal abrupt decreases in concentration, $(\Delta C_A)_I = (\Delta C_A)_2$. Each drop in concentration corresponds to a decrease in rate, Δr_A, on the r_A-axis. But, for the same ΔC_A, it is $(\Delta r_A)_I > (\Delta r_A)_2$. Therefore, since r_A is more sensitive to C_A when C_A is high, if there must be mixing (and, therefore, a sudden drop in concentration, which happens in a CSTR), it is better that this mixing is delayed as much as possible (when, that is, C_A is low), because this will penalise r_A less. Therefore: if the second derivative of the function $r_A(C_A)$ is positive, *it is convenient to postpone mixing*. For

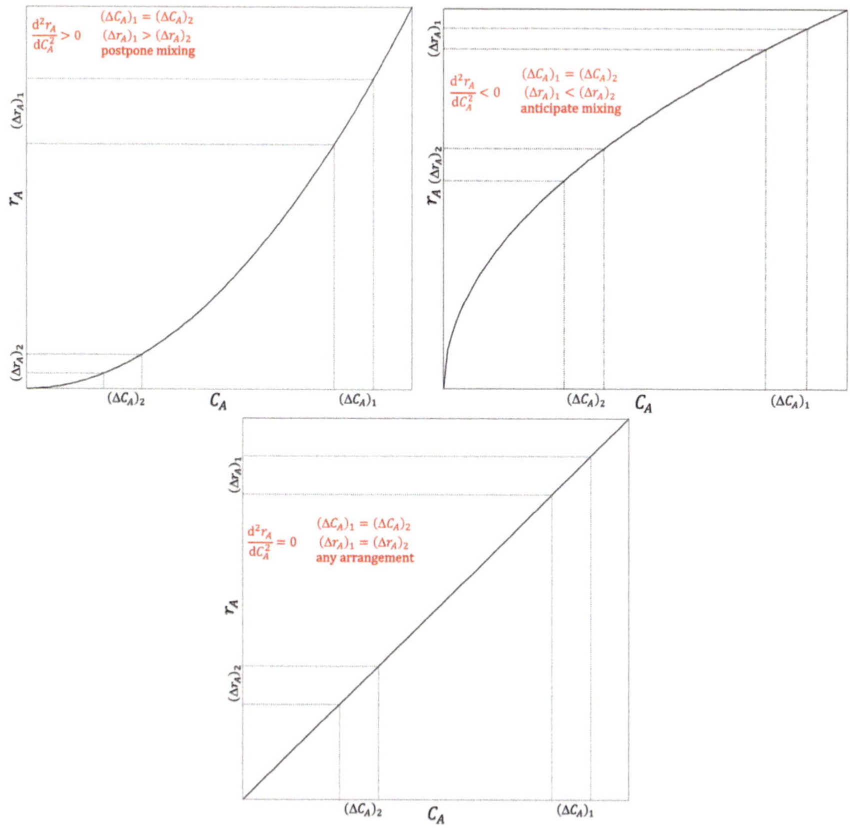

Fig. 12.12 Role of mixing. Up-left: $r_A(C_A)$ function concave upwards (convex). Up-right: $r_A(C_A)$ function concave downwards (concave). Down: $r_A(C_A)$ linear function

example, by placing the PFR first and then the CSTR. Or, more in general (recalling that a series of different PFR is always fluid dynamically equivalent to a single PFR, regardless of the arrangement):

$$\boxed{\frac{d^2 r_A}{dC_A^2} > 0 \Rightarrow \text{postpone mixing (first one/more PFR, then CSTR from the smallest to the largest)}} \tag{12.56}$$

Obviously, the opposite is true if the second derivative of $r_A(C_A)$ is <0 (function concave downwards or, simply, concave). Now, it is $(\Delta r_A)_1 < (\Delta r_A)_2$. As r_A is here more sensitive to C_A when C_A is low, if there must be mixing, it is better that this mixing is anticipated as much as possible (namely, when C_A is high), because this will penalise r_A less. Therefore: if the second derivative of the function $r_A(C_A)$ is negative, *it is convenient to anticipate mixing* (CSTR+PFR), generally stating that:

$$\frac{d^2 r_A}{dC_A^2} < 0 \Rightarrow \text{anticipate mixing (first CSTR from the largest to the smallest, then one/more PFR)} \tag{12.57}$$

To conclude the analysis, if we have a linear plot (the second derivative of $r_A(C_A)$ is 0, for example linear kinetics), anticipating or postponing the mixing has no advantage/disadvantage, since the sensitivity of r_A to C_A does not vary between high and low concentrations.

Example: Role of Mixing for Canonical Power-Law Kinetics

Assume a chemical reaction governed by:

$$r_A(C_A) = kC_A^n \tag{12.58}$$

The second derivative can be expressed as:

$$\frac{dr_A}{dC_A} = nkC_A^{n-1} \Rightarrow \frac{d^2 r_A}{dC_A^2} = n(n-1)kC_A^{n-2} \tag{12.59}$$

Thus:

$$\begin{cases} n > 1 \Rightarrow \dfrac{d^2 r_A}{dC_A^2} > 0 \text{ (postpone mixing)} \\ 0 < n < 1 \Rightarrow \dfrac{d^2 r_A}{dC_A^2} < 0 \text{ (anticipate mixing)} \\ n = 1 \text{ (or, } n = 0) \Rightarrow \dfrac{d^2 r_A}{dC_A^2} = 0 \text{ (any arrangement)} \end{cases} \tag{12.60}$$

Example: Arranging Two Different CSTR in Series

We consider the reaction A + 2B⟹3P. The reacting stream is composed by A (C_{A0} = 1 mol/L) and excess B, and has flow rate of Q_0 = 1 L/h. The kinetic law is:

$$\begin{cases} r_A = \dfrac{k_1 C_A}{k_2 + C_P} = \dfrac{k_1 C_{A0}(1 - x_A)}{k_2 + 3C_{A0}x_A} \\ k_1 = 1 \dfrac{\text{mol}}{\text{L} \cdot \text{h}} \\ k_2 = 0.5 \text{ mol/L} \end{cases} \tag{12.61}$$

(continued)

We have available two CSTR, one smaller (1 L) and the other larger (10 L). First, we observe that, when x_A increases (as C_A decreases and C_P increases), r_A decreases, resulting in a canonical kinetic plot ($1/r_A$ vs. x_A, see e.g. Figure 12.8) that would suggest the use of a PFR. Not having a PFR, we decide to put the two CSTR in series. The best arrangement will give the highest value for the conversion degree and, thus, for the productivity (being known Q_0 and C_{A0}). Note that the space-time for the two reactors (1 h and 10 h, respectively) is already determined. We can now calculate the second derivative for the $r_A(C_A)$ function:

$$C_P = 3(C_{A0} - C_A) \Rightarrow r_A(C_A) = \frac{k_1 C_A}{k_2 + 3(C_{A0} - C_A)} \tag{12.62}$$

$$\frac{dr_A}{dC_A} = \frac{k_1[k_2 + 3(C_{A0} - C_A)] + 3k_1 C_A}{[k_2 + 3(C_{A0} - C_A)]^2} = \frac{k_1 k_2 + 3k_1 C_{A0}}{[k_2 + 3(C_{A0} - C_A)]^2} \tag{12.63}$$

$$\frac{d^2 r_A}{dC_A^2} = \frac{6(k_1 k_2 + 3k_1 C_{A0})}{[k_2 + 3(C_{A0} - C_A)]^3} > 0 \tag{12.64}$$

The sign in Eq. 12.64 has been argued since $C_{A0} > C_A$. Thus, Eq. 12.56, we postpone mixing and propose the configuration in Fig. 12.13. Apply the design equation for the two CSTR:

$$\tau_1 = \frac{V_1}{Q_0} = 1\ \text{h} = \frac{C_{A0} x_{A1}}{r_A(x_{A1})} = \frac{x_{A1}(k_2 + 3C_{A0} x_{A1})}{k_1(1 - x_{A1})} \tag{12.65}$$

$$\tau_2 = \frac{V_2}{Q_0} = 10\ \text{h} = \frac{C_{A0}(x_{A2} - x_{A1})}{r_A(x_{A2})} = \frac{(x_{A2} - x_{A1})(k_2 + 3C_{A0} x_{A2})}{k_1(1 - x_{A2})} \tag{12.66}$$

Solving, respectively:

$$3C_{A0}(x_{A1})^2 + (k_1\,\tau_1 + k_2)x_{A1} - k_1\,\tau_1 = 0 \Rightarrow x_{A1} = 0.379 \tag{12.67}$$

$$3C_{A0}(x_{A2})^2 + (k_1\,\tau_2 + k_2 - 3C_{A0} x_{A1})x_{A2} - (k_1\,\tau_2 + k_2 x_{A1}) = 0 \Rightarrow x_{A2} = 0.854 \tag{12.68}$$

which gives the maximum productivity of $P_P = 3Q_0 C_{A0} x_{A2} = 2.562$ mol/h.

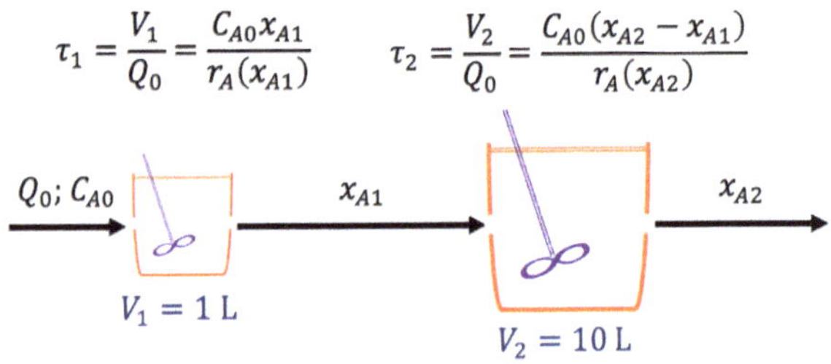

Fig. 12.13 Arrangement of 2 CSTR of different volume in series when mixing needs to be postponed

Example: Arranging in Series a PFR and a CSTR

A hydrolysis reaction A⟹P is governed by:

$$\begin{cases} r_A = \dfrac{k_1 C_A}{C_A + k_2 C_P + k_3} \\ k_1 = 1\ \dfrac{\text{mol}}{\text{L} \cdot \text{min}} \\ k_2 = 0.25 \\ k_3 = 0.5\ \text{mol/L} \end{cases} \tag{12.69}$$

The reacting stream is composed by A (C_{A0} = 1 mol/L), with flow rate Q_0 = 1 L/min. We have available one PFR and one CSTR, both having volume of 2 L, to arrange in series. Again, the best arrangement will give the highest value for the conversion degree and, thus, for the productivity. Note that, also in this case, the space-time for the two reactors (2 min each) is already determined. Start calculating the first derivative for the $r_A(C_A)$ function:

$$C_P = C_{A0} - C_A \Rightarrow r_A(C_A) = \frac{k_1 C_A}{C_A + k_2(C_{A0} - C_A) + k_3} = \frac{k_1 C_A}{(1 - k_2)C_A + k_2 C_{A0} + k_3} \tag{12.70}$$

$$\frac{dr_A}{dC_A} = \frac{k_1[(1 - k_2)C_A + k_2 C_{A0} + k_3] - k_1 C_A(1 - k_2)}{[(1 - k_2)C_A + k_2 C_{A0} + k_3]^2} = \frac{k_1 k_2 C_{A0} + k_1 k_3}{[(1 - k_2)C_A + k_2 C_{A0} + k_3]^2} \tag{12.71}$$

We observe that the first derivative is positive, i.e., when C_A decreases (x_A increases), r_A decreases—this results into a canonical kinetic behaviour although the expression for the reaction rate is not of standard form. We can now calculate the second derivative for the $r_A(C_A)$ function:

$$\frac{d^2 r_A}{dC_A^2} = \frac{-2(1 - k_2)(k_1 k_2 C_{A0} + k_1 k_3)}{[(1 - k_2)C_A + k_2 C_{A0} + k_3]^3} < 0 \tag{12.72}$$

(continued)

$$\tau_{CSTR} = \frac{V_{CSTR}}{Q_0} = \frac{C_{A0}x_{A1}}{r_A(x_{A1})} \qquad \tau_{PFR} = \frac{V_{PFR}}{Q_0} = C_{A0}\int_{x_{A1}}^{x_{A2}} \frac{1}{r_A(x_A)}\mathrm{d}x_A$$

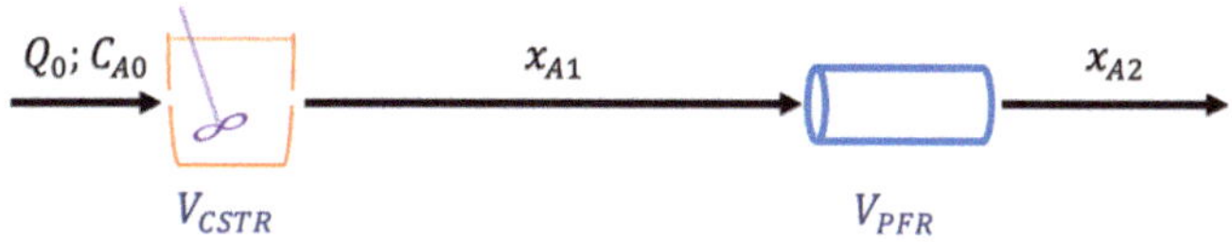

Fig. 12.14 Arrangement of 1 CSTR and 1 PFR in series when mixing needs to be anticipated

The sign in Eq. 12.72 has been argued since $1 - k_2 > 0$. Thus, Eq. 12.57, we anticipate mixing and propose the configuration in Fig. 12.14.

To tackle the design equations, we now need to obtain $r_A(x_A)$; start from Eq. 12.69:

$$r_A(x_A) = \frac{k_1 C_{A0}(1 - x_A)}{C_{A0}(1 - x_A) + k_2 C_{A0} x_A + k_3} = \frac{k_1 C_{A0}(1 - x_A)}{(k_2 C_{A0} - C_{A0})x_A + C_{A0} + k_3} \tag{12.73}$$

Solve the CSTR:

$$\tau_{CSTR} = \frac{V_{CSTR}}{Q_0} = 2 \text{ min} = \frac{C_{A0}x_{A1}}{r_A(x_{A1})} = \frac{x_{A1}[(k_2 C_{A0} - C_{A0})x_{A1} + C_{A0} + k_3]}{k_1(1 - x_{A1})} \tag{12.74}$$

$$(k_2 C_{A0} - C_{A0})(x_{A1})^2 + (C_{A0} + k_3 + k_1 \tau_{CSTR})x_{A1} - k_1\, \tau_{CSTR} = 0 \Longrightarrow x_{A1} = 2/3 \tag{12.75}$$

Solve now the PFR:

$$\tau_{PFR} = \frac{V_{PFR}}{Q_0} = 2 \text{ min} = C_{A0}\int_{x_{A1}}^{x_{A2}} \frac{1}{r_A(x_A)}\mathrm{d}x_A = \frac{1}{k_1}\int_{x_{A1}}^{x_{A2}} \frac{(k_2 C_{A0} - C_{A0})x_A + C_{A0} + k_3}{1 - x_A}\mathrm{d}x_A \tag{12.76}$$

Apply PFD method:

$$\frac{(k_2 C_{A0} - C_{A0})x_A + C_{A0} + k_3}{1 - x_A} = C_1 + \frac{C_2}{1 - x_A} \tag{12.77}$$

(continued)

Thus:

$$\tau_{PFR} = 2\ \text{min} = \frac{1}{k_1}\left(\int_{x_{A1}}^{x_{A2}} C_1 \mathrm{d}x_A + \int_{x_{A1}}^{x_{A2}} \frac{C_2}{1-x_A}\mathrm{d}x_A\right) = \frac{1}{k_1} \times \left[C_1(x_{A2}-x_{A1}) - C_2 \ln\left(\frac{1-x_{A2}}{1-x_{A1}}\right)\right] \quad (12.78)$$

Manipulate Eq. 12.77 to obtain the constants C_1 and C_2:

$$\frac{(k_2C_{A0} - C_{A0})x_A + C_{A0} + k_3}{1-x_A} = \frac{-C_1x_A + (C_1+C_2)}{1-x_A} \quad (12.79)$$

Valid if:

$$\begin{cases} -C_1 = k_2C_{A0} - C_{A0} \\ C_1 + C_2 = C_{A0} + k_3 \end{cases} \Rightarrow \begin{cases} C_1 = C_{A0} - k_2C_{A0} \\ C_2 = k_2C_{A0} + k_3 \end{cases} \quad (12.80)$$

Substituting the expressions for the constants, Eq. 12.78 is solved (by trial-and-error) for $x_{A2} = 0.969$, which gives the maximum productivity of $P_P = Q_0C_{A0}x_{A2} = 0.969$ mol/min.

Example: Indifference in the Arrangement of a CSTR and a PFR for Linear Kinetics

Consider A ⇒ P governed by linear kinetics, $r_A = kC_A$, and the availability of two reactors, a PFR (τ_{PFR}) and a CSTR (τ_{CSTR}), with known space-time. We want to demonstrate that the exit conversion degree is the same, either we place PFR first and then CSTR, or vice versa. Start with the series PFR + CSTR. The solution for PFR is that in Table 10.8 (where x_{A1} is the conversion degree at PFR outlet):

$$\tau_{PFR} = -\frac{1}{k}\ln(1-x_{A1}) \Rightarrow x_{A1} = 1 - \exp(-k\tau_{PFR}) \quad (12.81)$$

While, for CSTR following PFR and with exit conversion degree x_{A2}, it is:

$$\tau_{CSTR} = \frac{C_{A0}(x_{A2}-x_{A1})}{r_A(x_{A2})} = \frac{(x_{A2}-x_{A1})}{k(1-x_{A2})} \Rightarrow x_{A2} = \frac{x_{A1} + k\tau_{CSTR}}{1+k\tau_{CSTR}} = \frac{1-\exp(-k\tau_{PFR}) + k\tau_{CSTR}}{1+k\tau_{CSTR}} \quad (12.82)$$

(continued)

Now reverse (CSTR+PFR). The solution for CSTR is that in Table 10.9 (where x_{A1} is the conversion degree at CSTR outlet):

$$\tau_{CSTR} = \frac{1}{k}\frac{x_{A1}}{1-x_{A1}} \Rightarrow x_{A1} = \frac{k\tau_{CSTR}}{1+k\tau_{CSTR}} \tag{12.83}$$

While, for PFR following CSTR and with exit conversion degree x_{A2}, it is:

$$\tau_{PFR} = C_{A0}\int_{x_{A1}}^{x_{A2}} \frac{1}{r_A(x_A)}\mathrm{d}x_A = \frac{1}{k}\int_{x_{A1}}^{x_{A2}} \frac{1}{1-x_A}\mathrm{d}x_A =$$
$$-\frac{1}{k}\ln\left(\frac{1-x_{A2}}{1-x_{A1}}\right) \Rightarrow x_{A2} = 1-(1-x_{A1})\exp(-k\tau_{PFR}) \tag{12.84}$$

Substituting Eq. 12.83 into Eq. 12.84:

$$x_{A2} = 1-\left(1-\frac{k\tau_{CSTR}}{1+k\tau_{CSTR}}\right)\exp(-k\tau_{PFR}) = 1-\exp(-k\tau_{PFR})$$
$$+\frac{k\tau_{CSTR}}{1+k\tau_{CSTR}}\exp(-k\tau_{PFR}) = \frac{1-\exp(-k\tau_{PFR})+k\tau_{CSTR}}{1+k\tau_{CSTR}} \tag{12.85}$$

i.e., the same as in Eq. 12.82. The equivalence is demonstrated.

12.5 PFR with Recycle

12.5.1 Illustration

The *PFR with recycle* is an axially segregated reactor, in which a certain degree of back-mixing is induced by partial recycling of the exit stream. The extent of recycle is regulated by what we will call *recycle ratio*. This reactor solution can be useful in particular cases, such as for example to set the desired temperature, or to adjust the composition of the reactant mixture if a network of reactions takes place, topics that concern the following chapters of this textbook. Remaining instead in the context of single and isothermal reactions, the reactor with recycle is useful, e.g., in the case of processes governed by non-canonical kinetic laws, as it will be discussed in a following paragraph.

Let us assign the scheme in Fig. 12.15, where:

- the feed stream is characterised by volumetric flow rate Q_O, concentration of the limiting reagent C_{AO}, molar flow rate of the limiting reagent $F_{AO} = Q_O C_{AO}$;
- the stream exiting the PFR has characteristics Q_2, C_{A2}, $F_{A2} = Q_2 C_{A2}$, with conversion degree x_{A2};

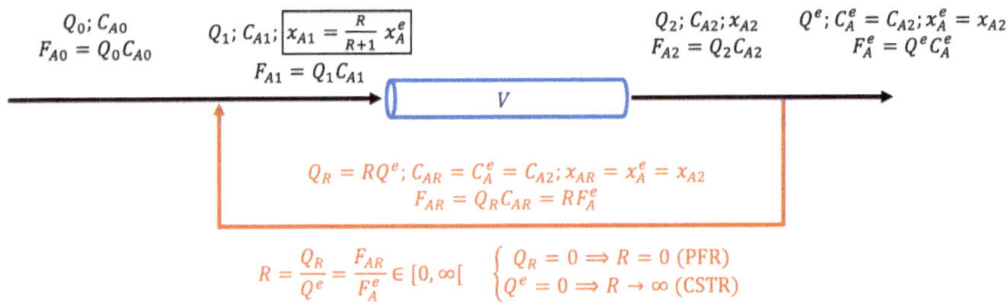

Fig. 12.15 PFR with recycle

- of this stream exiting the PFR, a part is collected at the exit (Q^e, $C_A^e = C_{A2}$, $F_A^e = Q^eC_A^e$, $x_A^e = x_{A2}$), and a part is *recycled* (Q_R, $C_{AR} = C_A^e = C_{A2}$, $F_{AR} = Q_RC_{AR}$, $x_{AR} = x_A^e = x_{A2}$);
- the recycle stream is mixed with the feed, via a mixing node immediately upstream of the PFR. This determines that, entering the PFR, is not the "0" stream but a "1" stream with characteristics Q_1, C_{A1}, $F_{A1} = Q_1C_{A1}$, and conversion degree x_{A1} *different from 0*, since the "0" stream has been mixed with the recycle stream, which is already converted.

The *recycle ratio R* is defined as:

$$\boxed{R = \frac{Q_R}{Q^e} = \frac{\frac{F_{AR}}{C_{AR}}}{\frac{F_A^e}{C_A^e}} = \frac{F_{AR}}{F_A^e} \in [0, \infty[\quad \begin{cases} R = 0 \text{ (PFR)} \\ R \to \infty \text{ (CSTR)} \end{cases}} \tag{12.86}$$

i.e. the ratio between the volumetric flow rate (or molar flow rate of A, it is equivalent) of the recycle stream and that of the stream exiting the whole system. If the exit stream is not recycled, $R = 0$ and we are back to the case of a standard PFR. If the exit stream is entirely recycled, then $Q^e = 0$ and $R \to \infty$, which recalls the case of complete mixing (CSTR). Therefore, R is a variable with which *the designer can explore the infinite ideal reactor behaviours between PFR and CSTR* (more generally, it is like saying that the PFR with recycle is an ideal reactor that includes, as particular cases, also PFR and CSTR), Fig. 12.16.

In a next paragraph, we will derive the design equation for PFR with recycle. To do so, it is here necessary to express the value of x_{A1} in more detail. We start with (cf. Eq. 10.21):

$$x_{A1} = \frac{C_{A0} - C_{A1}}{C_{A0} + \varepsilon C_{A1}} = \frac{1 - \frac{C_{A1}}{C_{A0}}}{1 + \varepsilon \frac{C_{A1}}{C_{A0}}} \tag{12.87}$$

where $C_{A1} = F_{A1}/Q_1$. Recalling that (cf. Eq. 10.10):

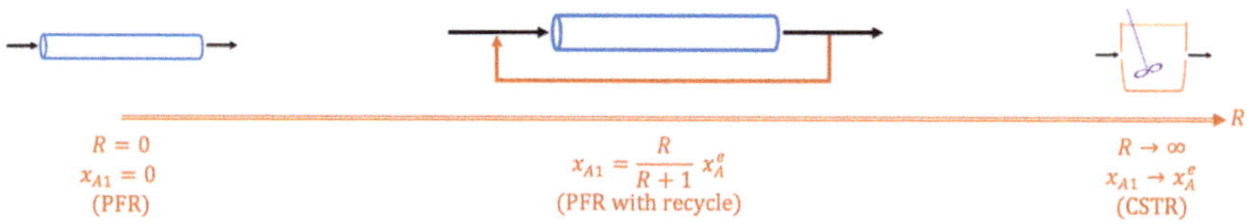

Fig. 12.16 Point of view: PFR with recycle can explore the infinite ideal reactor behaviours between PFR and CSTR

$$x_A^e = \frac{F_{A0} - F_A^e}{F_{A0}} \Rightarrow F_A^e = F_{A0}\left(1 - x_A^e\right) \tag{12.88}$$

the numerator of C_{A1} is found (using the definition for R):

$$F_{A1} = F_{A0} + F_{AR} = F_{A0} + RF_A^e = F_{A0} + RF_{A0}\left(1 - x_A^e\right) = F_{A0}\left(R + 1 - Rx_A^e\right) \tag{12.89}$$

Being, Eq. 10.15:

$$Q^e = Q_0\left(1 + \varepsilon x_A^e\right) \tag{12.90}$$

the denominator of C_{A1} is found (using again the definition for R):

$$Q_1 = Q_0 + Q_R = Q_0 + RQ^e = Q_0 + RQ_0\left(1 + \varepsilon x_A^e\right) = Q_0\left(R + 1 + R\varepsilon x_A^e\right) \tag{12.91}$$

We can now express C_{A1}:

$$C_{A1} = \frac{F_{A1}}{Q_1} = \frac{F_{A0}\left(R + 1 - Rx_A^e\right)}{Q_0\left(R + 1 + R\varepsilon x_A^e\right)} = C_{A0}\frac{R + 1 - Rx_A^e}{R + 1 + R\varepsilon x_A^e} \Rightarrow \frac{C_{A1}}{C_{A0}} = \frac{R + 1 - Rx_A^e}{R + 1 + R\varepsilon x_A^e} \tag{12.92}$$

Substituting in Eq. 12.87:

$$x_{A1} = \frac{1 - \frac{C_{A1}}{C_{A0}}}{1 + \varepsilon\frac{C_{A1}}{C_{A0}}} = \frac{1 - \frac{R+1-Rx_A^e}{R+1+R\varepsilon x_A^e}}{1 + \varepsilon\frac{R+1-Rx_A^e}{R+1+R\varepsilon x_A^e}} = \frac{R\varepsilon x_A^e + Rx_A^e}{R + 1 + R\varepsilon + \varepsilon} = \frac{Rx_A^e(1 + \varepsilon)}{(R + 1)(1 + \varepsilon)} \tag{12.93}$$

we land on the fundamental:

$$\boxed{x_{A1} = \frac{R}{R + 1}x_A^e} \tag{12.94}$$

Please observe that x_{A1} is a fraction (governed by R; the higher R, the higher this fraction) of x_A^e, and that, to strengthen the concept visualised in Fig. 12.16:

$$\begin{cases} R = 0 \Rightarrow x_{A1} = 0 \text{ (PFR)} \\ R \rightarrow \infty \Rightarrow x_{A1} \rightarrow x_A^e \text{ (CSTR)} \end{cases} \tag{12.95}$$

12.5.2 Design Equation

In the scheme illustrated in Fig. 12.15, as usual we define the space-time of our reactor as $\tau = V/Q_O$, i.e. the ratio between the volume of the reactor and the volumetric flow rate *entering the entire system*. We now postulate the following. Let the PFR with recycle be equivalent to the scheme in Fig. 12.17: a series of two reactors, of which the first is a "virtual reactor" with the role that, in reality, assumes the mixing node upstream of the PFR with recycle. That is, this virtual reactor pre-converts the stream up to the conditions "1" (Q_1, C_{A1}, $F_{A1} = Q_1 C_{A1}$, and x_{A1}), which are the same as those referred, in the real system, at the actual entrance in the PFR with recycle (Fig. 12.15). This virtual reactor is fed with a stream having A-concentration equal to C_{AO} (the same as in the real case), but volumetric flow rate Q_0' different from Q_O (and molar flow rate of A, therefore, $F_{A0}' = Q_0' C_{AO}$, different from F_{AO}). Then, the second reactor in the series of Fig. 12.17 is a "pure" PFR, of volume V (the same as in the real case), which converts the stream from x_{A1} up to x_A^e (as in the real case). From what was learned in a previous chapter, the space-time of this PFR (Fig. 12.17) is:

$$\tau' = \frac{V}{Q_0'} = C_{A0} \int_{x_{A1}}^{x_A^e} \frac{1}{r_A(x_A)} \mathrm{d}x_A \tag{12.96}$$

Please note that we have termed τ' this space-time, being it conceptually different from τ, the space-time of the PFR with recycle. The degree of freedom is, therefore, Q_0'. We ask: what must be the value of Q_0', so that the two systems (the real one in Fig. 12.15 and the virtual one in Fig. 12.17) are actually equivalent?

The molar flow rate of A in conditions "1", F_{A1}, is the same (as per postulated condition) in both configurations. In the real scheme, it is (see Eq. 12.89):

$$F_{A1} = F_{A0}\left(R + 1 - Rx_A^e\right) \tag{12.97}$$

In the virtual configuration, being:

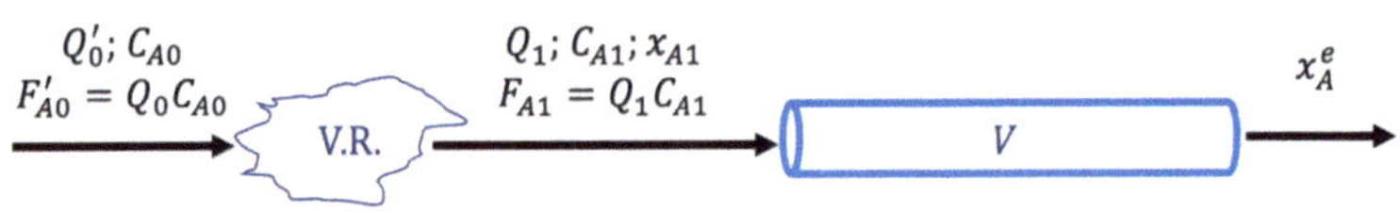

Fig. 12.17 Virtual reactor configuration equivalent to PFR with recycle in Fig. 12.15: series of a virtual reactor (termed V.R.) followed by a PFR. Comparing schemes in Figs. 12.15 and 12.17, equal symbols correspond to equal physico-chemical properties

$$x_{A1} = \frac{F'_{A0} - F_{A1}}{F'_{A0}} \tag{12.98}$$

it is, considering also Eq. 12.94:

$$x_{A1} = \frac{F'_{A0} - F_{A1}}{F'_{A0}} \Rightarrow F_{A1} = F'_{A0}(1 - x_{A1}) = F'_{A0}\left(1 - \frac{R}{R+1}x^e_A\right) = F'_{A0}\frac{R+1-Rx^e_A}{R+1} \tag{12.99}$$

Dividing Eq. 12.97 with Eq. 12.99:

$$1 = \frac{F_{A0}(R+1)}{F'_{A0}} \Rightarrow \frac{F'_{A0}}{F_{A0}} = \frac{Q'_0 C_{A0}}{Q_0 C_{A0}} = \frac{Q'_0}{Q_0} = R+1 \Rightarrow Q'_0 = (R+1)Q_0 \tag{12.100}$$

i.e., so that the two systems are equivalent, the volumetric flow rate fed to the virtual system, Q'_0, must be $(R + 1)$-times the real value, Q_0. Inserting the relation in Eq. 12.100 into Eq. 12.96:

$$\frac{V}{(R+1)Q_0} = C_{A0}\int_{x_{A1}}^{x^e_A} \frac{1}{r_A(x_A)}\mathrm{d}x_A \tag{12.101}$$

we finally recognise $\tau = V/Q_0$ so to obtain the *design equation for PFR with recycle*:

$$\boxed{\tau = \frac{V}{Q_0} = (R+1)C_{A0}\int_{x_{A1}}^{x^e_A} \frac{1}{r_A(x_A)}\mathrm{d}x_A} \tag{12.102}$$

The scrupulous reader will note that, first, it is no longer, as it was for STR, PFR, CSTR, a bijective relationship between τ and x^e_A, since τ is also a function of R (which explicitly appears outside the integral, and implicitly in the lower bound of integration, x_{A1}). This means that, fixed x^e_A, the operation can be conducted at various (infinite) values of R, posing the problem of optimisation for R, which will be addressed later. Moreover, reinforcing the point of view illustrated in Fig. 12.16 (updated in Fig. 12.18), for $R = 0$, Eq. 12.102 clearly leads back to the design equation for simple PFR (as $R + 1 = 1$ and $x_{A1} = 0$). Furthermore, for $R \to \infty$, Eq. 12.102 assumes an indeterminate form (as $R + 1 \to \infty$ and $x_{A1} \to x^e_A$, i.e. the

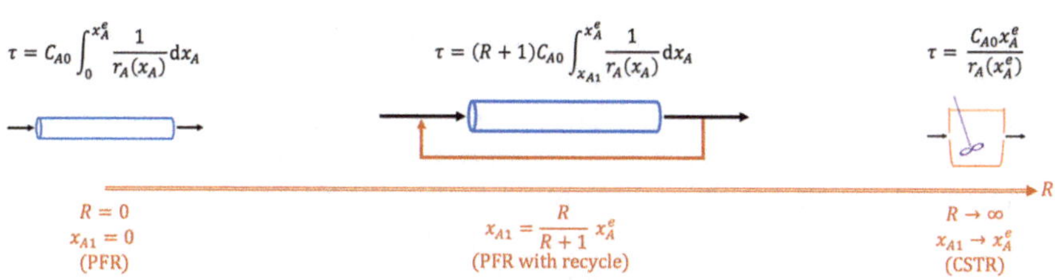

Fig. 12.18 Point of view (reprise): PFR with recycle can explore the infinite ideal reactor behaviours between PFR and CSTR

integral tends to zero). However, reasoning on the circumstance that, for very large R, the two integration bounds are very close each other, it is easy to proceed in this way:

$$\tau(R \rightarrow \infty) = (R+1)C_{A0}\frac{1}{r_A\left(x_A^e\right)}\int_{x_{A1}}^{x_A^e} \mathrm{d}x_A \tag{12.103}$$

where the reaction rate, since it does not vary much between two very close integration extremes, has been set outside the integral and equal to $r_A(x_A^e)$; in fact, now there is no difference between x_{A1} and x_A^e. Maths dictates:

$$\tau(R \rightarrow \infty) = (R+1)C_{A0}\frac{x_A^e - x_{A1}}{r_A\left(x_A^e\right)} \tag{12.104}$$

Being:

$$x_{A1} = \frac{R}{R+1}x_A^e \Rightarrow x_A^e - x_{A1} = x_A^e - \frac{R}{R+1}x_A^e = \frac{x_A^e}{R+1} \tag{12.105}$$

it is:

$$\tau(R \rightarrow \infty) = \frac{C_{A0}x_A^e}{r_A\left(x_A^e\right)} \tag{12.106}$$

the design equation for CSTR (as expected). Therefore (Fig. 12.18), Eq. 12.102 can be considered, in a certain sense, a "universal" design equation for ideal continuous reactors.

12.5.3 Optimisation of the Recycle Ratio

Excluding the case of reactions governed by canonical kinetics (for which, in the absence of other reasons, the optimal value of R is clearly $R = 0$, i.e. PFR without recycling; in this case, in fact, we do not want back-mixing, see Fig. 11.1), *the case of interest for using a PFR with recycle is that of autocatalytic reactions*. We can refer to the general kinetic law presented in Eqs. 11.22 and 11.24:

$$r_A = kC_AC_P = kC_{A0}(1 - x_A)(C_{P0} + C_{A0}x_A) \tag{12.107}$$

If we aim at $x_A^e \leq x_A^*$, Fig. 11.6, the preferred solution is clearly CSTR. But if, on the other hand, we aim at a sufficiently high value for the conversion degree ($x_A^e > x_A^*$), the best reactor solution is the one shown in Fig. 12.19 (series CSTR+PFR): in the zone where r_A increases as the reaction progresses (i.e., up to x_A^*), mixed flow will be preferred (reactor that works uniformly at x_A^*, for which r_A assumes a maximum value); in the zone where r_A decreases, segregated flow will be convenient (from x_A^* up to x_A^e). Moreover, recalling the graphical interpretation of the design equations,

Fig. 12.19 Optimal configuration (series CSTR +PFR) for autocatalytic reaction and sufficiently high degree of conversion

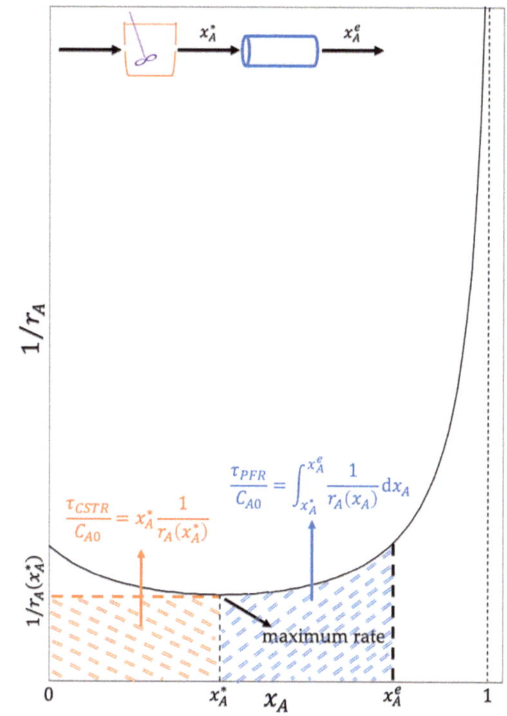

we see that the sum of the two areas in Fig. 12.19 (which indicates the overall space-time of the configuration we are proposing) is the best (smallest) possible.

In the case in which it is not feasible to use the series of two reactors as proposed in Fig. 12.19, having understood that, in reality, a combined "mixed/segregated" system would be convenient, the best solution remains that of the PFR with recycle, *which contains the two reactor paradigms*. Therefore, the question now is to find the condition of optimal R *such that the space-time of PFR with recycle is minimum*. To do this, known x_A^e, derive τ with respect to R, setting the partial derivative to zero:

$$\left.\frac{\partial \tau}{\partial R}\right|_{x_A^e} = 0 \tag{12.108}$$

$$\frac{d\left[(R+1)\int_{x_{A1}(R)}^{x_A^e} \frac{1}{r_A(x_A)} dx_A\right]}{dR} = \frac{d(R+1)}{dR}\int_{x_{A1}(R)}^{x_A^e} \frac{1}{r_A(x_A)} dx_A + (R+1)\frac{d\left[\int_{x_{A1}(R)}^{x_A^e} \frac{1}{r_A(x_A)} dx_A\right]}{dR} = 0 \tag{12.109}$$

Note that we have here explicitly recalled that, in the lower bound of integration, x_{A1} is a function of R as per Eq. 12.94 (as we are deriving respect to R). This is a case where the function to be derived (with respect to R) presents an integral that contains

R in the integration extremes, with the function under integral dependent instead on x_A, with respect to which the integral is calculated. Rearranging a theorem known from mathematical analysis (cf. Levenspiel 1999), it is:

$$\frac{\mathrm{d}\left[\int_{a(R)}^{b(R)} f(x_A)\mathrm{d}x_A\right]}{\mathrm{d}R} = f(b)\frac{\mathrm{d}b}{\mathrm{d}R} - f(a)\frac{\mathrm{d}a}{\mathrm{d}R} \tag{12.110}$$

Applying to our case (where $b = x_A^e$ does not depend on R, $a(R) = x_{A1}(R)$, and $f = 1/r_A$):

$$\frac{\mathrm{d}\left[\int_{x_{A1}(R)}^{x_A^e} \frac{1}{r_A(x_A)}\mathrm{d}x_A\right]}{\mathrm{d}R} = -\frac{1}{r_A(x_{A1})}\frac{\mathrm{d}\left(\frac{R}{R+1}x_A^e\right)}{\mathrm{d}R} = -\frac{1}{r_A(x_{A1})}\frac{x_A^e}{(R+1)^2} \tag{12.111}$$

Substituting into Eq. 12.109:

$$\frac{\mathrm{d}\left[(R+1)\int_{x_{A1}(R)}^{x_A^e} \frac{1}{r_A(x_A)}\mathrm{d}x_A\right]}{\mathrm{d}R} = \int_{x_{A1}(R)}^{x_A^e} \frac{1}{r_A(x_A)}\mathrm{d}x_A - \frac{R+1}{r_A(x_{A1})}\frac{x_A^e}{(R+1)^2} = 0 \tag{12.112}$$

$$\int_{x_{A1}(R)}^{x_A^e} \frac{1}{r_A(x_A)}\mathrm{d}x_A = \frac{1}{r_A(x_{A1})}\frac{x_A^e}{R+1} \tag{12.113}$$

Please recall Eq. 12.105:

$$\frac{x_A^e}{R+1} = x_A^e - x_{A1} \tag{12.114}$$

Therefore, Eq. 12.113 becomes:

$$\boxed{\frac{1}{r_A(x_{A1})} = \frac{\int_{x_{A1}}^{x_A^e} \frac{1}{r_A(x_A)}\mathrm{d}x_A}{x_A^e - x_{A1}} = \overline{\frac{1}{r_A(x_A)}}\Bigg|_{x_{A1}}^{x_A^e}} \tag{12.115}$$

The optimal recycle ratio, i.e. the one that minimises the space-time of a PFR with recycle, assigned x_A^e, is the value of R such that, in practice, the reaction rate at PFR inlet (i.e., at x_{A1} conversion, see Fig. 12.15) is equal to the average value of the reaction rate between PFR inlet (x_{A1}) and outlet (x_A^e). Although the fundamental Eq. 12.115 does not explicitly show R, we recall that R and x_{A1} are linked by a bijective relation (see Eq. 12.94):

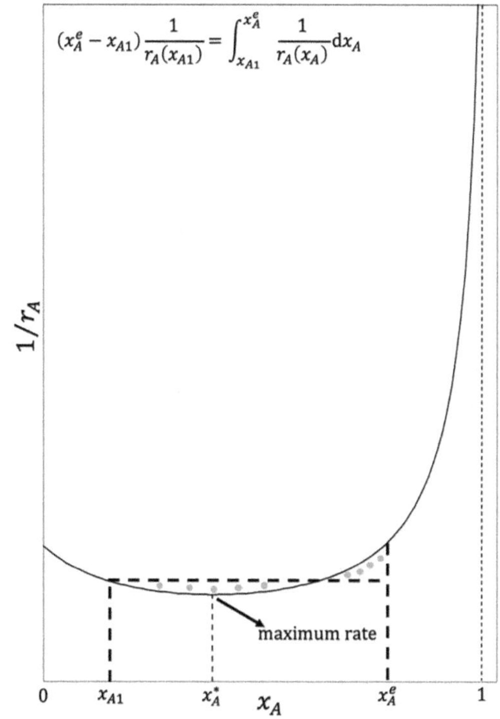

Fig. 12.20 Condition of optimisation for PFR with recycle

$$\boxed{R = \frac{x_{A1}}{x_A^e - x_{A1}}} \tag{12.116}$$

Thus, finding the value for x_{A1} that satisfies the optimisation condition in Eq. 12.115 means finding the optimal value for R.

Resuming the trend for the reciprocal of the reaction rate vs. conversion degree, it is evident that, if the function were canonical (Fig. 11.1), the condition in Eq. 12.115 could never occur (and, in fact, in this case no recycle is desired). The only possibility for Eq. 12.115 to materialise is that, at the same time: (i) the plot $1/r_A(x_A)$ vs. x_A is non-monotonic (this is the autocatalytic case in question, Fig. 12.20); (ii) x_{A1} and x_A^e are located, respectively, *before and after* x_A^*. Only in this way, in fact, the elements of fluid between x_{A1} and x_A^e will experience values of r_A both better and worse with respect to $r_A(x_{A1})$, and therefore it will be possible to verify that $r_A(x_{A1})$ and the average value of r_A between x_{A1} and x_A^e, substantially, are equivalent.

Equation 12.115 can be rewritten as:

$$\boxed{\left(x_A^e - x_{A1}\right)\frac{1}{r_A(x_{A1})} = \int_{x_{A1}}^{x_A^e} \frac{1}{r_A(x_A)}\mathrm{d}x_A} \tag{12.117}$$

which has an interesting graphic counterpart. The optimisation condition occurs if x_{A1} and x_A^e are such that the following areas are equal each other (Fig. 12.20): (i) the area of the rectangle with base $x_A^e - x_{A1}$ and height $1/r_A(x_{A1})$, i.e. the left-hand side of Eq. 12.117; (ii) the area under the curve between x_{A1} and x_A^e, i.e. the right-hand side of Eq. 12.117. Again, for this to happen (i.e., for the two areas indicated with green dots to be equal), x_{A1} and x_A^e must necessarily be found, respectively, upstream and downstream of x_A^*.

Concluding, from an operational point of view, the optimisation procedure can be schematised as follows:

- it must be verified that $x_A^e > x_A^*$ (otherwise, a CSTR would be more suitable, or a PFR with recycling at the infinite theoretical limit; practically this means R very large, f.i. $R = 50$–100, for which, Eq. 12.94, x_{A1} is about 98–99% of x_A^e);
- we proceed, also by trial-and-error if necessary, finding the value of x_{A1} (i.e., of R by Eq. 12.116) that satisfies the condition in Eq. 12.115 (or Eq. 12.117, this is equivalent)—it is remembered that x_{A1} must be less than x_A^* (this facilitates, i.e. restricts, the field of existence for optimal R);
- having found the optimal value for x_{A1} (and, therefore, for R), the design equation Eq. 12.102 can be used, to find the space-time (which will be the minimum possible) of the PFR with recycle.

Example: Optimisation of the Recycle Ratio

Consider, as reference case, the autocatalytic reaction illustrated in Chap. 11: A + P⟹2P, $r_A = kC_AC_P$, $C_{A0} = 1$ mol/L, $C_{P0} = 0.2$ mol/L, $k = 0.5$ L/(mol·min). We already know that $x_A^* = 0.4$ and that (rearranging Eq. 11.38):

$$\int_{x_{A1}}^{x_A^e} \frac{1}{r_A(x_A)}\mathrm{d}x_A = \frac{1}{kC_{A0}(C_{A0}+C_{P0})}\left[\ln\left(\frac{C_{P0}+C_{A0}x_A}{1-x_A}\right)\right]_{x_{A1}}^{x_A^e}$$

$$= \frac{1}{kC_{A0}(C_{A0}+C_{P0})}\ln\left(\frac{C_{P0}+C_{A0}x_A^e}{1-x_A^e}\,\frac{1-x_{A1}}{C_{P0}+C_{A0}x_{A1}}\right) \tag{12.118}$$

Take $x_A^e = 0.9$ and a PFR ($V = 0.5$ L) with recycle, which is a good choice for this system since $x_A^e > x_A^*$, as in Fig. 12.20. To minimise the space-time, we need to find the optimal value for R. The optimisation condition is (Eq. 12.117):

(continued)

$$\left(x_A^e - x_{A1}\right)\frac{1}{r_A(x_{A1})} = \int_{x_{A1}}^{x_A^e}\frac{1}{r_A(x_A)}\mathrm{d}x_A \Leftrightarrow \frac{x_A^e - x_{A1}}{(1-x_{A1})(C_{P0}+C_{A0}x_{A1})}$$
$$= \frac{1}{C_{A0}+C_{P0}}\ln\left(\frac{C_{P0}+C_{A0}x_A^e}{1-x_A^e}\frac{1-x_{A1}}{C_{P0}+C_{A0}x_{A1}}\right) \tag{12.119}$$

By trial-and-error (for x_{A1} less than x_A^*), this condition is verified for $x_{A1} = 0.108$. This results into (Eq. 12.116) an optimal recycle ratio of $R = 0.136$, that returns the following (minimum) value for the space-time (Eq. 12.102):

$$\tau = \frac{V}{Q_0} = (R+1)C_{A0}\int_{x_{A1}}^{x_A^e}\frac{1}{r_A(x_A)}\mathrm{d}x_A = (R+1)$$
$$\times \frac{1}{k(C_{A0}+C_{P0})}\ln\left(\frac{C_{P0}+C_{A0}x_A^e}{1-x_A^e}\frac{1-x_{A1}}{C_{P0}+C_{A0}x_{A1}}\right) \tag{12.120}$$

It results $\tau = 6.55$ min. Finally, working under these conditions allows us to use the maximum value for $Q_0 = 0.076$ L/min and, therefore, for the productivity $P_P = Q_0 C_{A0} x_A^e = 0.068$ mol/min.

12.6 Background Information

- A system of reactors in parallel (e.g., N CSTR or N PFR) is optimised by making all reactors operate with the same value of space-time (i.e., the same degree of conversion); in this case, the reactor system can be treated as a single reactor, with a volume equal to the sum of the volumes of the N reactors, which processes the entire volumetric flow rate up to the exit conversion degree.
- For reactor schemes in series, the space-time of a reactor is the ratio between the reactor volume and the flow rate entering the entire system and, in the design equation, the actual range of conversion degrees (between reactor inlet and outlet) must be considered (Eqs. 12.41 and 12.47).
- A combination of any number of PFR in series works as a single PFR with a volume equal to the sum of the volumes, and no optimisation is needed (Fig. 12.7).
- Having CSTR in series means, at least partially, segregating the flow (tending to a PFR case), Fig. 12.8.
- If a PFR-type behaviour is desired, and two (or more) PFR are available, they can be arranged in series, or in optimised parallel; if a PFR-type behaviour is desired, and two (or more) CSTR are available, they can be arranged in (optimised) series;

if a CSTR-type behaviour is desired, and two (or more) CSTR are available, they can be arranged in optimised parallel.

- For reactors in series (Fig. 12.12): if the second derivative of the function $r_A(C_A)$ is positive, it is convenient to postpone mixing—for example, by placing a PFR first and then a CSTR; if the second derivative is negative, it is convenient to anticipate mixing—for example, by placing a CSTR first and then a PFR; if the second derivative is zero, anticipating or postponing the mixing has no advantage/disadvantage.
- PFR with recycle (Figs. 12.15 and 12.18) can explore the infinite ideal reactor behaviours between PFR and CSTR, with design Eq. 12.102.
- The recycle ratio (Eqs. 12.86 and 12.116) in a PFR with recycle is optimised (Eq. 12.115) when the value of the reaction rate at PFR inlet is substantially equal to the average value of the reaction rate between PFR inlet and outlet—this can be verified for autocatalytic reactions, Fig. 12.20.

References

Levenspiel O (1999) Chemical reaction engineering. John Wiley & Sons, New York

Westerterp KR, van Swaaij WPM, Beenackers AACM (1988) Chemical reactor design and operation. John Wiley & Sons, Hoboken

Extensions to Ideal Reactors: Non-Isothermal Reactions

13

Abstract

This chapter explores the aspects associated to ideal reactors, when they are operated in non-isothermal mode. First, the essential arguments about the kinetics and thermodynamics of chemical reactions in relation to temperature are presented, highlighting how the design equations must be now declined. Then, the thermokinetic maps for irreversible and reversible, endothermic and exothermic reactions are built and commented. For each case, the optimal temperature profiles are discussed. Subsequently, the energy and material balances for adiabatic reactors are illustrated, with details on the optimisation conditions in terms of both temperature profiles and reactor choice. An extensive use of graphical methods complements the theoretical discussion.

Nomenclature

C_i	Concentration of *i* species [moles/volume]
C_{i0}	Concentration of *i* species at reaction beginning [moles/volume]
C_i^{EQ}	Equilibrium concentration of *i* species [moles/volume]
$\mathbb{C}$	Specific heat [energy/(mass·temperature)]
E	Activation energy [energy/moles]
E_d	Activation energy for direct reaction [energy/moles]
E_i	Activation energy for inverse reaction [energy/moles]
K_{EQ}	Equilibrium thermodynamic constant [–]
k	Kinetic parameter [(1/time)·(moles/volume)$^{1-n}$]
k_0	Arrhenius' pre-exponential [(1/time)·(moles/volume)$^{1-n}$]
k_{0d}	Arrhenius' pre-exponential for direct reaction [(1/time)·(moles/volume)$^{1-n}$]
k_{0i}	Arrhenius' pre-exponential for inverse reaction [(1/time)·(moles/volume)$^{1-n}$]
k_d	Kinetic parameter for direct reaction [(1/time)·(moles/volume)$^{1-n}$]

F. Montagnaro et al., *Elements of Unit Operations and Chemical Reactors*,
https://doi.org/10.1007/978-3-032-18046-9_13

k_i	Kinetic parameter for inverse reaction [(1/time)·(moles/volume)$^{1-n}$]
m	Mass [mass]
n	Total reaction order [–]
n_{i0}	Moles of i species at reaction beginning [moles]
P_P	Productivity of the desired species [moles/time]
Q_0	Volumetric flow rate at reaction beginning [volume/time]
Q_j	Volumetric flow rate entering a reactor "j" (in parallel) [volume/time]
q	Power of the heat exchanger [energy/time]
R	Recycle ratio [–]
$\mathfrak{R}$	Gas constant [energy/(moles·temperature)]
r_A	Rate of reaction [moles/(volume·time)]
S	Reactor cross-section area [length^2]
T	Temperature [temperature]
T^e	Exit temperature [temperature]
T^*	Inflection temperature [temperature]
T_0	Temperature at reaction beginning [temperature]
T_{IN}	Inlet temperature to adiabatic reactor [temperature]
T_{MAX}	Maximum temperature [temperature]
T_{in}	Temperature at process inlet [temperature]
T_k	Temperature in position "k" [temperature]
T_{out}	Temperature at process outlet [temperature]
T_{ref}	Reference temperature [temperature]
t	Time [time]
V	Volume [volume]
V_0	Volume at reaction beginning [volume]
V_j	Volume of a reactor "j" [volume]
W	Mass flow rate [mass/time]
x_A	Degree of conversion [–]
x_A^{EQ}	Equilibrium degree of conversion [–]
x_A^e	Exit degree of conversion [–]
x_A^*	Degree of conversion for maximum reaction rate [–]
x_{Aj}	Degree of conversion for a reactor "j" [–]
x_{Ak}	Degree of conversion in position "k" [–]
$x_{A,in}$	Degree of conversion at process inlet [–]
$x_{A,out}$	Degree of conversion at process outlet [–]
z	Reactor axial coordinate [length]
ΔG	Change in free energy [energy/moles]
ΔH	Change in enthalpy [energy/moles]
ΔT_{ad}	Adiabatic temperature change [temperature]
$\Delta\varsigma$	Change in entropy [energy/(moles· temperature)]
ε	Expansion coefficient [–]
ρ	Density [mass/volume]
τ	Space-time for continuous reactor [time]
τ_{CSTR}	Space-time for CSTR [time]

τ_{PFR}	Space-time for PFR [time]
τ_{TOT}	Sum of space-times [time]
τ_j	Space-time for a reactor "*j*" [time]

13.1 Introductory Considerations on Non-Isothermal Reacting Systems

While the "index" of this chapter essentially concerns non-isothermal reactors, it is specified that, given the nature of this textbook (which aims to provide elements), several simplifying assumptions will be made (e.g., liquid-phase systems with constant density are considered). Therefore, there are several topics that belong to the "non-index" of this chapter. For extensions, scholars may refer to more specialised textbooks (e.g., Levenspiel 1999; Westerterp et al. 1988).

13.1.1 Design Equations

When we consider the effect of temperature *T*, *the rate of reaction can be seen as a function of both conversion degree and T*, $r_A(x_A,T)$. In STR, the intensive property *T*, like x_A, is uniform in space for a fixed time. The general design equation will be the usual one, but now indicating the dependence $r_A(x_A,T)$. Similarly in a PFR: *T*, like x_A, presents a perfectly segregated profile along the axis, and it is instead uniform along the radial coordinate. In a CSTR, *T* assumes, at every instant, a value uniform in space and equal to the output value, T^e. The set of design equations (simplified for systems at constant density, and generally defining x_A^e as the degree of conversion at the end of the process) is, therefore:

$$\begin{cases} \boxed{t \ (\text{or } \tau) = C_{A0} \displaystyle\int_0^{x_A^e} \frac{1}{r_A(x_A,T)} \mathrm{d}x_A} & \text{for STR (or PFR)} \\ \boxed{\tau = \dfrac{C_{A0} x_A^e}{r_A\left(x_A^e, T^e\right)}} & \text{for CSTR} \end{cases} \tag{13.1}$$

We observe that: for STR/PFR, it is necessary to know the *profile* $T(x_A)$ to solve the design equation. For CSTR, it is necessary to know T^e, in addition to x_A^e, to evaluate *the unique value of the reaction rate* $r_A(x_A^e,T^e)$.

13.1.2 Kinetics

As highlighted, the reaction rate r_A is now a function $r_A(x_A,T)$, since *T* influences the *kinetic parameter k*. As discussed when presenting Eq. 10.3, in fact, *k* is no more a "constant"; the classical dependence of *k* on *T* is dictated by the *Arrhenius law*

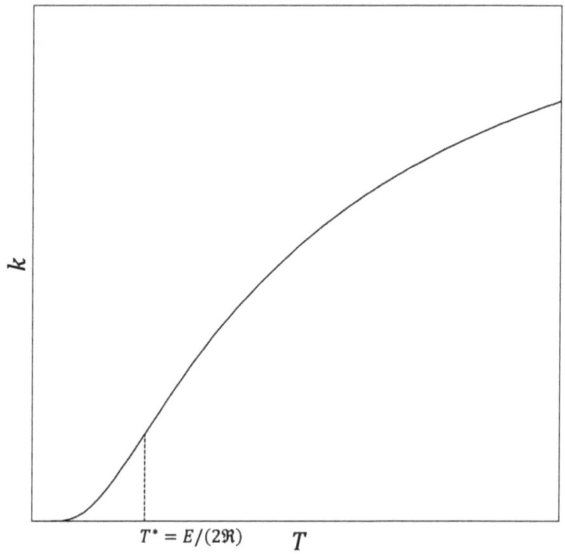

Fig. 13.1 Generic trend for $k(T)$, with indication of the inflection point

[named after Svante August Arrhenius (1859–1927), Swedish scientist, Professor at University of Stockholm; see Arrhenius 1889]:

$$k(T) = k_0 \exp\left(-\frac{E}{\Re T}\right) \tag{13.2}$$

Note that we are, however, considering that the kinetic parameter does not change with the composition of the reacting mixture. The implicit assumption, here, is that the reacting process evolves in a dilute system with homogenous spatial distribution of reactants. Particular cases arise when the diffusional self-stirring mechanisms may cause spatial gradients and segregation of the reacting species. The non-re-randomisation of the reactants location determines that the random walkers mostly fluctuate around their original positions, giving rise to a "fractal-like behaviour". The reader interested in *fractal-like kinetics* (k will change along with the reaction progress, also in isothermal systems) can extend these concepts in, e.g., Kopelman (1988); Balsamo and Montagnaro (2022).

Back to Eq. 13.2, the typical behavior of $k(T)$ is that of an increasing sigmoid, Fig. 13.1. Once the inflection point T^* is defined, it can be calculated by cancelling the second derivative of $k(T)$:

$$\frac{\mathrm{d}k}{\mathrm{d}T} = k_0 \frac{E}{\Re T^2} \exp\left(-\frac{E}{\Re T}\right) \tag{13.3}$$

$$\frac{\mathrm{d}^2 k}{\mathrm{d}T^2} = k_0 \frac{E}{\Re T^2} \frac{E}{\Re T^2} \exp\left(-\frac{E}{\Re T}\right) - 2k_0 \frac{E}{\Re T^3} \exp\left(-\frac{E}{\Re T}\right) \tag{13.4}$$

$$\frac{d^2k}{dT^2}=0 \Rightarrow \frac{E}{\Re T^2}\frac{E}{\Re T^2}=2\frac{E}{\Re T^3} \Rightarrow \frac{E}{\Re T}=2 \tag{13.5}$$

$$T^* = \frac{E}{2\Re} \tag{13.6}$$

Taking some industrially relevant reactions (in their "forward" expressions), e.g.: $N_2 + 3H_2 \Rightarrow 2NH_3$ (activation energy E ca. 46,000 cal/mol; by simplifying $\Re$ at about 2 cal/(mol·K), from Eq. 13.6 it is T^* about 11,500 K), $CO + H_2O \Rightarrow CO_2 + H_2$ (E ca. 27,000 cal/mol, $T^* = 6750$ K), $SO_2 + 0.5O_2 \Rightarrow SO_3$ (E ca. 9000 cal/mol; $T^* = 2250$ K), we note that T^* is much greater than 1000 K. To generalise (Westerterp et al. 1988), E lies in the range ca. 10,000–60,000 cal/mol ($T^* = 2500$–15,000 K). Therefore, in industrial practice, and apart from rare cases, reactors operate to the *left of the point of inflection* in Fig. 13.1, i.e. in a zone where *k is strongly dependent on T*—which is why the argument of this chapter is of interest. Moreover, processes with a higher activation energy have a higher value for T^* (Eq. 13.6), that is, the temperature range in which k is very sensitive to T is wider. In fact, a way to see E is that *it measures the sensitivity of k to T*, as we will discuss more in detail in the following chapter.

13.1.3 Thermodynamics

If the reaction being considered is characterised by *chemical equilibrium*, it will be necessary to consider the role of the equilibrium thermodynamic constant (K_{EQ}): the higher it is, the more relevant the concentration of the products is with respect to the reagents. Simplifying, we can assume:

$$K_{EQ}(T) = \exp\left(-\frac{\Delta G}{\Re T}\right) \tag{13.7}$$

Being the change in free energy (ΔG) related to the change in reaction enthalpy (ΔH) and the change in reaction entropy ($\Delta \varsigma$), on molar basis (these three parameters are assumed constant):

$$\Delta G = \Delta H - T\Delta \varsigma \tag{13.8}$$

it is:

$$K_{EQ}(T) = \exp\left(-\frac{\Delta H - T\Delta \varsigma}{\Re T}\right) = \exp\left(-\frac{\Delta H}{\Re T} + \frac{\Delta \varsigma}{\Re}\right) \tag{13.9}$$

Going through the logarithms:

$$\ln[K_{EQ}(T)] = -\frac{\Delta H}{\Re T} + \frac{\Delta \varsigma}{\Re} \tag{13.10}$$

Deriving with respect to T:

$$\frac{\mathrm{dln}[K_{EQ}(T)]}{\mathrm{d}T} = \frac{\Delta H}{\Re T^2} \tag{13.11}$$

and integrating by the Fourier method (separation of variables) between a reference temperature (T_{ref}) and a generic T:

$$\int_{T_{ref}}^{T} \mathrm{dln}[K_{EQ}(T)] = \frac{\Delta H}{\Re}\int_{T_{ref}}^{T}\frac{1}{T^2}\mathrm{d}T \Rightarrow \ln\left[\frac{K_{EQ}(T)}{K_{EQ}(T_{ref})}\right] = \frac{\Delta H}{\Re} \times \left(\frac{1}{T_{ref}} - \frac{1}{T}\right) \tag{13.12}$$

i.e., a formulation of the *van 't Hoff* law [named after Jacobus Henricus van 't Hoff (1852–1911), Dutch scientist, *alma mater* at Delft University of Technology; see e.g. Meijer 2001]. Therefore, according to van 't Hoff's law and considering $T > T_{ref}$:

- when the reaction is *endothermic* ($\Delta H > 0$; the reaction needs energy to proceed, as the enthalpy of the products is higher than the enthalpy of the reactants), the right-hand side of Eq. 13.12 (and, then, the logarithm at left-hand side) is >0, therefore the logarithm argument is >1—*the equilibrium thermodynamic constant increases as the temperature increases, and the reaction is thermodynamically favoured as T increases*;
- when the reaction is *exothermic* ($\Delta H < 0$; the reaction releases energy as it proceeds, as the enthalpy of the products is lower than the enthalpy of the reactants), the right-hand side of Eq. 13.12 (and, then, the logarithm at left-hand side) is <0, therefore the logarithm argument is <1—*the equilibrium thermodynamic constant decreases as the temperature increases, and the reaction is thermodynamically penalised as T increases.*

Example: Relation Between Equilibrium Degree of Conversion and Equilibrium Constant

Consider the simplified case offered by a liquid-phase A⇔P reaction. Under equilibrium, the *law of mass action* states that the equilibrium constant can be seen as the ratio between the concentration of the product and that of the reactant:

$$K_{EQ}(T) = \frac{C_P^{EQ}(T)}{C_A^{EQ}(T)} = \frac{x_A^{EQ}(T)}{1 - x_A^{EQ}(T)} \tag{13.13}$$

Thus, Eq. 13.13 gives a relation between the degree of conversion at equilibrium (x_A^{EQ}) and the equilibrium constant:

(continued)

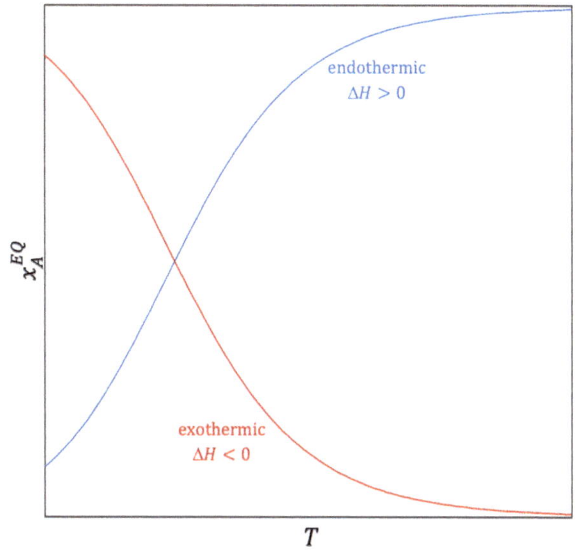

Fig. 13.2 Typical sigmoidal trend of equilibrium conversion vs. temperature for endo- and exo-thermic reactions

$$x_A^{EQ}(T) = \frac{K_{EQ}(T)}{K_{EQ}(T) + 1} \propto K_{EQ}(T) \qquad (13.14)$$

We can therefore re-phrase the van 't Hoff formulation:

- when the reaction is *endothermic* ($\Delta H > 0$), as K_{EQ} increases as T increases (the reaction is thermodynamically favoured as T increases), we see from Eq. 13.14 that x_A^{EQ} *increases with temperature*;
- when the reaction is *exothermic* ($\Delta H < 0$), as K_{EQ} decreases as T increases (the reaction is thermodynamically penalised as T increases), x_A^{EQ} *decreases with temperature.*

A typical trend for $x_A^{EQ}(T)$ is illustrated in Fig. 13.2: it has a sigmoidal shape, increasing or decreasing as function of the reaction enthalpy.

13.2 Thermokinetic Maps

In addition to r_A and x_A, therefore, a third character has appeared on the stage, namely T. While, previously, the relationships between r_A and x_A had been highlighted on graphs of the type $1/r_A$ vs. x_A (example in Fig. 10.3), now, to represent the interrelationships between r_A, x_A and T, we will use *thermokinetic maps*; here we choose to report x_A (on the ordinate) vs. T (on the abscissa), with *isokinetic curves*, i.e. parametric in r_A. This to avoid more complex representation of the three variables on 3D-graphs.

13.2.1 Irreversible Reactions

Consider the case of an irreversible reaction, A⟹P. The general aspect of a thermokinetic map is the one illustrated in Fig. 13.3. We will refer to a simple example, to theoretically explain the map. Be the reaction described by linear kinetics:

$$r_A(x_A, T) = k(T)C_A = k(T)C_{A0}(1 - x_A) \tag{13.15}$$

On a x_A vs. T plane, the equation of a generic isokinetic is, therefore:

$$x_A = 1 - \frac{r_A}{k(T)C_{A0}} \quad \text{at fixed } r_A \tag{13.16}$$

Of course, thermokinetic maps are constructed for a known value of C_{A0}. We observe:

- the isokinetic $r_A = 0$ is represented by the horizontal $x_A = 1$ (for any T, $r_A = 0$ when the reactant A disappears, see also Eq. 13.15);

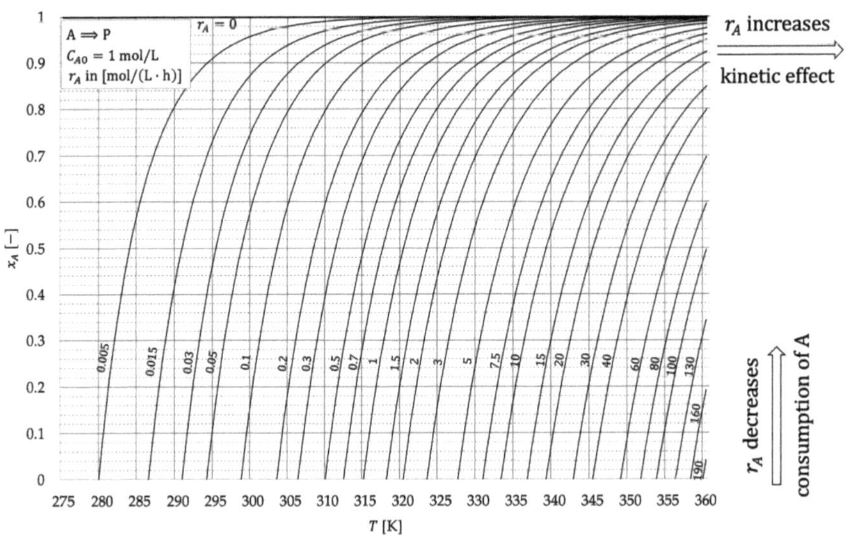

Fig. 13.3 Thermokinetic map for an irreversible reaction. The map has been built for A⟹P, $r_A = k(T)C_A$, $k_0 = 1.81 \cdot 10^{18}$ 1/h, $E = 110$ kJ/mol, $C_{A0} = 1$ mol/L. Values on the curves represent r_A in [mol/(L·h)]. The range of T is typical for a liquid-phase reaction

- for all other isokinetic curves, $x_A < 1$ (Eq. 13.16), increasing with T (as k increases with T) and approaching 1 for high values of T;
- at fixed x_A, as T increases, isokinetic curves with increasing r_A are encountered (r_A is proportional to T, Eq. 13.15);
- at fixed T, we fall back into the isothermal case (k is constant): as x_A increases, A is consumed and r_A decreases (Eq. 13.15).

13.2.2 Reversible Reactions

Consider the case of a reversible reaction, A⇔P. Be both the direct and the inverse reaction described by linear kinetics and governed by kinetic parameters k_d and k_i, respectively (as usual, we assume $C_{P0} = 0$; the approach can be easily extended to other cases, such as feed streams already containing P and/or different stoichiometric coefficients):

$$r_A(x_A, T) = k_d(T)C_A - k_i(T)C_P = k_d(T)C_{A0}(1 - x_A) - k_i(T)C_{A0}x_A \tag{13.17}$$

$$k_d(T) = k_{0d}\exp\left(-\frac{E_d}{\Re T}\right) \tag{13.18}$$

$$k_i(T) = k_{0i}\exp\left(-\frac{E_i}{\Re T}\right) \tag{13.19}$$

Equation 13.17 is rewritten as:

$$x_A = \frac{k_d(T)C_{A0} - r_A}{[k_d(T) + k_i(T)]C_{A0}} = \frac{k_d(T)}{k_d(T) + k_i(T)} - \frac{r_A}{[k_d(T) + k_i(T)]C_{A0}} \tag{13.20}$$

Under equilibrium, the net reaction rate must be 0. Therefore, Eq. 13.17 reads:

$$\begin{aligned} 0 &= k_d(T)C_{A0}\left[1 - x_A^{EQ}(T)\right] - k_i(T)C_{A0}x_A^{EQ}(T) \Rightarrow x_A^{EQ}(T) \\ &= \frac{k_d(T)}{k_d(T) + k_i(T)} \end{aligned} \tag{13.21}$$

Then, on a x_A vs. T plane, the equation of a generic isokinetic is:

$$x_A = x_A^{EQ}(T) - \frac{r_A}{[k_d(T) + k_i(T)]C_{A0}} \quad \text{at fixed } r_A \tag{13.22}$$

As a general consideration, we note that $x_A < x_A^{EQ}$, approaching it at high T (when, fixed r_A, k_d and k_i are both very large). The particular case is for $r_A = 0$, when, of course, $x_A = x_A^{EQ}$. But (Fig. 13.2), as the trend for $x_A^{EQ}(T)$ depends on whether the reaction is endo- or exo-thermic, we must now develop two distinct cases.

Assume A⇔P *endothermic*. The general aspect of a thermokinetic map is shown in Fig. 13.4. The isokinetic curve $r_A = 0$ ($x_A = x_A^{EQ}$) has the sigmoidal increasing

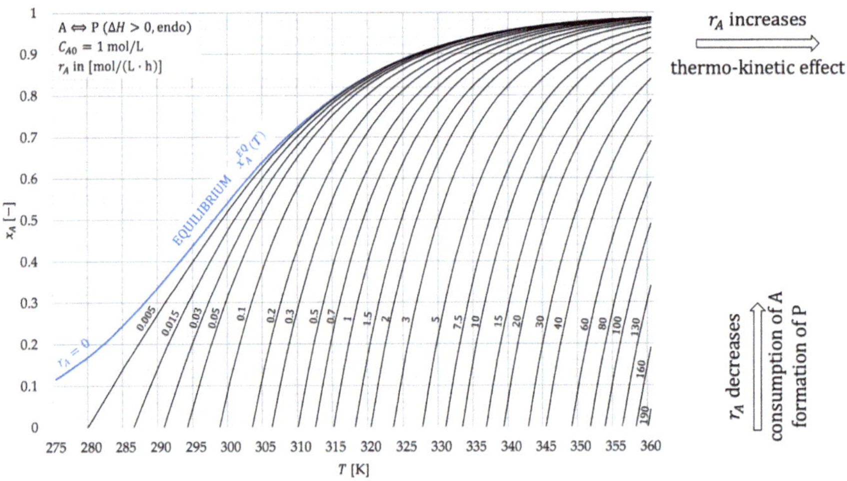

Fig. 13.4 Thermokinetic map for a reversible endothermic reaction. The map has been built for A⟺P, $r_A = k_d(T)C_A - k_i(T)C_P$, $k_{0d} = 1.81 \cdot 10^{18}$ 1/h, $E_d = 110$ kJ/mol, $k_{0i} = 3.39 \cdot 10^{7}$ 1/h, $E_i = 48.9$ kJ/mol, $C_{A0} = 1$ mol/L, $C_{P0} = 0$. Values on the curves represent r_A in [mol/(L·h)]. The blue sigmoid illustrates the equilibrium conditions. The range of T is typical for a liquid-phase reaction

shape of Fig. 13.2. In fact, the equilibrium is favoured when T is increased. Furthermore:

- for any isokinetic, as T increases, both x_A^{EQ} and k_d and k_i increase, then x_A must increase (Eq. 13.22);
- at fixed x_A, as T increases, isokinetic curves with increasing r_A are encountered (being the reaction endothermic, r_A is positively affected by T, in the sense that the direct reaction is more sensitive than the inverse reaction to T);
- at fixed T, we fall back into the isothermal case (k_d and k_i are constant): as x_A increases, A is consumed, P is formed, and r_A decreases (Eq. 13.17);
- this thermokinetic map, apart from the presence of the equilibrium sigmoid, is similar to the case of an irreversible process (Fig. 13.3). In fact, even reversible endothermic reactions will be (*in this case: both kinetically and thermodynamically*) favoured by increases in temperature, as happens for non-reversible reactions.

Now assume A⟺P *exothermic*. The general aspect of a (more complex, actually) thermokinetic map is shown in Fig. 13.5. The isokinetic curve $r_A = 0$ ($x_A = x_A^{EQ}$) has now the sigmoidal decreasing shape of Fig. 13.2. In fact, the equilibrium is penalised when T is increased. Furthermore:

- as T increases, in the isokinetic expression Eq. 13.22, x_A^{EQ} decreases while k_d and k_i increase, then (driving on an isokinetic) x_A presents a *non-monotonic trend with*

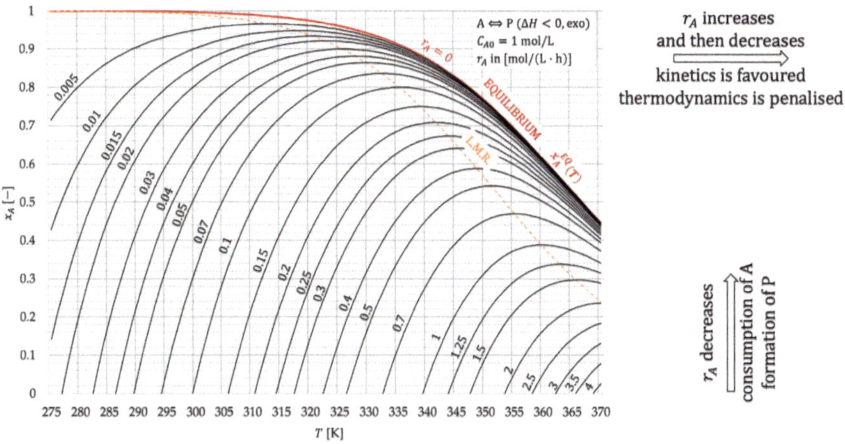

Fig. 13.5 Thermokinetic map for a reversible exothermic reaction. The map has been built for A⇔P, $r_A = k_d(T)C_A - k_i(T)C_P$, $k_{0d} = 3.39 \cdot 10^7$ 1/h, $E_d = 48.9$ kJ/mol, $k_{0i} = 1.81 \cdot 10^{18}$ 1/h, $E_i = 124.2$ kJ/mol, $C_{A0} = 1$ mol/L, $C_{P0} = 0$. Values on the curves represent r_A in [mol/(L·h)]. The red sigmoid illustrates the equilibrium conditions. The orange "L.M.R." dashed line is the locus of maximum rates. The range of T is typical for a liquid-phase reaction

T (for a generic isokinetic curve, when temperature increases, x_A first increases and then decreases). To explain this, please note that we are in the presence of a process in which the *kinetic and thermodynamic indications contrast with each other*. While kinetics is, conceptually, favoured by T, the thermodynamic equilibrium is penalised by T. Given an isokinetic curve, for low values of T (*far from equilibrium*, which is therefore not yet "seen" by the system), x_A is favoured by increases in T (the kinetic effect prevails). We then pass through a maximum, after which, by continuing to increase T, we approach thermodynamic equilibrium and, therefore, x_A is penalised by increases in T (now the system "sees" the equilibrium and the thermodynamic effect prevails). To summarise, when T is increased and having fixed an isokinetic curve: *at low T, the positive kinetic effect prevails and x_A increases; at high T, the negative thermodynamic effect prevails and x_A decreases approaching the equilibrium value*;

- for similar reasons, at fixed x_A, as T increases, we encounter first isokinetic curves with increasing r_A (the positive kinetic effect prevails) and then isokinetic curves with decreasing r_A (the negative thermodynamic effect prevails). In other words, we pass through a *maximum rate* and then we meet back, in decreasing order, the same isokinetic curves encountered before the point of maximum rate;
- for what stated above, the map presents a *locus of maximum rates*: for a given value of x_A, this represents the value of T at which the maximum reaction rate is obtained (for higher values of T, negative thermodynamic effects prevail over positive kinetic ones, and r_A begins to decrease). This locus is represented in Fig. 13.5;

- at fixed T, we fall back into the isothermal case (k_d and k_i are constant): as x_A increases, A is consumed, P is formed, and r_A decreases (Eq. 13.17).

Example: Equation for the Locus of Maximum Rates, Reversible Exothermic Reaction Governed by Linear Kinetics

We here show how, analytically, the locus of maximum rates for the kinetics under investigation has been evaluated (and, then, plotted in Fig. 13.5). Start with Eq. 13.17. To find the analytical relation between x_A and T that represents the locus of maximum rates, the partial derivative of r_A with respect to T (fixed x_A) must be set to zero. In this way, in fact, we represent the circumstance that, fixed x_A, as T varies, r_A passes through a maximum. Thus:

$$\left.\frac{\partial r_A}{\partial T}\right|_{x_A} = 0 \tag{13.23}$$

$$\begin{aligned} &\frac{\mathrm{d}\left[k_{0d}\exp\left(-\frac{E_d}{\Re T}\right)C_{A0}(1-x_A) - k_{0i}\exp\left(-\frac{E_i}{\Re T}\right)C_{A0}x_A\right]}{\mathrm{d}T} \\ &= C_{A0}\left[k_{0d}(1-x_A)\frac{E_d}{\Re T^2}\exp\left(-\frac{E_d}{\Re T}\right) - k_{0i}x_A\frac{E_i}{\Re T^2}\exp\left(-\frac{E_i}{\Re T}\right)\right] \\ &= 0 \end{aligned} \tag{13.24}$$

$$k_{0d}(1-x_A)\frac{E_d}{\Re T^2}\exp\left(-\frac{E_d}{\Re T}\right) = k_{0i}x_A\frac{E_i}{\Re T^2}\exp\left(-\frac{E_i}{\Re T}\right) \tag{13.25}$$

$$\exp\left(\frac{E_i - E_d}{\Re T}\right) = \frac{k_{0i}}{k_{0d}}\frac{x_A}{1-x_A}\frac{E_i}{E_d} \Rightarrow \frac{E_i - E_d}{\Re T} = \ln\left(\frac{k_{0i}}{k_{0d}}\frac{x_A}{1-x_A}\frac{E_i}{E_d}\right) \tag{13.26}$$

$$T(x_A) = \frac{E_i - E_d}{\Re \ln\left[\frac{k_{0i}E_i}{k_{0d}E_d}\frac{x_A}{(1-x_A)}\right]} \tag{13.27}$$

Equation 13.27 is the analytical expression for the locus of maximum rates. It is then possible to obtain, for each value of x_A, the optimal value of T so that r_A is maximum. Once realised that, for exothermic reactions, $E_i > E_d$, Eq. 13.27 illustrates that, when x_A increases, the temperature to reach the locus of maximum is lower (Fig. 13.5).

13.3 Optimal Temperature Profiles

In the case of STR/PFR, once the desired value of x_A^e has been set, choosing the optimal profile of T means determining, for each x_A, the value of T such that r_A *is the maximum possible* (see Eq. 13.1). This guarantees the minimum value for t (or τ) and, in general, optimal operating conditions. Known the desired $T(x_A)$ profile, it will be possible to substitute it into $r_A(x_A,T) \Rightarrow r_A(x_A)$ and solve the STR/PFR design equation by usual approaches. In the case of CSTR, once the desired value of x_A^e has been set, it will be necessary to choose the optimal operating temperature T^e (which coincides with the output temperature). It will be such that, for the desired conversion, *the maximum possible value* is obtained for the only operating reaction rate, that will allow to solve, Eq. 13.1, the CSTR design equation. Also in this case, we will minimise the CSTR space-time or, more in general, we will optimise the system.

13.3.1 Irreversible Reactions; Endothermic Reversible Reactions

For both irreversible and reversible endothermic reactions, both kinetic and (in the case of a reversible reaction) thermodynamic considerations agree in suggesting *working at the highest possible* T values. The situation is, for example, illustrated in Fig. 13.6. After having set the desired conversion degree, x_A^e, and identified the maximum value of temperature, T_{MAX}, at which it is possible to operate without incurring unwanted phenomena (examples: evaporation of the solvent, onset of undesired side-reactions, decomposition of chemicals, safety issues), for STR/PFR we will keep the *reactor isothermal* at $T = T_{MAX}$, and for CSTR it will be necessary to ensure that $T^e = T_{MAX}$. In this way, in all cases, maximum reaction rates will be encountered. This will be done by equipping the reactor with a heat exchange system (with design principles illustrated in Chap. 5; examples: internal coil/pipe, or jacketed reactor), with a service fluid that allows the desired T profile to be achieved (this will depend on the enthalpy change of the reaction, on the heat exchanges of the reactor with the external environment, on the feed temperature, etc.).

Example: Optimisation of a STR, Endothermic Reversible Reaction

Consider the $A \Leftrightarrow P$ endothermic reaction in Fig. 13.4. We have a STR, $x_A^e = 0.8$, $T_{MAX} = 350$ K, $C_{A0} = 1$ mol/L, $C_{P0} = 0$. The graphic approach in Fig. 13.6 makes us sure that, to optimise t, we need to keep the STR isothermal at T_{MAX} (the kinetic parameters will be constant). Then, it is (Eqs. from 13.17 to 13.19):

(continued)

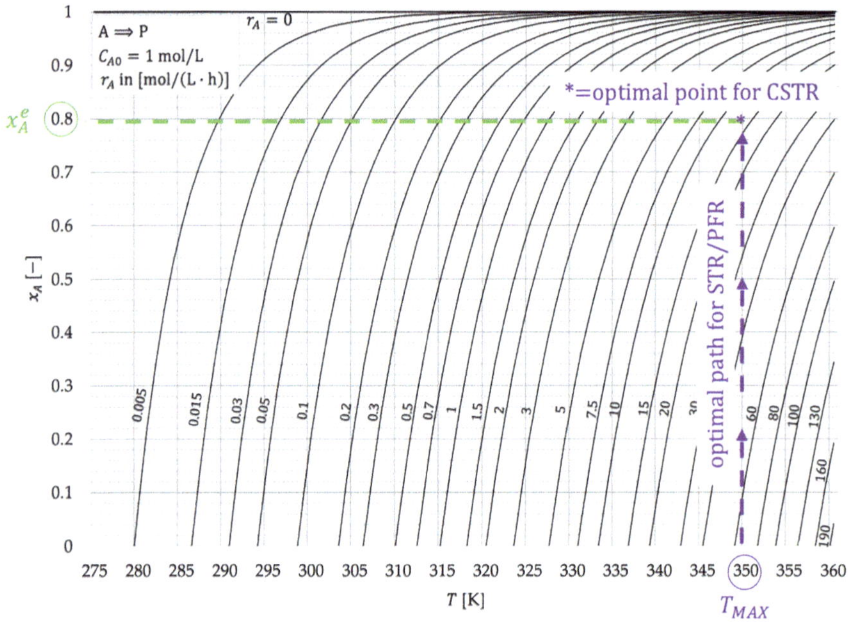

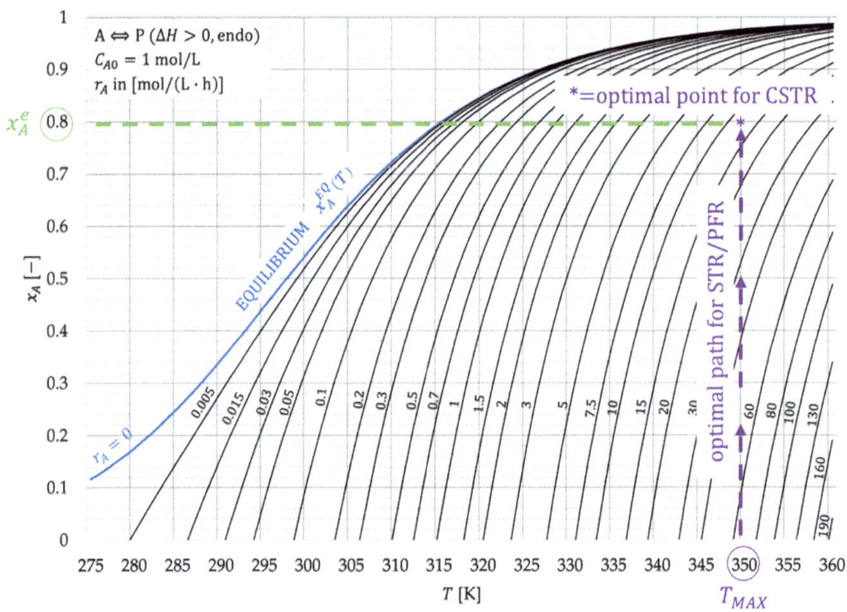

Fig. 13.6 Optimal operating conditions on thermokinetic maps, once set exit conversion degree and maximum temperature. Up: irreversible reaction. Down: reversible endothermic reaction

$$\begin{cases} r_A(x_A, T_{MAX}) = k_d(T_{MAX})C_{A0}(1 - x_A) - k_i(T_{MAX})C_{A0}x_A \\ k_d(T_{MAX}) = k_{0d}\exp\left(-\frac{E_d}{\Re T_{MAX}}\right) = 69.26\ 1/\text{h} \\ k_i(T_{MAX}) = k_{0i}\exp\left(-\frac{E_i}{\Re T_{MAX}}\right) = 1.71\ 1/\text{h} \end{cases} \tag{13.28}$$

We can now approach the design Eq. 13.1:

$$t = C_{A0}\int_0^{x_A^e} \frac{1}{r_A(x_A, T)}\mathrm{d}x_A = \int_0^{x_A^e} \frac{1}{k_d - (k_d + k_i)x_A}\mathrm{d}x_A =$$
$$-\frac{1}{k_d + k_i}\ln\{[k_d - (k_d + k_i)x_A]\}_0^{x_A^e}$$
$$= \frac{1}{k_d + k_i}\ln\left[\frac{k_d}{k_d - (k_d + k_i)x_A^e}\right] = 0.024\ \text{h} \tag{13.29}$$

13.3.2 Exothermic Reversible Reactions

For reversible exothermic reactions, kinetic and thermodynamic considerations *contrast each other*. We have seen that, far from equilibrium (i.e., at low conversion degree), kinetics prevails (temperature must be high) and, close to equilibrium (i.e., at high conversion degree), thermodynamics prevails (temperature must be low). We can sense that the optimal temperature profile, in this case, must be that of T decreasing as long as conversion degree increases (referring to a STR/PFR) or, in other words, *the optimal temperature profile must adhere to the locus of maximum rates*. The situation is illustrated in Fig. 13.7. Set the desired conversion degree, x_A^e, and T_{MAX}. Follow the path from $x_A = 0$ up to x_A^e. In general, for low values of x_A, the locus of maximum rates would suggest $T > T_{MAX}$. This can be tackled by maintaining $T = T_{MAX}$ until the locus of maximum rates is encountered, and then settling on the locus. Of course, we will need heat exchange systems able to fine-tune the temperature (by subtracting heat, as temperature must be decreased and as the exothermic reaction releases heat) along the conversion (in time for a STR, in space for a PFR). For practical purposes, it will be sufficient to operate in a band around the locus. In any case, the aim is to encounter the maximum possible reaction rates for each conversion degree. For a CSTR, it will be necessary to design the system so to ensure that T^e (the operating temperature) is the value dictated, for x_A^e, by the locus of maximum rates (in this way, the CSTR will operate at the maximum value of reaction rate for the given conversion).

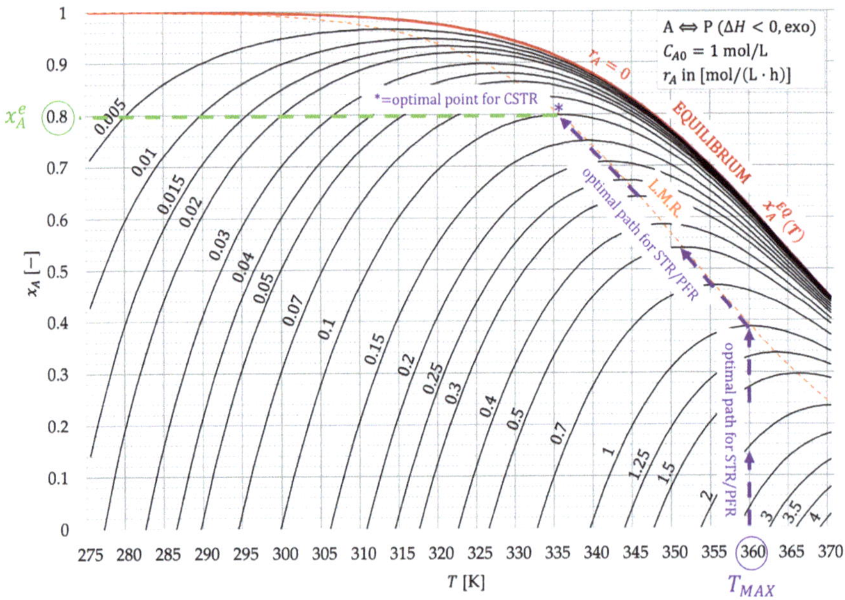

Fig. 13.7 Optimal operating conditions on thermokinetic maps, once set exit conversion degree and maximum temperature, for reversible exothermic reaction

Example: Optimisation of a PFR, Exothermic Reversible Reaction

Consider the A⇔P exothermic reaction in Fig. 13.5. We have a PFR, $x_A^e = 0.8$, $T_{MAX} = 360$ K, $C_{A0} = 1$ mol/L, $C_{P0} = 0$. We follow the graphic approach in Fig. 13.7. The optimal temperature profile is:

$$\begin{cases} T = T_{MAX} & \text{up to } x_A = 0.38 \\ T(x_A) = \dfrac{E_i - E_d}{\Re \ln\left[\dfrac{k_{0i}E_i}{k_{0d}E_d}\dfrac{x_A}{(1-x_A)}\right]} & \text{for } x_A > 0.38 \text{ as per Eq.13.27} \end{cases} \tag{13.30}$$

with $x_A = 0.38$ the point where we meet the locus of maximum rates for $T_{MAX} = 360$ K. The analytical solution of the design equation:

$$\tau = C_{A0}\int_0^{x_A^e} \frac{1}{r_A(x_A, T)}\mathrm{d}x_A \tag{13.31}$$

should impose to consider Eqs. from 13.17 to 13.19, substituting the T profile given in Eq. 13.30. This is a complex approach, that can be circumvented by adopting a mixed (graphical-analytical) procedure, presented here and used

(continued)

also elsewhere in this chapter (when needed). The basic idea is to follow the optimal path on the map (Fig. 13.7), and to approximate the integral in Eq. 13.31 to a summation with constant interval Δx_A (for example, take $\Delta x_A = 0.05$):

$$\tau \cong C_{A0}\Delta x_A \sum_0^{x_A^e} \frac{1}{r_A} \tag{13.32}$$

We can build a table (Table 13.1) where, for each x_A interval (for example, choose to calculate at the higher bound of the interval), we list the value of r_A. Then we will sum the reciprocals of each r_A value, and, by Eq. 13.32, we will calculate τ. To analytically follow this approach, first we need, for each x_A interval, to evaluate T (Eq. 13.30), and then we can calculate r_A (Eqs. from 13.17 to 13.19). Graphically, although it is a less precise procedure, we could read the values of r_A directly on the map in Fig. 13.7, without the need of calculating the temperatures.

By application of the design equation, it is:

$$\tau \cong C_{A0}\Delta x_A \left(\frac{1}{2.499} + \ldots + \frac{1}{0.098}\right) = 1.90\ \text{h} \tag{13.33}$$

Table 13.1 Data for the approximated solution of the design equation

x_A [–]	T [K] Eq. 13.30	r_A [mol/(L·h)] Eqs. 13.17–13.19
0.05	360	2.499
0.10	360	2.277
0.15	360	2.055
0.20	360	1.832
0.25	360	1.610
0.30	360	1.388
0.35	360	1.166
0.40	359.0	0.946
0.45	356.1	0.759
0.50	353.3	0.606
0.55	350.6	0.479
0.60	347.8	0.373
0.65	345.0	0.284
0.70	342.0	0.210
0.75	338.8	0.148
0.80	335.2	0.098

13.4 Mass and Energy Balances for Adiabatic Reactors

As already highlighted, to solve the design Eq. 13.1 for an ideal reactor, we need the profile $T(x_A)$, to be substituted into $r_A(x_A,T) \Rightarrow r_A(x_A)$. In the most general terms, the mass balance (on the limiting reactant species) must be coupled with the *energy balance*. Specifically, neglecting the changes in other energy contributions, the energy balance will be approximated to an enthalpy balance, although this is not strictly rigorous. That is, in a chemical reactor, we consider the changes in enthalpy as the main contributors when we write the energy balance (for example, we can neglect changes in potential, kinetic, etc., energy in most actual chemical reactors).

We will delve into the case of *adiabatic reactors*. Albeit this is a particular case, it is of interest in industrial practice: a reactor in which the temperature variations are important, due to the thermal effects connected to the course of the reaction, but thermally insulated with respect to the external environment. Strictly speaking, we consider the reactor purely adiabatic, i.e. with no internal heat exchangers. Furthermore, we will assume *reactions in the liquid phase* (and species involved in the reaction not much concentrated in the solvent), for which the density is constant and so it is the average specific heat ($\mathbb{C}$, here expressed on mass basis); typically, these values can be assumed equal to those of the solvent. This also implies that the variation of specific heat during the chemical reaction can be neglected and, therefore, allows us to assume ΔH as constant. Extensions (when these assumptions are not valid) can be found in, e.g., Levenspiel (1999); Westerterp et al. (1988).

Start with STR. The considerations on A mass balance (and control volume) have been developed from Eqs. 10.28 to 10.31. We rewrite Eq. 10.31, noting that $n_{AO} = C_{AO}V_O$ and that $\varepsilon = 0$ (that is, the reacting volume $V = V_O$):

$$n_{A0}\frac{\mathrm{d}x_A}{\mathrm{d}t} = r_A(x_A,T)V_0[1+\varepsilon x_A] \Rightarrow C_{A0}\frac{\mathrm{d}x_A}{\mathrm{d}t} = r_A(x_A,T) \tag{13.34}$$

Now we build the energy balance on STR, as ACC = GEN (cf. Eq. 10.29), in [energy/time], and taking the whole STR as control volume (as we did for the mass balance). In the accumulation term, m is the reacting mass, expressed as product between the density (ρ) and the volume V (both constant) of the reacting solution; the generation term is related to the change in molar reaction enthalpy (the minus sign is present as, for a reaction releasing energy, i.e. exothermic, $\Delta H < 0$ but the generation is positive, and vice versa):

$$m\mathbb{C}\frac{\mathrm{d}T}{\mathrm{d}t} = -\Delta H r_A(x_A,T)V \Rightarrow \rho\mathbb{C}\frac{\mathrm{d}T}{\mathrm{d}t} = -\Delta H r_A(x_A,T) \tag{13.35}$$

Proceed with PFR. The considerations on A mass balance (the control volume is illustrated in Fig. 10.10) have been developed from Eqs. 10.67 to 10.70. The latter is here recalled:

$$Q_0 C_{A0} \frac{\mathrm{d}x_A}{\mathrm{d}z} = r_A(x_A, T)S \tag{13.36}$$

Now we write the energy balance on the PFR differential control volume, as OUT–IN = GEN (cf. Eq. 10.67), where W is the (constant) reacting mass flow rate, expressed as product between the density and the volumetric flow rate (Q_0 = constant) of the reacting solution:

$$W\mathbb{C}[T(z + \mathrm{d}z) - T(z)] = -\Delta H r_A(x_A, T)S\mathrm{d}z \Rightarrow Q_0 \rho \mathbb{C} \frac{\mathrm{d}T}{\mathrm{d}z} = -\Delta H r_A(x_A, T)S \tag{13.37}$$

End with CSTR. Rewrite the mass balance in Eq. 10.88:

$$Q_0 C_{A0} x_A^e = r_A\left(x_A^e, T^e\right)V \tag{13.38}$$

The energy balance on the entire CSTR (OUT–IN = GEN) is (with T_0 the temperature of the stream at reactor inlet, and W expressed as usual):

$$W\mathbb{C}(T^e - T_0) = -\Delta H r_A\left(x_A^e, T^e\right)V \Rightarrow Q_0 \rho \mathbb{C}(T^e - T_0) = -\Delta H r_A\left(x_A^e, T^e\right)V \tag{13.39}$$

Summarising ("B.E." and "B.M." stand for balance of energy and balance of mass, respectively):

$$\begin{cases} \rho \mathbb{C} \dfrac{\mathrm{d}T}{\mathrm{d}t} = -\Delta H r_A(x_A, T) & \underline{\text{B.E.}} \text{ for STR} \\ C_{A0} \dfrac{\mathrm{d}x_A}{\mathrm{d}t} = r_A(x_A, T) & \underline{\text{B.M.}} \text{ for STR} \end{cases} \tag{13.40}$$

$$\begin{cases} Q_0 \rho \mathbb{C} \dfrac{\mathrm{d}T}{\mathrm{d}z} = -\Delta H r_A(x_A, T)S & \underline{\text{B.E.}} \text{ for PFR} \\ Q_0 C_{A0} \dfrac{\mathrm{d}x_A}{\mathrm{d}z} = r_A(x_A, T)S & \underline{\text{B.M.}} \text{ for PFR} \end{cases} \tag{13.41}$$

$$\begin{cases} Q_0 \rho \mathbb{C}(T^e - T_0) = -\Delta H r_A\left(x_A^e, T^e\right)V & \underline{\text{B.E.}} \text{ for CSTR} \\ Q_0 C_{A0} x_A^e = r_A\left(x_A^e, T^e\right)V & \underline{\text{B.M.}} \text{ for CSTR} \end{cases} \tag{13.42}$$

For STR/PFR, we end up with a system of two differential equations. The system is non-linear (think to the Arrhenius expression for r_A), whose solution gives the profiles (in t for STR, in z for PFR) for x_A and T. This yields a relationship $T(x_A)$, which is what we are aiming at. Similar is the situation for CSTR, where the non-linear equations are algebraic and are solved for the variables x_A^e and T^e. It can be observed that (in each case) the non-linearity of the equations is embedded in the reaction rate expression. One way to exclude this effect is to divide the equations member by member:

$$\begin{cases} \dfrac{\rho\mathbb{C}}{C_{A0}}\dfrac{dT}{dx_A} = -\Delta H & \text{for STR/PFR} \\ \dfrac{\rho\mathbb{C}}{C_{A0}}\dfrac{T^e - T_0}{x_A^e} = -\Delta H & \text{for CSTR} \end{cases} \quad (13.43)$$

Equation 13.43 represents, for each reactor, the *combined energy/mass balance*. Solving Eq. 13.43 for STR/PFR by separation of variables:

$$\frac{\rho\mathbb{C}}{C_{A0}}\int_{T_0}^{T^e} dT = -\Delta H\int_0^{x_A^e} dx_A \quad (13.44)$$

we land on the same equation obtained for CSTR (see Eq. 13.43). Therefore, the following represents, under our assumptions, the *generalised solution for any adiabatic reactor* (where we have put a generic T corresponding to a generic x_A):

$$\boxed{T - T_0 = \frac{-\Delta H C_{A0}}{\rho\mathbb{C}} x_A} \quad (13.45)$$

Rewrite Eq. 13.45 as:

$$\boxed{\begin{cases} T - T_0 = \Delta T_{ad} x_A \\ \Delta T_{ad} = \dfrac{-\Delta H C_{A0}}{\rho\mathbb{C}} \begin{cases} \Delta T_{ad} < 0 \text{ for endothermic reactions } (\Delta H > 0) \\ \Delta T_{ad} > 0 \text{ for exothermic reactions } (\Delta H < 0) \end{cases} \end{cases}} \quad (13.46)$$

defining ΔT_{ad} (*adiabatic temperature change*) as the *maximum temperature variation* that the reacting mixture can undergo (and this happens for $x_A = 1$, since it is recalled that, in the adiabatic case, the only reason why T can vary is linked to the progress of the chemical reaction). Note the following:

- Equation 13.46 *represents the desired relationship between T and x_A*;
- ΔT_{ad} is >0 for exothermic reactions ($\Delta H < 0$, the progress of the reaction makes T increase), and is <0 for endothermic reactions ($\Delta H > 0$, the progress of the reaction makes T decrease);
- ΔT_{ad}, in absolute value, is greater for reactions with greater variations in enthalpy and when there is more reactant, see the numerator of Eq. 13.46;
- ΔT_{ad}, in absolute value, is smaller when the mixture "absorbs energy variations better", that is, when density and specific heat are higher, see the denominator of Eq. 13.46;
- Equation 13.46 points out that, given a certain value for x_A, the effective temperature change ($T - T_0$) that the mixture undergoes between the beginning and the end of the process is a fraction (precisely, x_A) of ΔT_{ad}. For example, if $\Delta T_{ad} = 20$ °C (=20 K), and $x_A = 70\%$, the effective temperature change would be 14 °C (=14 K).

We also note the general nature of Eq. 13.46, i.e. it is not only valid for a single adiabatic reactor, but also for more complex schemes. For example:

- consider a reactant mixture (temperature T_0, concentration C_{A0}) sent to a series of two reactors. The first is an "any" reactor, which converts the mixture up to x_{A1} and temperature T_1. The second is an adiabatic reactor, where the mixture is converted up to x_{A2} and temperature T_2. It can be demonstrated, by rewriting the balances, that the adiabatic relation is: $T_2 - T_1 = \Delta T_{ad}(x_{A2} - x_{A1})$, defining ΔT_{ad} again as in Eq. 13.46, i.e. always referring to the concentration C_{A0} *at the inlet to the entire system* (because it is with respect to this concentration that the values of x_A are calculated, as discussed in Chap. 12);
- consider an adiabatic PFR with recycle (Chap. 12). Assuming that also the mixing node (upstream of the PFR) and the recycling node (downstream of the PFR) do not exchange energy with the outside, *the system is globally adiabatic*. Therefore, given any two points of the scheme, the adiabatic relation will hold between them. For example (scheme in Fig. 12.15), if T_0 is the temperature of the stream entering the entire system (at $x_A = 0$), and x_A^e and T^e are the characteristics of the stream exiting the entire system, it will be: $T^e - T_0 = \Delta T_{ad} x_A^e$. Or, if x_{A1} and T_1 are the characteristics of the stream after mixing with the recycle current (i.e., immediately before entering the PFR), it will be: $T_1 - T_0 = \Delta T_{ad} x_{A1}$ (because, here, the mixing node behaves as an "adiabatic reactor"). Again, if x_{A2} $(=x_A^e)$ and T_2 $(=T^e)$ are the characteristics of the stream immediately exiting the PFR (i.e., before the recycle node), it will be: $T_2 - T_1 = \Delta T_{ad}\,(x_{A2} - x_{A1})$, and so on, remembering that ΔT_{ad} is defined as in Eq. 13.46, i.e. with reference to C_{A0}.

Therefore, it can be said that, given an adiabatic "box" in any complex scheme, the temperature difference in the adiabatic "box" $(T_{out} - T_{in})$ is equal to the product between ΔT_{ad} and the variation in degree of conversion achieved in this part of the scheme $(x_{A,out} - x_{A,in})$:

$$\boxed{T_{out} - T_{in} = \Delta T_{ad}\left(x_{A,out} - x_{A,in}\right)} \tag{13.47}$$

Graphically, the adiabatic relation (e.g., Eq. 13.45) can be represented as in Fig. 13.8. Take a canonical thermokinetic map (x_A vs. T). We do not add isokinetic curves, since Eq. 13.45 is universal, i.e. valid for every reaction, whether exothermic or endothermic, reversible or irreversible. Eq. 13.45, on a thermokinetic map, is represented by an *operating line* (similarly to the concepts seen earlier in this textbook, concerning unit operations). This line shows the following features:

- starts at T_0, for $x_A = 0$;
- its slope is equal to $1/\Delta T_{ad}$, positive for exothermic reactions (T increases with x_A) and negative for endothermic reactions (T decreases with x_A);
- as ΔT_{ad} increases in absolute value, the slope, in absolute value, decreases; this means that, for the same x_A, the temperature change undergone by the reacting

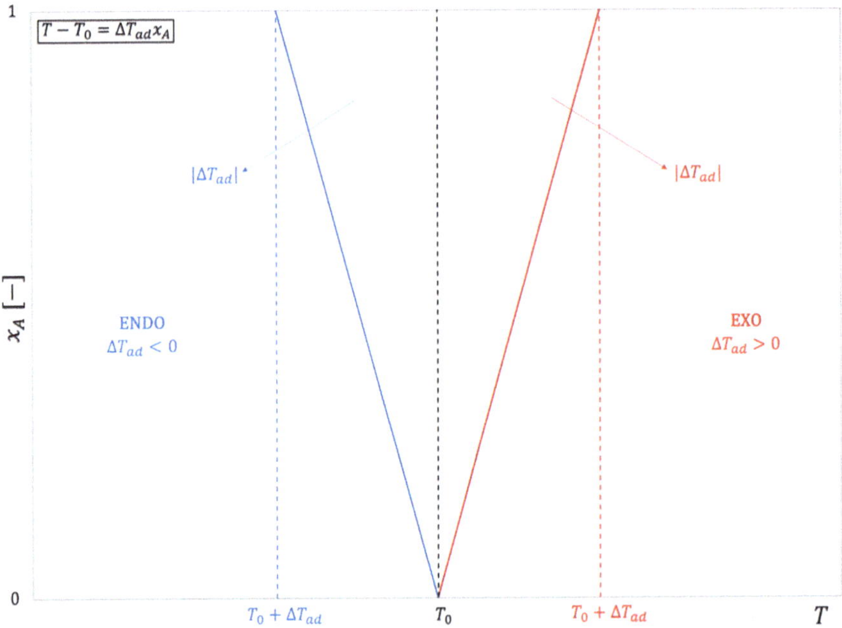

Fig. 13.8 Operating lines under the hypotheses for liquid-phase reactions in adiabatic reactors, drawn on a generic thermokinetic map

mixture is greater if ΔT_{ad} is, in absolute value, greater (i.e. if the enthalpy change of the reaction and/or C_{A0} are greater, and if the density and/or specific heat are lower).

Example: Multiplicity of Stationary States for Exothermic Irreversible Reaction in CSTR

Consider an A⟹P exothermic irreversible reaction, with thermokinetic data in Fig. 13.3. Be the reaction carried out in an adiabatic CSTR with known τ. We assign the following data: $T_0 = 278$ K, $C_{A0} = 2$ mol/L, $C_{P0} = 0$, $\Delta H = -40$ kcal/mol, aqueous phase for which the group $\rho \mathbb{C}$ approximately values (1000 g/L)·(1 cal/(g·K)) = 1000 cal/(L·K), $\tau = 1$ h. Being the kinetics described by:

$$r_A(x_A, T) = k(T)C_A = k(T)C_{A0}(1 - x_A) \tag{13.48}$$

the design equation reads:

(continued)

$$\tau = \frac{C_{A0}x_A^e}{r_A\left(x_A^e, T^e\right)} = \frac{x_A^e}{k(T^e)\left(1 - x_A^e\right)} \Rightarrow x_A^e = \frac{k(T^e)\tau}{1 + k(T^e)\tau} \tag{13.49}$$

Although τ is known, x_A^e is not, as it depends also on T^e. Rewrite Eq. 13.43 as:

$$\rho\mathbb{C}(T^e - T_0) = -\Delta H C_{A0} x_A^e \tag{13.50}$$

namely:

$$\left[\frac{\text{energy}}{\text{volume}}\right] \text{absorbed by the mixture} = \left[\frac{\text{energy}}{\text{volume}}\right] \text{released by the mixture upon reaction} \tag{13.51}$$

We have two equations, Eq. 13.49 and Eq. 13.50, in two variables (x_A^e and T^e), but the system is not linear (see the Arrhenius Eq. 13.2). Therefore, the solution vector is not necessarily unique (in terms of physically acceptable solutions). We may have multiple solutions (in terms of x_A^e and T^e) satisfying the balance Eqs. 13.49 and 13.50, so the CSTR may work under a *multiplicity of stationary states* (not necessarily stable, as we will see). This specific example illustrates the issue, that can be extended in, e.g., Levenspiel (1999); Westerterp et al. (1988). Rewriting Eq. 13.50, with the aid of Eq. 13.49, as:

$$\rho\mathbb{C}(T^e - T_0) = -\Delta H C_{A0} \frac{k(T^e)\tau}{1 + k(T^e)\tau} \tag{13.52}$$

we plot in Fig. 13.9 both the left-hand ("absorption" term) and the right-hand side ("release" term) vs. CSTR operating temperature. The absorption term is obviously linear and increasing with T^e; the release term has a sigmoidal shape, increasing with temperature as the conversion degree x_A^e (for an irreversible process) is favoured by temperature.

Each intersection of the two curves (the two sides of Eq. 13.52) represents a stationary solution for the adiabatic CSTR. We can observe, in this case, the existence of three solutions. Apart from a first one located at approximately T_0 (from Eq. 13.49, x_A^e is seen to be negligible), there is a second one at $T^e = 313$ K ($x_A^e = 0.443$) and a third one at $T^e = 357$ K ($x_A^e = 0.993$).

The intermediate stationary state is marked with a "U", meaning that it is *unstable*. Even minimal (accidental) variations of temperature have the following effect: if temperature increases, the release term is greater than the absorption term, and the temperature is destined to further increase, up to the stationary state "S2"; if temperature decreases, the release term is less than the

(continued)

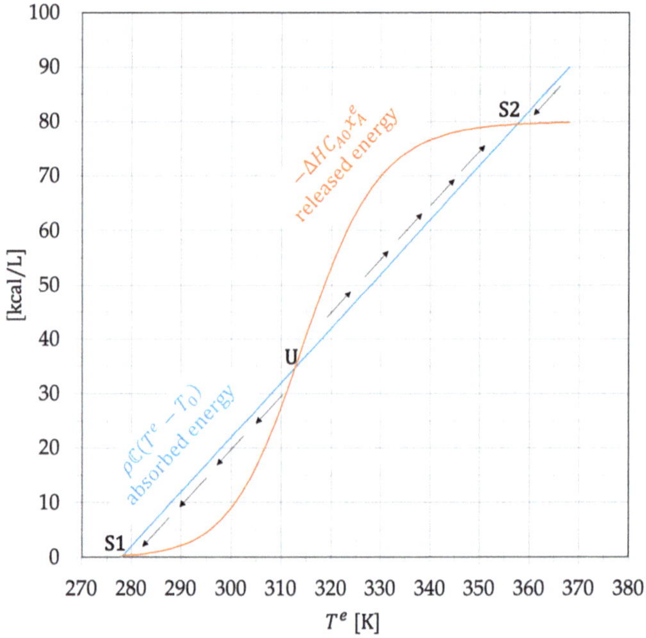

Fig. 13.9 Multiple steady states (U = unstable; S1 = stable at low conversion; S2 = stable at high conversion) for adiabatic CSTR, irreversible exothermic reaction, linear kinetics. The arrows starting from the U state indicate that, for small departures at higher temperature, the system will automatically move towards the S2 state, and for small departures at lower temperature, the system will automatically move towards the S1 state. Instead, starting from the S2 state, any small departure at higher or lower temperature will bring the system back towards the same S2 state (similarly for S1)

absorption term, and the temperature is destined to further decrease, up to the stationary state "S1".

For the stationary state "S2", variations of temperature make it a *stable* operating point. In fact: if temperature increases, the release term is less than the absorption term, and the temperature is destined to decrease again, falling back to "S2"; if temperature decreases, the release term is greater than the absorption term, and the temperature is destined to increase again, falling back to "S2". Although less relevant, the same arguments apply to the stable operating point "S1", given the relative position of the sigmoid vs. straight line. In other words: when, given a steady state, the release term is lower than the absorption term for higher temperatures, and is higher than the absorption term for lower temperatures, the state can be considered stable, and vice versa.

Since, clearly, "S2" is preferable to "S1" (obtaining, by making the CSTR work at "S2", a greater degree of conversion), it will be sufficient to ensure

(continued)

that, in the transient phase, the CSTR works at a temperature slightly higher than that corresponding to the unstable "ignition" point "U": the CSTR will thus automatically go to "S2" (that is, in our example, T^e = 357 K and $x_A^e = 0.993$).

13.5 Optimisation for Adiabatic Reactors

13.5.1 Optimal Temperature Profiles (Operating Lines)

Let us take, as an example, the thermokinetic map of Fig. 13.10 (*irreversible* reaction), and assign an *endothermic* reaction. The following discussion is also valid for a *reversible* reaction, provided it is endothermic (only the thermokinetic map would change, which in this case would show the sigmoid as in Fig. 13.4). Let T_0 be the initial temperature of the mixture, and T_{MAX} the maximum operating temperature. Let the desired value for x_A^e be assigned. Given the value for ΔT_{ad} (<0), there exist infinite operating lines with slope $1/\Delta T_{ad}$. Which is the best, i.e. the one that allows for the shortest possible space-time? Clearly, since the process is

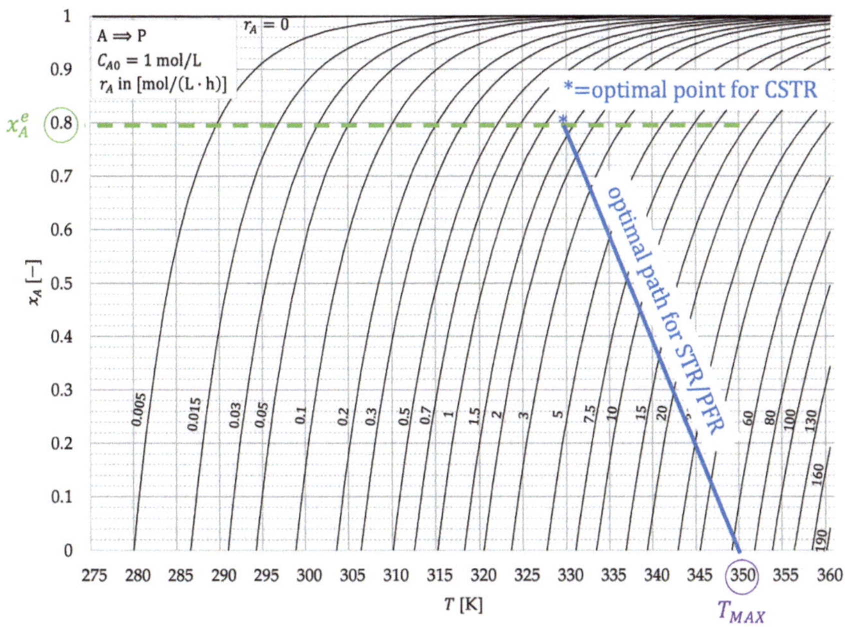

Fig. 13.10 Optimal operating line for irreversible endothermic reaction, adiabatic conditions, once set exit conversion degree and maximum temperature; similar considerations apply for reversible endothermic reaction

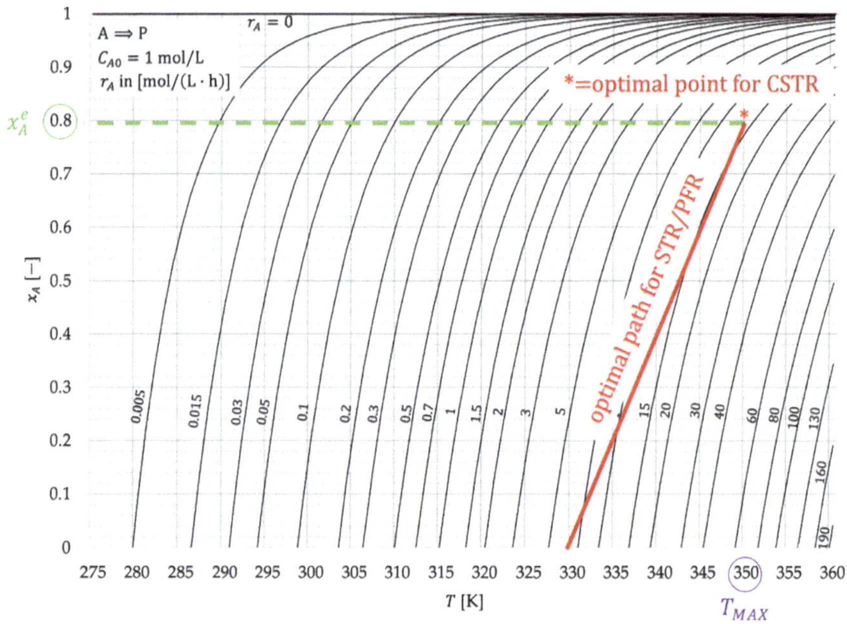

Fig. 13.11 Optimal operating line for irreversible exothermic reaction, adiabatic conditions, once set exit conversion degree and maximum temperature

favoured by high temperatures both on the kinetic side and, if applicable, the thermodynamic side, it will be convenient to pre-heat (with a heat exchanger external to the reactor) the stream from T_0 up to T_{MAX}, and then let it react. In this case, the operating line will be "as right as possible", that is, intercepting, for each x_A, the highest possible values for r_A. In fact, for each x_A, it would not be possible to have higher values for r_A, because this would involve using a line further to the right than the one indicated, and this line would be, in part, further to the right than T_{MAX} (therefore, in a non-operating zone). Working on the indicated line will consequently involve the lowest possible value for τ. Naturally, the temperature profile marked by the operating line is valid in the case of STR/PFR. In the case of CSTR, only the final point of the operating line must be considered.

Let us now consider the thermokinetic map of Fig. 13.11 (*irreversible* reaction), and assign an *exothermic* reaction. Again, given the value for ΔT_{ad} (>0), there exist infinite operating lines with slope $1/\Delta T_{ad}$. To reason on the line that allows to have the shortest possible space-time, we need to consider again that the process is favoured by high temperatures (since there is no thermodynamic limitation). Then, it will be convenient to feed the mixture (by pre-heating or pre-cooling) at a temperature such as, reacting the mixture up to x_A^e, the exit temperature is T_{MAX}. Also in this case, the operating line will be "as right as possible", that is, intercepting, for each x_A, the highest possible values for r_A, and determining the lowest possible value for τ. This is valid for STR/PFR (for CSTR, only the final point of the operating line must be considered).

Example: Optimisation of Adiabatic CSTR, Endothermic Irreversible Reaction

Consider the A⟹P endothermic reaction in Fig. 13.12. We have an adiabatic CSTR, $x_A^e = 0.7$, $T_O = 300$ K, $T_{MAX} = 330$ K, $Q_O = 1$ L/h, $C_{AO} = 1$ mol/L, $C_{PO} = 0$, $\Delta H = 40$ kcal/mol. The reaction is carried out in aqueous phase, so $\rho\mathbb{C} = 1$ kcal/(L·K). Thus, the value for $\Delta T_{ad} = -40$ K is easily calculated by Eq. 13.46. The optimal temperature profile is (from Eq. 13.46):

$$T = T_{MAX} + \Delta T_{ad} x_A \tag{13.53}$$

So, first, we need to place a pre-heater external to the reactor, to bring the mixture (with volumetric flow rate Q_O) from T_O to the inlet temperature, which is T_{MAX}. The scheme is illustrated in Fig. 13.13. While the design of the heat exchanger (with the calculus of the heat exchange area) has been discussed elsewhere in this textbook (Chap. 5), we here just evaluate the thermal power needed to the heat exchanger:

(continued)

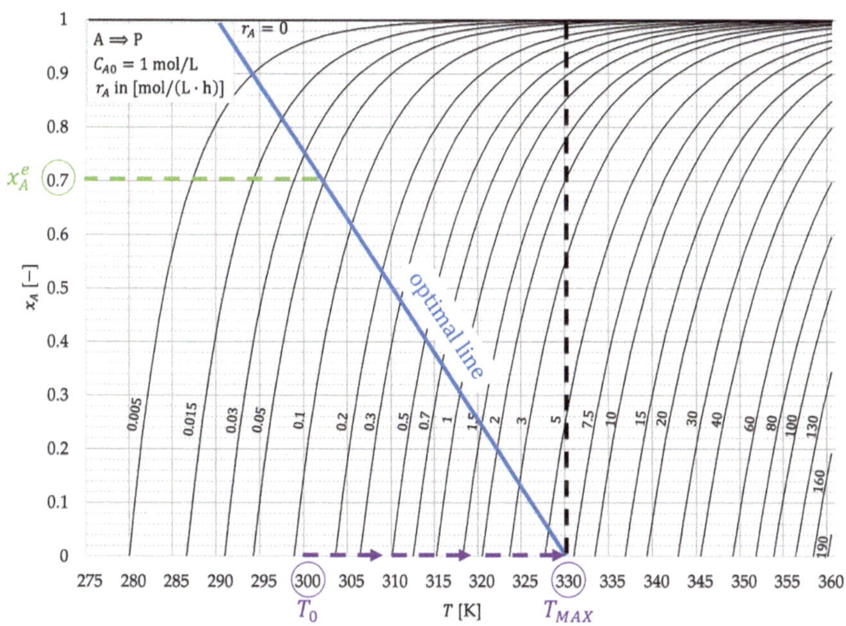

Fig. 13.12 Optimisation for irreversible endothermic reaction in adiabatic CSTR: first pre-heat from T_O to T_{MAX}, then feed to the CSTR

$$q = Q_0 \rho \mathbb{C}(T_{MAX} - T_0) \tag{13.54}$$

It is $q = 30$ kcal/h. In Fig. 13.13, we also see the value for the exit temperature ($T^e = 302$ K). This value can be analytically calculated by Eq. 13.53, or conceptually derived considering that 40 K is the maximum temperature variation (absolute value) that the reacting mixture can undergo (at unit conversion); being the desired conversion equal to 70%, the mixture will cool by 70% of 40 K, i.e. 28 K lower than the inlet temperature (330 K). To draw the operating line on the thermokinetic map in Fig. 13.12, once set the point having coordinates ($0; T_{MAX} = 330$ K), one can choose the other point: ($x_A^e = 0.7$; $T^e = 302$ K), or a virtual one with unit conversion and temperature 330–40 = 290 K, or any other respecting Eq. 13.53.

We can now obtain (see data in Fig. 13.3 and visually check with Fig. 13.12):

$$r_A\left(x_A^e, T^e\right) = k_0 \exp\left(-\frac{E}{\Re T^e}\right) C_{A0}\left(1 - x_A^e\right) = 0.051 \frac{\text{mol}}{\text{L} \cdot \text{h}} \tag{13.55}$$

This is the highest possible value for the reaction rate at exit conversion, that determines the minimum possible value for the CSTR space-time:

$$\tau = \frac{C_{A0} x_A^e}{r_A\left(x_A^e, T^e\right)} = 13.7 \text{ h} \tag{13.56}$$

That is, we design an adiabatic CSTR having optimal volume $V = Q_0 \tau = 13.7$ L.

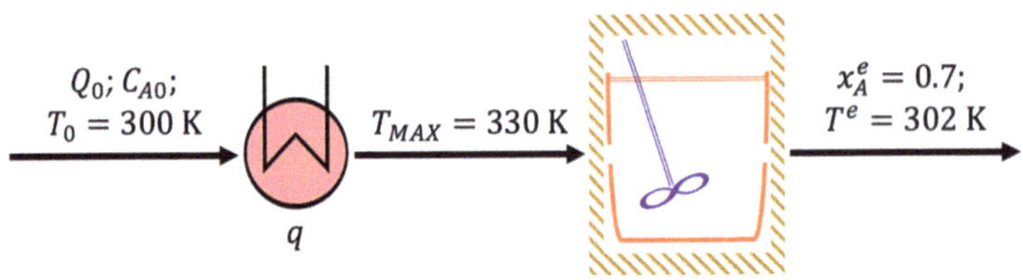

Fig. 13.13 Optimal design, with pre-heater, for irreversible endothermic reaction in adiabatic CSTR

Example: Optimisation of Adiabatic PFR, Exothermic Irreversible Reaction

Consider the A⟹P exothermic reaction in Fig. 13.14. We have an adiabatic PFR, $x_A^e = 0.9$, $T_O = 330$ K, $T_{MAX} = 340$ K, $Q_O = 10$ L/h, $C_{AO} = 1$ mol/L, $C_{PO} = 0$, $\Delta H = -20$ kcal/mol. The reaction is carried out in aqueous phase, therefore $\Delta T_{ad} = 20$ K. The optimal temperature profile is obtained by considering that, at $x_A^e = 0.9$, we will prefer $T^e = T_{MAX} = 340$ K. As the maximum T variation is 20 K (for unit conversion), in this case the T difference between PFR inlet and outlet will be the 90% of this value, i.e. 18 K. Thus, we need an inlet PFR temperature $T_{IN} = 340–18 = 322$ K. By joining the two points, we get our operating line. This can be also formally obtained by rewriting Eq. 13.46 as:

$$T_{MAX} - T_{IN} = \Delta T_{ad} x_A^e \Longrightarrow T_{IN} = T_{MAX} - \Delta T_{ad} x_A^e = 322\ \text{K} \tag{13.57}$$

Since $T_{IN} < T_O$, in this case we need a pre-cooler (Fig. 13.15), with q:

$$q = Q_0 \rho \mathbb{C}(T_0 - T_{IN}) \tag{13.58}$$

It is $q = 80$ kcal/h.

Formally, we should now insert the equation of the operating line:

$$T = T_{IN} + \Delta T_{ad} x_A \tag{13.59}$$

into the Arrhenius kinetic expression, and then solve the PFR integral. We instead choose the graphical procedure:

$$\tau \cong C_{A0} \Delta x_A \sum_0^{x_A^e} \frac{1}{r_A} \tag{13.60}$$

Data are listed in Table 13.2 (as usual, they can be also approximated by visual inspection of Fig. 13.14). By application of the design equation, it is:

$$\tau \cong C_{A0} \Delta x_A \left(\frac{1}{2.792} + \ldots + \frac{1}{2.278} \right) = 0.25\ \text{h} \tag{13.61}$$

That is, we design an adiabatic PFR having optimal volume $V = Q_0 \tau = 2.5$ L.

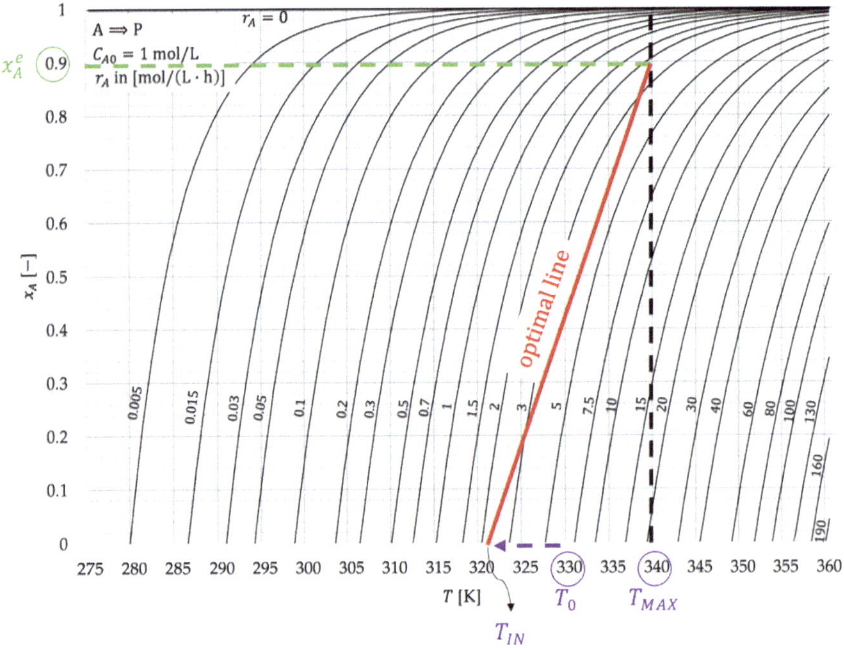

Fig. 13.14 Optimisation for irreversible exothermic reaction in adiabatic PFR: first pre-cool from T_0 to T_{IN}, then feed to the PFR

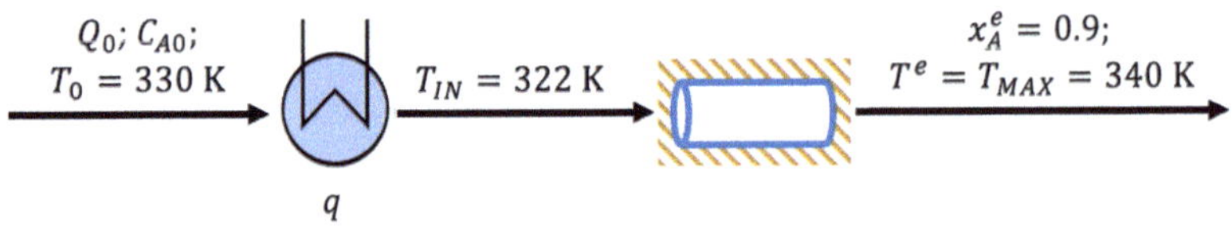

Fig. 13.15 Optimal design, with pre-cooler, for irreversible exothermic reaction in adiabatic PFR

We now consider the thermokinetic map of Fig. 13.16 (*reversible exothermic* reaction), and a CSTR. Given the value of ΔT_{ad} (>0), there exist infinite lines with slope $1/\Delta T_{ad}$. Which is the best, the one that allows to have the shortest possible space-time? Clearly, it will be convenient to exercise the CSTR at the operating point (T^e;x_A^e) that is on the locus of the maximum rates for x_A^e (in this condition, in fact, the only reaction rate will be the maximum possible). Having identified T^e in this way, the value of the CSTR inlet temperature will be found imposing: $T^e - T_{IN} = \Delta T_{ad} x_A^e$, and an external heat exchanger will be generally needed to adapt the inlet temperature to the desired value.

If we, instead, have a PFR (or STR), follow Fig. 13.17. To find the optimal operating line, we should remember that, in this case, it is not only the output value for r_A that is important, but all the reaction rate values between 0 and x_A^e. Draw the line "1", which meets the locus of the maximum rates in x_A^e. Line "1" cannot be

Table 13.2 Data for the approximated solution of the design equation (other details in Fig. 13.3)

x_A [–]	T [K] Eq. 13.59	r_A [mol/(L·h)] Eq. 13.15
0.05	323	2.792
0.10	324	3.001
0.15	325	3.214
0.20	326	3.427
0.25	327	3.638
0.30	328	3.841
0.35	329	4.032
0.40	330	4.204
0.45	331	4.350
0.50	332	4.460
0.55	333	4.525
0.60	334	4.530
0.65	335	4.461
0.70	336	4.301
0.75	337	4.028
0.80	338	3.620
0.85	339	3.047
0.90	340	2.278

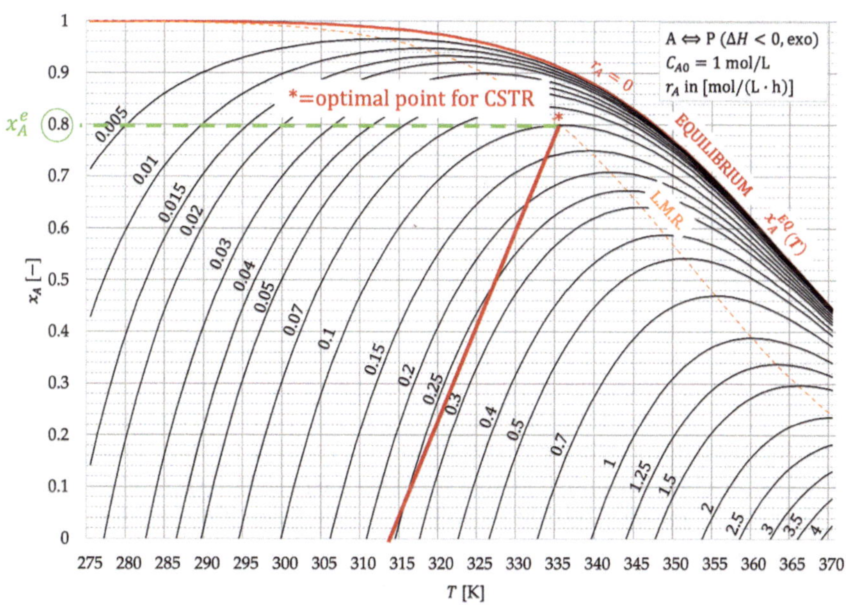

Fig. 13.16 Optimal operating line for reversible exothermic reaction, adiabatic CSTR, once set exit conversion degree

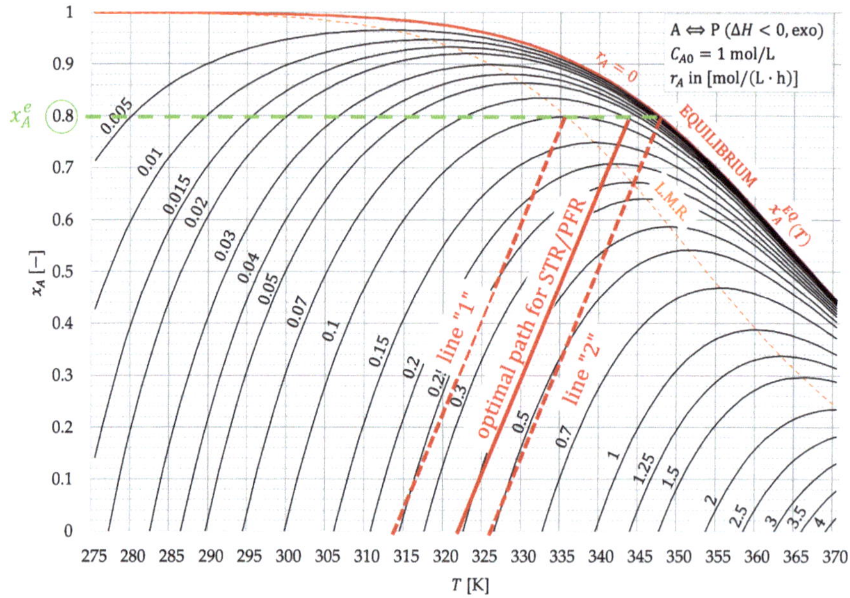

Fig. 13.17 Optimal operating line for reversible exothermic reaction, adiabatic PFR/STR, once set exit conversion degree

considered as optimal, since the best value of r_A is intercepted only at the exit, but the mixture is, on average, too "cold"; therefore, for low values of x_A (far from equilibrium, where higher temperatures would be more convenient), the values for r_A are too low. Operating lines further "to the left" of the line "1" will obviously be even less optimal. Now draw the line "2", which meets the equilibrium sigmoid in x_A^e. It cannot be considered as an optimal operating line, too, since it is true that now the values of r_A are favoured when we are far from equilibrium, but a pinch condition is imposed at the exit (for x_A^e, it would be $r_A = 0$, that is, the reactor is asked to work under equilibrium conditions, and this would be achieved with $\tau \longrightarrow \infty$). Lines further "to the right" of the line "2" will obviously be unfeasible. In doing so, we have drawn an *operating parallelogram*: the optimal line, that can be found by trial-and-error starting from various values of T_{IN} (a possible solution is represented in Fig. 13.17), will be inside the parallelogram, because we have to combine the need to have high temperatures far from equilibrium, and not to be too close to the sigmoid when we are close to equilibrium (not very high temperatures when we are close to equilibrium, since the reversible reaction is exothermic). The fact that this best operating line does not end on the locus (for x_A^e) should not worry: in fact, it will be "astride" the locus, which will be encountered at an intermediate point (time) of the PFR (STR), which is certainly no less important than the exit (final) condition. Again, having individuated the operating line, pre-heaters will help in adjusting the inlet temperature.

13.5.2 Autothermal Reactors

If we are dealing with an *adiabatic* reactor and an *exothermic* reaction, and if, to operate on the chosen line, it is necessary to *pre-heat* the feed stream (from T_0 to T_{IN}), an elegant solution is represented by the *autothermal reactor* (continuous), Fig. 13.18 (where the thermokinetic map is not specified, since the discussion is valid for both irreversible and reversible, exothermic, processes). The stream will exit the adiabatic reactor at a temperature:

$$T^e = T_{IN} + \Delta T_{ad} x_A^e \tag{13.62}$$

once x_A^e has been set and the operating line has been chosen, i.e. the one that goes from T_{IN} (at zero conversion) to T^e (at exit conversion), for which it was necessary to pre-heat the stream from T_0 to T_{IN} using an exchanger external to the reactor (the discussion that is being carried out is valid, in fact, in this case). In the heat exchanger, a hot service fluid will be needed. *This role can be assumed by the stream exiting the reactor.* The thermal power to be transferred to the cold fluid at the inlet is, under our hypotheses (with T_2 the temperature of the reacted stream downstream the heat exchanger):

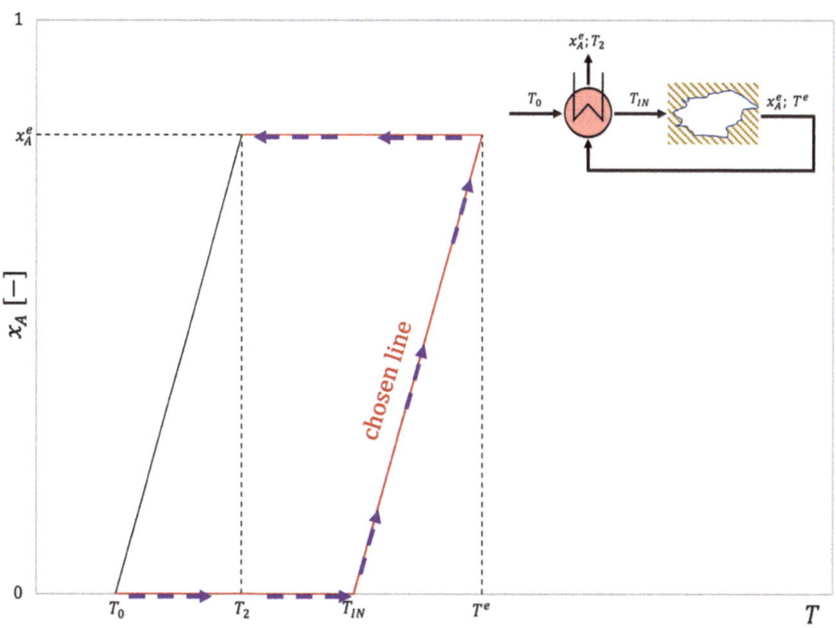

Fig. 13.18 Autothermal continuous adiabatic reactor (exothermic reaction, pre-heat of the feed, once set exit conversion degree): the arrows indicate the path of the fluid

$$q = Q_0 \rho \mathbb{C}(T_{IN} - T_0) = Q_0 \rho \mathbb{C}(T^e - T_2) \tag{13.63}$$

Namely, the thermal power gained by the inlet fluid is the same as that lost by the reacted mixture, that sacrifices its enthalpy content. It follows that the two temperature variations are equal:

$$T_{IN} - T_0 = T^e - T_2 \tag{13.64}$$

Remembering Eq. 13.62:

$$T^e - T_{IN} = T_2 - T_0 = \Delta T_{ad} x_A^e \tag{13.65}$$

Between T_2 and T_0 there is the *same adiabatic relation* that exists between T^e and T_{IN}. This is reasonable, because the entire exchanger+reactor system (and not only the reactor) can be considered *globally adiabatic*. Graphically, what has been said translates into a fluid that, starting from T_0, reaches at the end of its life T_2 (x_A^e conversion) not along the operating line from T_0 to T_2, but rather along the one (chosen, maybe optimal) from T_{IN} to T^e, and then returning to T_2 (at x_A^e conversion). This means to follow the sides of a parallelogram (the "left" side from T_0 to T_2, and the "right" side from T_{IN} to T^e, sides obviously parallel each other, given Eq. 13.65).

Example: Optimisation of Adiabatic CSTR, Exothermic Reversible Reaction

Consider an A⇔P exothermic reaction, with illustrations in Figs. 13.19 and 13.20. The analytical steps can be followed on the thermokinetic map. We have an adiabatic CSTR with $V = 1\ m^3$, $x_A^e = 0.8$, $T_0 = 290$ K, $C_{A0} = 1$ mol/L, $C_{P0} = 0$, $\Delta H = -30$ kcal/mol. It is highlighted that, here and sometimes in the following, the ΔH value that we will assign changes, to illustrate different problems, and therefore it does not coincide with the actual value for which the thermokinetic map has been built. For the sake of the clarity, the reader will forgive this approximation. Back to our problem, in aqueous phase it is $\Delta T_{ad} = 30$ K. The optimal value for T^e at 80% conversion is obtained by Eq. 13.27 (data in Fig. 13.5), and it is $T^e = 335$ K. Therefore, as the mixture must enter 24 K colder:

$$T_{IN} = T^e - \Delta T_{ad} x_A^e = 311\ \text{K} \tag{13.66}$$

This is a case in which we can use the autothermal solution. The pre-heater will have thermal power:

$$q = Q_0 \rho \mathbb{C}(T_{IN} - T_0) \tag{13.67}$$

The reaction rate will be:

(continued)

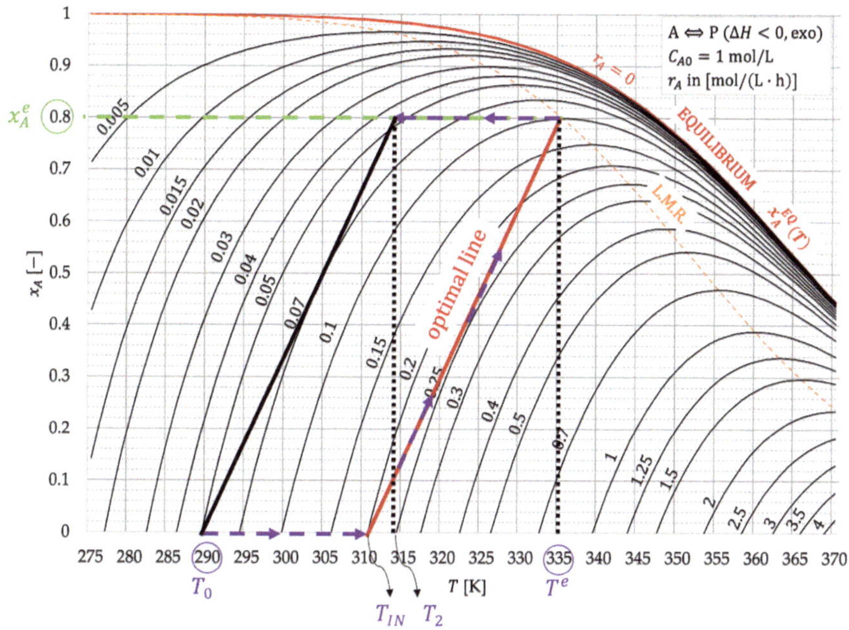

Fig. 13.19 Optimisation for reversible exothermic reaction in adiabatic autothermal CSTR: the arrows indicate the path of the fluid

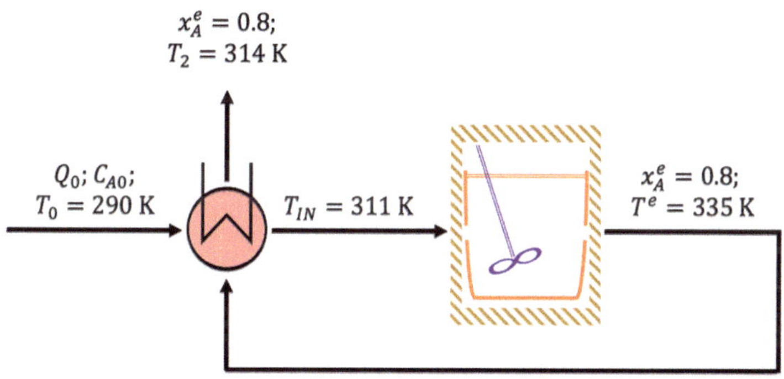

Fig. 13.20 Optimal autothermal design, for reversible exothermic reaction in adiabatic CSTR

$$r_A\left(x_A^e, T^e\right) = k_{0d}\exp\left(-\frac{E_d}{\Re T^e}\right)C_{A0}\left(1 - x_A^e\right) - k_{0i}\exp\left(-\frac{E_i}{\Re T^e}\right)C_{A0}x_A = 0.098\ \frac{\text{mol}}{\text{L}\cdot\text{h}} \tag{13.68}$$

This is the highest possible value for the reaction rate at exit conversion, that determines the minimum possible value for the CSTR space-time:

(continued)

$$\tau = \frac{C_{A0}x_A^e}{r_A\left(x_A^e, T^e\right)} = 8.16\ \text{h} \tag{13.69}$$

That is, we design an adiabatic CSTR with maximum $Q_0 = 122.5$ L and maximum productivity $P_P = Q_0 C_{A0} x_A^e = 98$ mol/h. This is possible provided that a maximum thermal heat is assigned to the external heat exchanger (Eq. 13.67, $q = 2572$ kcal/h; of course, more refined economic analyses should also consider this aspect, as it affects the exchange area). In the autothermal configuration, the stream will exit at $T_2 = 314$ K. This can be argued in many ways: one could observe that the inlet stream must be heated by 21 K (and, then, the outlet stream must be cooled from 335 K by 21 K as well), or observe that the adiabatic relation between T_0 and T_2 imposes a temperature jump of 24 K (Eq. 13.65), or obtain it graphically.

Example: Optimisation of Adiabatic PFR, Exothermic Reversible Reaction

Same case as above, illustrated now in Figs. 13.21, 13.22 and 13.23. We have an adiabatic PFR with $V = 1$ L, $x_A^e = 0.8$, $T_0 = 300$ K, $C_{A0} = 1$ mol/L, $C_{P0} = 0$, $\Delta H = -40$ kcal/mol. In aqueous phase, it is $\Delta T_{ad} = 40$ K. Following the instructions in Fig. 13.17, line "1" ends at 335 K (see example above). Line "2", instead, ends at the equilibrium temperature for x_A^e. Use Eqs. 13.18, 13.19 and 13.21:

$$x_A^{EQ} = \frac{k_d(T)}{k_d(T) + k_i(T)} = \frac{1}{1 + \frac{k_i(T)}{k_d(T)}} = \frac{1}{1 + \frac{k_{0i}}{k_{0d}} \exp\left(\frac{E_d - E_i}{\mathfrak{R}T}\right)} \tag{13.70}$$

Satisfied for equilibrium temperature:

$$T = \frac{E_d - E_i}{\mathfrak{R} \ln\left[\frac{k_{0d}}{k_{0i}} \frac{\left(1 - x_A^{EQ}\right)}{x_A^{EQ}}\right]} = 347\ \text{K} \tag{13.71}$$

The optimal line will end, for x_A^e, at a temperature between 335 and 347 K. By trial-and-error (we have tried some values for T^e, but more refined solutions are possible), we see that the optimal operating line (i.e., that giving the minimum value for the space-time, i.e. the maximum average reaction rate) is the one with exit temperature at 344 K. As the mixture must enter 32 K colder, it is:

$$T_{IN} = T^e - \Delta T_{ad} x_A^e = 312\ \text{K} \tag{13.72}$$

Choose the graphical approach:

(continued)

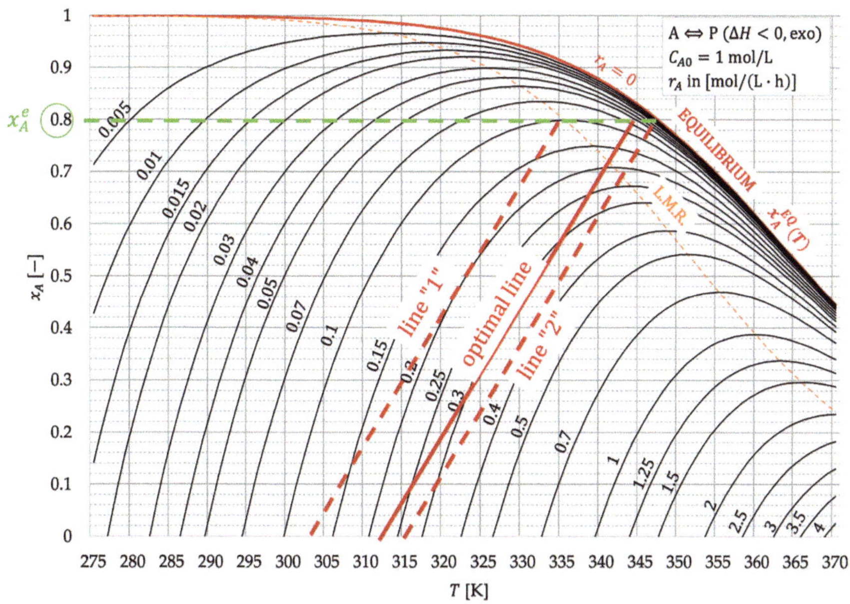

Fig. 13.21 Individuation of the operating parallelogram for reversible exothermic reaction in adiabatic PFR

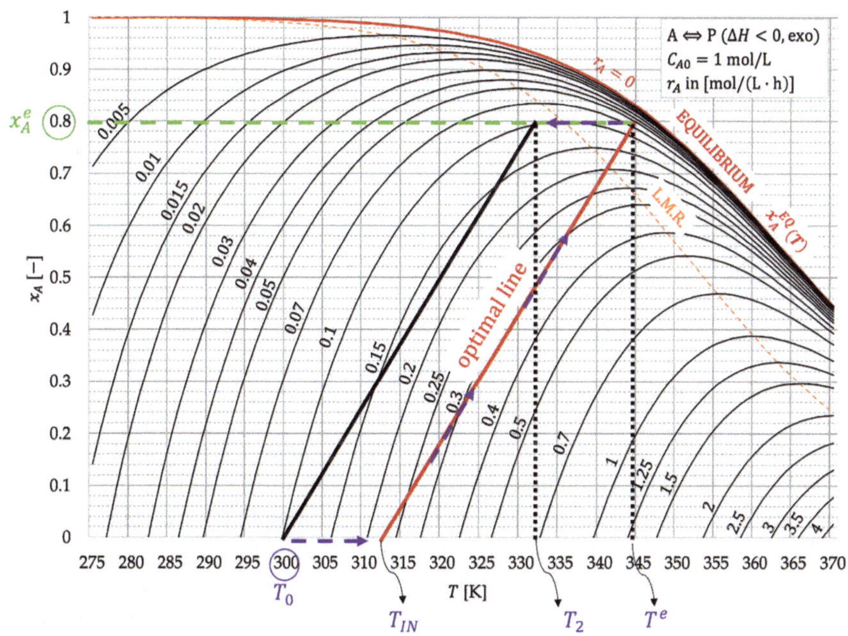

Fig. 13.22 Optimisation for reversible exothermic reaction in adiabatic autothermal PFR: arrows indicate the path of the fluid

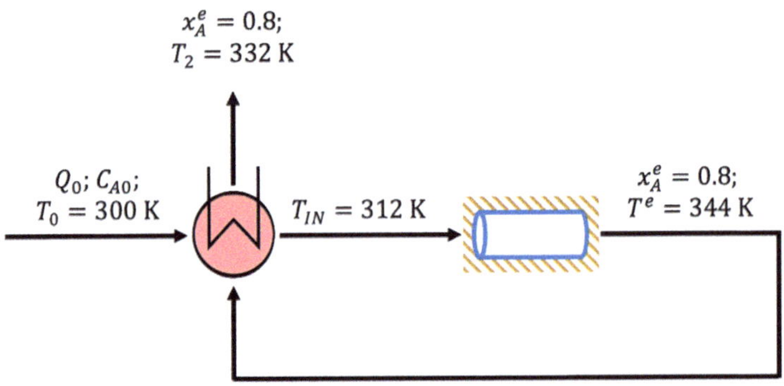

Fig. 13.23 Optimal autothermal design, for reversible exothermic reaction in adiabatic PFR

$$T = T_{IN} + \Delta T_{ad} x_A \tag{13.73}$$

$$\tau \cong C_{A0} \Delta x_A \sum_0^{x_A^e} \frac{1}{r_A} \tag{13.74}$$

Data are listed in Table 13.3. By application of the design equation, it is:

$$\tau \cong C_{A0} \Delta x_A \left(\frac{1}{0.236} + \ldots + \frac{1}{0.055} \right) = 3.78 \text{ h} \tag{13.75}$$

That is, we design an adiabatic PFR with maximum $Q_O = 0.26$ L and maximum productivity $P_P = Q_O C_{AO} x_A^e = 0.208$ mol/h.

Now go for an autothermal configuration. The pre-heater will have thermal power:

$$q = Q_O \rho \mathbb{C} (T_{IN} - T_O) \tag{13.76}$$

i.e., $q = 3.12$ kcal/h. In the autothermal configuration, the stream will exit at $T_2 = 332$ K, following the usual considerations.

13.5.3 Reactor Choice

Take the thermokinetic map of Fig. 13.24 (*irreversible* reaction), and assign an *endothermic* reaction. The following discussion is also valid for a *reversible* reaction, provided it is *endothermic*. The process is *adiabatic*. Given the value of x_A^e, let the line in the figure be the operating one (it may or may not be the optimal one; in this discussion, it is immaterial). The following question arises: what is the best reactor configuration, once the operating line has been defined? It is noted that, as x_A

Table 13.3 Data for the approximated solution of the design equation

x_A [–]	T [K] Eq. 13.73	r_A [mol/(L·h)] Eqs. 13.17–13.19
0.05	314	0.236
0.10	316	0.251
0.15	318	0.266
0.20	320	0.280
0.25	322	0.294
0.30	324	0.305
0.35	326	0.314
0.40	328	0.320
0.45	330	0.321
0.50	332	0.317
0.55	334	0.306
0.60	336	0.285
0.65	338	0.254
0.70	340	0.207
0.75	342	0.142
0.80	344	0.055

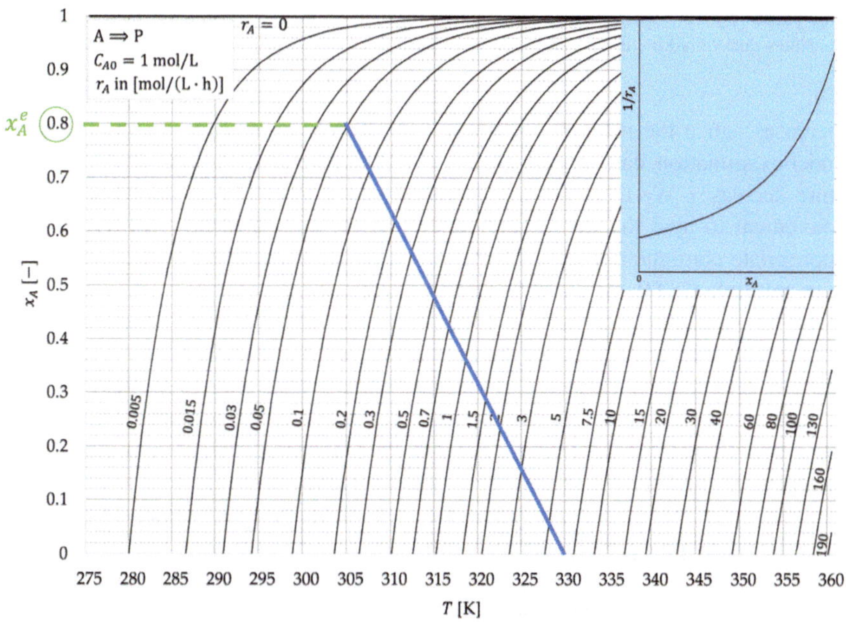

Fig. 13.24 Generic operating line for adiabatic process, endothermic irreversible reaction; the box translates data on a kinetic plot (canonical trend)

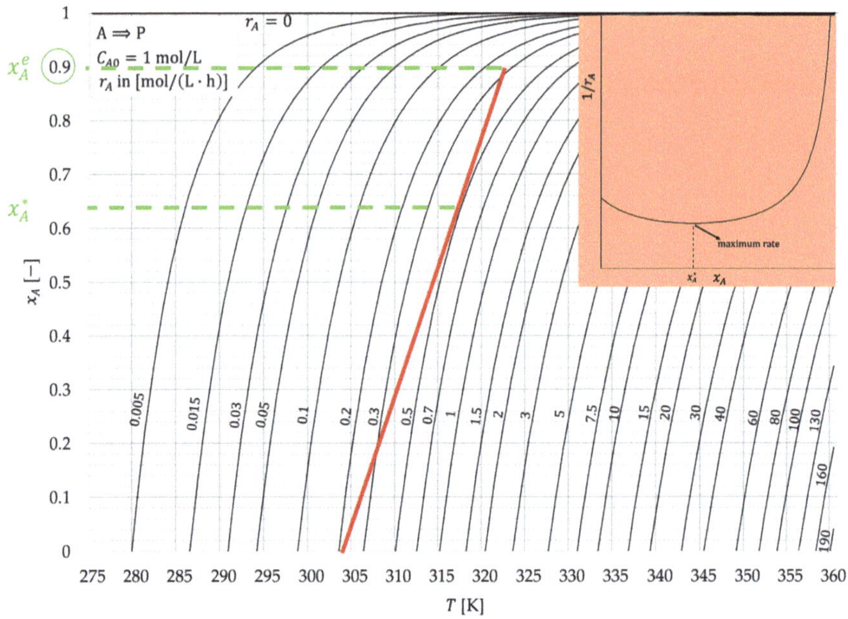

Fig. 13.25 Generic operating line for adiabatic process, exothermic irreversible reaction; the box translates data on a kinetic plot (autocatalytic trend)

increases, since the reaction is endothermic, T decreases. This, for the processes under examination, can only penalise r_A (for both reactant consumption and temperature decrease). We then fall in a canonical case for which, if possible, it will be convenient to tend towards a configuration with segregated flow (PFR), or other appropriate configurations will be chosen based on availability: for example, if we have multiple CSTR, they will be arranged in series.

If we have the case in Fig. 13.25 (*irreversible exothermic* reaction), as x_A increases, T increases. This, for the irreversible processes under examination, obviously accelerates r_A. However, it must be remembered that, as x_A increases, the concentration of A decreases and therefore r_A slows down. The opposition between these two effects can determine a *point of tangency* (of ordinate x_A^* in Fig. 13.25) between the operating line and an isokinetic curve. This, in turn, resembles an "autocatalytic" behaviour (*the autocatalyst, here, is the temperature itself*, whose increase is self-produced by the reaction):

- for low values of x_A, the positive effect of the increase of T on r_A prevails, as x_A increases: in fact, the mixture is still rich in A (it does not suffer much the consumption of A, while its temperature is still low);
- for high values of x_A, the negative effect of the consumption of A on r_A prevails, as x_A increases: in fact, the mixture is already sufficiently hot (while A is scarce);

- the value of x_A^* has the same meaning as in autocatalytic reactions, see, e.g., Fig. 11.4; at this value, corresponding here to the point of tangency, the reaction rate is maximum;
- the reactor choice will follow the usual clichés for autocatalytic plots (CSTR vs. PFR, or CSTR+PFR series, or PFR with recycle) depending on the availability and the value of x_A^e.

Similar is the situation (Fig. 13.26) for *reversible exothermic* reaction. As x_A increases, T increases. This accelerates r_A if far from equilibrium (positive kinetic effects prevail) and vice versa (negative thermodynamic effects prevail). Furthermore, as x_A increases, the concentration of A decreases and therefore r_A slows down. Again, we can have a tangency condition and an autocatalytic-type behaviour (with similar considerations on the reactor configuration):

- for low values of x_A, the positive effect of the increase of T on r_A prevails, as x_A increases: in fact, the mixture is still rich in A, not too hot, and we are far from equilibrium;
- for high values of x_A, the two negative effects (consumption of A and increase of T) on r_A prevail, as x_A increases: in fact, the mixture is poor in A, already sufficiently hot and we are close to equilibrium.

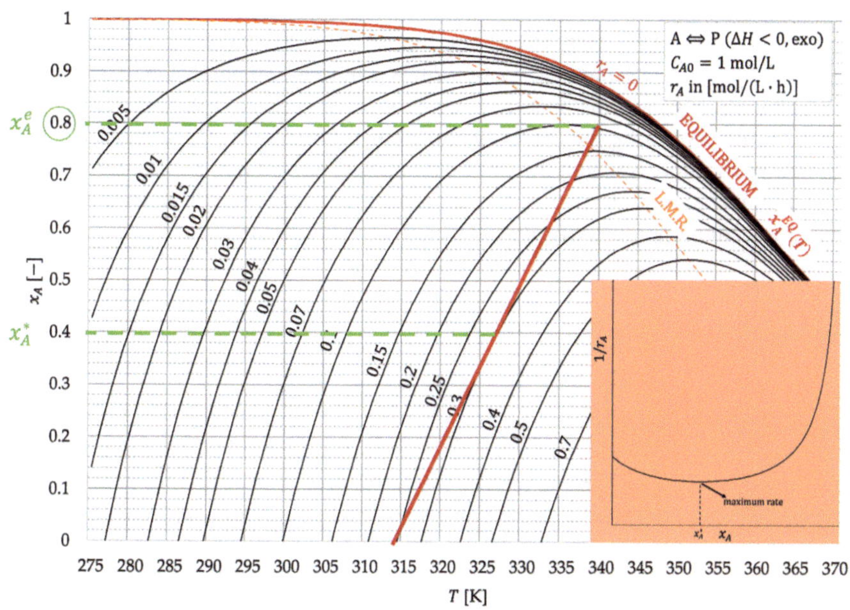

Fig. 13.26 Generic operating line for adiabatic process, exothermic reversible reaction; the box translates data on a kinetic plot (autocatalytic trend)

Example: Optimal Adiabatic Reactor Choice, Exothermic Irreversible Reaction

A⟹P exothermic reaction, Fig. 13.27, thermokinetic data in Fig. 13.3. Be $x_A^e = 0.6$, $T_O = 300$ K (without possibility of external heat exchangers, so $T_{IN} = T_O$), $C_{AO} = 1$ mol/L, $C_{PO} = 0$, $\Delta H = -24$ kcal/mol. Be the reaction conducted in a solvent with $\rho\mathbb{C} = 0.8$ kcal/(L·K). That is, $\Delta T_{ad} = 30$ K. In absence of heat exchangers, our operating line is fixed. We can draw it starting from the point (300;0) and ending, for example, at 330 K for unit conversion. For our conversion degree (60%), the exit temperature T^e will be 18 K higher than T_O. We have, available, two CSTR ($V_1 = 2$ L and $V_2 = 6$ L). First, analyse data as per Table 13.4. The combined provisions given by inspection of Table 13.4 and Fig. 13.27 make us observe that we are in a conversion range where the positive effect (related to increase in T) prevails, so we are well before the point of tangency (in this case, x_A^* is about 75%). In such a situation (Fig. 11.6, $x_A^e < x_A^*$), CSTR is better. Then, we will arrange the two CSTR in optimised parallel, so to make this configuration be equivalent to one CSTR, with a volume equal to the sum of the volumes (8 L), which processes the entire volumetric flow rate Q_O up to conversion degree x_A^e.

So, for the equivalent virtual (adiabatic) CSTR:

$$\tau = \frac{V_1 + V_2}{Q_0} = \frac{C_{A0} x_A^e}{r_A\left(x_A^e, T^e\right)} = 0.97\ \text{h} \Rightarrow Q_0 = 8.25\ \text{L/h} \qquad (13.77)$$

In reality, we need to design a system (Fig. 13.28) where a Q_O-fraction equal to $V_1/(V_1 + V_2) = 25\%$, i.e. $Q_1 = 2.06$ L/h, is sent to the adiabatic CSTR having volume V_1, and the remainder $Q_O - Q_1 = 6.19$ L/h is sent to the other adiabatic CSTR. In this way, as the reader already knows, the two CSTR will work at the same space-time, reaching the same conversion ($x_{A1} = x_{A2} = x_A^e$) and the same exit temperature (T^e). That is, we are designing a system with maximum Q_O and maximum productivity $P_P = Q_O C_{AO} x_A^e = 4.95$ mol/h.

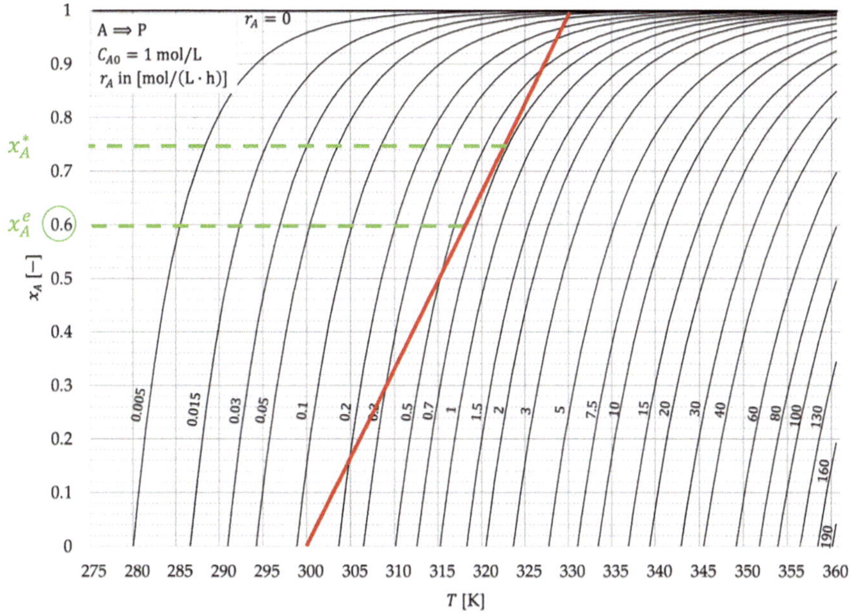

Fig. 13.27 Operating line for the system under investigation (irreversible exothermic reaction in adiabatic system)

Table 13.4 Data along the operating line

x_A [–]	T [K] Eq. 13.73	r_A [mol/(L·h)] Eq. 13.15
0.05	301.5	0.150
0.10	303	0.177
0.15	304.5	0.207
0.20	306	0.242
0.25	307.5	0.280
0.30	309	0.322
0.35	310.5	0.367
0.40	312	0.416
0.45	313.5	0.467
0.50	315	0.519
0.55	316.5	0.570
0.60	318	0.617

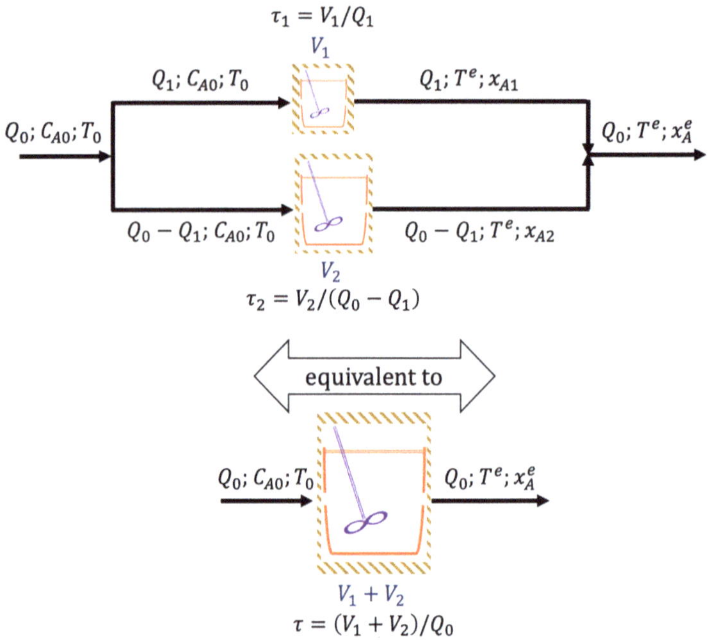

Fig. 13.28 Design for two adiabatic CSTR arranged in parallel

Example: Optimal Adiabatic Reactor Choice, Exothermic Reversible Reaction

A⇔P exothermic reaction, Fig. 13.29, thermokinetic data in Fig. 13.5. Be $x_A^e = 0.8$, $T_O = 310$ K (without possibility of external heat exchangers, so $T_{IN} = T_O$), $C_{AO} = 2$ mol/L, $C_{PO} = 0$, $\Delta H = -20$ kcal/mol, aqueous solution. Then, $\Delta T_{ad} = 40$ K. This is the first time we encounter a value for C_{AO} different from the reference one for the thermokinetic map we are using. But, once assigned the kinetic law (Eq. 13.17), it is easy to verify the dependency of r_A on C_{AO} (linear in this case). Therefore, the r_A-values read on the map must be multiplied by a factor of 2, that is the ratio between the real C_{AO} value and the reference one. In absence of heat exchangers, our operating line is fixed. We can draw it starting from the point (310;0) and ending, for example, at 350 K for unit conversion. For our conversion degree (80%), the exit temperature T^e will be 32 K higher than T_O. First, analyse data as per Table 13.5. We have a typical autocatalytic behaviour, with x_A^* of about 45% and $x_A^e > x_A^*$. We know (cf. Fig. 12.19) that there is no better option than a series CSTR+PFR, with CSTR working up to x_A^*, and then PFR. Our solution is illustrated in Fig. 13.30.

(continued)

Thus, we have first a CSTR working at 45% conversion and temperature of 328 K (i.e., that corresponding to the point of tangency, the maximum reaction rate):

$$\tau_{CSTR} = \frac{C_{A0}x_A^*}{r_A\left(x_A^*, T\left(x_A^*\right)\right)} = 1.55\ \text{h} \tag{13.78}$$

and then a PFR, working from x_A^* to x_A^e:

$$\tau_{PFR} = C_{A0}\int_{x_A^*}^{x_A^e} \frac{1}{r_A(x_A, T)}\,\mathrm{d}x_A \cong C_{A0}\Delta x_A \sum_{x_A^*}^{x_A^e} \frac{1}{r_A}$$
$$\cong C_{A0}\Delta x_A\left(\frac{1}{0.577} + \ldots + \frac{1}{0.150}\right) = 2.00\ \text{h} \tag{13.79}$$

The value for $\tau_{TOT} = \tau_{CSTR} + \tau_{PFR} = 3.55$ h is the minimum possible.

If, instead, we do not have the possibility of disposing two reactors in series, we could opt for a PFR with optimised recycle (Fig. 13.31). The optimisation condition (Eq. 12.117) is rewritten as:

$$\left(x_A^e - x_{A1}\right)\frac{1}{r_A(x_{A1})} \cong \Delta x_A \sum_{x_{A1}}^{x_A^e} \frac{1}{r_A} \tag{13.80}$$

By trial-and-error, with data in Table 13.5, the optimal x_{A1} value ($x_{A1} < x_A^*$) is around 5% (more refined solutions are possible). So, the optimal recycle ratio is (Eq. 12.116):

$$R = \frac{x_{A1}}{x_A^e - x_{A1}} = 0.067 \tag{13.81}$$

and the minimum space-time is:

$$\tau = (R+1)C_{A0}\int_{x_{A1}}^{x_A^e} \frac{1}{r_A(x_A, T)}\,\mathrm{d}x_A \cong (R+1)C_{A0}\Delta x_A \sum_{x_{A1}}^{x_A^e} \frac{1}{r_A}$$
$$\cong (R+1)C_{A0}\Delta x_A\left(\frac{1}{0.446} + \ldots + \frac{1}{0.150}\right) = 3.77\ \text{h} \tag{13.82}$$

a good option, only 6% higher than τ_{TOT} in the previous configuration. Finally, note that, as already discussed, if we have complex configurations (CSTR +PFR, PFR with recycle, …), the operating line remains the same for the whole scheme, once set inlet temperature and line slope.

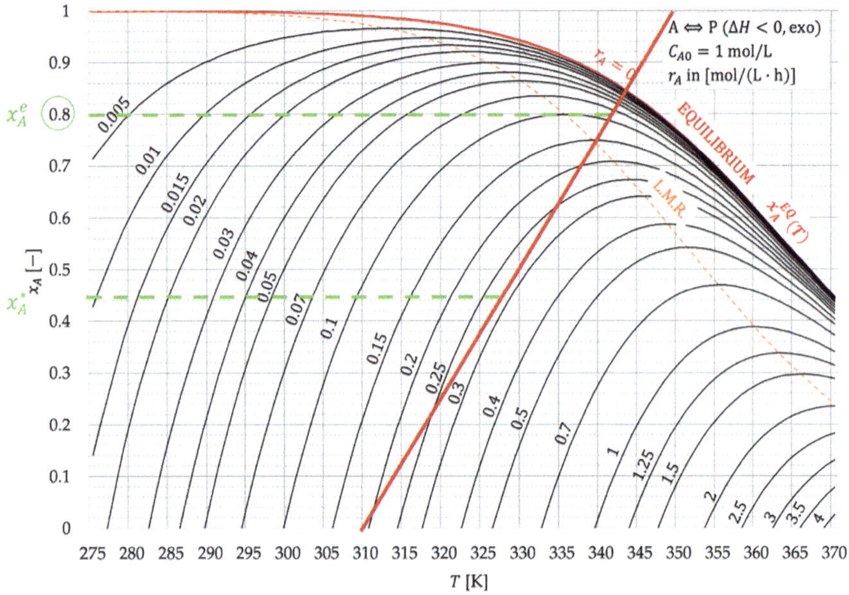

Fig. 13.29 Operating line for the system under investigation (reversible exothermic reaction in adiabatic system)

Table 13.5 Data along the operating line

x_A [−]	T [K] Eq. 13.73	r_A [mol/(L·h)] Eqs. 13.17–13.19
0.05	312	0.418
0.10	314	0.446
0.15	316	0.474
0.20	318	0.500
0.25	320	0.525
0.30	322	0.546
0.35	324	0.564
0.40	326	0.576
0.45	328	0.581
0.50	330	0.577
0.55	332	0.560
0.60	334	0.528
0.65	336	0.477
0.70	338	0.402
0.75	340	0.295
0.80	342	0.150

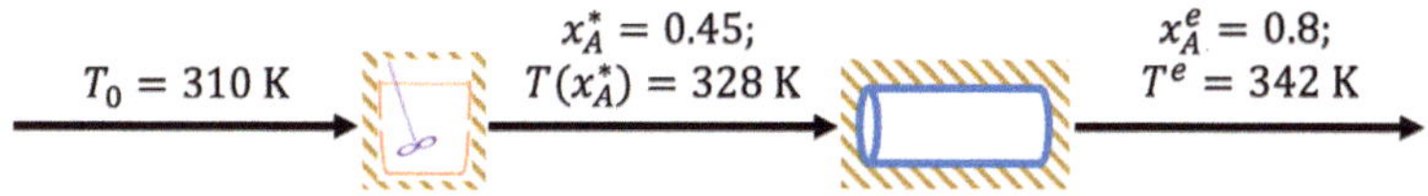

Fig. 13.30 Design for a series of adiabatic CSTR followed by adiabatic PFR in series

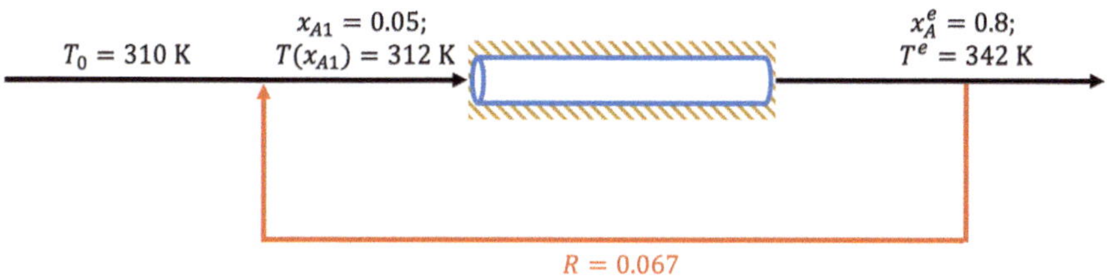

Fig. 13.31 Design for a PFR (adiabatic) with recycle

13.6 Background Information

- The design equations must be written considering the dependence of the reaction rate on both the degree of conversion and temperature, Eq. 13.1.
- For reactions governed by thermodynamic equilibrium, Fig. 13.2 shows how the equilibrium degree of conversion varies with temperature, for endothermic or exothermic reactions.
- The thermokinetic maps illustrate the mutual dependence between reaction rate, degree of conversion and temperature (Fig. 13.3 for irreversible reactions, Fig. 13.4 for reversible endothermic reactions, Fig. 13.5 for reversible exothermic reactions).
- Being able to choose any temperature profile, it is advisable to work at the highest possible temperatures for irreversible reactions and for reversible endothermic reactions (Fig. 13.6). While, for reversible exothermic reactions, it is advisable to adhere as much as possible to the path indicated by the locus of maximum rates (Fig. 13.7).
- For adiabatic reactors, the combined balance of energy and mass returns a universal equation (Eq. 13.46), which linearly links temperature and degree of conversion. This relationship can be expressed, on a thermokinetic map, by a linear trend (operating line) as in Fig. 13.8. Figures 13.10, 13.11, 13.16 and 13.17 show the optimal operating lines depending on the various cases.
- Given a temperature profile, the analysis of the thermokinetic map allows to translate the data on a kinetic graph that reports the reciprocal of reaction rate vs. degree of conversion, from which it is possible to derive the best reactor configuration (Figs. 13.24, 13.25 and 13.26).

References

Arrhenius S (1889) Über die Dissociationswärme und den Einfluss der Temperatur auf den Dissociationsgrad der Elektrolyte. Z Phys Chem 4:96–116. https://doi.org/10.1515/zpch-1889-0408

Balsamo M, Montagnaro F (2022) Fractal-like random pore model applied to CO_2 capture by CaO sorbent. Chem Eng Sci 254:117649. https://doi.org/10.1016/j.ces.2022.117649

Kopelman R (1988) Fractal reaction kinetics. Science 241:1620–1626. https://doi.org/10.1126/science.241.4873.1620

Levenspiel O (1999) Chemical reaction engineering. John Wiley & Sons, New York

Meijer EW (2001) Jacobus Henricus van 't Hoff; hundred years of impact on stereochemistry in The Netherlands. Angew Chem Int Ed 40:3783–3789. https://doi.org/10.1002/1521-3773(20011015)40:20<3783::AID-ANIE3783>3.0.CO;2-J

Westerterp KR, van Swaaij WPM, Beenackers AACM (1988) Chemical reactor design and operation. John Wiley & Sons, Hoboken

Extensions to Ideal Reactors: Reaction Networks

14

Abstract

So far, we have referred to single reactions. This chapter deals with the analysis in presence of multiple reactions (reaction networks). The chosen scenario is to provide, in a streamlined way, a general picture (in which the series and parallel schemes are presented, and the fundamental quantities such as yield and selectivity are defined) and a general approach (in which the methods of writing the material balances are illustrated, as a function of the reactor). All other considerations are left to the development of case studies that, although without the claim of generality, present the key characteristics of reaction networks. The chapter ends with considerations on the role of temperature.

Nomenclature

C_i	Concentration of i species [moles/volume]
C_{i0}	Concentration of i species at reaction beginning [moles/volume]
E	Activation energy [energy/moles]
I_F	Integrating factor [–]
k	Kinetic parameter [(1/time)·(moles/volume)$^{1-n}$]
k_0	Arrhenius' pre-exponential [(1/time)·(moles/volume)$^{1-n}$]
k_i	Kinetic constant for reaction i [(1/time)·(moles/volume)$^{1-n}$]
n	Total reaction order [–]
n_i	Moles of i species or reaction order for reaction i [moles] or [–]
n_{i0}	Moles of i species at reaction beginning [moles]
P_P	Productivity of the desired species [moles/time]
Q	Volumetric flow rate [volume/time]
Q_0	Volumetric flow rate at reaction beginning [volume/time]
$\mathfrak{R}$	Gas constant [energy/(moles·temperature)]
r_A	Rate of A disappearance [moles/(volume·time)]

F. Montagnaro et al., *Elements of Unit Operations and Chemical Reactors*,
https://doi.org/10.1007/978-3-032-18046-9_14

r_i	Rate of reaction i, referred to the reactant [moles/(volume·time)]
S	Reactor cross-section area [length^2]
T	Temperature [temperature]
t	Time [time]
t_m	Dead time [time]
V	Volume [volume]
x_A	Degree of conversion [–]
z	Reactor axial coordinate [length]
α	Stoichiometric coefficient for the limiting species [–]
β	Stoichiometric coefficient for the desired species [–]
γ	Normalised stoichiometric coefficient [–]
η	Yield [–]
η_{CSTR}	Yield for CSTR [–]
η_{PFR}	Yield for PFR [–]
ρ	Stoichiometric coefficient [–]
σ	Stoichiometric coefficient [–]
τ	Space-time for continuous reactor [time]
Φ	Global selectivity [–]
Φ_{CSTR}	Global selectivity for CSTR [–]
Φ_{PFR}	Global selectivity for PFR [–]
ϕ	Instantaneous selectivity [–]

14.1 General Picture

We will refer to isothermal systems with constant density. Extensions can be found in specialised textbooks (e.g., Levenspiel 1999; Westerterp et al. 1988). We start considering the reactive network illustrated in Fig. 14.1. While practical examples will be provided later, here, for simplicity, we refer only to the limiting reagent (species A) and to only one species per stage (in our case, P, R, S). We will call "P" *as the desired product*. Differently from what we have seen so far, now unwanted species ("R" and "S") are also present. We call this scheme a *series-parallel reaction network*, which simplifies in a series in absence of S, and in a parallel in absence of R. Starting from this general picture, any more complex reaction network can be conceptualised.

While, for a single reaction, a classic objective has been to minimise the space-time to obtain a given conversion, in the case of reaction networks this may not be

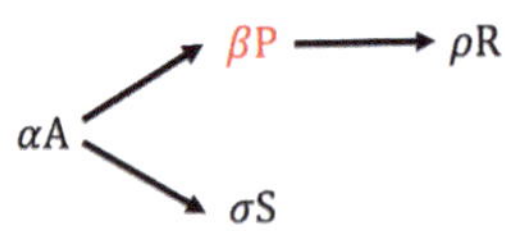

Fig. 14.1 Generic scheme for a series-parallel reaction network (P is the desired species)

sufficient. For example, a reactor could be designed with a fairly small space-time, but with an unfavourable distribution of products (i.e., shifted towards unwanted species and not P). Therefore, it is necessary to define additional quantities with respect to the conversion degree (Denbigh 1961). The first one is the *yield*. The yield (in the desired product P) represents the amount of formed P with respect to the initial amount of the limiting species, A. The expression for the yield in P considers: (i) the normalisation taking into account the stoichiometric coefficients (as usual, $\gamma = \beta/\alpha$ is the ratio of the stoichiometric coefficients related to the desired product and the limiting reactant); (ii) the possibility that, at the beginning of the process, there can be some P, and this must be subtracted from the amount of P formed in the network; (iii) the circumstance that we can refer to concentrations, and not moles (STR) or molar flow rates (PFR/CSTR), since we are assuming systems with constant density. Under these assumptions, the yield is:

$$\boxed{\eta = \frac{\frac{C_P - C_{P0}}{\beta}}{\frac{C_{A0}}{\alpha}} = \frac{1}{\gamma}\frac{C_P - C_{P0}}{C_{A0}}} \tag{14.1}$$

The second relevant parameter is the *global selectivity* (also, simply called "selectivity") of A in P, representing the amount of formed P with respect to the amount of reacted A. Under similar considerations as developed above, it is:

$$\boxed{\Phi = \frac{\frac{C_P - C_{P0}}{\beta}}{\frac{C_{A0} - C_A}{\alpha}} = \frac{1}{\gamma}\frac{C_P - C_{P0}}{C_{A0} - C_A}} \tag{14.2}$$

We observe that both η and Φ have been defined *so to belong to the interval [0, 1]*. While, for a single reaction, it is $\Phi = 1$ and $\eta = x_A$, in a network the selectivity is not unitary, since *not all A that reacts goes towards P* at the end of the process. Moreover, it is clearly (being $C_{A0} – C_A < C_{A0}$):

$$\Phi > \eta \tag{14.3}$$

and, given the degree of conversion:

$$x_A = \frac{C_{A0} - C_A}{C_{A0}} \tag{14.4}$$

we obtain the fundamental relation between the three quantities, that confirms Eq. 14.3 as $x_A < 1$:

$$\boxed{\eta = \Phi x_A} \tag{14.5}$$

The concept of productivity in the desired species for a continuous reactor (see Eq. 10.103) is here expressed as, using Eq. 14.1:

$$P_P = Q_0(C_P - C_{P0}) = \gamma Q_0 C_{A0} \eta \tag{14.6}$$

This is similar to the expression we use for single reaction (Eq. 10.103), but substituting the degree of conversion with the yield (the two quantities are equal only in case of a single reaction). For a discontinuous process, similarly, we end up with η instead of x_A starting from Eq. 10.53:

$$P_P = \frac{n_P - n_{P0}}{t + t_m} = \frac{V(C_P - C_{P0})}{t + t_m} = \frac{\gamma V C_{A0} \eta}{t + t_m} = \frac{\gamma n_{A0} \eta}{t + t_m} \tag{14.7}$$

The concept of global selectivity is accompanied by that of *instantaneous selectivity*, representing here the instantaneous variation of P concentration with respect to the instantaneous variation of A concentration:

$$\phi = -\frac{1}{\gamma} \frac{\mathrm{d}C_P}{\mathrm{d}C_A} \tag{14.8}$$

With simple mathematical steps, we observe that global selectivity can be seen as the average of the infinite values of instantaneous selectivity between input and output conditions. In fact:

$$\Phi = \frac{1}{\gamma} \frac{C_P - C_{P0}}{C_{A0} - C_A} = \frac{1}{\gamma} \frac{\int_{C_{P0}}^{C_P} \mathrm{d}C_P}{C_{A0} - C_A} \tag{14.9}$$

but, being (Eq. 14.8):

$$\mathrm{d}C_P = -\gamma \phi \mathrm{d}C_A \tag{14.10}$$

it is:

$$\Phi = -\frac{\int_{C_{A0}}^{C_A} \phi \mathrm{d}C_A}{C_{A0} - C_A} = \frac{\int_{C_A}^{C_{A0}} \phi \mathrm{d}C_A}{C_{A0} - C_A} = \overline{\phi}\Big|_{\mathrm{out}}^{\mathrm{in}} \tag{14.11}$$

Example: Medley of Reaction Networks

Figure 14.2 illustrates a series-parallel network in which *o*-dichlorobenzene, in an environment of Cl_2 and releasing HCl, reacts towards a desired species (1,2,3-trichlorobenzene) and an undesired species (1,2,4-trichlorobenzene). Furthermore, in the same environment, the desired species can be further converted to 1,2,3,4-tetrachlorobenzene, another undesired product.

(continued)

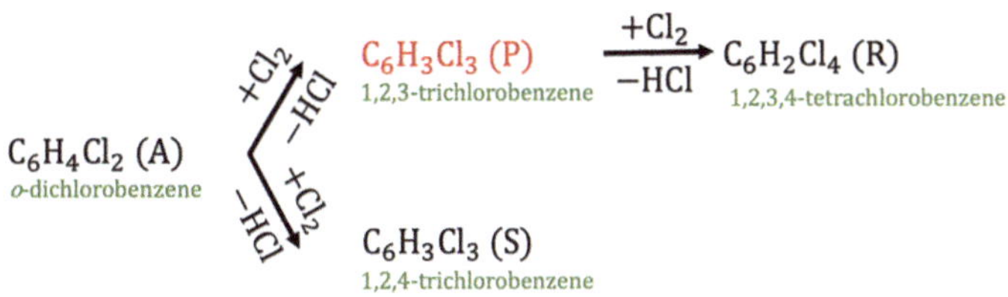

Fig. 14.2 Series-parallel reaction network involving *o*-dichlorobenzene (A-reactant) and 1,2,3-trichlorobenzene (desired product)

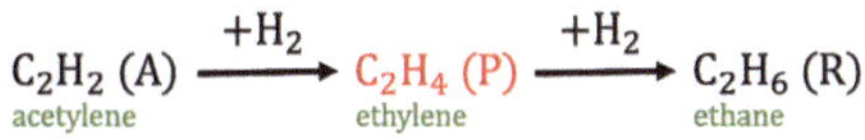

Fig. 14.3 Series reaction network involving acetylene (A-reactant) and ethylene (desired product)

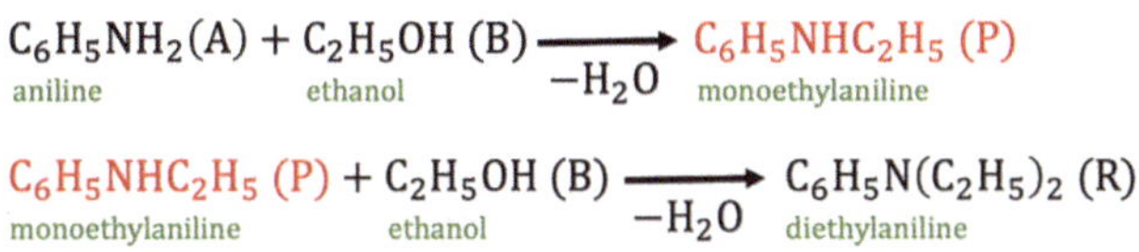

Fig. 14.4 Series-parallel reaction network involving aniline (A-reactant) and monoethylaniline (desired product)

A network in series is represented in Fig. 14.3: here we have acetylene in H_2, that is converted towards the desired product (ethylene), but the latter can further react with H_2 to give the undesired (in this case) ethane.

A more complex case is the one in Fig. 14.4, where A (aniline) reacts with B (ethanol) to give the desired P (monoethylaniline). At the same time, in presence of B, P can be converted to R (diethylaniline, undesired). Therefore, the process can be seen as a series respect to A and P, and as a parallel if we refer to B that gives both P and R. Please observe that both reactions release water.

A case of a parallel network with three lines is reported in Fig. 14.5: 1,2,4-trichlorobenzene, in presence of sodium methoxide, gives (releasing NaCl) both the desired species (1,4-dichloro-2-methoxybenzene) and the undesired 1,2-dichloro-4-methoxybenzene and 2,4-dichloro-1-methoxybenzene.

We end this medley with (Fig. 14.6) a series-parallel network where ethylene oxide and water react to give the desired ethylene glycol which, in turn, in presence of ethylene oxide, gives diethylene glycol (undesired).

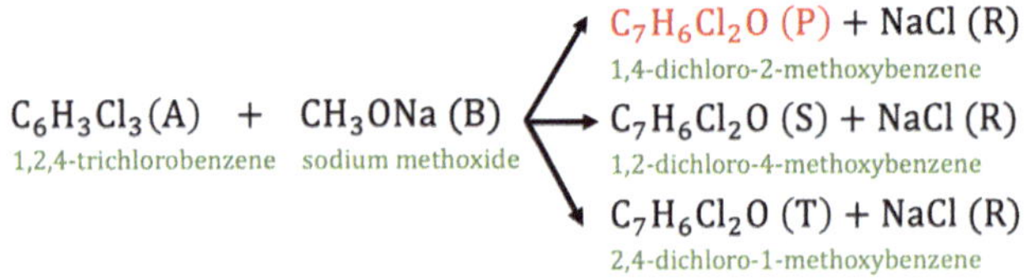

Fig. 14.5 Parallel reaction network involving 1,2,4-trichlorobenzene (A-reactant) and 1,4-dichloro-2-methoxybenzene (desired product)

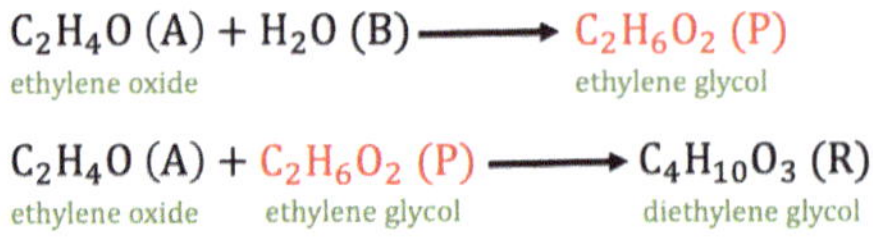

Fig. 14.6 Series-parallel reaction network involving ethylene oxide (A-reactant) and ethylene glycol (desired product)

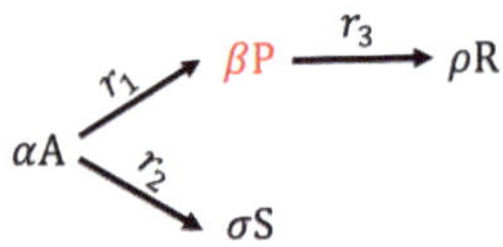

Fig. 14.7 Generic scheme for a series-parallel reaction network (P is the desired species), with indication of the reaction rates. The reaction rates are referred to the reactant: r_1 refers to the moles of A converted to P, r_2 refers to the moles of A converted to S, r_3 refers to the moles of P converted to R

14.2 Approach

We will refer to the general scheme of Fig. 14.7, where we define r_i as the rate of reaction "i" (it represents, as usual, the rate of consumption of the reactant). Any extension will be then possible to any reaction network, adopting the same approach. Let us begin by observing that the mass balance equations on the limiting species (A), which led to the drafting of the design equations (Eq. 10.33 for STR, Eq. 10.73 for PFR, Eq. 10.89 for CSTR), *are also valid for any network*, provided that r_A (rate of disappearance of A for a single reaction) is replaced with *the sum of all the reaction rates involving species A* (in the case of Fig. 14.7, it is $r_A = r_1 + r_2$, both positive as A is consumed by both reactions). Therefore, *design equations do not change*.

Now, follow the approach on mass balance presented for STR, Eq. 10.30. The balances on A, P, R, and S, respectively read:

$$\begin{cases} -\dfrac{dC_A}{dt} = r_1 + r_2 \\ \dfrac{dC_P}{dt} = \dfrac{\beta}{\alpha} r_1 - r_3 \\ \dfrac{dC_R}{dt} = \dfrac{\rho}{\beta} r_3 \\ \dfrac{dC_S}{dt} = \dfrac{\sigma}{\alpha} r_2 \end{cases} \quad (14.12)$$

Similarly in the case of PFR. Starting from Eq. 10.69 (mass balance on A):

$$-\frac{QdC_A}{Sdz} = -\frac{QdC_A}{dV} = -\frac{dC_A}{d\tau} = r_1 + r_2 \quad (14.13)$$

it is seen that, as expected, the set of equations is equal to that for STR, provided that t is substituted by τ.

For CSTR, start from Eq. 10.88 (mass balance on A):

$$Q(C_{A0} - C_A) = (r_1 + r_2)V \quad (14.14)$$

where, for the sake of simplicity, we omit the superscript "exit" for concentrations, and we remember that each reaction rate is expressed at exit conditions. Extending to the other species, we end up with:

$$\begin{cases} \dfrac{C_{A0} - C_A}{\tau} = r_1 + r_2 \\ \dfrac{C_P - C_{P0}}{\tau} = \dfrac{\beta}{\alpha} r_1 - r_3 \\ \dfrac{C_R - C_{R0}}{\tau} = \dfrac{\rho}{\beta} r_3 \\ \dfrac{C_S - C_{S0}}{\tau} = \dfrac{\sigma}{\alpha} r_2 \end{cases} \quad (14.15)$$

The approach is as follows. For STR (PFR), solve the system of differential equations. For CSTR, solve the system of algebraic equations. The solution will give the t-profiles (or τ-profiles) for the concentrations of all species. Once these profiles are substituted into the equations for yield (Eq. 14.1) and selectivity (Eq. 14.2), we can obtain the η and Φ trends, and identify the operating conditions for maximising yield or selectivity, depending on what is required.

We will now present a number of case studies, highlighting their conceptual aspects. This, of course, cannot saturate the number of possible cases, but it offers ideas for the analysis of any case the scholar will have to deal with.

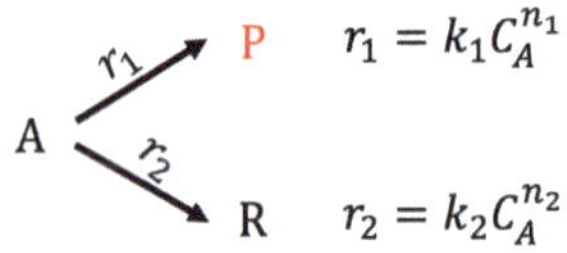

Fig. 14.8 Parallel network, power law kinetics

Example: Optimisation of Selectivity for a Parallel Network, Power Law Kinetics

Consider the simple parallel network depicted in Fig. 14.8. Let the desired reaction "1" be represented by the reaction rate r_1 (power law with order n_1). Let the unwanted reaction "2" be represented by the reaction rate r_2 (power law with order n_2).

If we want the distribution of the products to be as favourable to P as possible compared to R, obviously we want the ratio:

$$\frac{r_1}{r_2} = \frac{k_1}{k_2} C_A^{n_1 - n_2} \tag{14.16}$$

to be as high as possible. For a fixed temperature, given that the values of the kinetic constants are fixed, we see that the ratio r_1/r_2 depends on the difference between the reaction orders, $n_1 - n_2$:

- *if the desired reaction has a higher order than the unwanted reaction* ($n_1 > n_2$), r_1/r_2 increases as C_A increases. The desired reaction is more sensitive to changes in C_A than the undesired reaction, so increases in C_A will favour r_1 more than r_2. Therefore, we want to keep C_A *as high as possible*. This can be done by choosing a PFR over a CSTR and/or, in case, by removing any inert materials, working under pressure (if in gas phase), of course within the limits imposed by restrictions such as, e.g., economics, safety, availability;
- *if the desired reaction has a lower order than the unwanted reaction* ($n_1 < n_2$), r_1/r_2 increases as C_A decreases. The desired reaction is less sensitive to changes in C_A than the undesired reaction, so increases in C_A will favour r_2 more than r_1. Therefore, we want to keep C_A *as low as possible* to favour the product distribution. This can be done by choosing a CSTR over a PFR and/or, in case, by adding inert materials, working under low pressure (if in gas phase). However, keeping low C_A values, the reactions will all proceed more slowly, requiring a greater value for the space-time. Therefore, in this case, the conditions that favour the distribution of the products are not those that favour the minimisation of the space-time. It will be necessary to optimise considering the two aspects;
- if $n_1 = n_2$, the ratio r_1/r_2 = constant = k_1/k_2. Therefore, with nothing changing in terms of distribution of the products, a PFR will be chosen, for reasons of space-time (see Fig. 11.1 and considerations around).

Fig. 14.9 Parallel network, linear kinetics

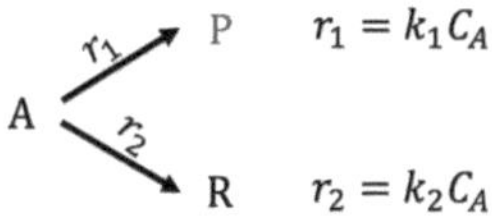

Example: Study of a Parallel Network, PFR (or STR), Linear Kinetics

Consider the parallel network represented in Fig. 14.9: the two reactions, both desired and undesired, are governed by linear kinetics. We examine a PFR (but, since there is no variation in density, all applies to STR as well, provided that τ is substituted by t), and we assign k_1 and k_2. Be $C_{PO} = C_{RO} = 0$.

The set of equations is:

$$\begin{cases} -\dfrac{\mathrm{d}C_A}{\mathrm{d}\tau} = r_1 + r_2 = (k_1 + k_2)C_A \\ \dfrac{\mathrm{d}C_P}{\mathrm{d}\tau} = r_1 = k_1 C_A \\ \dfrac{\mathrm{d}C_R}{\mathrm{d}\tau} = r_2 = k_2 C_A \end{cases} \tag{14.17}$$

Solve the profile for A:

$$\frac{\mathrm{d}C_A}{C_A} = -(k_1 + k_2)\mathrm{d}\tau \Rightarrow \int_{C_{A0}}^{C_A} \frac{\mathrm{d}C_A}{C_A} = -(k_1 + k_2)\int_0^{\tau} \mathrm{d}\tau \Rightarrow \ln\left(\frac{C_A}{C_{A0}}\right) = -(k_1 + k_2)\tau \tag{14.18}$$

that is:

$$\frac{C_A}{C_{A0}} = \exp[-(k_1 + k_2)\tau] \tag{14.19}$$

The normalised profile for C_A is reported in Fig. 14.10: the concentration of A, starting from C_{AO}, decreases to tend asymptotically to 0 (condition of $x_A = 1$ for infinitely long space-time). The balance on P reads:

(continued)

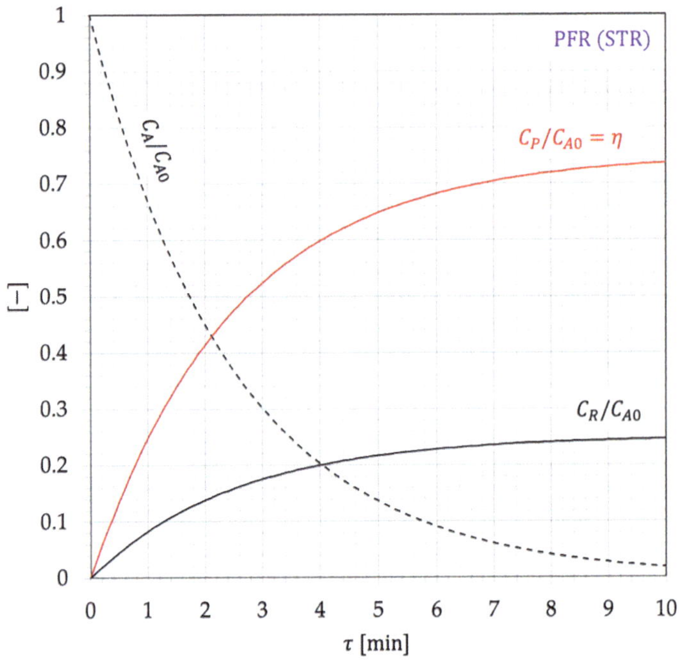

Fig. 14.10 Profiles (normalised) for the concentration of all species, network of Fig. 14.9, PFR (or STR) reactor. $k_1 = 0.3$ 1/min; $k_2 = 0.1$ 1/min

$$\frac{dC_P}{d\tau} = k_1 C_{A0} \exp[-(k_1 + k_2)\tau] \Rightarrow \int_0^{C_P} dC_P = C_P$$
$$= k_1 C_{A0} \int_0^{\tau} \exp[-(k_1 + k_2)\tau] d\tau$$
$$= \frac{k_1 C_{A0}}{k_1 + k_2} \{\exp[-(k_1 + k_2)\tau]\}_{\tau}^{0} = \frac{k_1 C_{A0}}{k_1 + k_2}$$
$$\times \{1 - \exp[-(k_1 + k_2)\tau]\} \tag{14.20}$$

$$\frac{C_P}{C_{A0}} = \eta = \frac{\frac{k_1}{k_2}}{\frac{k_1}{k_2} + 1} \{1 - \exp[-(k_1 + k_2)\tau]\} \tag{14.21}$$

The concentration of the desired species, P, increases as space-time increases, tending to the asymptotic value:

(continued)

$$\lim_{\tau \to \infty} \frac{C_P}{C_{A0}} = \frac{\frac{k_1}{k_2}}{\frac{k_1}{k_2} + 1} \tag{14.22}$$

This value (with the given kinetic constants in Fig. 14.10, it is 0.75) is greater as the ratio between the two kinetic constants (desired/undesired) increases (the normalised value would be at the limit =1 if k_1 were much greater than k_2, i.e. if we returned to the case of a single reaction for which A entirely converts to P at infinite space-time). The reader has observed, in Eq. 14.21, that the normalised (per C_{A0}) profile of C_P is the alter ego of the yield (cf. Eq. 14.1). Therefore, we can say that the yield is maximum (and it values the limit in Eq. 14.22) when the space-time is conceptually infinite.

The balance on R reads:

$$\frac{dC_R}{d\tau} = k_2 C_{A0} \exp[-(k_1 + k_2)\tau] \Rightarrow \int_0^{C_R} dC_R = C_R$$
$$= k_2 C_{A0} \int_0^{\tau} \exp[-(k_1 + k_2)\tau] d\tau$$
$$= \frac{k_2 C_{A0}}{k_1 + k_2} \{\exp[-(k_1 + k_2)\tau]\}_{\tau}^{0} = \frac{k_2 C_{A0}}{k_1 + k_2}$$
$$\times \{1 - \exp[-(k_1 + k_2)\tau]\} \tag{14.23}$$

$$\frac{C_R}{C_{A0}} = \frac{1}{\frac{k_1}{k_2} + 1} \{1 - \exp[-(k_1 + k_2)\tau]\} \tag{14.24}$$

The concentration of the undesired species, R, increases as space-time increases, tending to the asymptotic value:

$$\lim_{\tau \to \infty} \frac{C_R}{C_{A0}} = \frac{1}{\frac{k_1}{k_2} + 1} \tag{14.25}$$

This value (with the given kinetic constants, it is 0.25) obviously complements the previous one: for infinite space-time, A disappears, and $(C_P + C_R)/C_{A0} = 1$, as the mixture will be composed by P and R only. This is confirmed by summing the two limits in Eqs. 14.22 and 14.25. More generally, for any value of τ:

(continued)

$$\frac{C_A}{C_{A0}}+\frac{C_P}{C_{A0}}+\frac{C_R}{C_{A0}}=\exp[-(k_1+k_2)\tau]+\frac{\frac{k_1}{k_2}}{\frac{k_1}{k_2}+1}$$
$$\times\{1-\exp[-(k_1+k_2)\tau]\}+\frac{1}{\frac{k_1}{k_2}+1}\{1-\exp[-(k_1+k_2)\tau]\}$$
$$=1 \tag{14.26}$$

which is a way to satisfy the condition valid for constant density systems (to be modified adding C_{P0} and C_{R0} at right-hand side, if needed), namely the total concentration must be constant (apart from stoichiometric corrections when needed):

$$C_A+C_P+C_R=C_{A0} \tag{14.27}$$

The selectivity must be constant along with τ, as the two reactions have the same reaction order. Starting from Eq. 14.2:

$$\Phi=\frac{\frac{C_P}{C_{A0}}}{1-\frac{C_A}{C_{A0}}}=\frac{\frac{\frac{k_1}{k_2}}{\frac{k_1}{k_2}+1}\{1-\exp[-(k_1+k_2)\tau]\}}{1-\exp[-(k_1+k_2)\tau]}=\frac{\frac{k_1}{k_2}}{\frac{k_1}{k_2}+1} \tag{14.28}$$

We observe that the global selectivity constantly assumes the value that the yield has at its maximum, i.e. at unit conversion (remember Eqs. 14.3 and 14.5).

Example: Study of a Parallel Network, CSTR, Linear Kinetics

Consider the same parallel network as above, Fig. 14.9, but now carried out in a CSTR. The set of pertinent equations is:

$$\begin{cases}\frac{C_{A0}-C_A}{\tau}=r_1+r_2=(k_1+k_2)C_A\\ \frac{C_P}{\tau}=r_1=k_1C_A\\ \frac{C_R}{\tau}=r_2=k_2C_A\end{cases} \tag{14.29}$$

Solve the profile for A:

(continued)

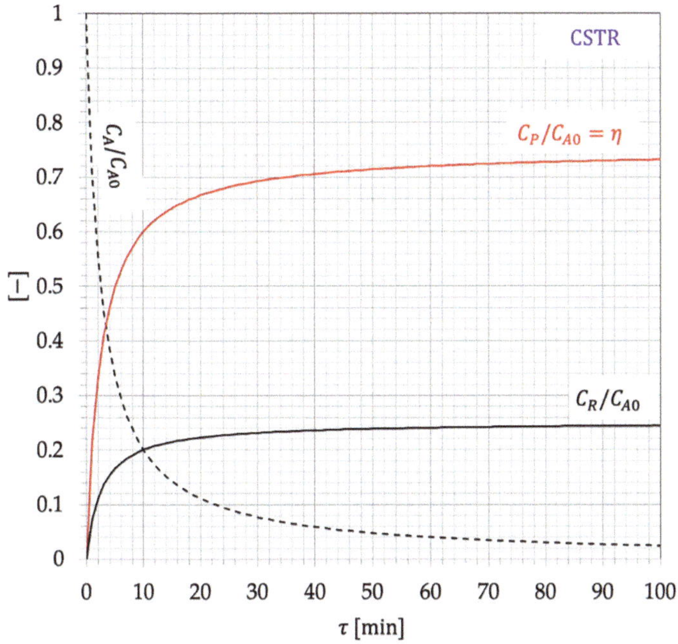

Fig. 14.11 Profiles (normalised) for the concentration of all species, network of Fig. 14.9, CSTR reactor. $k_1 = 0.3$ 1/min; $k_2 = 0.1$ 1/min

$$C_{A0} - C_A = (k_1 + k_2)\tau C_A \Rightarrow C_A[1 + (k_1 + k_2)\tau] = C_{A0} \tag{14.30}$$

$$\frac{C_A}{C_{A0}} = \frac{1}{1 + (k_1 + k_2)\tau} \tag{14.31}$$

The normalised profile for C_A is reported in Fig. 14.11: also in this case, the concentration of A, starting from C_{A0}, decreases to tend asymptotically to 0 (condition of $x_A = 1$ for infinitely long space-time). The balance on P reads:

$$C_P = k_1\tau C_A \Rightarrow \frac{C_P}{C_{A0}} = \eta = \frac{k_1\tau}{1 + (k_1 + k_2)\tau} \tag{14.32}$$

Again, the concentration of the desired species, P, increases as space-time increases, tending to the asymptotic value:

(continued)

$$\lim_{\tau \to \infty} \frac{C_P}{C_{A0}} = \frac{k_1}{k_1 + k_2} = \frac{\frac{k_1}{k_2}}{\frac{k_1}{k_2} + 1} \tag{14.33}$$

i.e., the same limit as the one observed for PFR in Eq. 14.22. Equation 14.32 gives η, again maximum (the limit in Eq. 14.33) when the space-time is conceptually infinite.

The balance on R reads:

$$C_R = k_2 \tau C_A \Rightarrow \frac{C_R}{C_{A0}} = \frac{k_2 \tau}{1 + (k_1 + k_2)\tau} \tag{14.34}$$

The concentration of the undesired species, R, increases as space-time increases, tending to the asymptotic value:

$$\lim_{\tau \to \infty} \frac{C_R}{C_{A0}} = \frac{k_2}{k_1 + k_2} = \frac{1}{\frac{k_1}{k_2} + 1} \tag{14.35}$$

i.e., the same limit as the one observed for PFR in Eq. 14.25. Again, for infinite space-time, A disappears, and $(C_P + C_R)/C_{A0} = 1$. More generally, for any value of τ:

$$\frac{C_A}{C_{A0}} + \frac{C_P}{C_{A0}} + \frac{C_R}{C_{A0}} = \frac{1}{1 + (k_1 + k_2)\tau} + \frac{k_1 \tau}{1 + (k_1 + k_2)\tau} + \frac{k_2 \tau}{1 + (k_1 + k_2)\tau} = 1 \Rightarrow C_A + C_P + C_R = C_{A0} \tag{14.36}$$

Of course, also here the selectivity is constant along with τ:

$$\Phi = \frac{\frac{C_P}{C_{A0}}}{1 - \frac{C_A}{C_{A0}}} = \frac{\frac{k_1 \tau}{1+(k_1+k_2)\tau}}{1 - \frac{1}{1+(k_1+k_2)\tau}} = \frac{k_1 \tau}{(k_1 + k_2)\tau} = \frac{\frac{k_1}{k_2}}{\frac{k_1}{k_2} + 1} \Rightarrow \Phi_{CSTR} = \Phi_{PFR} \tag{14.37}$$

and it assumes the same value as for PFR. In fact, as observed in a previous example, if the desired and the undesired reactions have the same order, the selectivity is not affected by reactor fluid dynamics. Nonetheless, we have already noticed that in this case (canonical kinetic laws) a PFR will be chosen, for reasons of space-time: at fixed τ, $\eta_{PFR} > \eta_{CSTR}$ or, equivalently, a lower value of τ will be needed to reach a given yield. This is demonstrated in Fig. 14.12.

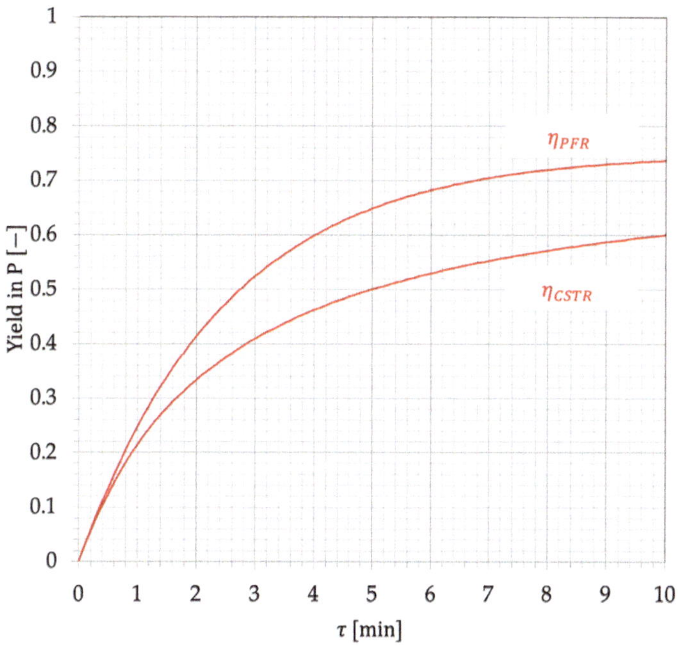

Fig. 14.12 Profiles for the yield in PFR (Eq. 14.21) vs. CSTR (Eq. 14.32), network of Fig. 14.9, for space-time up to 10 min. $k_1 = 0.3$ 1/min; $k_2 = 0.1$ 1/min

Example: Solution of a Differential Equation Via the Method of the Integrating Factor

Before tackling examples of series network, we illustrate here the *method of the integrating factor* for the solution of a canonic differential equation (expressed in time) in the form:

$$\frac{df(t)}{dt} + g(t)f(t) = h(t) \tag{14.38}$$

This method will result useful for the determination of relevant profiles in case of a series network carried out in PFR/STR. Once individuated the function ($g(t)$) that multiplies the function ($f(t)$) respect to which the differential equation must be solved, we define *integrating factor*:

(continued)

$$I_F = \exp\left[\int g(t)\mathrm{d}t\right] \tag{14.39}$$

Now, multiply Eq. 14.38 per I_F:

$$\frac{\mathrm{d}f(t)}{\mathrm{d}t}\exp\left[\int g(t)\mathrm{d}t\right] + g(t)f(t)\exp\left[\int g(t)\mathrm{d}t\right]$$
$$= h(t)\exp\left[\int g(t)\mathrm{d}t\right] \tag{14.40}$$

The left-hand side of Eq. 14.40 is the derivative of the product between $f(t)$ and I_F, invoking the Leibniz product rule. So, we can write:

$$\frac{\mathrm{d}[f(t)I_F]}{\mathrm{d}t} = h(t)I_F \tag{14.41}$$

We can now approach the solution by application of the Fourier method (separation of variables):

$$\mathrm{d}[f(t)I_F] = h(t)I_F\mathrm{d}t \tag{14.42}$$

that can be integrated between two extremes dictated by the case under analysis.

Example: Study of a Series Network, PFR (or STR), Linear Kinetics

Consider now the series network represented in Fig. 14.13: the two reactions, both desired and undesired, are governed by linear kinetics. We examine a PFR (or, equivalently, an STR), and we assign k_1 and k_2. Be $C_{P0} = C_{R0} = 0$.

The set of equations reads:

$$\begin{cases} -\dfrac{\mathrm{d}C_A}{\mathrm{d}\tau} = r_1 = k_1 C_A \\ \dfrac{\mathrm{d}C_P}{\mathrm{d}\tau} = r_1 - r_2 = k_1 C_A - k_2 C_P \\ \dfrac{\mathrm{d}C_R}{\mathrm{d}\tau} = r_2 = k_2 C_P \end{cases} \tag{14.43}$$

Solve the profile for A:

(continued)

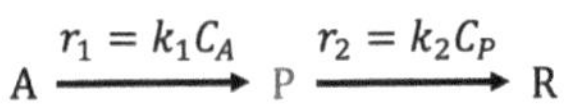

Fig. 14.13 Series network, linear kinetics

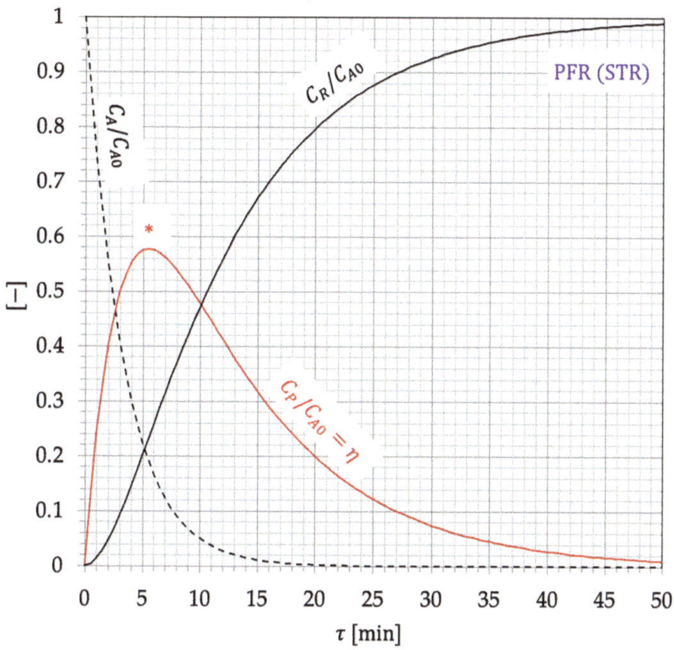

Fig. 14.14 Profiles (normalised) for the concentration of all species, network of Fig. 14.13, PFR (or STR) reactor. $k_1 = 0.3$ 1/min; $k_2 = 0.1$ 1/min. The asterisk marks the condition for maximum yield

$$\frac{dC_A}{C_A} = -k_1 d\tau \Rightarrow \int_{C_{A0}}^{C_A} \frac{dC_A}{C_A} = -k_1 \int_0^\tau d\tau \Rightarrow \ln\left(\frac{C_A}{C_{A0}}\right) = -k_1\tau \tag{14.44}$$

$$\frac{C_A}{C_{A0}} = \exp(-k_1\tau) \tag{14.45}$$

The normalised profile for C_A is reported in Fig. 14.14: the concentration of A, starting from C_{A0}, decreases to tend asymptotically to 0 (condition of $x_A = 1$ for infinitely long space-time). The balance on P reads:

$$\frac{dC_P}{d\tau} = k_1 C_{A0} \exp(-k_1\tau) - k_2 C_P \tag{14.46}$$

To solve this equation, put in the canonical form of Eq. 14.38:

(continued)

$$\frac{dC_P}{d\tau} + k_2 C_P = k_1 C_{A0} \exp(-k_1 \tau) \tag{14.47}$$

We recognise that the function $g(t)$ is, actually, a constant (k_2), while the function $h(t)$ is the right-hand side of Eq. 14.47. The integrating factor is (Eq. 14.39):

$$I_F = \exp\left(k_2 \int d\tau\right) = \exp(k_2 \tau) \tag{14.48}$$

and Eq. 14.42 becomes:

$$\begin{aligned} d[C_P \exp(k_2\tau)] &= k_1 C_{A0} \exp(-k_1\tau) \exp(k_2\tau) d\tau \\ &= k_1 C_{A0} \exp[(k_2 - k_1)\tau] d\tau \end{aligned} \tag{14.49}$$

$$\int_{\tau=0, C_P=0}^{\tau, C_P} d[C_P \exp(k_2\tau)] = k_1 C_{A0} \int_0^\tau \exp[(k_2 - k_1)\tau] d\tau \tag{14.50}$$

$$\begin{aligned} C_P \exp(k_2\tau) &= \frac{k_1 C_{A0}}{k_2 - k_1} \{\exp[(k_2 - k_1)\tau]\}_0^\tau = \frac{k_1 C_{A0}}{k_2 - k_1} \\ &\times \{\exp[(k_2 - k_1)\tau] - 1\} \end{aligned} \tag{14.51}$$

with solution:

$$\frac{C_P}{C_{A0}} = \eta = \frac{k_1}{k_2 - k_1} [\exp(-k_1\tau) - \exp(-k_2\tau)] \tag{14.52}$$

Very interestingly, the profile of C_P (i.e., the yield profile) starts from zero, ends to zero and *passes through a maximum*. This is an intrinsic feature of a series network where the desired product is an intermediate (Fig. 14.13): at the beginning, the formation of P prevails over its consumption, as C_A is high and C_P is low; therefore, C_P increases. If the reaction is allowed to proceed further, the consumption of P will prevail over its formation, as C_A is low (the desired r_1 is low) and C_P is high (the undesired r_2 is high); therefore, C_P decreases. The *point of maximum yield* can be now determined:

$$\frac{d\left(\frac{C_P}{C_{A0}}\right)}{d\tau} = \frac{k_1}{k_2 - k_1} [-k_1 \exp(-k_1\tau) + k_2 \exp(-k_2\tau)] = 0 \tag{14.53}$$

$$k_1 \exp(-k_1\tau) = k_2 \exp(-k_2\tau) \Rightarrow \exp[(k_2 - k_1)\tau] = \frac{k_2}{k_1} \tag{14.54}$$

(continued)

$$\tau = \frac{\ln\left(\frac{k_2}{k_1}\right)}{k_2 - k_1} \tag{14.55}$$

This condition, marked with an asterisk in Fig. 14.14, is reflected by a space-time corresponding to the *reciprocal of the logarithmic mean* between undesired and desired process kinetic constants. The balance on R reads:

$$\frac{dC_R}{d\tau} = k_2 \frac{k_1 C_{A0}}{k_2 - k_1}[\exp(-k_1\tau) - \exp(-k_2\tau)] \tag{14.56}$$

$$\int_0^{C_R} dC_R = k_2 \frac{k_1 C_{A0}}{k_2 - k_1}\int_0^{\tau}[\exp(-k_1\tau) - \exp(-k_2\tau)]d\tau \tag{14.57}$$

$$C_R = k_2 \frac{k_1 C_{A0}}{k_2 - k_1}\left\{\frac{1}{k_1}[\exp(-k_1\tau)]_{\tau}^{0} - \frac{1}{k_2}[\exp(-k_2\tau)]_{\tau}^{0}\right\} = k_2 \frac{k_1 C_{A0}}{k_2 - k_1} \times \left[\frac{1}{k_1} - \frac{1}{k_1}\exp(-k_1\tau) - \frac{1}{k_2} + \frac{1}{k_2}\exp(-k_2\tau)\right] \tag{14.58}$$

$$\frac{C_R}{C_{A0}} = \frac{k_1 k_2}{k_2 - k_1}\left[\frac{1}{k_1} - \frac{1}{k_1}\exp(-k_1\tau) - \frac{1}{k_2} + \frac{1}{k_2}\exp(-k_2\tau)\right] \tag{14.59}$$

The concentration of the undesired species, R, increases as space-time increases (R is the final product of the series), tending to the asymptotic value:

$$\lim_{\tau \to \infty} \frac{C_R}{C_{A0}} = \frac{k_1 k_2}{k_2 - k_1}\left(\frac{1}{k_1} - \frac{1}{k_2}\right) = 1 \tag{14.60}$$

which is obvious as, for infinite space-time, both A and P disappear in favour of R. More generally, for any value of τ:

$$\frac{C_A}{C_{A0}} + \frac{C_P}{C_{A0}} + \frac{C_R}{C_{A0}} = \exp(-k_1\tau) + \frac{k_1}{k_2 - k_1} \times [\exp(-k_1\tau) - \exp(-k_2\tau)] + \frac{k_1 k_2}{k_2 - k_1} \times \left[\frac{1}{k_1} - \frac{1}{k_1}\exp(-k_1\tau) - \frac{1}{k_2} + \frac{1}{k_2}\exp(-k_2\tau)\right] = 1 \Rightarrow C_A + C_P + C_R = C_{A0} \tag{14.61}$$

Another interesting aspect concerning the profile of the concentration of R species is that it shows an *inflection point where the yield is maximum*:

(continued)

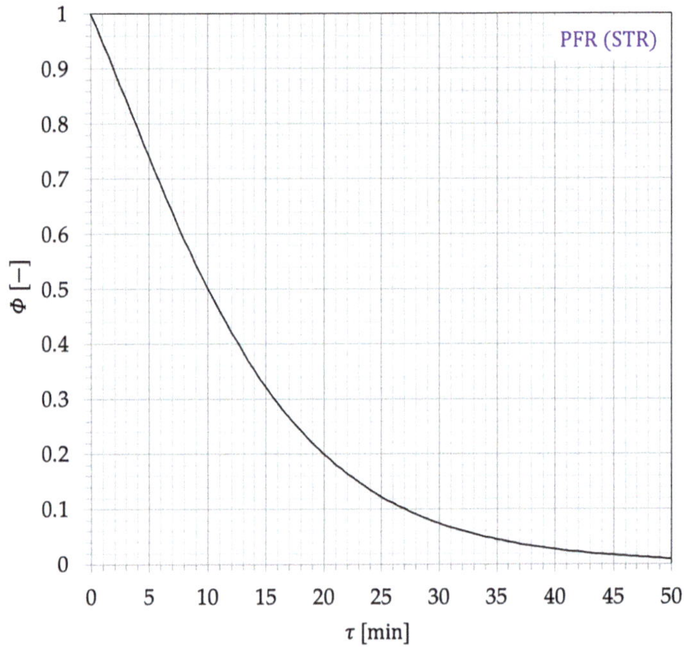

Fig. 14.15 Profile for the global selectivity, network of Fig. 14.13, PFR (or STR) reactor. $k_1 = 0.3$ 1/min; $k_2 = 0.1$ 1/min

$$\frac{dC_R}{d\tau} = k_2 C_P \Rightarrow \frac{d^2 C_R}{d\tau^2} = k_2 \frac{dC_P}{d\tau} \tag{14.62}$$

$$\frac{d^2 C_R}{d\tau^2} = 0 \Leftrightarrow \frac{dC_P}{d\tau} = 0 \tag{14.63}$$

Now we can discuss about global selectivity (Fig. 14.15):

$$\Phi = \frac{\frac{C_P}{C_{A0}}}{1 - \frac{C_A}{C_{A0}}} = \frac{\frac{k_1}{k_2 - k_1}\left[\exp(-k_1\tau) - \exp(-k_2\tau)\right]}{1 - \exp(-k_1\tau)} \tag{14.64}$$

It is maximum (100%) for differential reactor. In fact, applying the de l'Hôpital's rule:

$$\lim_{\tau\to 0} \Phi = \frac{k_1}{k_2 - k_1} \lim_{\tau\to 0} \frac{-k_1 \exp(-k_1\tau) + k_2 \exp(-k_2\tau)}{k_1 \exp(-k_1\tau)} = 1 \tag{14.65}$$

(continued)

This fundamental result was predictable from the trends in Fig. 14.14, and from physical and mathematical considerations. In differential conditions, the global selectivity conceptually becomes equal to the instantaneous selectivity (Eq. 14.8 and balances in Eq. 14.43):

$$\phi = -\frac{\mathrm{d}C_P}{\mathrm{d}C_A} = \frac{k_1 C_A - k_2 C_P}{k_1 C_A} \tag{14.66}$$

For space-time close to zero, the undesired reaction has not yet taken place, because the process begins with only the desired reaction. Therefore, $C_P = 0$ in Eq. 14.66, $\phi = 1 = \Phi$ or, in other words, A that reacts can only give P. Practically, it will be necessary to work at sufficiently low space-time (and, therefore, conversion degrees), if we want to maximise the process selectivity (this means that *the conditions for optimal yield and optimal selectivity are different*). As the space-time increases, the selectivity is destined to decrease (see Fig. 14.15), because the undesired reaction will become preponderant, reaching the limit of determining the complete disappearance of P (and therefore, zero selectivity with all A that has globally converted towards R, see also the profiles in Fig. 14.14). Saying that, in differential conditions, $\phi = 1$, means saying (Eq. 14.66) that, in absolute value, the slope of C_A and C_P profiles is the same: this can be observed by the visual inspection of Fig. 14.14. It also implies that, under differential conditions, $\mathrm{d}C_R/\mathrm{d}\tau = k_2 C_P = 0$: as a matter of fact, the profile for C_R starts with zero slope, and then increases passing through the inflection point of Eq. 14.63.

Example: Study of a Series Network, CSTR, Linear Kinetics

Consider the same series network as above, Fig. 14.13, but now carried out in a CSTR. The set of pertinent equations reads:

$$\begin{cases} \dfrac{C_{A0} - C_A}{\tau} = r_1 = k_1 C_A \\ \dfrac{C_P}{\tau} = r_1 - r_2 = k_1 C_A - k_2 C_P \\ \dfrac{C_R}{\tau} = r_2 = k_2 C_P \end{cases} \tag{14.67}$$

Solve the profile for A:

$$C_{A0} - C_A = k_1 \tau C_A \Rightarrow C_A(1 + k_1 \tau) = C_{A0} \tag{14.68}$$

(continued)

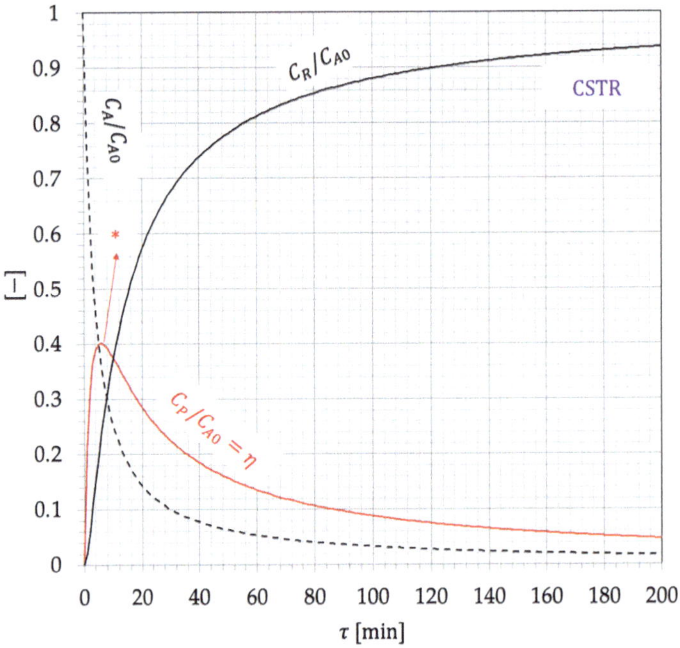

Fig. 14.16 Profiles (normalised) for the concentration of all species, network of Fig. 14.13, CSTR reactor. $k_1 = 0.3$ 1/min; $k_2 = 0.1$ 1/min. The asterisk marks the condition for maximum yield

$$\frac{C_A}{C_{A0}} = \frac{1}{1 + k_1\tau} \tag{14.69}$$

with the usual trend in Fig. 14.16. The balance on P reads:

$$C_P = k_1\tau C_A - k_2\tau C_P \Rightarrow C_P(1 + k_2\tau) = \frac{k_1\tau C_{A0}}{1 + k_1\tau} \Rightarrow \frac{C_P}{C_{A0}} = \eta$$
$$= \frac{k_1\tau}{(1 + k_1\tau)(1 + k_2\tau)} \tag{14.70}$$

As conceptually expected, C_P is zero for $\tau = 0$ and for infinite τ, passing through the point of maximum yield:

$$\frac{\mathrm{d}\left(\frac{C_P}{C_{A0}}\right)}{\mathrm{d}\tau} = \frac{k_1(1 + k_1\tau)(1 + k_2\tau) - k_1\tau(k_1 + k_2 + 2k_1k_2\tau)}{[(1 + k_1\tau)(1 + k_2\tau)]^2} = 0 \tag{14.71}$$

or, equivalently:

(continued)

$$(1+k_1\tau)(1+k_2\tau)-\tau(k_1+k_2+2k_1k_2\tau)=1-k_1k_2\tau^2=0 \tag{14.72}$$

$$\tau=\frac{1}{\sqrt{k_1k_2}} \tag{14.73}$$

This condition, marked with an asterisk in Fig. 14.16, is reflected by a space-time corresponding to the *reciprocal of the geometric mean* between undesired and desired process kinetic constants.

The balance on R reads:

$$C_R=k_2\tau C_P \Rightarrow \frac{C_R}{C_{A0}}=\frac{k_1k_2\tau^2}{(1+k_1\tau)(1+k_2\tau)} \tag{14.74}$$

The concentration of the undesired species, R, increases as space-time increases, tending to the (obvious) asymptotic value:

$$\lim_{\tau\to\infty}\frac{C_R}{C_{A0}}=\lim_{\tau\to\infty}\frac{2k_1k_2\tau}{k_1+k_2+2k_1k_2\tau}=1 \tag{14.75}$$

More generally, for any value of τ:

$$\frac{C_A}{C_{A0}}+\frac{C_P}{C_{A0}}+\frac{C_R}{C_{A0}}=\frac{1}{1+k_1\tau}+\frac{k_1\tau}{(1+k_1\tau)(1+k_2\tau)} + \frac{k_1k_2\tau^2}{(1+k_1\tau)(1+k_2\tau)}=1\Rightarrow C_A+C_P+C_R=C_{A0} \tag{14.76}$$

Now we can discuss about global selectivity (Fig. 14.17):

$$\Phi=\frac{\frac{C_P}{C_{A0}}}{1-\frac{C_A}{C_{A0}}}=\frac{\frac{k_1\tau}{(1+k_1\tau)(1+k_2\tau)}}{1-\frac{1}{1+k_1\tau}}=\frac{1}{1+k_2\tau} \tag{14.77}$$

with the expected trend, starting from 100% (differential reactor) and then decreasing to zero at infinite space-time (where $C_P=0$).

To conclude, while both configurations (PFR and CSTR) allow for unit selectivity (provided the system is operated under differential conditions), the maximum yield for PFR (with our values, it is 57.7% at $\tau=5.49$ min, cf. Figure 14.14, and Eqs. 14.52 and 14.55) is higher than for CSTR (40.2% at $\tau=5.77$ min, cf. Figure 14.17, and Eqs. 14.70 and 14.73). This is related to the general superior performance of a PFR when canonical kinetics (as those we are considering) apply.

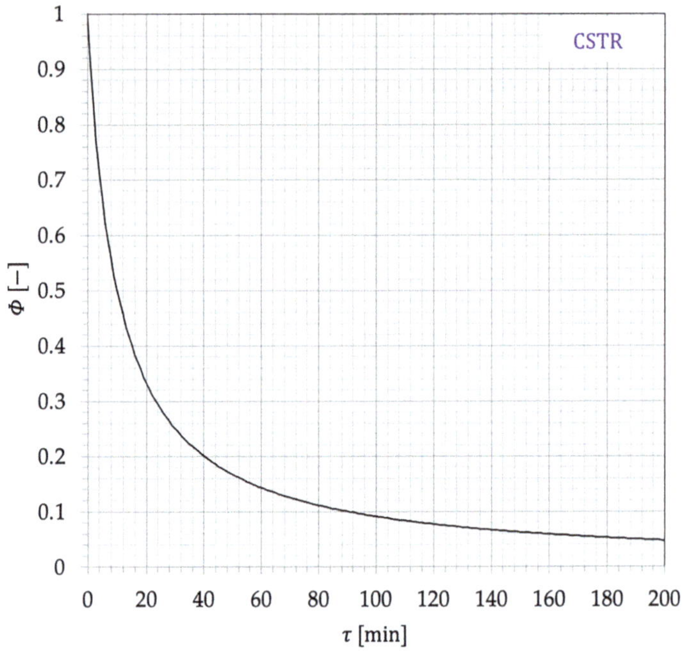

Fig. 14.17 Profile for the global selectivity, network of Fig. 14.13, CSTR reactor. $k_1 = 0.3$ 1/min; $k_2 = 0.1$ 1/min

14.3 Considerations on the Role of Temperature

Although we have considered isothermal systems so far, we now briefly discuss the effect of temperature T on reaction networks. Consider the general Arrhenius expression (Eq. 13.2):

$$k(T) = k_0 \exp\left(-\frac{E}{\mathfrak{R}T}\right) \tag{14.78}$$

$$\ln\left[\frac{k(T)}{k_0}\right] = -\frac{E}{\mathfrak{R}T} \tag{14.79}$$

The previous relation translates a linearity condition, if on a plane (Fig. 14.18) we report the left-hand side of Eq. 14.79 as a function of the reciprocal of the absolute temperature. Therefore, we have a straight line with slope $-E/\mathfrak{R}$, indicating that, when T increases, k increases. In Chap. 13, we have observed that a way to see E is that *the activation energy measures the sensitivity of k to T*: reactions characterised by higher values of E will present variations of k that are more sensitive to variations of T. In other words (Fig. 14.18): if two reactions are considered, with the same increase in T, the reaction that presents the highest activation energy will be the one

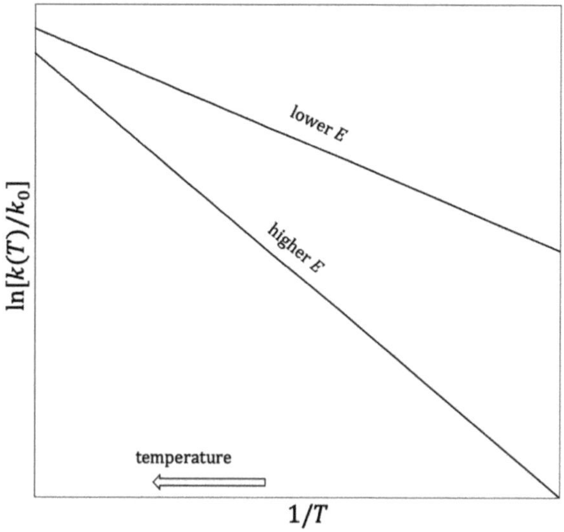

Fig. 14.18 Arrhenius plot: increasing temperature "from right to left" (while accelerates both reactions) favours the reaction with higher activation energy more than the other reaction (the relative position of the two lines is arbitrary)—the kinetic parameter increases more rapidly if the activation energy is higher. The slope of the lines is $-E/\mathfrak{R}$

whose rate will increase the most (in fact, in absolute value, the slope is higher if E is higher and, therefore, k will increase faster).

Given these premises, in a network of reactions:

- *if the desired reaction has a higher activation energy than the other reactions*, it is preferable to work at the highest possible temperature, of course within the limits imposed by restrictions such as, e.g., economics, safety, possible evaporation of a solvent, onset of unwanted processes, resistance of the materials. In this way, in fact, all reactions will be favoured, but the desired reaction will be favoured the most;
- *if the desired reaction has a lower activation energy than the other reactions*, it is preferable to work at the lowest possible temperature in terms of product distribution. In this way, in fact, all reactions will be unfavoured, but the desired reaction will be unfavoured the least. However, even if the network is oriented towards the desired product, the absolute temperatures will be low, and therefore each reaction will be, in absolute sense, slowed down. This will require higher (space-)time values, with consequent penalties. It will be necessary to optimise.

14.4 Background Information

- The fundamental quantities for a reaction network are the yield (Eq. 14.1) and the selectivity (Eq. 14.2), linked to each other and to the degree of conversion by Eq. 14.5.

- The design equations for the reactors remain valid, provided that the rate of reaction present in the design equation is the sum of all the reaction rates involving species A.
- The general writing of the balances is given in Eq. 14.12 (PFR/STR) and in Eq. 14.15 (CSTR).
- The solution of the balances allows to know the trend of yield and selectivity.
- In a parallel network with reactions governed by power laws, in terms of orientation of the network towards the desired product, if the desired reaction has the highest order, it will be convenient to use a PFR (vice versa, a CSTR, but at the expense of a greater space-time).
- In a parallel network with reactions governed by canonical laws, the conditions of maximum yield are obtained for infinite space-time at the limit.
- In a series network with reactions governed by canonical laws, the conditions of maximum yield are obtained for intermediate space-time, while the selectivity is conceptually unitary for differential operating conditions.
- The activation energy measures the sensitivity of the kinetic constant to the temperature. If the desired reaction has a higher activation energy, in terms of orientation of the network towards the desired product, it will be convenient to work at the highest possible temperature (vice versa, it will be convenient to work at the lowest possible temperature, but at the expense of a greater space-time).

References

Denbigh KG (1961) Instantaneous and overall reaction yields. Chem Eng Sci 14:25–35. https://doi.org/10.1016/0009-2509(61)85049-5

Levenspiel O (1999) Chemical reaction engineering. John Wiley & Sons, New York

Westerterp KR, van Swaaij WPM, Beenackers AACM (1988) Chemical reactor design and operation. John Wiley & Sons, Hoboken

Non-Ideal Reactors: Fluid Dynamic Aspects 15

Abstract

We begin the last part of the textbook, which refers to chemical reactors when they cannot be considered systems characterised by ideal fluid dynamics. The topic is divided into two chapters: in this one, we will focus on fluid dynamics aspects, without the need to consider systems with chemical reaction occurring. The main causes of non-ideality will be illustrated, as well as the function that can characterise a non-ideal system, i.e. the distribution of residence times (with its properties). The tracing methods for obtaining this function will then be shown and, when necessary, reference will be made to ideal systems. This is because, in some cases of interest, it is possible to apply compartment models that make the real system approachable through block diagrams consisting of ideal reactors.

Nomenclature

C_T	Concentration of tracer [mass (or moles)/volume]
C_T^{max}	Maximum (steady state) tracer concentration [mass (or moles)/volume]
E	Residence time distribution [1/time]
E_{CSTR}	Residence time distribution for CSTR [1/time]
E_{PFR}	Residence time distribution for PFR [1/time]
F	Cumulative residence time distribution [–]
F_{CSTR}	Cumulative residence time distribution for CSTR [–]
F_{PFR}	Cumulative residence time distribution for PFR [–]
m_T	Quantity of tracer [mass (or moles)]
$\dot{m}_T$	Flow rate of tracer [mass (or moles)/time]
Q	Volumetric flow rate [volume/time]
Q_a	Active volumetric flow rate [volume/time]
Q_b	By-pass volumetric flow rate [volume/time]
Q_j	Volumetric flow rate entering a reactor "*j*" (in parallel) [volume/time]

F. Montagnaro et al., *Elements of Unit Operations and Chemical Reactors*,
https://doi.org/10.1007/978-3-032-18046-9_15

t	Time [time]
$\bar{t}$	Mean residence time [time]
t_0	Support of the Dirac delta function [time]
t_i	Generic time value [time]
V	Volume [volume]
V_{CSTR}	Volume of CSTR [volume]
V_{PFR}	Volume of PFR [volume]
V_a	Active volume [volume]
V_j	Volume of a reactor "j" [volume]
V_m	Dead volume [volume]
δ	Dirac delta function [1/time]
τ	Space-time for continuous ideal reactor [time]

15.1 Premise

In continuation of what has been seen in the previous chapters for ideal reactors, even in the cases of systems that deviate from conditions of, e.g., ideal mixing (CSTR) or ideal axial segregation (PFR), we will consider fluid phases (in particular, with *constant density and isothermal*). Extensions on the treatment of reactors in which solid particles are present can be found in more specialised books (including Levenspiel 1999; Montagnaro 2024).

What are the fluid dynamic conditions that can lead a reactor system to behave in a *non-ideal* way? The deviations from ideal behaviours can be caused, f.i., by dead volumes (*stagnant regions*) or by-passes (*short circuits*), as in the simple example illustrated in Fig. 15.1 (where a tank reactor equipped with non-perfect stirring is represented; but dead volumes and by-passes can be present in any system), or by *non-turbulent motion* in a tubular reactor (which determines a radial velocity profile, parabolic in the laminar limit, with fluid elements that have different velocity and therefore residence times, and therefore conversion degrees, depending on their radial position). Many other examples could be made, on the basis of the specific system under consideration.

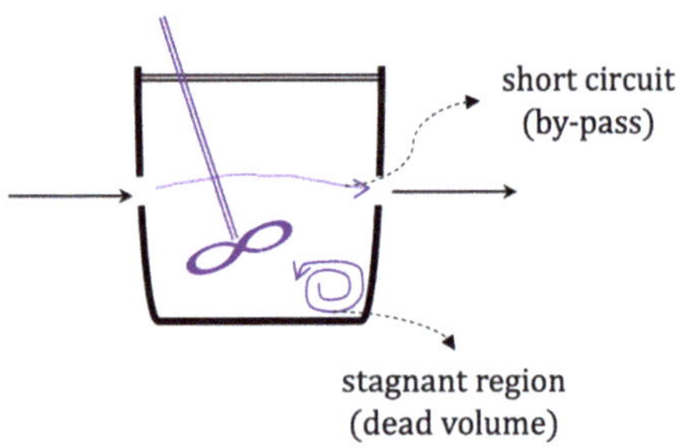

Fig. 15.1 Simplifying scheme of two non-ideal situations in a mixed tank reactor: stagnant region and by-pass

15.2 Residence Time Distribution

15.2.1 General Presentation and Properties

In a non-ideal system, for what discussed above, the fluid elements *will not all spend the same residence time* in the reactor. For example, those that fall into stagnant regions will "live" much longer, and those that enter a by-pass flow much shorter than expected. This will determine a *distribution of residence times*, depicted by a function called RTD (residence time distribution) and indicated with $E(t)$, expressed in [1/time]. Without necessarily having to solve the entire flow field, to design the reactor it is often sufficient to know the RTD function. The function $E(t)$ is defined such that the integral between t_1 and t_2 of $E(t)\mathrm{d}t$ represents the fraction of fluid elements that reside in the system for a time $t \in [t_1, t_2]$, as indicated in Fig. 15.2 where a generic (arbitrary) example of $E(t)$ is illustrated. In particular, values for $E(t)$ at short residence times indicate the presence of by-pass (fluid elements that reside in the reactor shorter than expected), while values for $E(t)$ in the long tail indicate the presence of dead volumes (fluid elements that reside in the reactor longer than expected). Bringing the discussion to the differential limit:

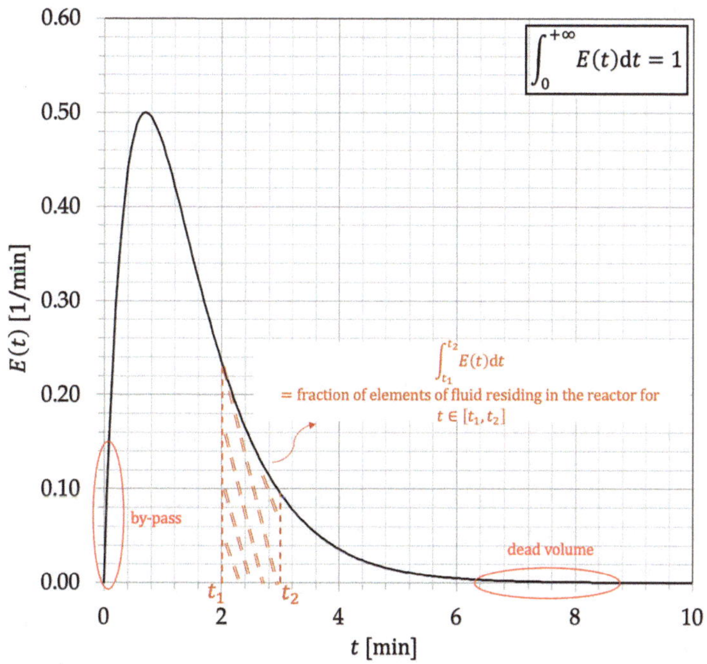

Fig. 15.2 General appearance of an RTD curve for a non-ideal reactor, with the condition of congruence and the presence of zones indicating by-pass or dead volume

$$\int_t^{t+\mathrm{d}t} E(t)\mathrm{d}t \cong E(t)\int_t^{t+\mathrm{d}t} \mathrm{d}t = E(t)\mathrm{d}t \tag{15.1}$$

i.e., $E(t)\mathrm{d}t$ represents the fraction of fluid elements that reside in the reactor for a time comprised between t and $t + \mathrm{d}t$, namely that reside in the reactor for that time t. Obviously, extending the integral between all times between 0 and infinite:

$$\boxed{\int_0^{+\infty} E(t)\mathrm{d}t = 1} \tag{15.2}$$

i.e. the *condition of congruence* for $E(t)$. The RTD can be also characterised by the *mean residence time*:

$$\boxed{\bar{t} = \int_0^{+\infty} tE(t)\mathrm{d}t} \tag{15.3}$$

where the value of each residence time t is "weighted" by the relevance of the fluid elements that live in the reactor for that time t, relevance given by $E(t)\mathrm{d}t$ as per Eq. 15.1; the procedure is then extended, via integral, to all possible values of t.

Example: Learning to Manage an RTD Curve

The RTD curve in Fig. 15.2 is represented by:

$$E(t) = a[\exp(-t) - \exp(-at)] \tag{15.4}$$

with, in particular, $a = 2$. Subsequently in the chapter, it will become clear how to obtain an $E(t)$ function. The first thing is to verify the condition of congruence, Eq. 15.2:

$$\begin{aligned}\int_0^{+\infty} E(t)\mathrm{d}t &= a\int_0^{+\infty} [\exp(-t) - \exp(-at)]\mathrm{d}t \\ &= a\left[\int_0^{+\infty} \exp(-t)\mathrm{d}t - \int_0^{+\infty} \exp(-at)\mathrm{d}t\right] \\ &= a\left\{[-\exp(-t)]_0^{+\infty} + \frac{1}{a}[\exp(-at)]_0^{+\infty}\right\} = a\left(1 - \frac{1}{a}\right) = 1\end{aligned} \tag{15.5}$$

If, for example, we want to know the fraction of fluid elements residing in our non-ideal reactor for times between $t_1 = 2$ min and $t_2 = 3$ min (cf. Figure 15.2):

(continued)

$$\int_{t_1}^{t_2} E(t)\mathrm{d}t = a\int_{t_1}^{t_2} [\exp(-t) - \exp(-at)]\mathrm{d}t$$
$$= a\left[\int_{t_1}^{t_2} \exp(-t)\mathrm{d}t - \int_{t_1}^{t_2} \exp(-at)\mathrm{d}t\right]$$
$$= a\left\{[-\exp(-t)]_{t_1}^{t_2} + \frac{1}{a}[\exp(-at)]_{t_1}^{t_2}\right\}$$
$$= a\left[-\exp(-t_2) + \exp(-t_1) + \frac{1}{a}\exp(-at_2) - \frac{1}{a}\exp(-at_1)\right]$$
$$= 15.5\% \qquad (15.6)$$

If we want to know the fraction of fluid elements that live in the reactor for a time not longer than a given time t (say, for simplicity, $t = t_2 = 3$ min):

$$\int_{0}^{t_2} E(t)\mathrm{d}t = a\int_{0}^{t_2} [\exp(-t) - \exp(-at)]\mathrm{d}t$$
$$= a\left[\int_{0}^{t_2} \exp(-t)\mathrm{d}t - \int_{0}^{t_2} \exp(-at)\mathrm{d}t\right]$$
$$= a\left\{[-\exp(-t)]_{0}^{t_2} + \frac{1}{a}[\exp(-at)]_{0}^{t_2}\right\}$$
$$= a\left[-\exp(-t_2) + 1 + \frac{1}{a}\exp(-at_2) - \frac{1}{a}\right] = 90.3\% \qquad (15.7)$$

Of course, from the condition of congruence, the fraction of fluid elements residing in the reactor for a time not shorter than this given time $t = t_2$ is:

$$\int_{t_2}^{+\infty} E(t)\mathrm{d}t = \int_{0}^{+\infty} E(t)\mathrm{d}t - \int_{0}^{t_2} E(t)\mathrm{d}t = 1 - \int_{0}^{t_2} E(t)\mathrm{d}t = 9.7\% \qquad (15.8)$$

To conclude, the mean residence time is (Eq. 15.3):

$$\bar{t} = \int_{0}^{+\infty} tE(t)\mathrm{d}t = a\int_{0}^{+\infty} t[\exp(-t) - \exp(-at)]\mathrm{d}t$$
$$= a\left\{\int_{0}^{+\infty} t\exp(-t)\mathrm{d}t - \int_{0}^{+\infty} t\exp(-at)\mathrm{d}t\right\} \qquad (15.9)$$

Recall the method of integration by parts:

$$\int_{t_1}^{t_2} \frac{\mathrm{d}f(t)}{\mathrm{d}t} g(t)\mathrm{d}t = [f(t)g(t)]_{t_1}^{t_2} - \int_{t_1}^{t_2} f(t)\frac{\mathrm{d}g(t)}{\mathrm{d}t}\mathrm{d}t \qquad (15.10)$$

Take $g(t) = t$ and apply separately to the two integrals in Eq. 15.9:

(continued)

$$\int_0^{+\infty} t\exp(-t)\mathrm{d}t = [-t\exp(-t)]_0^{+\infty} + \int_0^{+\infty}\exp(-t)\mathrm{d}t =$$
$$-[\exp(-t)]_0^{+\infty} = 1 \tag{15.11}$$

$$\int_0^{+\infty} t\exp(-at)\mathrm{d}t = \left[-\frac{t}{a}\exp(-at)\right]_0^{+\infty}$$
$$+\frac{1}{a}\int_0^{+\infty}\exp(-at)\mathrm{d}t = -\frac{1}{a^2}[\exp(-t)]_0^{+\infty} = \frac{1}{a^2} \tag{15.12}$$

Now substitute into Eq. 15.9:

$$\bar{t} = a\left(1 - \frac{1}{a^2}\right) = \frac{a^2-1}{a} = 1.5\ \text{min} \tag{15.13}$$

15.2.2 Residence Time Distribution for PFR

While we deal with non-ideal reactors, it is sometimes important to refer to what $E(t)$ would be for ideal systems. We will present further on the situation for a CSTR, while, here, we discuss the aspect of the RTD function for a PFR. First, define a generic function $f(t)$ as:

$$f(t) = \begin{cases} \dfrac{1}{t_2 - t_1} \forall t \in [t_1, t_2] \\ 0 \ \ \text{elsewhere} \end{cases} \tag{15.14}$$

This function is represented in Fig. 15.3. Between the two times t_1 and t_2, it is constantly equal to $1/(t_2 - t_1)$, and it is equal to 0 elsewhere. The overall area under the curve, being simply represented by the area of the rectangle, is clearly equal to 1. It is also noted that this function has dimensions [1/time].

The values of t_1 and t_2 have been arbitrarily chosen. If we narrow the interval $[t_1, t_2]$, the rectangle in Fig. 15.3 will become narrower and taller, but the function will keep the same features. In the limit of $t_2 \to t_1$, and defining $t_0 = t_1 = t_2$, we will obtain a special function:

$$\delta(t - t_0) = \lim_{t_2 \to t_1 = t_0} f(t) = \begin{cases} \dfrac{1}{0} \to +\infty & \text{for } t = t_0 \\ 0 \ \ \text{elsewhere} \end{cases} \tag{15.15}$$

This is called the *Dirac delta function* [named after Paul Adrien Maurice Dirac (1902–1984), English physicist, Professor at University of Cambridge; see Dirac 1930] and it is represented in Fig. 15.4. This function: deriving from $f(t)$ in Eq. 15.14, has dimensions [1/time] and its integral between 0 and $+\infty$ is still $=1$ (these are two

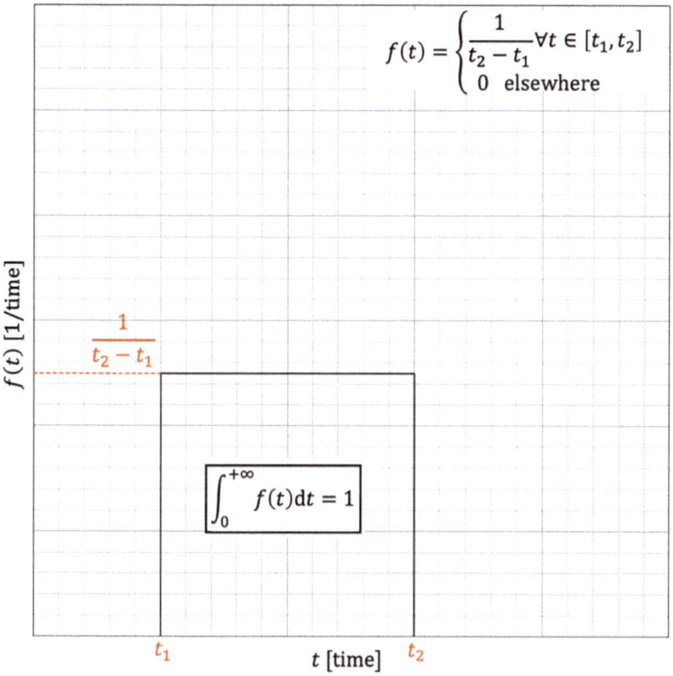

Fig. 15.3 Function defined in Eq. 15.14

characteristics typical of an RTD function); in its centre (the support point) t_O, values $+\infty$; is equal to zero everywhere except at t_O. Since such a function cannot be represented graphically, by convention we indicated it with an arrow at t_O, and this function is reported as $\delta(t - t_O)$.

In the case of a PFR, all the elements of fluid exit the reactor after a residence time that is equal to τ, the space-time. This means that the $E(t)$ function can be represented by the Dirac function centred at τ:

$$E_{PFR}(t) = \delta(t - \tau) \tag{15.16}$$

to indicate that 100% of the fluid elements live in the reactor for a time equal to the space-time, and that 0% of the elements will reside for a shorter, or longer, time (Fig. 15.5).

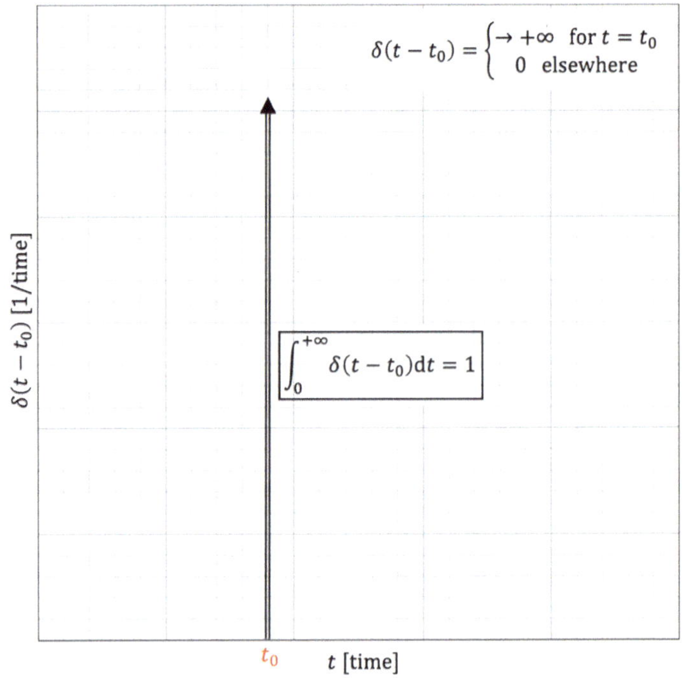

Fig. 15.4 The Dirac delta function

15.3 Tracing of Non-Ideal Reactors to Find the RTD Function

15.3.1 Tracing

The *tracing* of a non-ideal flow reactor represents the method to determine the $E(t)$ function. During the tracing test, a tracer is added to the fluid stream that passes through the system. A tracer is a material that can be recognised at the exit, e.g. since it is a dye or radioactive compound, or because it has a specific absorbance or refractive index. The *tracer must not alter the properties* (e.g., density, viscosity, flow rate) of the main fluid (and, therefore, must be introduced in small quantities). It must not separate from the main stream, sediment, be adsorbed on the internal walls, react, generate additional diffusive phenomena (apart from those intrinsically present in the fluid stream), evaporate, etc. Obviously, the tracer must be well soluble in the fluid and be easily detected (and quantified) even in small concentrations. The concept is to study, through the tracer (which will be present only during the tracing test and *not during the operation of the reactor*), what the behaviour of the fluid is in reality, and therefore what is the RTD function that characterises the system (reactor + fluid destined to be worked in the reactor) under examination, *not the tracer as such*. In this sense, the tracer must replace the fluid in its behaviour, by virtue of its

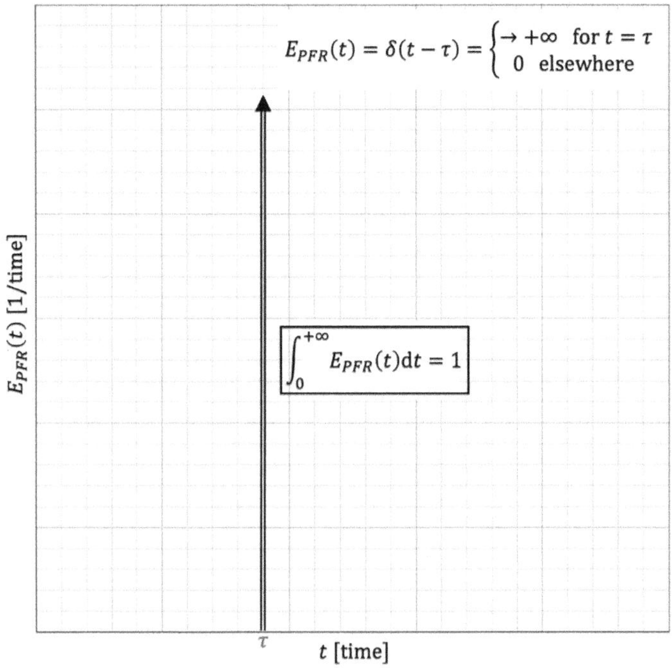

Fig. 15.5 The residence time distribution for PFR

recognisability at the exit (which would not be possible for molecules, e.g. water, of the carrier fluid: for them, we would not be able to know, at the exit of a continuous process, when such molecules have entered). We will now present two methods for reactor tracing.

15.3.2 Pulse Tracing

The argument is illustrated in Fig. 15.6. Take a non-ideal reactor of volume V. Add, to the operating volumetric flow rate Q (for constant-density systems, we will generally use this symbol), a quantity m_T of tracer. Let this addition be, as far as possible, *instantaneous* (from which the name "*pulse tracing*"), at a time that we will call $t = 0$. In this sense, we represent the tracer supply, graphically, as a sort of Dirac delta. The tracer will not be fed either before or after this instant $t = 0$. Let us place ourselves at the exit of the reactor, to measure the time-evolution of the tracer concentration, $C_T(t)$. Figure 15.6 represents a generic $C_T(t)$ curve: starting from $t = 0$, we will observe first some tracer at the exit (the tracer molecules that reside for a short time), then a peak (those that live an intermediate time value), and finally the "oldest" tracer. Of course, any other situation is possible, depending on the system. The mass balance on the tracer (IN = OUT) is closed if:

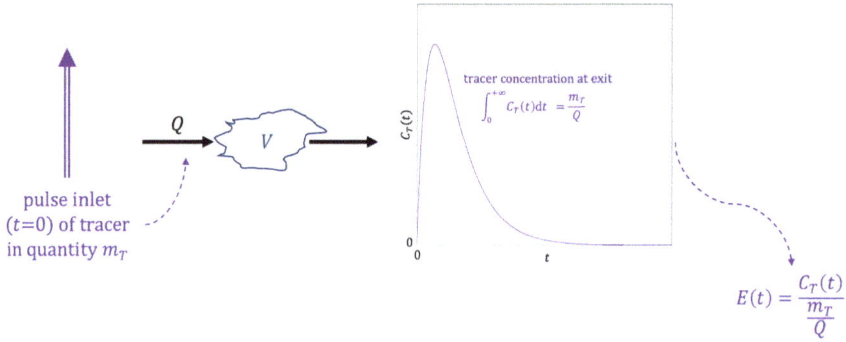

Fig. 15.6 The pulse experiment for reactor tracing (direct method to obtain the RTD)

$$m_T = Q\int_0^{+\infty} C_T(t)\mathrm{d}t \tag{15.17}$$

Conceptually, the function $C_T(t)$ is already a sort of $E(t)$, since the quantity of tracer, all injected at $t = 0$, is then recognised at the output at different times, from the shortest to the longest. However, $C_T(t)$ does not respond to the condition of congruence in Eq. 15.2. In fact, from Eq. 15.17, the area under the curve is not unitary, but equal to m_T/Q. Therefore, it is sufficient to divide the values of $C_T(t)$ by the known ratio m_T/Q to obtain what we need, i.e. a proper $E(t)$:

$$E(t) = \frac{C_T(t)}{\frac{m_T}{Q}} \tag{15.18}$$

$$\int_0^{+\infty} E(t)\mathrm{d}t = \frac{\int_0^{+\infty} C_T(t)\mathrm{d}t}{\frac{m_T}{Q}} = 1 \tag{15.19}$$

In Eq. 15.19, we have demonstrated the condition of congruence by virtue of Eq. 15.17. Moreover, please observe (from Eq. 15.18) that the dimension of $E(t)$ is, as needed, [1/time]. Pulse tracing represents the most direct way to obtain $E(t)$. In fact, we are in the hypotheses in which *the tracer behaves like the carrier fluid* and does not alter its properties; therefore, even in the absence of tracer, the carrier fluid will be characterised by the RTD function in Eq. 15.18.

Example: RTD from Pulse Tracing

In the system of Fig. 15.6, take $m_T = 1$ g and $Q = 10$ L/min. Analysing the profile of the tracer concentration at exit, we obtain the data of Table 15.1, retrieved each minute up to reading concentration of tracer equal to zero.

(continued)

Table 15.1 Data for pulse tracing experiment

t [min]	C_T [g/L]	$E(t)$ [1/min] Eq. 15.18
0	0	0
1	0.0264	0.264
2	0.0505	0.505
3	0.0115	0.115
4	0.0066	0.066
5	0.0023	0.023
6	0.0015	0.015
7	0.0007	0.007
8	0.0004	0.004
9	0.0001	0.001
10	0	0

First, check the closure of the mass balance on the tracer, Eq. 15.17, following the usual approximation ($\Delta t = 1$ min):

$$\frac{m_T}{Q} = \int_0^{+\infty} C_T(t)\mathrm{d}t \cong \Delta t \sum_0^{+\infty} C_T(t) = 0.0264 + \ldots + 0.0001$$
$$= 0.1\ \frac{\mathrm{g}\cdot\mathrm{min}}{\mathrm{L}} \tag{15.20}$$

Now, apply Eq. 15.18 (data in Table 15.1) to get $E(t)$. Then, verify Eq. 15.19. It is:

$$\int_0^{+\infty} E(t)\mathrm{d}t \cong \Delta t \sum_0^{+\infty} E(t) = 0.264 + \ldots + 0.001 = 1 \tag{15.21}$$

The mean residence time (Eq. 15.3) is:

$$\bar{t} = \int_0^{+\infty} tE(t)\mathrm{d}t \cong \Delta t \sum_0^{+\infty} tE(t) = 1 \times 0.264 + \ldots + 9 \times 0.001$$
$$= 2.178\ \mathrm{min} \tag{15.22}$$

Our RTD is represented in Fig. 15.7. Of course, fitting procedures of experimental data can help in describing the $E(t)$ curve with an analytical expression.

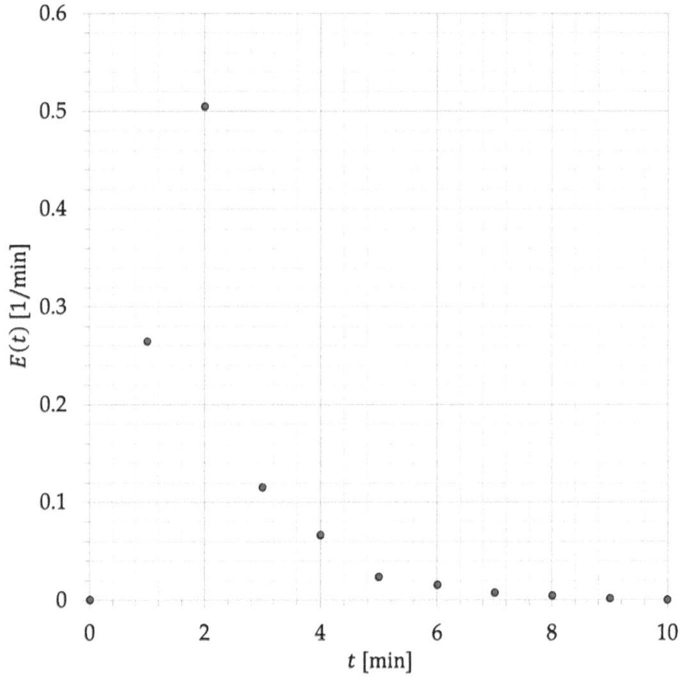

Fig. 15.7 Residence time distribution from pulse tracing experiment (data in Table 15.1)

15.3.3 Step Tracing

The argument is illustrated in Fig. 15.8. In *step tracing*, at time $t = 0$, we introduce a molar (or mass) flow rate of tracer ($\dot{m}_T$), which then remains constant over time. At reactor exit, the function $C_T(t)$ will increase in time until its steady-state value C_T^{max}:

$$\dot{m}_T = QC_T^{max} \Rightarrow C_T^{max} = \frac{\dot{m}_T}{Q} \tag{15.23}$$

since we are continuing, over all the step experiment, to feed the system with a constant tracer flow rate. The $C_T(t)$ curve can be normalised by C_T^{max} to obtain a dimensionless function, $F(t)$, which tends to 1:

$$F(t) = \frac{C_T(t)}{C_T^{max}} \tag{15.24}$$

What is the relationship between $F(t)$ and $E(t)$? How can we, through a step tracing that returns $F(t)$, obtain, albeit indirectly, what we need, namely $E(t)$? Let us look at the following example to answer. If, at a certain t (say, $t = 10$ min), $F = 0.4$, it means that, at 10 min after the beginning of the tracing test, in the exit stream there is 40% of the tracer concentration that should be there at steady state. This 40% is made up

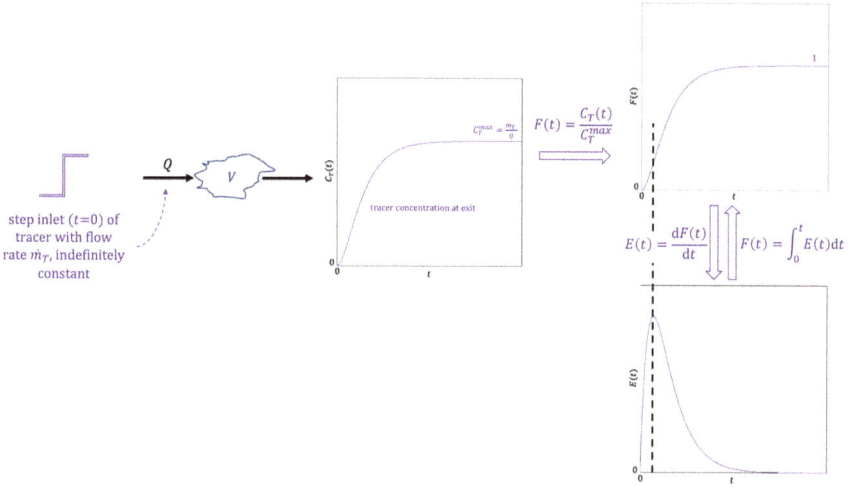

Fig. 15.8 The step experiment for reactor tracing (indirect method to obtain the RTD). The dashed line represents the condition where $F(t)$ has an inflection point, and $E(t)$ has a maximum

of tracer that may have entered at time 0 (and therefore, remained in the reactor for a time $t = 10$ min), or it may have entered at time >0 (and therefore, remained in the reactor for a time $<t = 10$ min), or it may have entered immediately before $t = 10$ min (and, therefore, remained in the reactor for a time approximately 0). In other words, 40% represents the *cumulative fraction* of fluid elements that have residence time between 0 and t, and therefore:

$$F(t) = \int_0^t E(t)\mathrm{d}t \tag{15.25}$$

as the right-hand side of Eq. 15.25 represents this cumulative fraction as well, given the RTD properties. In other words, $F(t)$ can be seen as a *cumulative residence time distribution*, and RTD can be obtained from $F(t)$ as:

$$\boxed{E(t) = \frac{\mathrm{d}F(t)}{\mathrm{d}t}} \tag{15.26}$$

We can observe that Eqs. 15.25 and 15.26 are congruent with the dimensions of $E(t)$ [1/time] and $F(t)$ (dimensionless).

As we can see in Fig. 15.8:

- $F(t)$ starts from 0 and increases up to 1, as the area under the $E(t)$ curve starts from 0 and increases up to 1 (Eq. 15.25);
- in a general case as the one depicted, the derivative of $F(t)$ is zero at $t = 0$ (and, in fact, $E(t) = 0$), then this derivative reaches a maximum at the inflection point (and, in fact, $E(t)$ has a maximum value in this point), and then the derivative

decreases, until it becomes zero for long times (and, in fact, $E(t)$ tends to 0 for long times), cf. Eq. 15.26.

Example: RTD from Step Tracing

In the system of Fig. 15.8, take $\dot{m}_T = 1.5$ g/min and $Q = 10$ L/min (then, Eq. 15.23, $C_T^{max} = 0.15$ g/L). Analysing the profile of the tracer concentration at exit, we obtain the data of Table 15.2, retrieved each 0.5 min up to reading concentration of tracer equal to C_T^{max}.

First, it is possible to obtain data for $F(t)$ according to Eq. 15.24. Then, we apply Eq. 15.26:

$$E(t) = \frac{\mathrm{d}F(t)}{\mathrm{d}t} \cong \frac{\Delta F(t)}{\Delta t} \tag{15.27}$$

Namely:

$$\begin{cases} E(t = 0.5\ \text{min}) = \dfrac{0.002 - 0}{0.5} = 0.004\ 1/\text{min} \\ E(t = 1\ \text{min}) = \dfrac{0.006 - 0.002}{0.5} = 0.008\ 1/\text{min} \\ \quad\vdots \\ E(t = 6\ \text{min}) = \dfrac{1 - 0.915}{0.5} = 0.170\ 1/\text{min} \end{cases} \tag{15.28}$$

It is:

(continued)

Table 15.2 Data for step tracing experiment

t [min]	C_T [g/L]	$F(t)$ [–] Eq. 15.24	$E(t)$ [1/min] Eq. 15.26
0	0	0	0
0.5	0.00030	0.002	0.004
1	0.00090	0.006	0.008
1.5	0.00270	0.018	0.024
2	0.00495	0.033	0.030
2.5	0.00930	0.062	0.058
3	0.02295	0.153	0.182
3.5	0.04425	0.295	0.284
4	0.06930	0.462	0.334
4.5	0.09885	0.659	0.394
5	0.11880	0.792	0.266
5.5	0.13725	0.915	0.246
6	0.15	1	0.170
6.5	0.15	1	0

$$\int_0^{+\infty} E(t)\mathrm{d}t \cong \Delta t \sum\nolimits_0^{+\infty} E(t) = 0.5(0.004 + \ldots + 0.170) = 1 \qquad (15.29)$$

$$\bar{t} = \int_0^{+\infty} tE(t)\mathrm{d}t \cong \Delta t \sum\nolimits_0^{+\infty} tE(t)$$
$$= 0.5(0.5 \times 0.004 + \ldots + 6 \times 0.170) = 4.301 \text{ min} \qquad (15.30)$$

Our $F(t)$ and $E(t)$ functions are represented in Fig. 15.9, with the usual trends.

If the reactor to be traced were a PFR, the response would be equal to the stimulus but translated by a time equal to the PFR space-time, τ. Thus, pulse tracing (a pulse at $t = 0$) translates into a pulse at $t = \tau$ (the function of Dirac in Fig. 15.5) and step tracing (starting from $t = 0$) gives, as $F(t)$ function, the same step but shifted in time by τ. As illustrated in Fig. 15.10: $F(t) = 0$ up to $t = \tau$ and, then, $F(t) = 1$ (as the value of the area under the Dirac $E(t)$, Eq. 15.25); $E(t) = 0$ for all times except τ, and values infinite at $t = \tau$ (like the derivative of $F(t)$, Eq. 15.26).

What if step tracing were executed on a CSTR, what $F(t)$ and, ultimately, $E(t)$ would characterise an ideal continuous stirred tank reactor? We can write the following mass balance on the tracer species, considering that there must be no chemical reaction (no generation or consumption of tracer), but the system (though continuous) is non-stationary with respect to the tracer (fed at constant flow rate $\dot{m}_T$ starting from $t = 0$, but exiting at concentration dependent on time, $C_T(t)$):

$$\mathrm{ACC} = \mathrm{IN} - \mathrm{OUT} \Rightarrow V\frac{\mathrm{d}C_T(t)}{\mathrm{d}t} = \dot{m}_T - QC_T(t) \qquad (15.31)$$

Being a CSTR, Eq. 15.31 presents the special feature that the tracer concentration inside the reactor (ACC term) is the same as that at outlet (OUT term). Now use Eq. 15.23 and solve:

$$V\frac{\mathrm{d}C_T(t)}{\mathrm{d}t} = Q\left[C_T^{max} - C_T(t)\right] \qquad (15.32)$$

$$\mathrm{d}t = \frac{V}{Q}\frac{1}{C_T^{max} - C_T(t)}\mathrm{d}C_T(t) \Rightarrow \int_0^t \mathrm{d}t = \tau\int_0^{C_T(t)} \frac{1}{C_T^{max} - C_T(t)}\mathrm{d}C_T(t) \qquad (15.33)$$

$$t = -\tau \ln\left[C_T^{max} - C_T(t)\right]_0^{C_T(t)} = -\tau \ln\left[\frac{C_T^{max} - C_T(t)}{C_T^{max}}\right] \qquad (15.34)$$

$$\frac{C_T^{max} - C_T(t)}{C_T^{max}} = 1 - \frac{C_T(t)}{C_T^{max}} = \exp\left(-\frac{t}{\tau}\right) \qquad (15.35)$$

Recalling the definition for $F(t)$ in Eq. 15.24:

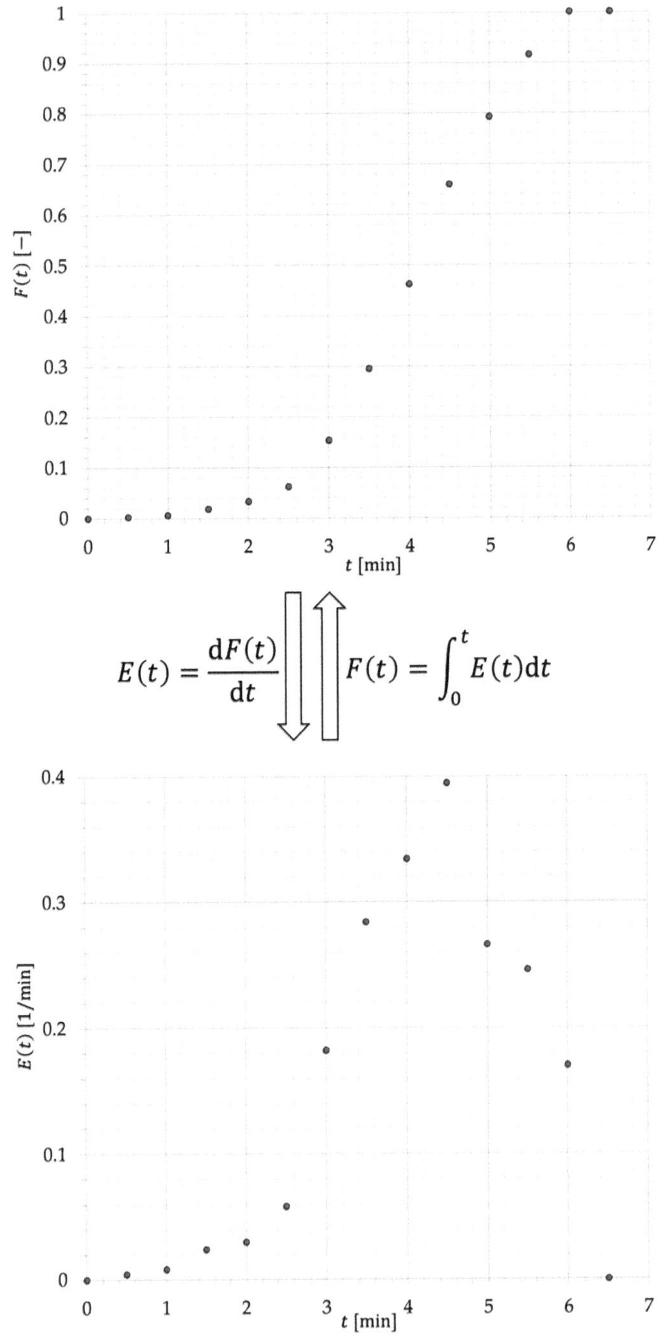

Fig. 15.9 Cumulative dimensionless residence time distribution (up) and residence time distribution (down) from step tracing experiment (data in Table 15.2)

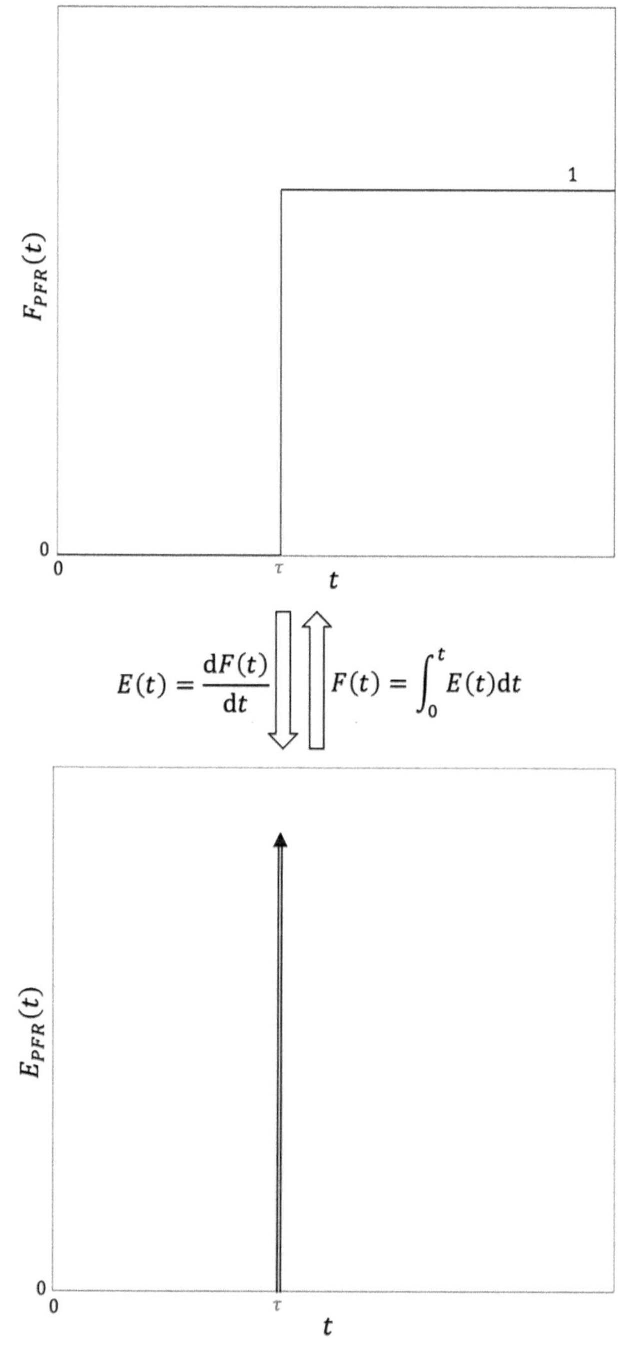

Fig. 15.10 Cumulative dimensionless residence time distribution (up) and residence time distribution (down) for PFR

$$F_{CSTR}(t) = 1 - \exp\left(-\frac{t}{\tau}\right) \tag{15.36}$$

and, finally, by Eq. 15.26:

$$E_{CSTR}(t) = \frac{\mathrm{d}F_{CSTR}(t)}{\mathrm{d}t} = \frac{1}{\tau}\exp\left(-\frac{t}{\tau}\right) \tag{15.37}$$

First, note that F_{CSTR} is non-dimensional and E_{CSTR} has the usual dimension [1/time]. Then, verify the condition of congruence:

$$\int_0^{+\infty} E_{CSTR}(t)\mathrm{d}t = \frac{1}{\tau}\int_0^{+\infty} \exp\left(-\frac{t}{\tau}\right)\mathrm{d}t = -\left[\exp\left(-\frac{t}{\tau}\right)\right]_0^{+\infty} = 1 \tag{15.38}$$

The two curves are represented in Fig. 15.11:

- $F_{CSTR}(t)$ starts from 0 and increases up to 1, as the area under the $E_{CSTR}(t)$ curve starts from 0 and increases up to 1;
- the derivative of $F_{CSTR}(t)$ is maximum at $t = 0$ (and, in fact, $E_{CSTR}(t)$ is maximum here and values the reciprocal of the space-time), then it decreases until zero (and, in fact, $E_{CSTR}(t)$ tends to 0).

To conclude, it can be verified that the mean residence time, in a CSTR, coincides with its space-time. Start from Eq. 15.3:

$$\bar{t} = \frac{1}{\tau}\int_0^{+\infty} t\exp\left(-\frac{t}{\tau}\right)\mathrm{d}t \tag{15.39}$$

Apply the method of integration by parts (Eq. 15.10) with $g(t) = t$:

$$\int_0^{+\infty} t\exp\left(-\frac{t}{\tau}\right)\mathrm{d}t = \left[-t\tau\exp\left(-\frac{t}{\tau}\right)\right]_0^{+\infty} + \tau\int_0^{+\infty} \exp\left(-\frac{t}{\tau}\right)\mathrm{d}t =$$
$$-\tau^2\left[\exp\left(-\frac{t}{\tau}\right)\right]_0^{+\infty} = \tau^2 \tag{15.40}$$

Therefore, we demonstrate from Eq. 15.39:

$$\bar{t} = \frac{1}{\tau}\tau^2 = \tau \tag{15.41}$$

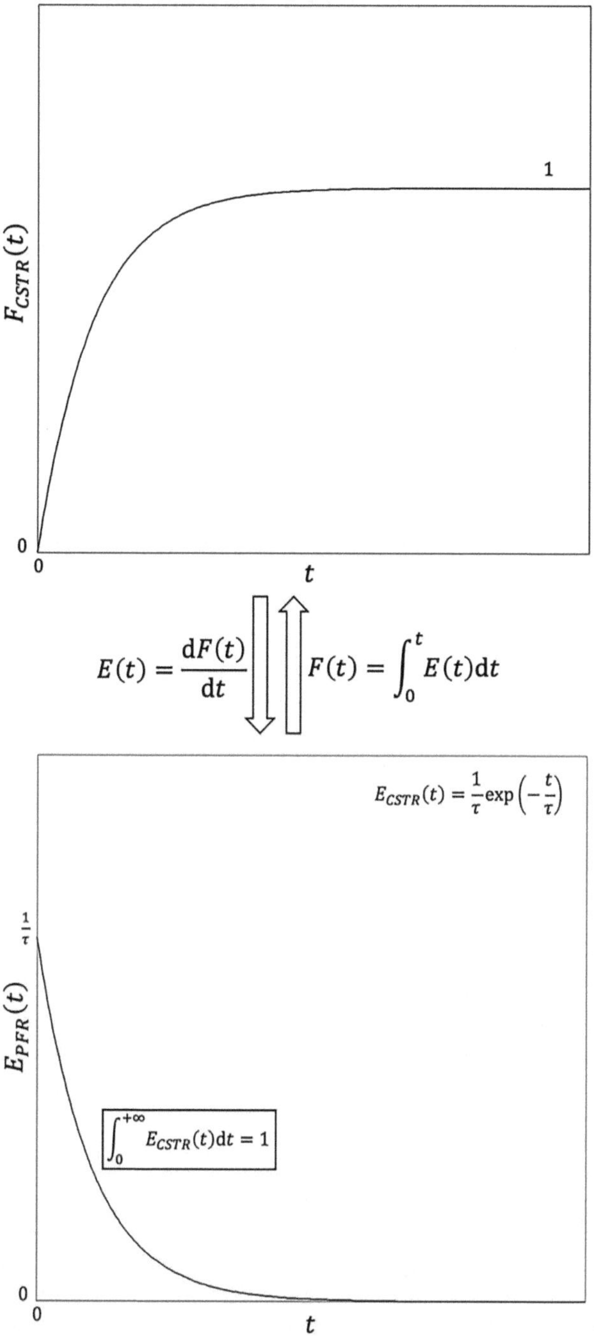

Fig. 15.11 Cumulative dimensionless residence time distribution (up) and residence time distribution (down) for CSTR

15.4 Interpretation of Non-Ideal Reactors Via Compartment Models

The reason why we have presented, in a chapter dedicated to non-ideal reactors, the RTD for the two ideal reactors (PFR and CSTR) is the following. Once the function $E(t)$ has been obtained by tracing our non-ideal system, there are situations in which it can be approximated to the known RTD for ideal reactors, with some variations (e.g., dead volumes and by-passes) that make our system non-ideal. In these cases, it is possible to treat the real system by means of a *compartment model* (further details in Levenspiel 1999), consisting of ideal blocks, which turns the system into a more easily approachable one. If this does not happen, the next chapter will address the methodologies to follow.

Generally speaking, we can say that:

- if the $E(t)$ function obtained by tracing resembles that for a PFR or CSTR, but indicates a *residence time* (volume/volumetric flow rate) *lower than expected*, since the volumetric flow rate cannot be higher than what we feed, this could mean that the active volume of the reactor is lower than the total volume ($V_a < V$), and therefore there is a *stagnant region of dead volume* $V_m = V - V_a$;
- if $E(t)$ resembles that for a PFR or CSTR, but indicates a *residence time* (volume/volumetric flow rate) *higher than expected*, since the volume cannot be higher than the total volume of the system, this could mean that the active volumetric flow rate circulating in the reactor is lower than the total flow rate ($Q_a < Q$), and therefore there is a *by-pass current* (of flow rate $Q_b = Q - Q_a$) that lives in the reactor for a virtually zero time. This is accompanied by the reading, during the tracing, of a peak (for virtually differential times) that accompanies (and completes) the RTD we obtained, to testify to the existence of a by-pass;
- if $E(t)$ derives from the superimposition of residence time distributions for ideal reactors, this could indicate the combination (in series and, or, in parallel) of PFR and/or CSTR.

Now follows a medley of cases to illustrate, by examples, what has been discussed. These cases do not claim to be exhaustive (examples of compartment models could be much more numerous), but they trace an approach that the scholar can then extend (e.g., combining the various situations presented here) depending on the cases they face.

Example: Presence of Dead Volume

In this and all the following cases in this chapter, we will imagine a non-ideal reference system consisting of a reactor of volume $V = 1$ L fed with a volumetric flow rate $Q = 0.5$ L/min. Therefore, in an ideal situation, it would be $\tau = 2$ min. Let us imagine that we have traced the reactor and

(continued)

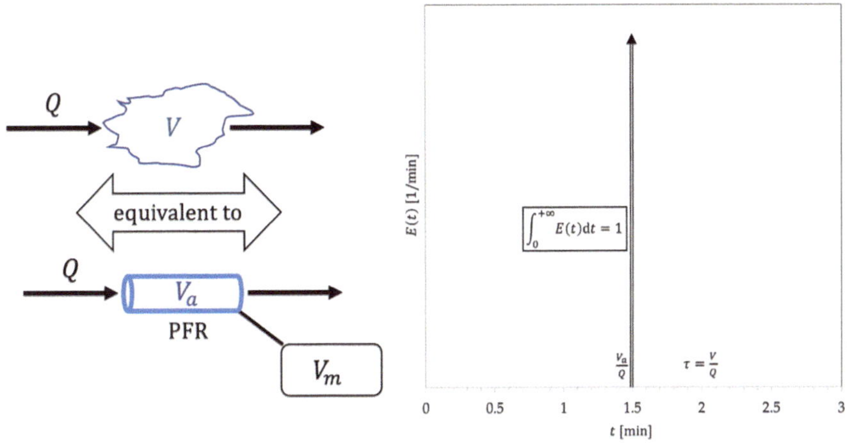

Fig. 15.12 Compartment model for PFR with dead volume

obtained the function $E(t)$. Let the function $E(t)$ be characterised by what is shown in Fig. 15.12: a Dirac delta (typical for a PFR) centred not in τ (as we would expect from Fig. 15.5), but in a time shorter than expected. We will call this time V_a/Q. This situation is characteristic of the presence of stagnant regions. We will obtain, from the information on $E(t)$, the value of V_a, and then the value of $V_m = V - V_a$. The equivalent compartment model is shown in Fig. 15.12: a PFR that processes the flow rate Q in a volume V_a (according to the tracing response we received), associated with a dead volume V_m virtually excluded from the reactive processes. This makes the real system easier to manage (we can treat it as an ideal reactor), and obviously we observe how the presence of dead volumes will worsen the system performance (since the equivalent PFR has volume $V_a < V$). For example, for $V_a/Q = 1.5$ min, we obtain $V_a = 0.75$ L and $V_m = 0.25$ L. This means that 25% of the total volume is made up of stagnant regions.

Let now the function $E(t)$ be characterised as shown in Fig. 15.13: a decreasing exponential (typical for a CSTR) which however intercepts the y-axis not at $1/\tau$ (as we would expect from Fig. 15.11), but at a value higher than expected (i.e., indicating an effective residence time lower than expected). We will call this intercept Q/V_a. Also this situation is characteristic of the presence of stagnant regions. We will obtain, from the information on $E(t)$, the value of V_a, and then the value of $V_m = V - V_a$. This approach is valid assuming that the function $E(t)$ can be approximated to the expression $E(t) = (Q/V_a)\exp.[-(Q/V_a)t]$, cf. Eq. 15.37 where $\tau = V/Q$ has been replaced by V_a/Q, with area under the curve clearly $=1$ (as per Eq. 15.38). The equivalent compartment model is shown in Fig. 15.13: a CSTR that processes the flow rate Q in a

(continued)

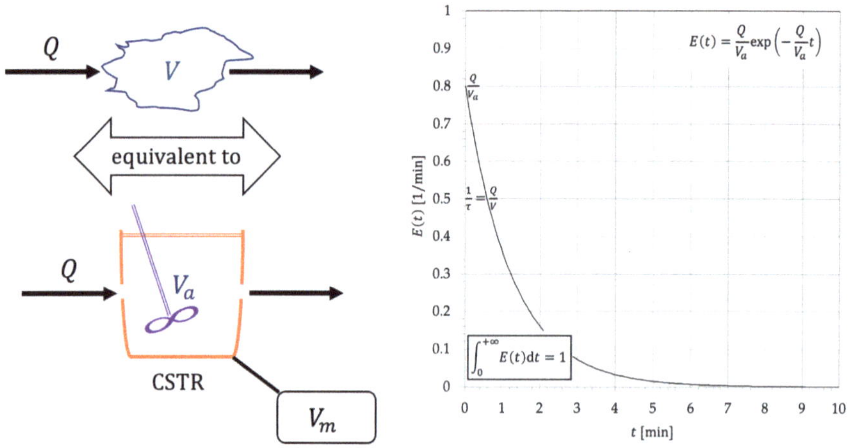

Fig. 15.13 Compartment model for CSTR with dead volume

volume V_a (according to the tracing response we received), associated with a dead volume V_m virtually excluded from the reactive processes. Also in this case, the real system will be easier to manage but with lower performance than an ideal reactor (the equivalent CSTR has volume $V_a < V$). For example, for intercept $Q/V_a = 0.8$ 1/min, we obtain $V_a = 0.625$ L and $V_m = 0.375$ L. This means that 37.5% of the total volume is made up of stagnant regions.

Example: Presence of by-Pass

Let the function $E(t)$ be characterised by what is shown in Fig. 15.14: two pulse functions, resembling Dirac deltas, the main one centred not in τ, but in a time longer than expected. We will call this time V/Q_a. This situation is characteristic of the presence of a by-pass, individuated by another pulse response, at time virtually zero. The RTD satisfies the congruence condition if the sum of the two areas under the two pulse functions is =1 (the area under the pulse at differential time must equal Q_b/Q as it takes into account the fraction of Q incurring in short circuit; the area under the pulse centred at V/Q_a must equal Q_a/Q as it takes into account the fraction of Q actually elaborated by the reactor). Thus, we can obtain the value of Q_a, and then the value of $Q_b = Q - Q_a$. The equivalent compartment model is shown in Fig. 15.14: a PFR (of volume V) that elaborates the flow rate Q_a (according to the tracing response we received), associated with a by-pass flow rate Q_b virtually excluded from the reactive processes. Again, the departure from ideality (the

(continued)

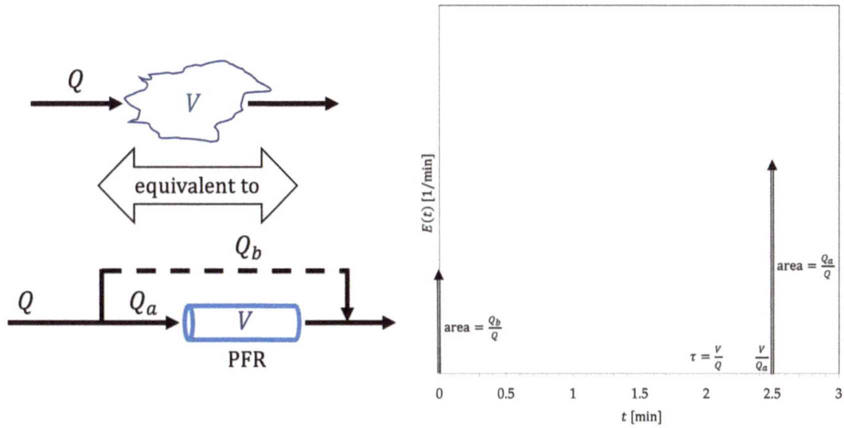

Fig. 15.14 Compartment model for PFR with by-pass

presence of short circuits) worsens the system performance (since, in the equivalent PFR, only the flow rate $Q_a < Q$ reacts; the exit conversion degree must consider the mass balance on the exit mixing node, where Q_b at zero conversion is added to the reacted Q_a). For example, for $V/Q_a = 2.5$ min, we obtain $Q_a = 0.4$ L/min and $Q_b = 0.1$ L/min. This means that 20% of the inlet flow rate incurs in short circuits.

Let now the function $E(t)$ be characterised as in Fig. 15.15: a decreasing exponential which intercepts the y-axis not at $1/\tau$, but at a value lower than expected. We will call this intercept $(Q_a/Q)(Q_a/V)$: please observe that, since $Q_a/Q < 1$ and $Q_a/V < Q/V = 1/\tau$, surely it is $(Q_a/Q)(Q_a/V) < 1/\tau$. Also this situation is characteristic of the presence of a by-pass, again individuated by a pulse response at time virtually zero that superimposes to the exponential curve and that is for all purposes part of $E(t)$. We can therefore obtain the value of Q_a, and then the value of $Q_b = Q - Q_a$. This approach is valid assuming that: (i) the decreasing exponential can be approximated to the expression $(Q_a/Q)(Q_a/V)\exp[-(Q_a/V)t]$, whose area under the curve is (this area must equal Q_a/Q as it takes into account the fraction of Q actually elaborated by the reactor):

$$\frac{Q_a}{Q}\frac{Q_a}{V}\int_0^{+\infty}\exp\left(-\frac{Q_a}{V}t\right)\mathrm{d}t = -\frac{Q_a}{Q}\left[\exp\left(-\frac{Q_a}{V}t\right)\right]_0^{+\infty} = \frac{Q_a}{Q} \quad (15.42)$$

and (ii) the area under the pulse at differential time is equal to Q_b/Q as it takes into account the fraction of Q incurring in short circuit. In this way, the overall RTD (that is the sum of the differential time pulse and the decreasing exponential) satisfies the congruence condition of Eq. 15.2. The equivalent

(continued)

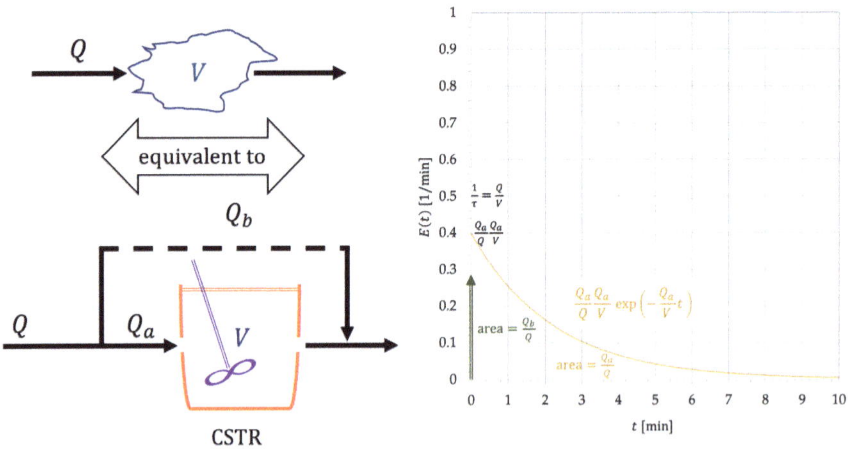

Fig. 15.15 Compartment model for CSTR with by-pass. Q_a/Q is the area under the decreasing exponential; Q_b/Q is the area under the pulse response

compartment model is shown in Fig. 15.15: a CSTR (of volume V) that elaborates the flow rate Q_a, associated with a by-pass flow rate Q_b virtually excluded from the reactive processes. Of course, the departure from ideality worsens the system performance. For example, for intercept $(Q_a/Q)(Q_a/V) = 0.4$ 1/min, we obtain $Q_a = 0.45$ L/min and $Q_b = 0.05$ L/min. This means that 10% of the inlet flow rate incurs in short circuits.

Example: Reactors in Parallel

If we observe, Fig. 15.16, an $E(t)$ function that results from the sum of two pulse functions centred at two different times (we can call these V_1/Q_1 and V_2/Q_2, respectively), this may be indicative from a channelling situation in which the real system internally behaves as 2 PFR in parallel: there is a zone (volume V_1) in which a part of volumetric flow rate (Q_1) is treated as plug flow were active and, similarly, a zone V_2 that elaborates (in PFR mode) the remainder flow rate (Q_2). The two time values represent two known data for 4 unknowns (V_1, V_2, Q_1, Q_2), but we can add the two conditions: $V_1 + V_2 = V$ (otherwise, we may have dead volumes) and $Q_1 + Q_2 = Q$ (otherwise, we may have short circuits). Moreover, an additional condition is that the RTD must satisfy the condition of congruence, so the pulses centred at V_1/Q_1 and V_2/Q_2 must be characterised by areas equal to Q_1/Q and Q_2/Q, respectively. Also this one is a situation worse than an ideal one, since there is no certainty that $V_1/Q_1 = V_2/$

(continued)

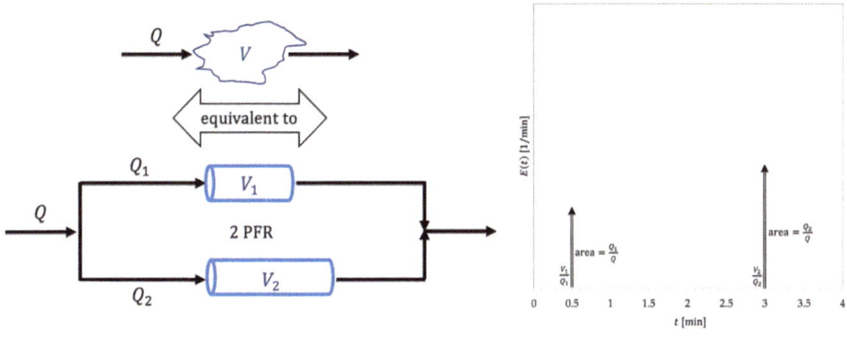

Fig. 15.16 Compartment model for 2 PFR in parallel

Q_2 as it would be instead required in an optimal system of reactors in parallel (see Eq. 12.6). For example, if the two characteristic times are 0.5 and 3 min, the 4×4 system is:

$$\begin{cases} \dfrac{V_1}{Q_1} = 0.5 \text{ min} \\ \dfrac{V_2}{Q_2} = 3 \text{ min} \\ V_1 + V_2 = V = 1 \text{ L} \\ Q_1 + Q_2 = Q = 0.5 \text{ L/min} \end{cases} \Rightarrow \begin{cases} V_1 = 0.1 \text{ L} \\ V_2 = 0.9 \text{ L} \\ Q_1 = 0.2 \text{ L/min} \\ Q_2 = 0.3 \text{ L/min} \end{cases} \tag{15.43}$$

and the system can be solved by considering the two PFR separately, obtaining the two degrees of conversion, and then calculating the exit degree of conversion by mass balance on the mixing node (Eq. 12.5).

An $E(t)$ resulting from the superimposition of two decreasing exponentials makes the non-ideal system equivalent to two CSTR in parallel, Fig. 15.17: there is a zone (volume V_1) in which a part of volumetric flow rate (Q_1) is treated as it were in a CSTR and, similarly, a zone V_2 that elaborates (in CSTR mode) the remainder flow rate (Q_2). This approach is valid assuming that the RTD can be approximated to the expression $E(t) = a\exp(-bt) + c\exp(-dt)$ with:

(continued)

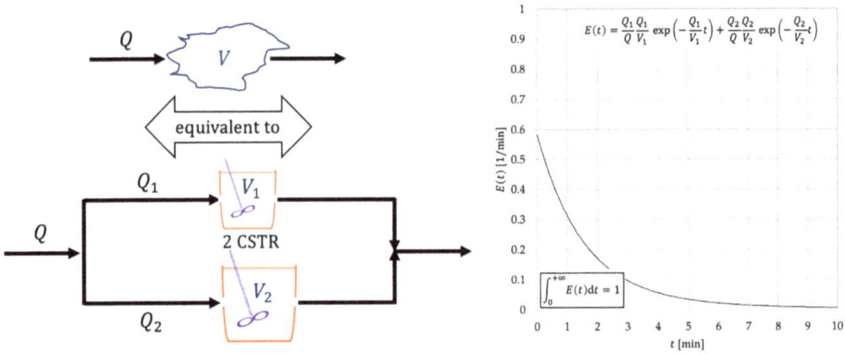

Fig. 15.17 Compartment model for 2 CSTR in parallel

$$\begin{cases} a = \frac{Q_1 Q_1}{Q\,V_1} \\ b = \frac{Q_1}{V_1} \\ c = \frac{Q_2 Q_2}{Q\,V_2} \\ d = \frac{Q_2}{V_2} \end{cases} \tag{15.44}$$

In other words, b and d are the reciprocal of the space-time of the two ideal CSTR (respectively), while a and c are the same of b and d (respectively) but multiplied per the fraction of flow rate circulating in the relative reactor. Once solved, the 4×4 system must also verify that $V_1 + V_2 = V$ and $Q_1 + Q_2 = Q$. The condition of congruence is ensured:

$$\int_0^{+\infty}[a\exp(-bt) + c\exp(-dt)]\mathrm{d}t = a\int_0^{+\infty}\exp(-bt)\mathrm{d}t + c\int_0^{+\infty}\exp(-dt)\mathrm{d}t = -\frac{a}{b}[\exp(-bt)]_0^{+\infty} - \frac{c}{d}[\exp(-dt)]_0^{+\infty} = \frac{a}{b} + \frac{c}{d} = \frac{Q_1}{Q} + \frac{Q_2}{Q} = 1 \tag{15.45}$$

Also in this case, the system does not behave as an optimal one (the two space-times are, in general, different). In an example, if $a = 0.05$ 1/min, $b = 0.25$ 1/min, $c = 8/15$ 1/min and $d = 2/3$ 1/min, it is $Q_1 = 0.1$ L/min, $Q_2 = 0.4$ L/min, $V_1 = 0.4$ L and $V_2 = 0.6$ L. Again, the system can be solved by considering the two CSTR separately, obtaining the two degrees of conversion, and then calculating the exit degree of conversion by mass balance on the mixing node (Eq. 12.5).

Example: Reactors in Series

An $E(t)$ function corresponding to a typical response for CSTR (decreasing exponential, unitary area under the curve, maximum value equal to the reciprocal of the space-time), but delayed by a certain time, tells us that the system is equivalent to a series CSTR+PFR (or vice versa; this is the first time we meet an $E(t)$ that explicitly corresponds to more than one configuration), where the delay time is the space-time of the PFR. Conceptually, if we are (as we do) the observer at the system exit, we will get the same RTD regardless the CSTR anticipates the PFR, or vice versa. In fact, here the effect of the PFR is to shift the response of the CSTR by a time equal to the PFR space-time. The scheme is illustrated in Fig. 15.18. Things work if the RTD can be expressed as:

$$E(t) = \frac{Q}{V_{CSTR}} \exp\left(-\frac{Q}{V_{CSTR}}t + \frac{V_{PFR}}{V_{CSTR}}\right) \quad \forall t \geq \frac{V_{PFR}}{Q} \tag{15.46}$$

i.e., a function that: (i) starts at the PFR space-time (for shorter times, it is null), where it values the reciprocal of the CSTR space-time; (ii) satisfies the congruence condition:

$$\frac{Q}{V_{CSTR}} \int_{\frac{V_{PFR}}{Q}}^{+\infty} \exp\left(-\frac{Q}{V_{CSTR}}t + \frac{V_{PFR}}{V_{CSTR}}\right) dt = -\left[\exp\left(-\frac{Q}{V_{CSTR}}t + \frac{V_{PFR}}{V_{CSTR}}\right)\right]_{\frac{V_{PFR}}{Q}}^{+\infty} = 1 \tag{15.47}$$

(continued)

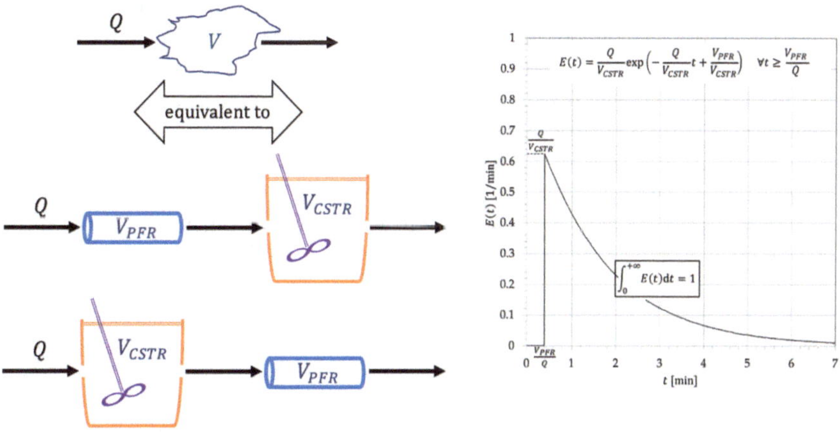

Fig. 15.18 Compartment model for CSTR and PFR in series

Of course, an additional condition is that $V_{PFR} + V_{CSTR} = V$ (otherwise, we should expect dead volumes), and Eq. 15.47 ensures the absence of short circuits (in presence of an area < 1, there should be a peak at differential times accounting for the remainder area and instructing us about by-passes). Since there is no proof that the series is optimised (as, instead, we have discussed in Chap. 12; see, e.g., Fig. 12.12), the system can have worst behaviour vs. an ideal one. Take the example of an $E(t)$ approximated by the function: 0.625exp $(-0.625\ t + 0.25)$, and starting at time = 0.4 min. From this last datum (the PFR space-time), we can get $V_{PFR} = 0.2$ L. From the value of the pre-exponential (0.625 1/min, the reciprocal of the CSTR space-time), we obtain $V_{CSTR} = 0.8$ L (by the way, $V_{PFR} + V_{CSTR} = V$).

The last observation, i.e. that the same $E(t)$ can correspond to multiple reactor configurations (for any $E(t)$, conceptually *there are infinite non-ideal systems that can return the same RTD*), acts as a hook to the arguments that we will discuss in the next chapter.

15.5 Background Information

- Non-ideal conditions can be related to various factors such as, for example, dead volumes and by-passes.
- The residence time distribution function gives indication on the fraction of fluid elements that reside in the system for a time comprised between two given times. Its overall subtended area must be unitary (Fig. 15.2).
- The mean residence time can be calculated as in Eq. 15.3.
- For ideal reactors, the RTD is illustrated in Fig. 15.5 for PFR (a Dirac delta) and in Fig. 15.11 for CSTR (a decreasing exponential).
- The pulse tracing allows to directly derive the RTD function (Eq. 15.18); the step tracing allows to indirectly derive the RTD function (Eq. 15.26) via the cumulative distribution function in Eq. 15.24.
- If the RTD resembles that for a PFR or CSTR, but indicates a residence time shorter (or longer) than expected, this can suggest the presence of dead volumes (or by-passes, respectively). If the RTD derives from the superimposition of RTD for ideal reactors, this could suggest the combination (in series and, or, in parallel) of PFR and/or CSTR. All these situations can be handled via compartment models.

References

Dirac PAM (1930) The principles of quantum mechanics. Oxford University Press
Levenspiel O (1999) Chemical reaction engineering. John Wiley & Sons, New York
Montagnaro F (2024) Sustainable energy production using solid materials. Springer Nature, Cham

Non-Ideal Reactors: Conceptualisation and Design 16

Abstract

In the last chapter of this textbook, which is also the second chapter devoted to non-ideal reactors, we will focus on the design equations in presence of a chemical reaction. The concepts of maximum segregation and maximum mixing reactor will be presented, developing the related characterising equations. Using the point of view that these two models represent the extremes of any non-ideal configuration with a given distribution of residence times, the expected conversion degree will be discussed. Subsequently, in the hypothesis of a tubular reactor with slight deviations from the ideality of the plug flow, the axial dispersion reactor will be presented and, finally, the case of fluid dynamics opposite to that of the plug flow (i.e., tubular reactor with laminar flow) will be discussed, with the consequent design equations.

Nomenclature

C_i	Concentration of i species [moles/volume]
C_{i0}	Concentration of i species at reaction beginning [moles/volume]
$\mathcal{D}$	Diffusion coefficient [length^2/time]
$\mathbb{D}$	Axial dispersion coefficient [length^2/time]
Da	Damköhler number [–]
d_T	Diameter of tubular reactor [length]
E	Residence time distribution [1/time]
E_{LFR}	Residence time distribution for laminar flow reactor [1/time]
F	Cumulative residence time distribution [–]
k	Kinetic constant [(1/time)·(moles/volume)$^{1-n}$]
L	Reactor length [length]
n	Total reaction order [–]
P_s	Specific mechanical power [energy/(time·mass)]

F. Montagnaro et al., *Elements of Unit Operations and Chemical Reactors*,
https://doi.org/10.1007/978-3-032-18046-9_16

Pe	Péclet number (axial) [–]
Q	Volumetric flow rate [volume/time]
Q_0	Volumetric flow rate at reaction beginning [volume/time]
r_A	Rate of reaction [moles/(volume·time)]
S	Reactor cross-section area [length^2]
t	Time [time]
$\bar{t}$	Mean residence time [time]
t_R	Characteristic reaction time [time]
$t_{d,a}$	Characteristic time for axial diffusive phenomena [time]
t_{min}	Minimum time [time]
t_{mm}	Micro-mixing time [time]
V	Volume [volume]
v	Fluid velocity [length/time]
$\bar{v}$	Mean fluid velocity [length/time]
v_{max}	Maximum fluid velocity [length/time]
x_A	Degree of conversion [–]
x_A^{ADR}	Degree of conversion for axial dispersion reactor [–]
x_A^{LFR}	Degree of conversion for laminar flow reactor [–]
x_A^{MMR}	Degree of conversion for maximum mixing reactor [–]
x_A^{MSR}	Degree of conversion for maximum segregation reactor [–]
x_A^{STR}	Degree of conversion for stirred tank reactor [–]
z	Reactor axial coordinate [length]
α	Age [time]
$\|\Delta \mathcal{P}\|$	Pressure drop [mass/(length·time^2)]
λ	Life expectancy [time]
μ	Dynamic viscosity [mass/(length·time)]
ν_k	Kinematic viscosity [length^2/time]
ξ	Non-dimensional axial coordinate [–]
τ	Space-time for continuous ideal reactor [time]

16.1 Maximum Segregation/Mixing Reactor—General Approach

At the end of the previous chapter, we have observed that a given $E(t)$ can correspond to multiple (conceptually, infinite) non-ideal reactors, since the fluid dynamic phenomena that determine a certain residence time distribution for the fluid elements can be as various as possible. Once obtained, for our non-ideal reactor, the RTD by tracing, if we can not reduce the system to a compartment model of ideal reactors (as seen in the previous chapter), it is necessary to address the issue of reactor conceptualisation and design with other approaches, whose general features are here presented.

First, we observe that a fluid can behave as:

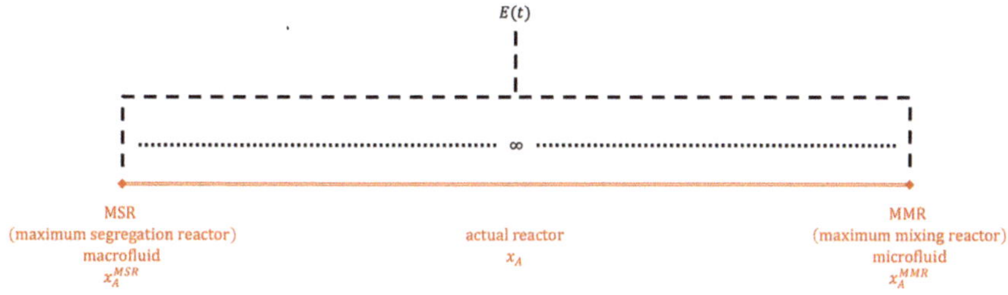

Fig. 16.1 Point of view: correspondence between RTD and infinite possible non-ideal reactors, delimited by maximum segregation reactor and maximum mixing reactor, each with its own conversion degree (the relative position of MMR and MSR is arbitrary)

- a *microfluid*, as generally in the case of, e.g., gases, or non-viscous liquids, with molecules that are free to intermix;
- a *macrofluid*, as generally in the case of, e.g., viscous liquids, or non-coalescing drops, with molecules grouped in packets or, at the extreme limit, to be considered individually. In this case, the groups of molecules maintain their identity in the system (physico-chemical properties), and do not show, virtually, transport phenomena (momentum/energy/mass) each other.

This distinction has been presented here as simple as possible (intermediate situations can apply), but it should not lead us to the trivial conclusion that ordinary gases and liquids are always microfluids, since their behaviour depends on the system under examination. For example: if we have gas bubbles inside an ordinary liquid phase, the latter remains a microfluid, but the gas bubbles can be treated as a macrofluid. Or, vice versa, if we have ordinary liquid drops inside a gas, the latter is a microfluid, but the liquid drops are closer to a macrofluid.
Apparently, infinite possible situations can occur between these two extremes, but indeed this distinction is useful to conceptualise as follows. For a given reactor and a given $E(t)$, the system can correspond to any of the infinite configurations between, included the extremes (Fig. 16.1):

- *maximum segregation reactor* (MSR), with fluid behaving as a perfect macrofluid;
- *maximum mixing reactor* (MMR), with fluid behaving as a perfect microfluid.

The approach we propose here, therefore, is to present how the conversion degree can be calculated for both MSR and MMR, relating it to the distribution of residence times (as we already know, the relationship between residence time and conversion degree represents, to all intents and purposes, the *design equation* for the reactor under consideration). Having done this, we will extend by commenting on how the real conversion degree (x_A) can be estimated, the value of which is limited by the values obtainable if the system were in MSR (x_A^{MSR}) or MMR (x_A^{MMR}) conditions, that therefore represent two ideal extremes but still pertaining to non-ideal reactors.

16.2 Maximum Segregation Reactor

In MSR (macrofluid behaviour), each "unit" is made up of molecules that remain associated with each other in the system and therefore follow the chemical reaction as if they were a single reactor not interacting with the other units (other grouped molecules). At the dimensional scale of the molecular package, the fluid elements are (inside the package) mixed each other: each package is, therefore, a sort of "micro-reactor", which behaves as if it were an STR. The degree of conversion, for a package of molecules that lives a certain time t in the reactor, will therefore be $x_A^{STR}(t)$, obtainable through the usual relations (see the design equation for STR, Eq. 10.33).

Since each package of molecules resides a given time t in the reactor, and since the fraction of fluid elements that lives a time t is $E(t)\mathrm{d}t$ (Eq. 15.1), the average conversion degree will result from the average (weighted via $E(t)\mathrm{d}t$) of the values for $x_A^{STR}(t)$ shown by the elements of fluid (grouped in units, each of them behaving as an STR separated from other units, and residing a time t in the reactor, Fig. 16.2):

$$\boxed{x_A^{MSR} = \int_0^{+\infty} x_A^{STR}(t)E(t)\mathrm{d}t} \tag{16.1}$$

We want to highlight how this actually represents the *design equation for MSR*, since it relates the degree of conversion with the residence times (which are taken into account by the RTD function) and, ultimately, with the mean residence time $\bar{t}$, that can always be calculated, given $E(t)$, by means of Eq. 15.3.

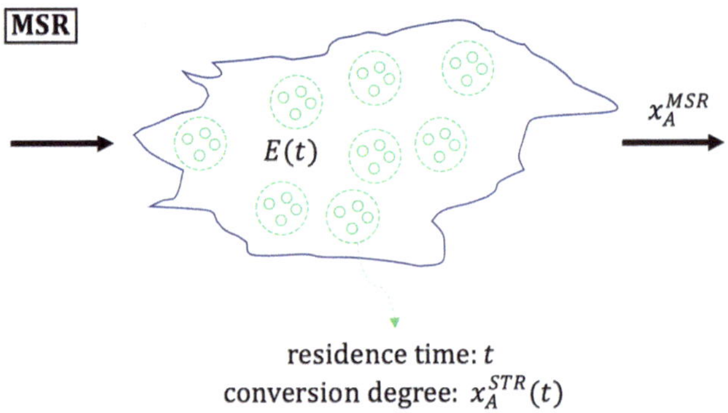

Fig. 16.2 Schematic of a maximum segregation reactor characterised by the presence of a macrofluid and a given RTD function ($E(t)$): each fluid package lives in the reactor for its own time t, is independent of the others, and presents a conversion degree as if it were an STR operated for that time t (the elements of fluid within the package are mixed each other). For the sake of clarity, only a small number of fluid packages has been depicted. The output stream presents a conversion degree that will result from the average of the conversion degrees for each fluid package, as illustrated in Eq. 16.1

Table 16.1 Data for residence time distribution and STR conversion degree

t [min]	$E(t)$ [1/min]	$x_A^{STR}(t)$ [–] Eq. 16.2
0	0	0
1	0.264	0.393
2	0.505	0.632
3	0.115	0.777
4	0.066	0.865
5	0.023	0.918
6	0.015	0.950
7	0.007	0.970
8	0.004	0.982
9	0.001	0.989
10	0	0.993

Example: MSR, Reaction with Linear Kinetics

Consider a non-ideal reactor as the one illustrated in Fig. 16.2 (macrofluid behaviour such as, e.g., a viscous liquid). Reactor tracing has given the RTD calculated in another example (cf. Table 15.1), and it is reported here in Table 16.1. Consider, in this system, the reference chemical reaction A⟹P driven by linear kinetics: $r_A = kC_A$ ($k = 0.5$ 1/min). In the hypothesis of MSR, first we can use the solution of the STR design equation for this case (cf. Table 10.5):

$$x_A^{STR}(t) = 1 - \exp(-kt) \tag{16.2}$$

Data are listed in Table 16.1. Then, we can apply the MSR design Eq. 16.1:

$$x_A^{MSR} \cong \Delta t \sum_0^{+\infty} x_A^{STR}(t)E(t) = 0.393 \times 0.264 + \ldots + 0.989 \times 0.001 = 0.616 \tag{16.3}$$

Therefore, for our MSR with mean residence time $\bar{t} = 2.178$ min (cf. Eq. 15.22), we get, under our conditions, an exit conversion degree of 61.6%.

16.3 Maximum Mixing Reactor

16.3.1 Premise

As a premise to the MMR, it is specified that we assume conditions under which diffusive phenomena at the reactor entrance and exit can be neglected (of course, they can be instead active inside the reactor). This is equivalent to saying that the

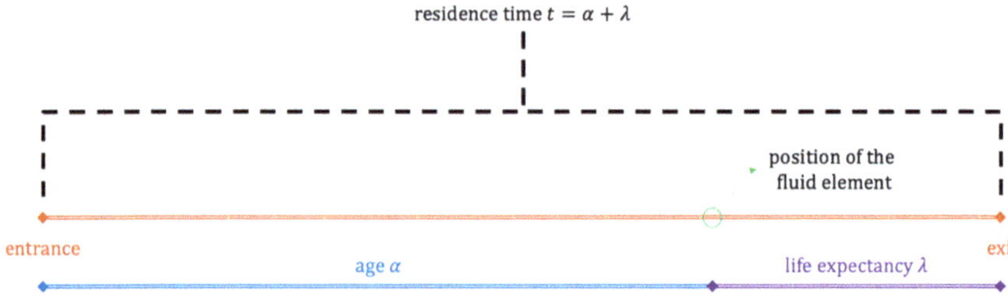

Fig. 16.3 Residence time, age and life expectancy for a fluid element in a non-ideal reactor

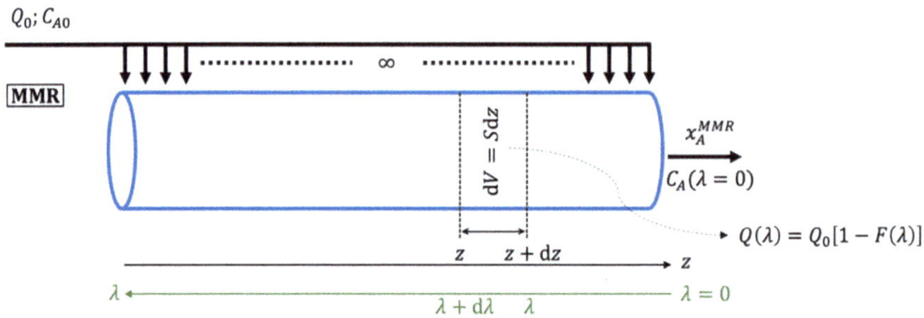

Fig. 16.4 Schematic of a maximum mixing reactor characterised by the presence of a microfluid, with infinite adduction channels for the inlet flow rate. The reference coordinate of the Zwietering reactor is the life expectancy, while the differential volume is defined as usual. The flow rate in the differential volume is expressed by Eq. 16.6. The output stream presents a conversion degree that will result from Eq. 16.7

fluid "*enters only once and exits only once*" without, e.g., backmixing in the entrance/exit sections that could allow the fluid to "exit and re-enter" the system.

Moreover, picturing an element of fluid at a certain moment of its life in the reactor (Fig. 16.3), it, destined to have a certain residence time t, has already spent a time α in the system (a symbol borrowed from the initial letter of the word "age"), while the time remaining until its exit is called λ ("*life expectancy*").

16.3.2 Presentation of the System

The MMR is schematised, elaborating approaches in literature (Westerterp et al. 1988; Zwietering 1959), as in Fig. 16.4. The feed stream, with volumetric flow rate Q_0 and limiting reagent concentration C_{A0}, is virtually subdivided into many (in the limit, infinite) adduction channels along the axial dimension z of a tubular system with a cross-section having area S. Of course: (i) each adduction is characterised by an infinitesimal flow rate, and the concentration of the reactant is C_{A0} for each adduction stream; (ii) the real system is not geometrically coincident with the one in Fig. 16.4, and the flow rate actually enters all together and not as depicted in

Fig. 16.4, whose scheme is used only as a virtual equivalent system to obtain the MMR design equation, as this equivalent system illustrates a situation in which maximum mixing occurs (due to the presence of infinite adduction channels).

The mass balance will not be written as a function of z, but as a function of the coordinate λ (the life expectancy of the microfluid elements). This coordinate starts from zero at reactor exit (there, the fluid elements will no longer have life expectancy, as their age coincides with their residence time), and increases in the opposite direction to z. We will define, as infinitesimal control volume, the one that extends between λ and $\lambda + d\lambda$.

Further considerations are:

- even though we are assuming systems with constant density, the volumetric flow rate (and, thus, the velocity) vary along λ;
- the velocity of the fluid elements at coordinate λ is, recalling the classic definition of a velocity and that λ has the dimensions of a time:

$$v(\lambda) = \frac{dz}{d\lambda} \tag{16.4}$$

- therefore, the volumetric flow rate travelling the reactor at coordinate λ is (defining the differential reactor volume, as usual, $dV = Sdz$):

$$Q(\lambda) = Sv(\lambda) = S\frac{dz}{d\lambda} = \frac{dV}{d\lambda} \tag{16.5}$$

- seen in other terms, the volumetric flow rate travelling the reactor at coordinate λ is the fraction of Q_0 made by fluid elements destined to have residence time *at least* equal to λ (fluid elements destined to have a residence time shorter than that value of λ must enter closer to the reactor exit). In fact, fluid elements entering the reactor at coordinate λ will have residence time exactly equal to λ, while all the other fluid elements that make up $Q(\lambda)$ entered closer to the reactor entrance and are, therefore, destined to have a residence time longer than λ. Recalling the general properties of the RTD and of the cumulative function $F(t)$ in Eq. 15.25, we can write:

$$Q(\lambda) = Q_0 \int_{\lambda}^{+\infty} E(\lambda)d\lambda = Q_0\left[1 - \int_0^{\lambda} E(\lambda)d\lambda\right] = Q_0[1 - F(\lambda)] \tag{16.6}$$

- please note that: (i) E and F have been equivalently written as a function of λ (and not of t) that is here our coordinate expressing the time values; (ii) E and F are two known functions, once traced the reactor;
- the solution of the mass balance will, in particular, return the concentration value of A at the exit, i.e. $C_A(\lambda = 0)$. This will allow to calculate:

$$\boxed{x_A^{MMR} = \frac{C_{A0} - C_A(\lambda = 0)}{C_{A0}}} \tag{16.7}$$

16.3.3 Design Equation

We start from the approach presented in Eq. 10.67 when dealing with the mass balance on the species A (the limiting reactant), around a control volume $S\mathrm{d}z$, for a stationary tubular reactor (terms are in [moles/time], as usual):

$$\mathrm{IN} - \mathrm{OUT} = -\mathrm{GEN} \tag{16.8}$$

where, for a MMR, the control volume is (see Eq. 16.5):

$$S\mathrm{d}z = Q(\lambda)\mathrm{d}\lambda \tag{16.9}$$

Now, we separately write the terms in Eq. 16.8:

$$\begin{cases} \text{axial IN} = Q(\lambda + \mathrm{d}\lambda)C_A(\lambda + \mathrm{d}\lambda) \\ \text{lateral IN} = Q_0\left[\int_{\lambda}^{\lambda+\mathrm{d}\lambda} E(\lambda)\mathrm{d}\lambda\right]C_{A0} \cong Q_0[E(\lambda)\mathrm{d}\lambda]C_{A0} \\ \text{axial OUT} = Q(\lambda)C_A(\lambda) \\ -\mathrm{GEN} = r_A(\lambda)Q(\lambda)\mathrm{d}\lambda \end{cases} \tag{16.10}$$

Let us illustrate the above terms (cf. Figure 16.4):

- axial inlet and outlet terms are written as usual, but considering that the inlet is positioned at coordinate λ + dλ, while the outlet at coordinate λ;
- the term of disappearance of A by chemical reaction in the differential control volume (the opposite of GEN) is written as usual, considering Eq. 16.9;
- the element of novelty is that, in the differential control volume, there is a lateral inlet which adds to the axial inlet. This lateral inlet stream has concentration of A equal to C_{A0}, while its volumetric flow rate is infinitesimal. In particular, it corresponds to the fraction of Q_0 made by fluid elements destined to live in the reactor a residence time equal to the expectancy of life coordinate (λ) corresponding to the differential volume we are considering. This is expressed via Eq. 15.1.

Now, please re-write Eq. 16.10 considering Eq. 16.6:

$$\begin{cases} \text{axial IN} = Q_0[1 - F(\lambda + \mathrm{d}\lambda)]C_A(\lambda + \mathrm{d}\lambda) \\ \text{lateral IN} = Q_0[E(\lambda)\mathrm{d}\lambda]C_{A0} \\ \text{axial OUT} = Q_0[1 - F(\lambda)]C_A(\lambda) \\ -\mathrm{GEN} = r_A(\lambda)Q_0[1 - F(\lambda)]\mathrm{d}\lambda \end{cases} \tag{16.11}$$

This approach has the advantage of erasing Q_0 when arranging the balance in Eq. 16.8:

$$[1-F(\lambda+\mathrm{d}\lambda)]C_A(\lambda+\mathrm{d}\lambda)+C_{A0}E(\lambda)\mathrm{d}\lambda-[1-F(\lambda)]C_A(\lambda)-r_A(\lambda)\times[1-F(\lambda)]\mathrm{d}\lambda=0 \tag{16.12}$$

$$\frac{[1-F(\lambda+\mathrm{d}\lambda)]C_A(\lambda+\mathrm{d}\lambda)-[1-F(\lambda)]C_A(\lambda)}{\mathrm{d}\lambda}+C_{A0}E(\lambda)-r_A(\lambda)\times[1-F(\lambda)]=0 \tag{16.13}$$

$$\frac{\mathrm{d}\{[1-F(\lambda)]C_A(\lambda)\}}{\mathrm{d}\lambda}+C_{A0}E(\lambda)-r_A(\lambda)[1-F(\lambda)]=0 \tag{16.14}$$

Solve using the Leibniz product rule, and use the renowned Eq. 15.26:

$$[1-F(\lambda)]\frac{\mathrm{d}C_A(\lambda)}{\mathrm{d}\lambda}-C_A(\lambda)\frac{\mathrm{d}F(\lambda)}{\mathrm{d}\lambda}+C_{A0}E(\lambda)-r_A(\lambda)[1-F(\lambda)]=0 \tag{16.15}$$

$$[1-F(\lambda)]\frac{\mathrm{d}C_A(\lambda)}{\mathrm{d}\lambda}-C_A(\lambda)E(\lambda)+C_{A0}E(\lambda)-r_A(\lambda)[1-F(\lambda)]=0 \tag{16.16}$$

$$[1-F(\lambda)]\frac{\mathrm{d}C_A(\lambda)}{\mathrm{d}\lambda}+E(\lambda)[C_{A0}-C_A(\lambda)]-r_A(\lambda)[1-F(\lambda)]=0 \tag{16.17}$$

Finally, we can simplify and obtain the *MMR design equation*:

$$\begin{cases} \boxed{\dfrac{\mathrm{d}C_A(\lambda)}{\mathrm{d}\lambda}+\dfrac{E(\lambda)}{1-F(\lambda)}[C_{A0}-C_A(\lambda)]-r_A(\lambda)=0} \\ \text{B.C.}: \left.\dfrac{\mathrm{d}C_A(\lambda)}{\mathrm{d}\lambda}\right|_{\lambda\to+\infty}\cong 0 \end{cases} \tag{16.18}$$

This is a (first-order) differential equation, generally non-linear (for example, the kinetic law embedded in r_A can show any dependence on C_A), with coefficients (the functions E and F) which are known (by tracing) but not constant. The solution of this equation gives the profile $C_A(\lambda)$ and, in particular, the value $C_A(\lambda=0)$ that, once used in Eq. 16.7, allows to calculate x_A^{MMR}. In this sense, this actually represents a design equation, since it relates the degree of conversion with the residence times, as the value of $C_A(\lambda=0)$ depends on the functions E and F (that, in turn, define the mean residence time $\bar{t}$, as per Eq. 15.3). Of course, the solution of Eq. 16.18 is possible only once the equation is solved via a boundary condition ("B.C."). This condition is placed at reactor inlet, i.e. when the life expectancy is sufficiently high (see Fig. 16.4; sometimes, this can be approximated by posing λ equal to 3 or 4 times the mean residence time $\bar{t}$). In this position, we have already assumed that diffusive phenomena (and, therefore, the derivative of C_A) can be neglected, see the discussion in premise. This allows to impose, as B.C., a *Neumann condition* (i.e., specifying the value assumed, at the boundary of the domain, by the derivative of the function and not by the function itself, as we instead do in a usual *Dirichlet condition*).

The solution of the Zwietering equation is not trivial, and will depend on the reactor RTD and on the reaction kinetics. An interesting hook to the arguments that we will discuss in the next section, is the solution of the MMR design equation for first-order kinetics, $r_A = kC_A$. Literature (Westerterp et al. 1988; Zwietering 1959) reports that, in this particular case, and re-expressing the life expectancy (which is a time) in terms of time *t*:

$$C_A(\lambda=0)=C_{A0}\int_0^{+\infty}\exp(-kt)E(t)\mathrm{d}t \tag{16.19}$$

Substituting in Eq. 16.7, we get:

$$x_A^{MMR}=1-\int_0^{+\infty}\exp(-kt)E(t)\mathrm{d}t \tag{16.20}$$

But we know that, Eq. 16.2:

$$x_A^{STR}(t)=1-\exp(-kt) \tag{16.21}$$

Then, and by applying the congruence condition in Eq. 15.2 and the MSR design Eq. 16.1:

$$\begin{aligned} x_A^{MMR}&=1-\int_0^{+\infty}\left[1-x_A^{STR}(t)\right]E(t)\mathrm{d}t=1-\int_0^{+\infty}E(t)\mathrm{d}t \\ &+\int_0^{+\infty}x_A^{STR}(t)E(t)\mathrm{d}t=\int_0^{+\infty}x_A^{STR}(t)E(t)\mathrm{d}t=x_A^{MSR} \end{aligned} \tag{16.22}$$

Namely, *in this particular case of first-order kinetics*, given a certain non-ideal reactor with its *E*(*t*), the two extreme cases (macrofluid and microfluid behaviour) return the same conversion degree and, therefore, any other of the infinite possible situations between MSR and MMR (see Fig. 16.1) will be characterised by the same value for x_A.

16.4 Actual Conversion Degree

It has been said, see Fig. 16.1, that given a non-ideal reactor characterised by its RTD, and given a chemical process, the *actual conversion degree* x_A will be included in a range whose extremes (not necessarily in this order) are the values obtained in a MMR (perfect microfluid behaviour, Eqs. 16.7 and 16.18) and in a MSR (perfect macrofluid behaviour, Eq. 16.1). Generally speaking, it is not possible to give more precise indications. It could be sufficient to know the range in which x_A can fall, considering that MSR and MMR are idealisations of a non-ideal behaviour, useful for defining the extremes of this range.

That said, reworking the concepts previously exposed about the role of mixing (see Fig. 12.12) can help in defining, depending on the case, which of the two (between x_A^{MMR} and x_A^{MSR}) represents the lower bound, and which the upper bound, of the interval where the actual value for x_A will fall. We know that, if the second derivative of $r_A(C_A)$ is >0, it is better to postpone the mixing. Therefore, the MMR condition will be the worst possible, because it anticipates the mixing, and MSR the best possible, because it delays the mixing. So:

$$\boxed{\frac{d^2 r_A}{dC_A^2} > 0 \Rightarrow \text{postpone mixing} \Rightarrow x_A \in \left[x_A^{MMR}, x_A^{MSR}\right]} \tag{16.23}$$

Vice versa, if the second derivative of $r_A(C_A)$ is <0, it is better to anticipate the mixing. Therefore, the MSR condition will be the worst possible, and MMR the best possible:

$$\boxed{\frac{d^2 r_A}{dC_A^2} < 0 \Rightarrow \text{anticipate mixing} \Rightarrow x_A \in \left[x_A^{MSR}, x_A^{MMR}\right]} \tag{16.24}$$

If instead the second derivative of $r_A(C_A)$ is 0, there is no role of mixing. Therefore, x_A^{MMR} and x_A^{MSR} will coincide (this concept has been discussed in Eq. 16.22 for a power-law kinetics with $n = 1$, which returns a function $r_A(C_A)$ with zero second derivative, see Fig. 12.12), and the calculation of only one of the two (it is, indeed, simpler to calculate x_A^{MSR}) allows us to know the actual value of x_A:

$$\boxed{\frac{d^2 r_A}{dC_A^2} = 0 \Rightarrow x_A = x_A^{MSR} = x_A^{MMR}} \tag{16.25}$$

Finally we observe that, in cases in which the solution of the differential design equation for MMR is not simple, the evaluation of x_A^{MSR} (which is, undoubtedly, an easier task) allows us, at least, to individuate the highest (or the lowest) value for the actual conversion degree when the second derivative of $r_A(C_A)$ is positive (or negative, respectively).

Example: Case Study Applying the Theory of Isotropic and Homogeneous Turbulence

The approach we present here allows us to understand whether our system is closer to MSR or MMR conditions. The case study we refer to is that of a stirred tank system. We assume that the turbulence (e.g., developed by mechanical blades that stir the fluid) dissipates uniformly throughout the fluid (i.e., without dead volumes: a condition of homogeneity), and that the turbulent fluctuations propagate with the same intensity in all directions (a condition of isotropy). According to *Kolmogorov's theory of isotropic and homogeneous turbulence* (Kolmogorov and Levin 1991, which is the translation of the original article dated 1941) we rework by recalling

(continued)

(Eq. 3.5) ν_k, the kinematic viscosity of the fluid, and by defining P_s [power/mass] as the specific mechanical power dissipated (by the action of the stirring system). We assume the following expression for the *micro-mixing time*:

$$t_{mm} = \sqrt{\frac{\nu_k}{P_s}} \quad \begin{cases} t_{mm} \ll \bar{t} \Rightarrow \text{MMR (microfluid)} \\ t_{mm} \gg \bar{t} \Rightarrow \text{MSR (macrofluid)} \end{cases} \tag{16.26}$$

This represents the time scale over which turbulence penetrates down to the smallest scales, making the fluid homogeneous (and, therefore, well-mixed) even on the molecular scale. The more viscous the fluid, and the lower the specific mechanical power, the longer the micro-mixing time. To reassure us of the dimensional correctness of the Kolmogorov's relation, we can observe:

$$t_{mm} = \left[\frac{\frac{\text{length}^2}{\text{time}}}{\frac{\text{energy}}{\text{mass} \cdot \text{time}}} = \frac{\frac{\text{length}^2}{\text{time}}}{\frac{\text{force} \cdot \text{length}}{\text{mass} \cdot \text{time}}} = \frac{\frac{\text{length}^2}{\text{time}}}{\frac{\text{mass} \cdot \frac{\text{length}}{\text{time}^2} \cdot \text{length}}{\text{mass} \cdot \text{time}}} = \frac{\frac{\text{length}^2}{\text{time}}}{\frac{\text{length}^2}{\text{time}^3}} = \text{time}^2 \right]^{0.5} = [\text{time}] \tag{16.27}$$

Back to Eq. 16.26:

- given the mean residence time $\bar{t}$ (known by reactor tracing), if $t_{mm} \ll \bar{t}$, then the fluid elements have had time to micro-mix, and the system will be closer to the MMR extreme;
- vice versa, if $t_{mm} \gg \bar{t}$, then the fluid elements did not have time to micro-mix, and the system will be closer to the MSR extreme;
- evidently, nothing can be said for systems for which this theory is not applicable, and for cases in which the micro-mixing time and the mean residence time are similar in terms of order of magnitude.

Let us consider a stirred reactor (industrial scale with volume 10 m^3) which, following tracing, returns $\bar{t} = 1.5$ min (as calculated in Eq. 15.13). We have a stirring system that delivers a power of 50 kW. If the reacting mixture has the characteristics of an aqueous solution maintained at room temperature (for simplicity of calculation), the mass occupying the volume of 10 m^3 is equal to 10,000 kg, from which we obtain $P_s = 5$ W/kg $= 5$ m^2/s^3. Since for an aqueous solution we can assume $\nu_k = 10^{-6}$ m^2/s (Table 3.1), the application of Eq. 16.26 returns t_{mm} with order of magnitude of milliseconds. Therefore $t_{mm} \ll \bar{t}$, and we can approximate the system to a microfluid (and apply the Zwietering MMR approach).

16.5 Axial Dispersion Reactor

16.5.1 Presentation of the System

The system we present here (*axial dispersion reactor*, ADR; Fig. 16.5) is a non-ideal model, applicable if the reactor, truly tubular like a PFR, presents *small deviations* from the ideal conditions of the PFR. Basically, we must remove here (for a non-ideal tubular reactor) the assumptions of perfect axial segregation and consider a (although not high) *degree of mixing* along the axial coordinate. The lack of knowledge of the phenomena, even complex ones, that can determine this situation (e.g., backmixing, diffusion) is incorporated in an *axial dispersion coefficient* ($\mathbb{D}$). This coefficient has the same dimensions as the molecular diffusion coefficient $\mathcal{D}$ [length^2/time], but does not coincide with it, even if depending on it. In fact, the coefficient $\mathbb{D}$ is a function of the properties of the fluid and the fluid dynamics (including the geometry) of the system under consideration. The literature reports various relationships depending on the cases (see Levenspiel 1999). Here we recall, as an example, a relation between $\mathbb{D}$ and $\mathcal{D}$ valid for a tubular reactor of diameter d_T in which the fluid circulates with (average) velocity v:

$$\mathbb{D} = \mathcal{D} + \frac{(d_T v)^2}{192\mathcal{D}} \tag{16.28}$$

When we discussed the ideality hypotheses that support the PFR, we noted in particular that the axial Péclet number (Eq. 10.63) must be much greater than 1 (at the conceptual limit, tending to infinity), which allows us to exclude axial dispersion phenomena. If we now want to consider their existence, redefine the axial Péclet number by inserting, more properly, $\mathbb{D}$ in place of $\mathcal{D}$ (cf. Eq. 10.63), where L is the reactor length:

$$Pe = \frac{Lv}{\mathbb{D}} \tag{16.29}$$

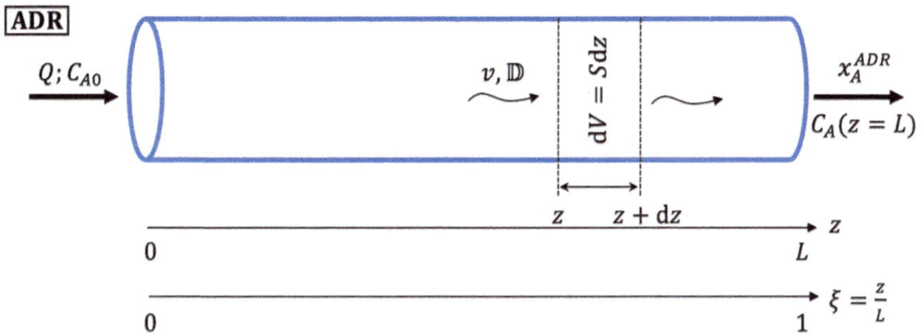

Fig. 16.5 Schematic of an axial dispersion reactor characterised by both convective (governed by the velocity v) and dispersive (governed by the axial dispersion coefficient $\mathbb{D}$) motion. The output stream presents a conversion degree that will result from Eq. 16.35

We will say that the ADR assumptions (*small deviations from ideality*) are valid if *Pe* is sufficiently large (e.g., $Pe > 10$), which is equivalent to say that $\mathbb{D}$ (although not zero as it should be in an ideal PFR) is sufficiently small. The assumption of deviations from PFR ideality that exist but are modest, will allow us to write the design equation by borrowing the approach presented for the PFR, with the appropriate modifications to account for the deviations from ideality, all enclosed in the coefficient $\mathbb{D}$.

16.5.2 Design Equation

We start from the approach presented in Eq. 10.67 when dealing with the mass balance on the species A (the limiting reactant), around a control volume $S\mathrm{d}z$, for a PFR (terms are in [moles/time], as usual), but considering (at inlet z and outlet $z + \mathrm{d}z$) not only the convective terms but also those related to dispersion phenomena (ruled by the Fick's law here governed by the axial dispersion coefficient), with $Q =$ constant as we are dealing with constant density systems:

$$\begin{cases} \text{IN (convective)} = QC_A(z) \\ \text{IN (dispersive)} = -\mathbb{D}\left.\dfrac{\mathrm{d}C_A(z)}{\mathrm{d}z}\right|_z S \\ \text{OUT (convective)} = QC_A(z+\mathrm{d}z) \\ \text{OUT (dispersive)} = -\mathbb{D}\left.\dfrac{\mathrm{d}C_A(z)}{\mathrm{d}z}\right|_{z+\mathrm{d}z} S \\ -\text{GEN} = r_A(z)S\mathrm{d}z \end{cases} \tag{16.30}$$

Then, the mass balance reads:

$$QC_A(z) - \mathbb{D}\left.\frac{\mathrm{d}C_A(z)}{\mathrm{d}z}\right|_z S - QC_A(z+\mathrm{d}z) + \mathbb{D}\left.\frac{\mathrm{d}C_A(z)}{\mathrm{d}z}\right|_{z+\mathrm{d}z} S = r_A(z)S\mathrm{d}z \tag{16.31}$$

$$\mathbb{D}\left[\left.\frac{\mathrm{d}C_A(z)}{\mathrm{d}z}\right|_{z+\mathrm{d}z} - \left.\frac{\mathrm{d}C_A(z)}{\mathrm{d}z}\right|_z\right] S + Q[C_A(z) - C_A(z+\mathrm{d}z)] - r_A(z)S\mathrm{d}z = 0 \tag{16.32}$$

$$\mathbb{D}\frac{\left.\frac{\mathrm{d}C_A(z)}{\mathrm{d}z}\right|_{z+\mathrm{d}z} - \left.\frac{\mathrm{d}C_A(z)}{\mathrm{d}z}\right|_z}{\mathrm{d}z} + \frac{Q}{S}\frac{C_A(z) - C_A(z+\mathrm{d}z)}{\mathrm{d}z} - r_A(z) = 0 \tag{16.33}$$

$$\begin{cases} \boxed{\mathbb{D}\dfrac{\mathrm{d}^2C_A(z)}{\mathrm{d}z^2} - v\dfrac{\mathrm{d}C_A(z)}{\mathrm{d}z} - r_A(z) = 0} \\ \text{B.C.1} : C_A(z=0) = C_{A0} \\ \text{B.C.2} : \left.\dfrac{\mathrm{d}C_A(z)}{\mathrm{d}z}\right|_{z=L} \cong 0 \end{cases} \tag{16.34}$$

We land at a second-order differential equation. We observe the two B.C. needed to solve this equation: the first one is a canonical Dirichlet condition imposing the inlet value to the concentration of A; the second one is a Neumann condition that arises from the assumption that, in our non-ideal systems, we are neglecting backmixing at reactor (entrance and) exit, therefore the Fick's flux must be nil in those points. The solution of the ADR design equation, once known the kinetic law, will allow to calculate the $C_A(z)$ profile and, in particular, the value for $C_A(z = L)$. Thus, finally:

$$\boxed{x_A^{ADR} = \frac{C_{A0} - C_A(z=L)}{C_{A0}}} \tag{16.35}$$

Example: ADR Design Equation for Power-Law Kinetics in Non-dimensional Form

We elaborate Eq. 16.34 with the aim of obtaining a non-dimensional differential equation if the kinetics can be expressed by a power law:

$$r_A(z) = kC_A^n(z) \tag{16.36}$$

First, put $C_A = C_{A0}(1 - x_A)$:

$$-\mathbb{D}C_{A0}\frac{d^2x_A(z)}{dz^2} + vC_{A0}\frac{dx_A(z)}{dz} - kC_{A0}^n[1 - x_A(z)]^n = 0 \tag{16.37}$$

Now, define a non-dimensional axial coordinate (spanning from 0 to 1, Fig. 16.5):

$$\xi = \frac{z}{L} \tag{16.38}$$

Therefore, it is:

$$z = \xi L \tag{16.39}$$

$$z^2 = \xi^2 L^2 \tag{16.40}$$

and Eq. 16.37 becomes:

$$-\frac{\mathbb{D}}{L^2}C_{A0}\frac{d^2x_A(\xi)}{d\xi^2} + \frac{v}{L}C_{A0}\frac{dx_A(\xi)}{d\xi} - kC_{A0}^n[1 - x_A(\xi)]^n = 0 \tag{16.41}$$

$$-\frac{\mathbb{D}}{L^2}\frac{d^2x_A(\xi)}{d\xi^2} + \frac{v}{L}\frac{dx_A(\xi)}{d\xi} - kC_{A0}^{n-1}[1 - x_A(\xi)]^n = 0 \tag{16.42}$$

The characteristic (axial) time for dispersive phenomena is, on the basis of Eq. 10.62:

(continued)

$$t_{d,a} = O\left(\frac{L^2}{\mathbb{D}}\right) \tag{16.43}$$

while, borrowing the concept of reactor space-time for PFR (Eq. 10.61), we can still use the symbol:

$$\tau = \frac{L}{v} \tag{16.44}$$

and the characteristic reaction time, Eq. 10.26, is:

$$t_R = O\left(\frac{1}{kC_{A0}^{n-1}}\right) \tag{16.45}$$

Let us include these three characteristic times in Eq. 16.42:

$$-\frac{1}{t_{d,a}}\frac{d^2x_A(\xi)}{d\xi^2} + \frac{1}{\tau}\frac{dx_A(\xi)}{d\xi} - \frac{1}{t_R}[1 - x_A(\xi)]^n = 0 \tag{16.46}$$

$$-\frac{\tau}{t_{d,a}}\frac{d^2x_A(\xi)}{d\xi^2} + \frac{dx_A(\xi)}{d\xi} - \frac{\tau}{t_R}[1 - x_A(\xi)]^n = 0 \tag{16.47}$$

But, Eq. 10.63:

$$Pe = \frac{t_{d,a}}{\tau} \tag{16.48}$$

and, Eq. 10.25:

$$Da = \frac{\tau}{t_R} \tag{16.49}$$

Thus, Eq. 16.47 becomes:

$$\frac{1}{Pe}\frac{d^2x_A(\xi)}{d\xi^2} - \frac{dx_A(\xi)}{d\xi} + Da[1 - x_A(\xi)]^n = 0 \tag{16.50}$$

That is:

$$\begin{cases} \boxed{\dfrac{d^2x_A(\xi)}{d\xi^2} - Pe\dfrac{dx_A(\xi)}{d\xi} + DaPe[1 - x_A(\xi)]^n = 0} \\ \text{B.C.1}: x_A(\xi = 0) = 0 \\ \text{B.C.2}: \left.\dfrac{dx_A(\xi)}{d\xi}\right|_{\xi=1} \cong 0 \end{cases} \tag{16.51}$$

(continued)

This is the desired ADR design equation in an elegant non-dimensional form. The solution of this equation will return, in particular, the exit conversion degree, i.e. $x_A(\xi = 1)$ which is strictly related to Da and Pe, that include the space-time (it can be seen, in the ADR, as a measure of the mean residence time). The reader has noted that we have translated the two boundary conditions present in Eq. 16.34: the first one is that, in $z = 0$ (that becomes $\xi = 0$ per Eq. 16.38), $C_A = C_{A0}$ (i.e., $x_A = 0$); the second is that, in $z = L$ (i.e., $\xi = 1$), the concentration of A does not vary, and this means that the gradient of the conversion degree must be nil as well.

Example: ADR Design Equation for Linear Kinetics, and Related Chart

We present here the solution for the particular case of $n = 1$. Equation 16.51 is rewritten as:

$$\frac{d^2 x_A(\xi)}{d\xi^2} - Pe\frac{dx_A(\xi)}{d\xi} + DaPe[1 - x_A(\xi)] = 0 \tag{16.52}$$

To make this second-order differential equation, which is already linear and with constant coefficients, also homogeneous, express it not as a function of x_A, but as a function of $1 - x_A$:

$$-\frac{d^2[1 - x_A(\xi)]}{d\xi^2} + Pe\frac{d[1 - x_A(\xi)]}{d\xi} + DaPe[1 - x_A(\xi)] = 0 \tag{16.53}$$

$$\begin{cases} \dfrac{d^2[1 - x_A(\xi)]}{d\xi^2} - Pe\dfrac{d[1 - x_A(\xi)]}{d\xi} - DaPe[1 - x_A(\xi)] = 0 \\ \text{B.C.1} : 1 - x_A(\xi = 0) = 1 \\ \text{B.C.2} : \left.\dfrac{d[1 - x_A(\xi)]}{d\xi}\right|_{\xi=1} \cong 0 \end{cases} \tag{16.54}$$

with the obvious translation (starting from Eq. 16.51) of the two B.C. in terms, now, of the function respect to which the differential equation must be solved $(1 - x_A)$. From mathematical background, the solution is:

$$1 - x_A(\xi) = C_1 \exp(\omega_1 \xi) + C_2 \exp(\omega_2 \xi) \tag{16.55}$$

where ω_1 and ω_2 represent the solutions of the characteristic polynomial:

$$\omega^2 - Pe\omega - DaPe = 0 \Rightarrow \begin{cases} \omega_1 = \dfrac{Pe + \sqrt{Pe^2 + 4DaPe}}{2} \\ \omega_2 = \dfrac{Pe - \sqrt{Pe^2 + 4DaPe}}{2} \end{cases} \tag{16.56}$$

Now, apply the two B.C. to Eq. 16.55, finding the constants C_1 and C_2:

(continued)

$$\begin{cases} 1-x_A(\xi=0)=C_1+C_2=1 \\ \left.\dfrac{d[1-x_A(\xi)]}{d\xi}\right|_{\xi=1}=C_1\omega_1\exp(\omega_1)+C_2\omega_2\exp(\omega_2)=0 \end{cases} \Rightarrow \begin{cases} C_1=-\dfrac{\omega_2\exp(\omega_2)}{\omega_1\exp(\omega_1)-\omega_2\exp(\omega_2)} \\ C_2=\dfrac{\omega_1\exp(\omega_1)}{\omega_1\exp(\omega_1)-\omega_2\exp(\omega_2)} \end{cases} \quad (16.57)$$

The solution can be specified for $\xi = 1$, that is, the exit conversion degree. Equation 16.55 becomes:

$$\begin{aligned} x_A(\xi=1) &= x_A^{ADR} = 1 - C_1\exp(\omega_1) - C_2\exp(\omega_2) = 1 \\ &+ \frac{\omega_2\exp(\omega_2)\exp(\omega_1)}{\omega_1\exp(\omega_1)-\omega_2\exp(\omega_2)} - \frac{\omega_1\exp(\omega_1)\exp(\omega_2)}{\omega_1\exp(\omega_1)-\omega_2\exp(\omega_2)} = 1 \\ &- \frac{(\omega_1-\omega_2)\exp(\omega_1+\omega_2)}{\omega_1\exp(\omega_1)-\omega_2\exp(\omega_2)} \end{aligned} \quad (16.58)$$

where we can also substitute:

$$\begin{aligned} \omega_1-\omega_2 &= \frac{Pe+\sqrt{Pe^2+4DaPe}}{2} - \frac{Pe-\sqrt{Pe^2+4DaPe}}{2} \\ &= \sqrt{Pe^2+4DaPe} \end{aligned} \quad (16.59)$$

$$\omega_1+\omega_2 = \frac{Pe+\sqrt{Pe^2+4DaPe}}{2} + \frac{Pe-\sqrt{Pe^2+4DaPe}}{2} = Pe \quad (16.60)$$

Figure 16.6 offers a map which represents a solution chart for this case: the ADR exit conversion degree, Eq. 16.58, is plotted vs. *Da* and *Pe*. The values of the Péclet number starts from 10, as discussed above, and ends at 100, where the results substantially coincide with those of an ideal PFR (negligible dispersion).

Example: Solving an ADR for Linear Kinetics

Consider a tubular reactor of length $L = 3$ m and diameter $d_T = 2/3$ m. First, observe that the ratio $L/d_T = 4.5$: that is, Eq. 10.66—which requires this ratio be much larger than 1 for an ideal PFR—is not satisfied. Be, in this illustrative example, the flow rate $Q = 0.01$ L/s $= 10^{-5}$ m^3/s. Being $S = \pi d_T^2 / 4 = 0.349$ m^2, the fluid velocity is $Q/S = v = 2.86\times10^{-5}$ m/s. For a fluid with molecular diffusivity $\mathcal{D} = 10^{-6}$ m^2/s, application of Eq. 16.28 returns $\mathbb{D} = 2.90\times10^{-6}$ m^2/s (it is not unusual that $\mathcal{D}$ and $\mathbb{D}$ share the same order of

(continued)

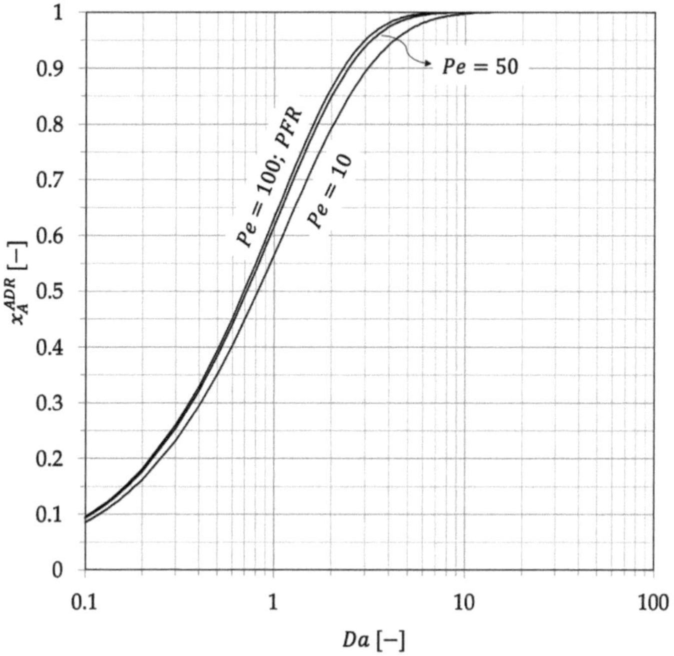

Fig. 16.6 Solution chart for ADR and linear kinetics; values for $Pe = 100$ almost coincide with results for ideal PFR (cf. Fig. 11.3)

magnitude, but conceptually, as discussed, they are not the same quantity). This means that, Eq. 16.29, *Pe* is 30. Although the tubular reactor can not be considered a PFR, we are within the contours of an ADR.

For a reaction A⟹P governed by $r_A = kC_A$ with $k = 1.9{\times}10^{-5}$ 1/s, it is, Eqs. 16.44, 16.45 and 16.49, $Da = k\tau = k(L/v) = 1.99$. From Eq. 16.56, it is $\omega_1 = 31.87$ and $\omega_2 = -\,1.87$. Applying Eq. 16.58, $x_A^{ADR} = 0.837$ (cf. chart in Fig. 16.6).

16.6 Laminar Flow Reactor

The last topic we will discuss is, in a certain sense, the opposite of the case of the axial dispersion reactor. While the latter refers to a case of small deviations from the ideality of the PFR, the *laminar flow reactor* (LFR, Fig. 16.7) is a concept that applies to tubular reactors characterised by laminar flow, which is the opposite of the plug flow characteristic of the PFR.

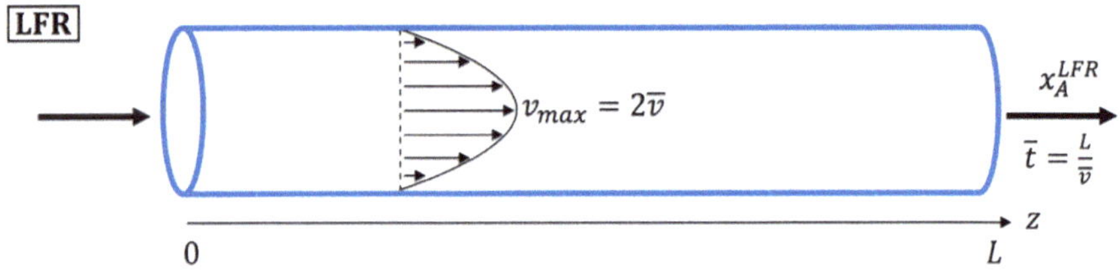

Fig. 16.7 Schematic of a laminar flow reactor with the characteristic parabolic profile for the radial velocity, with indication of mean and maximum velocity, and mean residence time. The output stream presents a conversion degree that will result from Eq. 16.61

As we have studied in another section of this textbook (Chap. 3, regarding the transport of momentum), for laminar flow, each fluid element follows its streamline along the main direction of flow (*without radial mixing*) and travels the system, depending on its radial position, with a certain velocity and, therefore, in a certain time. The fluid elements in axial position will be the fastest; those on the wall, the slowest (virtually, the fluid elements on the wall have zero velocity).

In reactor terminology, therefore, the behaviour is that of a macrofluid (maximum segregation reactor), and thus we can write (cf. Eq. 16.1):

$$\boxed{x_A^{LFR} = \int_0^{+\infty} x_A^{STR}(t) E_{LFR}(t) \mathrm{d}t} \tag{16.61}$$

In order to solve, we now need to indicate the distribution of residence times for LFR (Fig. 16.8):

$$E_{LFR} = \frac{\bar{t}^2}{2t^3} \quad \forall t \geq t_{min} = \frac{\bar{t}}{2} \tag{16.62}$$

Equation 16.62 (where E_{LFR} has dimensions of [1/time], as expected) can be analysed by recalling, first of all, that, for laminar flow, the radial velocity profile is parabolic, with the characteristic that the fluid elements on the axis (i.e., the fastest ones) have a velocity double the radial mean value of the velocity ($\bar{v}$). Therefore, the fluid elements on the axis will travel through the reactor (of length L) with a residence time (the minimum possible, t_{min}) equal to half of $\bar{t}$, the mean value of the residence time:

$$\bar{t} = \frac{L}{\bar{v}} \tag{16.63}$$

where, from our studies on the transport of momentum ($|\Delta \mathcal{P}|$ is the pressure drop between inlet and outlet of the LFR having internal diameter d_T in which a fluid with viscosity μ circulates, cf. Eq. 3.41):

$$\bar{v} = \frac{v_{max}}{2} = \frac{|\Delta \mathcal{P}| d_T^2}{32 \mu L} \tag{16.64}$$

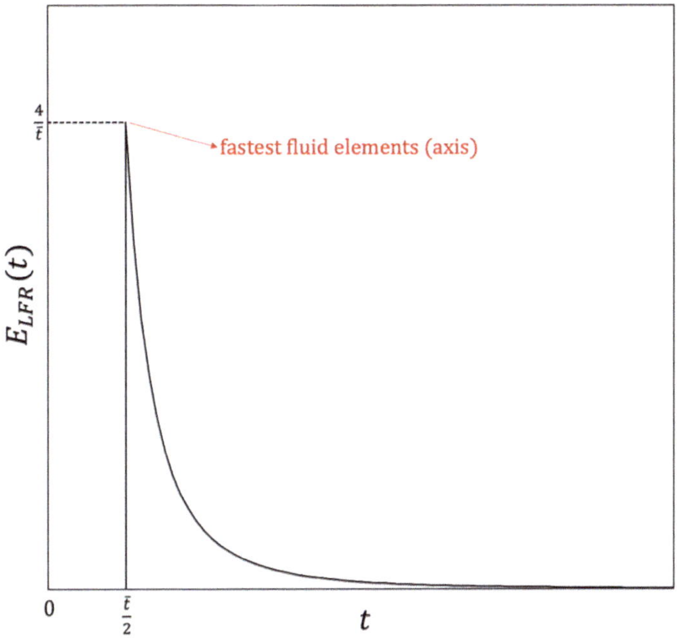

Fig. 16.8 General trend for the RTD characteristic of a laminar flow reactor (Eq. 16.62); elements of fluid with residence time shorter than t_{min} do not exist

So, up to t_{min} equal to 50% of $\bar{t}$, E_{LFR} must be nil (no fluid elements can live in the reactor for a time shorter than t_{min}, the residence time for the fastest fluid elements). Moreover, the conditions of congruence is respected as:

$$\int_{\frac{\bar{t}}{2}}^{+\infty} E_{LFR}(t)\mathrm{d}t = \frac{\bar{t}^2}{2}\int_{\frac{\bar{t}}{2}}^{+\infty} \frac{1}{t^3}\mathrm{d}t = \frac{\bar{t}^2}{2}\left[-\frac{1}{2t^2}\right]_{\frac{\bar{t}}{2}}^{+\infty} = 1 \tag{16.65}$$

The trend for E_{LFR} in Fig. 16.8 recalls, by mimicking it, the parabolic behaviour of the radial velocity profile: as we move away from the axis towards the wall, we will have increasingly slower fluid elements, which will therefore exit in an increasingly longer time; in the limit, the elements on the wall will never exit (stagnant region) because they are still.

Example: LFR, Reaction with Power Law Kinetics of Second Order
Consider a laminar flow reactor with $\bar{t} = 10$ min, and a chemical reaction A⟹P driven by:

$$\begin{cases} r_A = kC_A^2 \\ k = 10\ \dfrac{\mathrm{L}}{\mathrm{mol \cdot min}} \end{cases} \tag{16.66}$$

(continued)

Table 16.2 Data for residence time distribution and STR conversion degree (up to a point where E_{LFR} becomes vanishingly small)

t [min]	$E_{LFR}(t)$ [1/min] Eq. 16.62	$x_A^{STR}(t)$ [–] Eq. 16.67
0	0	0
2.5	0	0.714
5	0	0.833
7.5	0.118	0.882
10	0.050	0.909
12.5	0.026	0.926
15	0.015	0.938
17.5	0.009	0.946
20	0.006	0.952
22.5	0.004	0.957
25	0.003	0.962
27.5	0.002	0.965
30	0.002	0.968
32.5	0.001	0.970
35	0.001	0.972
37.5	0.001	0.974
40	0.001	0.976

with $C_{A0} = 0.1$ M. First, the solution of the STR design equation for this case (cf. Table 10.5) is:

$$x_A^{STR}(t) = \frac{kC_{A0}t}{1 + kC_{A0}t} \tag{16.67}$$

Then, recalling Eq. 16.62, we can apply the LFR design Eq. 16.61 with data in Table 16.2, by the usual simplified approach (please note that $t_{min} = \bar{t}/2 = 5$ min):

$$\begin{aligned} x_A^{LFR} &\cong \Delta t \sum_0^{+\infty} x_A^{STR}(t)E(t) \\ &= 2.5(0.118 \times 0.882 + \ldots + 0.001 \times 0.976) = 0.541 \end{aligned} \tag{16.68}$$

16.7 Background Information

- Given an RTD, there can be infinite corresponding reactor configurations.
- At the two extremes of these infinite configurations (Fig. 16.1), we find the maximum segregation reactor (design Eq. 16.1) typical in the case of macrofluid, and the maximum mixing reactor (design Eqs. 16.7 and 16.18) typical in the case of microfluid.

- The effective conversion degree, given an RTD and the infinite corresponding reactor configurations, will be in a range having as extremes the two conversion degrees of the two limiting configurations, i.e. MSR and MMR.
- A tubular reactor that presents small deviations from the ideal behaviour (PFR) can be treated as an ADR (design Eqs. 16.34 and 16.35).
- For laminar flow in a tubular reactor, the design equation for LFR (Eq. 16.61) is applied, with the characteristic RTD function (Eq. 16.62).
- Finally, Table 16.3 illustrates an abecedary on the reactors encountered with the respective design equations.

Table 16.3 Chemical reactors (isothermal): abecedary

Reactor	Design equation
STR = stirred tank r. 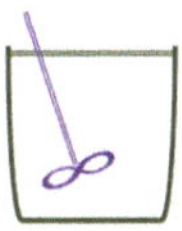	$t = C_{A0}\int_0^{x_A} \frac{1}{r_A(x_A)[1+\varepsilon x_A]}\mathrm{d}x_A$
PFR = plug flow r. 	$\tau = C_{A0}\int_0^{x_A^e} \frac{1}{r_A(x_A)}\mathrm{d}x_A$
CSTR = continuous stirred tank r. 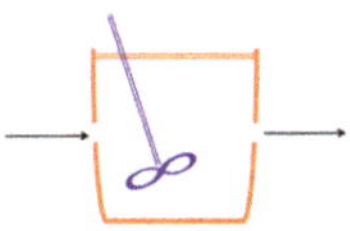	$\tau = \frac{C_{A0}x_A^e}{r_A\left(x_A^e\right)}$
PFR with recycle 	$\tau = (R+1)C_{A0}\int_{x_{A1}}^{x_A^e} \frac{1}{r_A(x_A)}\mathrm{d}x_A$
MSR = maximum segregation r. 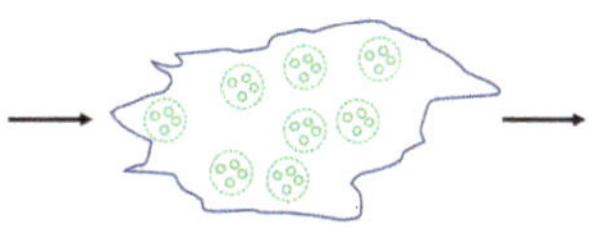	$x_A^{MSR} = \int_0^{+\infty} x_A^{STR}(t)E(t)\mathrm{d}t$
MMR = maximum mixing r. 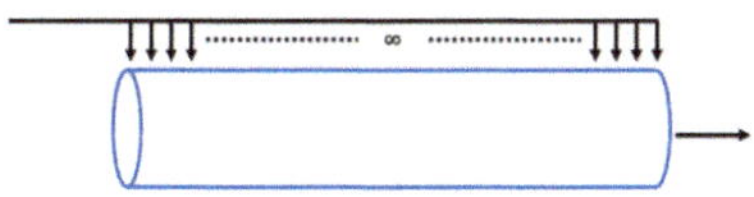	$\frac{\mathrm{d}C_A(\lambda)}{\mathrm{d}\lambda} + \frac{E(\lambda)}{1-F(\lambda)}[C_{A0} - C_A(\lambda)] - r_A(\lambda) = 0$
ADR = axial dispersion r. 	$\mathbb{D}\frac{\mathrm{d}^2C_A(z)}{\mathrm{d}z^2} - v\frac{\mathrm{d}C_A(z)}{\mathrm{d}z} - r_A(z) = 0$
LFR = laminar flow r. 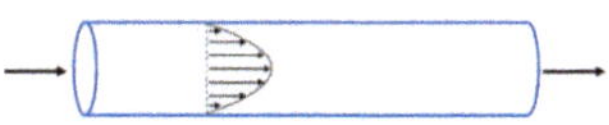	$x_A^{LFR} = \int_0^{+\infty} x_A^{STR}(t)E_{LFR}(t)\mathrm{d}t$

References

Kolmogorov AN, Levin V (1991) The local structure of turbulence in incompressible viscous fluid for very large Reynolds numbers. Proc R Soc Lond A 434:9–13. https://doi.org/10.1098/rspa.1991.0075

Levenspiel O (1999) Chemical reaction engineering. John Wiley & Sons, New York

Westerterp KR, van Swaaij WPM, Beenackers AACM (1988) Chemical reactor design and operation. John Wiley & Sons, Hoboken

Zwietering TN (1959) The degree of mixing in continuous flow systems. Chem Eng Sci 11:1–15. https://doi.org/10.1016/0009-2509(59)80068-3

MIX
Papier aus verantwortungsvollen Quellen
Paper from responsible sources
FSC® C105338

FSC
www.fsc.org

If you have any concerns about our products,
you can contact us on
ProductSafety@springernature.com

In case Publisher is established outside the EU,
the EU authorized representative is:
Springer Nature Customer Service Center GmbH
Europaplatz 3, 69115 Heidelberg, Germany

Printed by Libri Plureos GmbH
in Hamburg, Germany